Longitudinal and Panel Data

This book focuses on models and data that arise from repeated observations of a cross section of individuals, households, or firms. These models have found important applications within business, economics, education, political science, and other social science disciplines.

The author introduces the foundations of longitudinal and panel data analysis at a level suitable for quantitatively oriented graduate social science students as well as individual researchers. He emphasizes mathematical and statistical fundamentals but also describes substantive applications from across the social sciences, showing the breadth and scope that these models enjoy. The applications are enhanced by real-world data sets and software programs in SAS, Stata, and R.

EDWARD W. FREES is Professor of Business at the University of Wisconsin–Madison and is holder of the Fortis Health Insurance Professorship of Actuarial Science. He is a Fellow of both the Society of Actuaries and the American Statistical Association. He has served in several editorial capacities including Editor of the *North American Actuarial Journal* and Associate Editor of *Insurance: Mathematics and Economics*. An award-winning researcher, he has published in the leading refereed academic journals in actuarial science and insurance, other areas of business and economics, and mathematical and applied statistics.

Longitudinal and Panel Data

Analysis and Applications in the Social Sciences

EDWARD W. FREES
University of Wisconsin–Madison

CAMBRIDGE
UNIVERSITY PRESS

PUBLISHED BY THE PRESS SYNDICATE OF THE UNIVERSITY OF CAMBRIDGE
The Pitt Building, Trumpington Street, Cambridge, United Kingdom

CAMBRIDGE UNIVERSITY PRESS
The Edinburgh Building, Cambridge CB2 2RU, UK
40 West 20th Street, New York, NY 10011-4211, USA
477 Williamstown Road, Port Melbourne, VIC 3207, Australia
Ruiz de Alarcón 13, 28014 Madrid, Spain
Dock House, The Waterfront, Cape Town 8001, South Africa

http://www.cambridge.org

First published 2004

Printed in the United States of America

Typeface Times Roman 10/13 pt. *System* LaTeX 2ε [TB]

A catalog record for this book is available from the British Library.

Library of Congress Cataloging in Publication Data

Frees, Edward W.

Longitudinal and panel data : analysis and applications in the social sciences / Edward W. Frees.

p. cm.

Includes bibliographical references and index.

ISBN 0-521-82828-7 – ISBN 0-521-53538-7 (pb.)

1. Social sciences – Research – Statistical methods.
2. Longitudinal method. 3. Panel analysis.
4. Social sciences – Mathematical models.
5. Social sciences – Statistical methods. I. Title.

HA29.F6816 2004
300′.72′7 – dc22 2004043583

ISBN 0 521 82828 7 hardback
ISBN 0 521 53538 7 paperback

Contents

Preface

Intended Audience and Level

This text focuses on models and data that arise from repeated measurements taken from a cross section of subjects. These models and data have found substantive applications in many disciplines within the biological and social sciences. The breadth and scope of applications appears to be increasing over time. However, this widespread interest has spawned a hodgepodge of terms; many different terms are used to describe the same concept. To illustrate, even the subject title takes on different meanings in different literatures; sometimes this topic is referred to as "longitudinal data" and sometimes as "panel data." To welcome readers from a variety of disciplines, the cumbersome yet more inclusive descriptor "longitudinal and panel data" is used.

This text is primarily oriented to applications in the social sciences. Thus, the data sets considered here come from different areas of social science including business, economics, education, and sociology. The methods introduced in the text are oriented toward handling observational data, in contrast to data arising from experimental situations, which are the norm in the biological sciences.

Even with this social science orientation, one of my goals in writing this text is to introduce methodology that has been developed in the statistical and biological sciences, as well as the social sciences. That is, important methodological contributions have been made in each of these areas; my goal is to synthesize the results that are important for analyzing social science data, regardless of their origins. Because many terms and notations that appear in this book are also found in the biological sciences (where panel data analysis is known as longitudinal data analysis), this book may also appeal to researchers interested in the biological sciences.

Despite a forty-year history and widespread usage, a survey of the literature shows that the quality of applications is uneven. Perhaps this is because longitudinal and panel data analysis has developed in separate fields of inquiry; what is widely known and accepted in one field is given little prominence in a related field. To provide a treatment that is accessible to researchers from a variety of disciplines, this text introduces the subject using relatively sophisticated quantitative tools, including regression and linear model theory. Knowledge of calculus, as well as matrix algebra, is also assumed. For Chapter 8 on dynamic models, a time-series course at the level of Box, Jenkins, and Reinsel (1994G) would also be useful.

With this level of prerequisite mathematics and statistics, I hope that the text is accessible to my primary audience: quantitatively oriented graduate social science students. To help students work through the material, the text features several analytical and empirical exercises. Moreover, detailed appendices on different mathematical and statistical supporting topics should help students develop their knowledge of the topic as they work the exercises. I also hope that the textbook style, such as the boxed procedures and an organized set of symbols and notation, will appeal to applied researchers who would like a reference text on longitudinal and panel data modeling.

Organization

The beginning chapter sets the stage for the book. Chapter 1 introduces longitudinal and panel data as repeated observations from a subject and cites examples from many disciplines in which longitudinal data analysis is used. This chapter outlines important benefits of longitudinal data analysis, including the ability to handle the heterogeneity and dynamic features of the data. The chapter also acknowledges some important drawbacks of this scientific methodology, particularly the problem of attrition. Furthermore, Chapter 1 provides an overview of the several types of models used to handle longitudinal data; these models are considered in greater detail in subsequent chapters. This chapter should be read at the beginning and end of one's introduction to longitudinal data analysis.

When discussing heterogeneity in the context of longitudinal data analysis, we mean that observations from different subjects tend to be dissimilar when compared to observations from the same subject, which tend to be similar. One way of modeling heterogeneity is to use fixed parameters that vary by individual; this formulation is known as a *fixed-effects* model and is described in Chapter 2. A useful pedagogic feature of fixed-effects models is that they can be introduced using standard linear model theory. Linear model and regression theory is widely known among research analysts; with this solid foundation, fixed-effects models provide a desirable foundation for introducing longitudinal data models. This text is written assuming that readers are familiar with linear model and regression theory at the level of, for example, Draper and Smith (1981G) or Greene (2002E). Chapter 2 provides an overview of linear models with a heavy emphasis on analysis of covariance techniques that are useful for longitudinal and panel data analysis. Moreover, the Chapter 2 fixed-effects models provide a solid framework for introducing many graphical and diagnostic techniques.

Another way of modeling heterogeneity is to use parameters that vary by individual yet that are represented as random quantities; these quantities are known as *random effects* and are described in Chapter 3. Because models with random effects generally include fixed effects to account for the mean, models that incorporate both fixed and random quantities are known as *linear mixed-effects models*. Just as a fixed-effects model can be thought of in the linear model context, a linear mixed-effects model can be expressed as a special case of the mixed linear model. Because mixed linear model theory is not as widely known as regression, Chapter 3 provides more details on the estimation and other inferential aspects than the corresponding development in Chapter 2. Still, the good news for applied researchers is that, by writing linear mixed-effects models as mixed linear models, widely available statistical software can be used to analyze linear mixed-effects models.

By appealing to linear model and mixed linear model theory in Chapters 2 and 3, we will be able to handle many applications of longitudinal and panel data models. Still, the special structure of longitudinal data raises additional inference questions and issues that are not commonly addressed in the standard introductions to linear model and mixed linear model theory. One such set of questions deals with the problem of "estimating" random quantities, known as *prediction*. Chapter 4 introduces the prediction problem in the longitudinal data context and shows how to "estimate" residuals, conditional means, and future values of a process. Chapter 4 also shows how to use Bayesian inference as an alternative method for prediction.

To provide additional motivation and intuition for Chapters 3 and 4, Chapter 5 introduces *multilevel modeling*. Multilevel models are widely used in educational sciences and developmental psychology where one assumes that complex systems can be modeled hierarchically; that is, modeling is done one level at a time, with each level conditional on lower levels. Many multilevel models can be written as linear mixed-effects models; thus, the inference properties of estimation and prediction that we develop in Chapters 3 and 4 can be applied directly to the Chapter 5 multilevel models.

Chapter 6 returns to the basic linear mixed-effects model but adopts an econometric perspective. In particular, this chapter considers situations where the explanatory variables are stochastic and may be influenced by the response variable. In such circumstances, the explanatory variables are known as *endogenous*. Difficulties associated with endogenous explanatory variables, and methods for addressing these difficulties, are well known for cross-sectional data. Because not all readers will be familiar with the relevant econometric literature, Chapter 6 reviews these difficulties and methods. Moreover, Chapter 6 describes the more recent literature on similar situations for longitudinal data.

Chapter 7 analyzes several issues that are specific to a longitudinal or panel data study. One issue is the choice of the representation to model heterogeneity. The many choices include fixed-effects, random-effects, and serial correlation models. Chapter 7 also reviews important identification issues when trying to decide upon the appropriate model for heterogeneity. One issue is the comparison of fixed- and random-effects models, a topic that has received substantial attention in the econometrics literature. As described in Chapter 7, this comparison involves interesting discussions of the omitted-variables problem. Briefly, we will see that time-invariant omitted variables can be captured through the parameters used to represent heterogeneity, thus handling two problems at the same time. Chapter 7 concludes with a discussion of sampling and selectivity bias. Panel data surveys, with repeated observations on a subject, are particularly susceptible to a type of selectivity problem known as *attrition*, where individuals leave a panel survey.

Longitudinal and panel data applications are typically "long" in the cross section and "short" in the time dimension. Hence, the development of these methods stems primarily from regression-type methodologies such as linear model and mixed linear model theory. Chapters 2 and 3 introduce some dynamic aspects, such as serial correlation, where the primary motivation is to provide improved parameter estimators. For many important applications, the dynamic aspect is the primary focus, not an ancillary consideration. Further, for some data sets, the temporal dimension is long, thus providing opportunities to model the dynamic aspect in detail. For these situations, longitudinal data methods are closer in spirit to multivariate time-series analysis than to cross-sectional regression analysis. Chapter 8 introduces dynamic models, where the time dimension is of primary importance.

Chapters 2 through 8 are devoted to analyzing data that may be represented using models that are linear in the parameters, including linear and mixed linear models. In contrast, Chapters 9 through 11 are devoted to analyzing data that can be represented using nonlinear models. The collection of nonlinear models is vast. To provide a concentrated discussion that relates to the applications orientation of this book, we focus on models where the distribution of the response cannot be reasonably approximated by a normal distribution and alternative distributions must be considered.

We begin in Chapter 9 with a discussion of modeling responses that are dichotomous; we call these *binary dependent-variable* models. Because not all readers with a background in regression theory have been exposed to binary dependent models such as logistic regression, Chapter 9 begins with an introductory section under the heading of "homogeneous" models; these are simply the usual cross-sectional models without heterogeneity parameters. Then, Chapter 9 introduces the issues associated with random- and fixed-effects models to accommodate the heterogeneity. Unfortunately, random-effects model estimators are difficult to compute and the usual fixed-effects model estimators have undesirable properties. Thus, Chapter 9 introduces an alternative modeling strategy that is widely used in biological sciences based on a so-called marginal model. This model employs generalized estimating equations (GEE) or generalized method of moments (GMM) estimators that are simple to compute and have desirable properties.

Chapter 10 extends that Chapter 9 discussion to generalized linear models (GLMs). This class of models handles the normal-based models of Chapters 2–8, the binary models of Chapter 9, and additional important applied models. Chapter 10 focuses on count data through the Poisson distribution, although the general arguments can also be used for other distributions. Like Chapter 9, we begin with the homogeneous case to provide a review for readers who have not been introduced to GLMs. The next section is on marginal models that are particularly useful for applications. Chapter 10 follows with an introduction to random- and fixed-effects models.

Using the Poisson distribution as a basis, Chapter 11 extends the discussion to multinomial models. These models are particularly useful in economic "choice" models, which have seen broad applications in the marketing research literature. Chapter 11 provides a brief overview of the economic basis for these choice models and then shows how to apply these to random-effects multinomial models.

Statistical Software

My goal in writing this text is to reach a broad group of researchers. Thus, to avoid excluding large segments of the readership, I have chosen not to integrate any specific statistical software package into the text. Nonetheless, because of the applications orientation, it is critical that the methodology presented be easily accomplished using readily available packages. For the course taught at the University of Wisconsin, I use the statistical package SAS. (However, many of my students opt to use alternative packages such as Stata and R. I encourage free choice!) In my mind, this is the analog of an "existence theorem." If a procedure is important and can be readily accomplished by one package, then it is (or will soon be) available through its competitors. On the book Web site,

http://research.bus.wisc.edu/jfrees/Book/PDataBook.htm,

users will find routines written in SAS for the methods advocated in the text, thus demonstrating that they are readily available to applied researchers. Routines written for Stata and R are also available on the Web site. For more information on SAS, Stata, and R, visit their Web sites:

http://www.sas.com,
http://www.stata.com, and
http://www.r-project.org.

References Codes

In keeping with my goal of reaching a broad group of researchers, I have attempted to integrate contributions from different fields that regularly study longitudinal and panel data techniques. To this end, the references are subdivided into six sections. This subdivision is maintained to emphasize the breadth of longitudinal and panel data analysis and the impact that it has made on several scientific fields. I refer to these sections using the following coding scheme:

B: Biological Sciences Longitudinal Data,
E: Econometrics Panel Data,
EP: Educational Science and Psychology,
O: Other Social Sciences,
S: Statistical Longitudinal Data, and
G: General Statistics.

For example, I use "Neyman and Scott (1948E)" to refer to an article written by Neyman and Scott, published in 1948, that appears in the "Econometrics Panel Data" portion of the references.

Approach

This book grew out of lecture notes for a course offered at the University of Wisconsin. The pedagogic approach of the manuscript evolved from the course. Each chapter consists of an introduction to the main ideas in words and then as mathematical expressions. The concepts underlying the mathematical expressions are then reinforced with empirical examples; these data are available to the reader at the Wisconsin book Web site. Most chapters conclude with exercises that are primarily analytic; some are designed to reinforce basic concepts for (mathematically) novice readers. Others are designed for (mathematically) sophisticated readers and constitute extensions of the theory presented in the main body of the text. The beginning chapters (2–5) also include empirical exercises that allow readers to develop their data analysis skills in a longitudinal data context. Selected solutions to the exercises are also available from the author.

Readers will find that the text becomes more mathematically challenging as it progresses. Chapters 1–3 describe the fundamentals of longitudinal data analysis and are prerequisites for the remainder of the text. Chapter 4 is prerequisite reading for Chapters 5 and 8. Chapter 6 contains important elements necessary for reading Chapter 7. As already mentioned, a time-series analysis course would also be useful for mastering Chapter 8, particularly Section 8.5 on the Kalman filter approach.

Chapter 9 begins the section on nonlinear modeling. Only Chapters 1–3 are necessary background for this section. However, because it deals with nonlinear models, the requisite level of mathematical statistics is higher than Chapters 1–3. Chapters 10 and 11 continue the development of these models. I do not assume prior background on nonlinear models. Thus, in Chapters 9–11, the first section introduces the chapter topic in a nonlongitudinal context called a *homogeneous model.*

Despite the emphasis placed on applications and interpretations, I have not shied from using mathematics to express the details of longitudinal and panel data models. There are many students with excellent training in mathematics and statistics who need to see the foundations of longitudinal and panel data models. Further, there are now available a number of texts and summary articles (which are cited throughout the text) that place a heavier emphasis on applications. However, applications-oriented texts tend to be field-specific; studying only from such a source can mean that an economics student will be unaware of important developments in educational sciences (and vice versa). My hope is that many instructors will chose to use this text as a technical supplement to an applications-oriented text from their own field.

The students in my course come from the wide variety of backgrounds in mathematical statistics. To develop longitudinal and panel data analysis tools and achieve a common set of notation, most chapters contain a short appendix that develops mathematical results cited in the chapter. In addition, there are four appendices at the end of the text that expand mathematical developments used throughout the text. A fifth appendix, on symbols and notation, further summarizes the set of notation used throughout the text. The sixth appendix provides a brief description of selected longitudinal and panel data sets that are used in several disciplines throughout the world.

Acknowledgments

This text was reviewed by several generations of longitudinal and panel data classes here at the University of Wisconsin. The students in my classes contributed a tremendous amount of input into the text; their input drove the text's development far more than they realize.

I have enjoyed working with several colleagues on longitudinal and panel data problems over the years. Their contributions are reflected indirectly throughout the text. Moreover, I have benefited from detailed reviews by Anocha Ariborg, Mousumi Banerjee, Jee-Seon Kim, Yueh-Chuan Kung, and Georgios Pitelis. Thanks also to Doug Bates for introducing me to R.

Moreover, I am happy to acknowledge financial support through the Fortis Health Professorship in Actuarial Science.

Saving the most important for last, I thank my family for their support. Ten thousand thanks go to my mother Mary, my wife Deirdre, our sons Nathan and Adam, and the family source of amusement, our dog Lucky.

1
Introduction

Abstract. This chapter introduces the many key features of the data and models used in the analysis of longitudinal and panel data. Here, longitudinal and panel data are defined and an indication of their widespread usage is given. The chapter discusses the benefits of these data; these include opportunities to study dynamic relationships while understanding, or at least accounting for, cross-sectional heterogeneity. Designing a longitudinal study does not come without a price; in particular, longitudinal data studies are sensitive to the problem of attrition, that is, unplanned exit from a study. This book focuses on models appropriate for the analysis of longitudinal and panel data; this introductory chapter outlines the set of models that will be considered in subsequent chapters.

1.1 What Are Longitudinal and Panel Data?

Statistical Modeling

Statistics is about data. It is the discipline concerned with the collection, summarization, and analysis of data to make statements about our world. When analysts collect data, they are really collecting information that is quantified, that is, transformed to a numerical scale. There are many well-understood rules for reducing data, using either numerical or graphical summary measures. These summary measures can then be linked to a theoretical representation, or model, of the data. With a model that is calibrated by data, statements about the world can be made.

As users, we identify a basic entity that we measure by collecting information on a numerical scale. This basic entity is our *unit of analysis*, also known as the *research unit* or *observational unit*. In the social sciences, the unit of analysis is typically a person, firm, or governmental unit, although other applications can and do arise. Other terms used for the observational unit include *individual*, from the econometrics literature, as well as *subject*, from the biostatistics literature.

Regression analysis and *time-series analysis* are two important applied statistical methods used to analyze data. Regression analysis is a special type of multivariate analysis in which several measurements are taken from each subject. We identify one measurement as a *response*, or *dependent variable*; our interest is in making statements about this measurement, controlling for the other variables.

With regression analysis, it is customary to analyze data from a cross section of subjects. In contrast, with *time-series analysis*, we identify one or more subjects and observe them over time. This allows us to study relationships over time, the *dynamic* aspect of a problem. To employ time-series methods, we generally restrict ourselves to a limited number of subjects that have many observations over time.

Defining Longitudinal and Panel Data

Longitudinal data analysis represents a marriage of regression and time-series analysis. As with many regression data sets, *longitudinal data* are composed of a cross section of subjects. Unlike regression data, with longitudinal data we observe subjects over time. Unlike time-series data, with longitudinal data we observe many subjects. Observing a broad cross section of subjects over time allows us to study dynamic, as well as cross-sectional, aspects of a problem.

The descriptor *panel data* comes from surveys of individuals. In this context, a "panel" is a group of individuals surveyed repeatedly over time. Historically, panel data methodology within economics had been largely developed through labor economics applications. Now, economic applications of panel data methods are not confined to survey or labor economics problems and the interpretation of the descriptor "panel analysis" is much broader. Hence, we will use the terms "longitudinal data" and "panel data" interchangeably although, for simplicity, we often use only the former term.

Example 1.1: Divorce Rates Figure 1.1 shows the 1965 divorce rates versus AFDC (Aid to Families with Dependent Children) payments for the fifty states. For this example, each state represents an observational unit, the divorce rate is the response of interest, and the level of AFDC payment represents a variable that may contribute information to our understanding of divorce rates.

The data are observational; thus, it is not appropriate to argue for a causal relationship between welfare payments (AFDC) and divorce rates without invoking additional economic or sociological theory. Nonetheless, their relation is important to labor economists and policymakers.

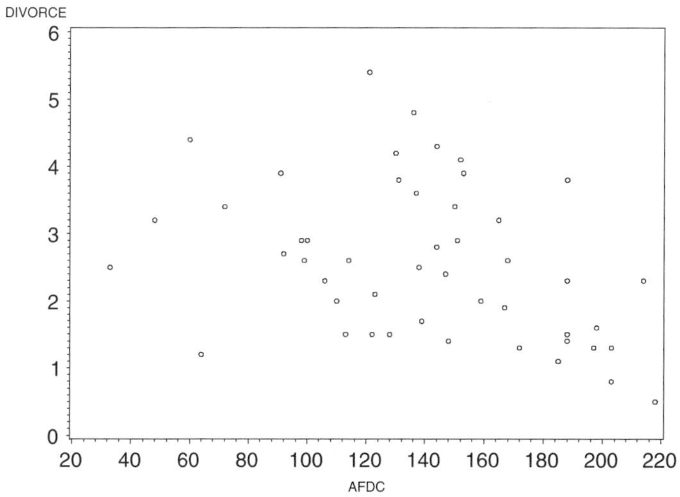

Figure 1.1. Plot of 1965 divorce rates versus AFDC payments.
(*Source: Statistical Abstract of the United States.*)

Figure 1.1 shows a negative relation; the corresponding correlation coefficient is −.37. Some argue that this negative relation is counterintuitive in that one would expect a positive relation between welfare payments and divorce rates; states with desirable economic climates enjoy both a low divorce rate and low welfare payments. Others argue that this negative relationship is intuitively plausible; wealthy states can afford high welfare payments and produce a cultural and economic climate conducive to low divorce rates.

Another plot, not displayed here, shows a similar negative relation for 1975; the corresponding correlation is −.425. Further, a plot with both the 1965 and 1975 data displays a negative relation between divorce rates and AFDC payments.

Figure 1.2 shows both the 1965 and 1975 data; a line connects the two observations within each state. These lines represent a change over time (dynamic), not a cross-sectional relationship. Each line displays a positive relationship; that is, as welfare payments increase so do divorce rates for each state. Again, we do not infer directions of causality from this display. The point is that the dynamic relation between divorce and welfare payments within a state differs dramatically from the cross-sectional relationship between states.

Some Notation

Models of longitudinal data are sometimes differentiated from regression and time-series data through their double subscripts. With this notation, we may

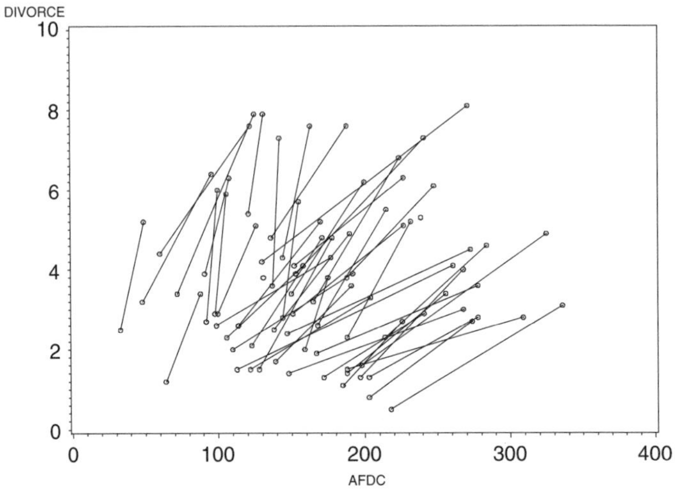

Figure 1.2. Plot of divorce rate versus AFDC payments from 1965 and 1975.

distinguish among responses by subject and time. To this end, define y_{it} to be the response for the ith subject during the tth time period. A longitudinal data set consists of observations of the ith subject over $t = 1, \ldots, T_i$ time periods, for each of $i = 1, \ldots, n$ subjects. Thus, we observe

$$\begin{gathered}\text{first subject} - \{y_{11}, y_{12}, \ldots, y_{1T_1}\}\\ \text{second subject} - \{y_{21}, y_{22}, \ldots, y_{2T_2}\}\\ \vdots\\ n\text{th subject} - \{y_{n1}, y_{n2}, \ldots, y_{nT_n}\}.\end{gathered}$$

In Example 1.1, most states have $T_i = 2$ observations and are depicted graphically in Figure 1.2 by a line connecting the two observations. Some states have only $T_i = 1$ observation and are depicted graphically by an open-circle plotting symbol. For many data sets, it is useful to let the number of observations depend on the subject; T_i denotes the number of observations for the ith subject. This situation is known as the *unbalanced data* case. In other data sets, each subject has the same number of observations; this is known as the *balanced data* case. Traditionally, much of the econometrics literature has focused on the balanced data case. We will consider the more broadly applicable unbalanced data case.

Prevalence of Longitudinal and Panel Data Analysis

Longitudinal and panel databases and models have taken on important roles in the literature. They are widely used in the social science literature, where panel data are also known as *pooled cross-sectional time series*, and in the natural sciences, where panel data are referred to as *longitudinal data*. To illustrate their prevalence, consider that an index of business and economic journals, ABI/INFORM, lists 326 articles in 2002 and 2003 that use panel data methods. Another index of scientific journals, the ISI Web of Science, lists 879 articles in 2002 and 2003 that use longitudinal data methods. Note that these are only the applications that were considered innovative enough to be published in scholarly reviews.

Longitudinal data methods have also developed because important databases have become available to empirical researchers. Within economics, two important surveys that track individuals over repeated surveys include the Panel Survey of Income Dynamics (PSID) and the National Longitudinal Survey of Labor Market Experience (NLS). In contrast, the Consumer Price Survey (CPS) is another survey conducted repeatedly over time. However, the CPS is generally not regarded as a panel survey because individuals are not tracked over time. For studying firm-level behavior, databases such as Compustat and CRSP (University of Chicago's Center for Research on Security Prices) have been available for over thirty years. More recently, the National Association of Insurance Commissioners (NAIC) has made insurance company financial statements available electronically. With the rapid pace of software development within the database industry, it is easy to anticipate the development of many more databases that would benefit from longitudinal data analysis. To illustrate, within the marketing area, product codes are scanned in when customers check out of a store and are transferred to a central database. These *scanner data* represent yet another source of data information that may inform marketing researchers about purchasing decisions of buyers over time or the efficiency of a store's promotional efforts. Appendix F summarizes longitudinal and panel data sets used worldwide.

1.2 Benefits and Drawbacks of Longitudinal Data

There are several advantages of longitudinal data compared with either purely cross-sectional or purely time-series data. In this introductory chapter, we focus on two important advantages: the ability to study dynamic relationships and to model the differences, or *heterogeneity*, among subjects. Of course, longitudinal data are more complex than purely cross-sectional or times-series data and so there is a price to pay in working with them. The most important drawback is the difficulty in designing the sampling scheme to reduce the problem of subjects leaving the study prior to its completion, known as *attrition*.

Dynamic Relationships

Figure 1.1 shows the 1965 divorce rate versus welfare payments. Because these are data from a single point in time, they are said to represent a *static* relationship.

For example, we might summarize the data by fitting a line using the method of least squares. Interpreting the slope of this line, we estimate a *decrease* of 0.95% in divorce rates for each \$100 increase in AFDC payments.

In contrast, Figure 1.2 shows changes in divorce rates for each state based on changes in welfare payments from 1965 to 1975. Using least squares, the overall slope represents an *increase* of 2.9% in divorce rates for each \$100 increase in AFDC payments. From 1965 to 1975, welfare payments increased an average of \$59 (in nominal terms) and divorce rates increased 2.5%. Now the slope represents a typical time change in divorce rates per \$100 unit time change in welfare payments; hence, it represents a *dynamic* relationship.

Perhaps the example might be more economically meaningful if welfare payments were in real dollars, and perhaps not (for example, deflated by the Consumer Price Index). Nonetheless, the data strongly reinforce the notion that dynamic relations can provide a very different message than cross-sectional relations.

Dynamic relationships can only be studied with repeated observations, and we have to think carefully about how we define our "subject" when considering dynamics. Suppose we are looking at the event of divorce on individuals. By looking at a cross section of individuals, we can estimate divorce rates. By looking at cross sections repeated over time (without tracking individuals), we can estimate divorce rates over time and thus study this type of dynamic movement. However, only by tracking repeated observations on a sample of individuals can we study the duration of marriage, or time until divorce, another dynamic event of interest.

Historical Approach

Early panel data studies used the following strategy to analyze pooled cross-sectional data:

- Estimate cross-sectional parameters using regression.
- Use time-series methods to model the regression parameter estimators, treating estimators as known with certainty.

Although useful in some contexts, this approach is inadequate in others, such as Example 1.1. Here, the slope estimated from 1965 data is −0.95%. Similarly, the slope estimated from 1975 data turns out to be −1.0%. Extrapolating these negative estimators from different cross sections yields very different results from the dynamic estimate: a positive 2.9%. Theil and Goldberger (1961E) provide an early discussion of the advantages of estimating the cross-sectional and time-series aspects simultaneously.

Dynamic Relationships and Time-Series Analysis

When studying dynamic relationships, univariate time-series analysis is a well-developed methodology. However, this methodology does not account for relationships among different subjects. In contrast, multivariate time-series analysis does account for relationships among a limited number of different subjects. Whether univariate or multivariate, an important limitation of time-series analysis is that it requires several (generally, at least thirty) observations to make reliable inferences. For an annual economic series with thirty observations, using time-series analysis means that we are using the same model to represent an economic system over a period of thirty years. Many problems of interest lack this degree of stability; we would like alternative statistical methodologies that do not impose such strong assumptions.

Longitudinal Data as Repeated Time Series

With longitudinal data we use several (repeated) observations of many subjects. Repeated observations from the same subject tend to be correlated. One way to represent this correlation is through dynamic patterns. A model that we use is the following:

$$y_{it} = \mathrm{E} y_{it} + \varepsilon_{it}, \quad t = 1, \ldots, T_i, \quad i = 1, \ldots, n, \tag{1.1}$$

where ε_{it} represents the deviation of the response from its mean; this deviation may include dynamic patterns. Further, the symbol E represents the expectation operator so that $\mathrm{E} y_{it}$ is the expected response. Intuitively, if there is a dynamic pattern that is common among subjects, then by observing this pattern over many subjects, we hope to estimate the pattern with fewer time-series observations than required of conventional time-series methods.

For many data sets of interest, subjects do not have identical means. As a first-order approximation, a linear combination of known, *explanatory* variables such as

$$\mathrm{E} y_{it} = \alpha + \mathbf{x}_{it}' \boldsymbol{\beta}$$

serves as a useful specification of the mean function. Here, $\mathbf{x}_{it}$ is a vector of explanatory, or *independent*, variables.

Longitudinal Data as Repeated Cross-Sectional Studies

Longitudinal data may be treated as a repeated cross section by ignoring the information about individuals that is tracked over time. As mentioned earlier, there are many important repeated surveys such as the CPS where subjects are not tracked over time. Such surveys are useful for understanding *aggregate* changes in a variable, such as the divorce rate, over time. However, if the interest is in studying the time-varying economic, demographic, or sociological characteristics *of an individual* on divorce, then tracking individuals over time is much more informative than using a repeated cross section.

Heterogeneity

By tracking subjects over time, we may model subject behavior. In many data sets of interest, subjects are unlike one another; that is, they are *heterogeneous*. In (repeated) cross-sectional regression analysis, we use models such as $y_{it} = \alpha + \mathbf{x}'_{it}\boldsymbol{\beta} + \varepsilon_{it}$ and ascribe the uniqueness of subjects to the disturbance term ε_{it}. In contrast, with longitudinal data we have an opportunity to model this uniqueness. A basic longitudinal data model that incorporates heterogeneity among subjects is based on

$$\mathrm{E}y_{it} = \alpha_i + \mathbf{x}'_{it}\boldsymbol{\beta}, \quad t = 1, \ldots, T_i, \; i = 1, \ldots, n. \qquad (1.2)$$

In cross-sectional studies where $T_i = 1$, the parameters of this model are unidentifiable. However, in longitudinal data, we have a sufficient number of observations to estimate $\boldsymbol{\beta}$ and $\alpha_1, \ldots, \alpha_n$. Allowing for *subject-specific parameters*, such as α_i, provides an important mechanism for controlling heterogeneity of individuals. Models that incorporate heterogeneity terms such as in Equation (1.2) will be called *heterogeneous models*. Models without such terms will be called *homogeneous models*.

We may also interpret heterogeneity to mean that observations from the same subject tend to be similar compared to observations from different subjects. Based on this interpretation, heterogeneity can be modeled by examining the sources of correlation among repeated observations from a subject. That is, for many data sets, we anticipate finding a positive correlation when examining $\{y_{i1}, y_{i2}, \ldots, y_{iT_i}\}$. As already noted, one possible explanation is the dynamic pattern among the observations. Another possible explanation is that the response shares a common, yet unobserved, subject-specific parameter that induces a positive correlation.

There are two distinct approaches for modeling the quantities that represent heterogeneity among subjects, $\{\alpha_i\}$. Chapter 2 explores one approach, where $\{\alpha_i\}$ are treated as fixed, yet unknown, parameters to be estimated. In this case, Equation (1.2) is known as a *fixed-effects* model. Chapter 3 introduces the second approach, where $\{\alpha_i\}$ are treated as draws from an unknown population and thus are random variables. In this case, Equation (1.2) may be expressed as

$$\mathrm{E}(y_{it} \mid \alpha_i) = \alpha_i + \mathbf{x}'_{it}\boldsymbol{\beta}.$$

This is known as a *random-effects* formulation.

Heterogeneity Bias

Failure to include heterogeneity quantities in the model may introduce serious bias into the model estimators. To illustrate, suppose that a data analyst mistakenly uses the function

$$\mathrm{E} y_{it} = \alpha + \mathbf{x}_{it}' \boldsymbol{\beta},$$

when Equation (1.2) is the true function. This is an example of heterogeneity bias, or a problem with data aggregation.

Similarly, one could have different (heterogeneous) slopes

$$\mathrm{E} y_{it} = \alpha + \mathbf{x}_{it}' \boldsymbol{\beta}_i$$

or different intercepts and slopes

$$\mathrm{E} y_{it} = \alpha_i + \mathbf{x}_{it}' \boldsymbol{\beta}_i.$$

Omitted Variables

Incorporating heterogeneity quantities into longitudinal data models is often motivated by the concern that important variables have been omitted from the model. To illustrate, consider the true model

$$y_{it} = \alpha_i + \mathbf{x}_{it}' \boldsymbol{\beta} + \mathbf{z}_i' \boldsymbol{\gamma} + \varepsilon_{it}.$$

Assume that we do not have available the variables represented by the vector $\mathbf{z}_i$; these *omitted variables* are also said to be *lurking*. If these omitted variables do not depend on time, then it is still possible to get reliable estimators of other model parameters, such as those included in the vector $\boldsymbol{\beta}$. One strategy is to consider the deviations of a response from its time-series average. This yields the derived model

$$\begin{aligned} y_{it}^* = y_{it} - \bar{y}_i &= (\alpha_i + \mathbf{x}_{it}' \boldsymbol{\beta} + \mathbf{z}_i' \boldsymbol{\gamma} + \varepsilon_{it}) - (\alpha_i + \bar{\mathbf{x}}_i' \boldsymbol{\beta} + \mathbf{z}_i' \boldsymbol{\gamma} + \bar{\varepsilon}_i) \\ &= (\mathbf{x}_{it} - \bar{\mathbf{x}}_i)' \boldsymbol{\beta} + \varepsilon_{it} - \bar{\varepsilon}_i = \mathbf{x}_{it}^{*\prime} \boldsymbol{\beta} + \varepsilon_{it}^*, \end{aligned}$$

where we use the response time-series average $\bar{y}_i = T_i^{-1} \sum_{t=1}^{T_i} y_{it}$ and similar quantities for $\bar{\mathbf{x}}_i$ and $\bar{\varepsilon}_i$. Thus, using ordinary least-square estimators based on regressing the deviations in $\mathbf{x}$ on the deviations in y yields a desirable estimator of $\boldsymbol{\beta}$.

This strategy demonstrates how longitudinal data can mitigate the problem of *omitted-variable bias*. For strategies that rely on purely cross-sectional data, it is well known that correlations of lurking variables, $\mathbf{z}$, with the model explanatory variables, $\mathbf{x}$, induce bias when estimating $\boldsymbol{\beta}$. If the lurking variable is time-invariant, then it is perfectly collinear with the subject-specific variables α_i. Thus, estimation strategies that account for subject-specific parameters also account for time-invariant omitted variables. Further, because of the collinearity between subject-specific variables and time-invariant omitted variables, we may interpret the subject-specific quantities α_i as proxies for omitted variables. Chapter 7 describes strategies for dealing with omitted-variable bias.

Efficiency of Estimators

A longitudinal data design may yield more efficient estimators than estimators based on a comparable amount of data from alternative designs. To illustrate, suppose that the interest is in assessing the average change in a response over time, such as the divorce rate. Thus, let $\bar{y}_{\bullet 1} - \bar{y}_{\bullet 2}$ denote the difference between divorce rates between two time periods. In a repeated cross-sectional study such as the CPS, we would calculate the reliability of this statistic assuming independence among cross sections to get

$$\mathrm{Var}\,(\bar{y}_{\bullet 1} - \bar{y}_{\bullet 2}) = \mathrm{Var}\,\bar{y}_{\bullet 1} + \mathrm{Var}\,\bar{y}_{\bullet 2}.$$

However, in a panel survey that tracks individuals over time, we have

$$\mathrm{Var}\,(\bar{y}_{\bullet 1} - \bar{y}_{\bullet 2}) = \mathrm{Var}\,\bar{y}_{\bullet 1} + \mathrm{Var}\,\bar{y}_{\bullet 2} - 2\,\mathrm{Cov}\,(\bar{y}_{\bullet 1}, \bar{y}_{\bullet 2})\,.$$

The covariance term is generally positive because observations from the same subject tend to be positively correlated. Thus, other things being equal, a panel survey design yields more efficient estimators than a repeated cross-section design.

One method of accounting for this positive correlation among same-subject observations is through the heterogeneity terms, α_i. In many data sets, introducing subject-specific variables α_i also accounts for a large portion of the variability. Accounting for this variation reduces the mean-square error and standard errors associated with parameter estimators. Thus, we are more efficient in parameter estimation than for the case without subject-specific variables α_i.

It is also possible to incorporate subject-invariant parameters, often denoted by λ_t, to account for period (temporal) variation. For many data sets, this does not account for the same amount of variability as $\{\alpha_i\}$. With small numbers of time periods, it is straightforward to use time dummy (binary) variables to incorporate subject-invariant parameters.

Other things equal, standard errors become smaller and efficiency improves as the number of observations increases. For some situations, a researcher may obtain more information by sampling each subject repeatedly. Thus, some advocate that an advantage of longitudinal data is that we generally have more observations, owing to the repeated sampling, and greater efficiency of estimators compared to a purely cross-sectional regression design. The danger of this philosophy is that generally observations from the same subject are related.

Thus, although more information is obtained by repeated sampling, researchers need to be cautious in assessing the amount of additional information gained.

Correlation and Causation

For many statistical studies, analysts are happy to describe associations among variables. This is particularly true of forecasting studies where the goal is to predict the future. However, for other analyses, researchers are interested in assessing causal relationships among variables.

Longitudinal and panel data are sometimes touted as providing "evidence" of causal effects. Just as with any statistical methodology, longitudinal data models in and of themselves are insufficient to establish causal relationships among variables. However, longitudinal data can be more useful than purely cross-sectional data in establishing causality. To illustrate, consider the three ingredients necessary for establishing causality, taken from the sociology literature (see, for example, Toon, 2000EP):

- A statistically significant relationship is required.
- The association between two variables must not be due to another, omitted, variable.
- The "causal" variable must precede the other variable in time.

Longitudinal data are based on measurements taken over time and thus address the third requirement of a temporal ordering of events. Moreover, as previously described, longitudinal data models provide additional strategies for accommodating omitted variables that are not available in purely cross-sectional data.

Observational data do not come from carefully controlled experiments where random allocations are made among groups. Causal inference is not directly accomplished when using observational data and only statistical models. Rather, one thinks about the data and statistical models as providing relevant empirical evidence in a chain of reasoning about causal mechanisms. Although longitudinal data provide stronger evidence than purely cross-sectional data, most of the work in establishing causal statements should be based on the theory of the substantive field from which the data are derived. Chapter 6 discusses this issue in greater detail.

Drawbacks: Attrition

Longitudinal data sampling design offers many benefits compared to purely cross-sectional or purely time-series designs. However, because the sampling structure is more complex, it can also fail in subtle ways. The most common failure of longitudinal data sets to meet standard sampling design assumptions is through difficulties that result from *attrition*. In this context, attrition refers to a gradual erosion of responses by subjects. Because we follow the same subjects over time, nonresponse typically increases through time. To illustrate, consider the U.S. Panel Study of Income Dynamics (PSID). In the first year (1968), the nonresponse rate was 24%. However, by 1985, the nonresponse rate grew to about 50%.

Attrition can be a problem because it may result in a *selection bias*. Selection bias potentially occurs when a rule other than simple random (or stratified) sampling is used to select observational units. Examples of selection bias often concern endogenous decisions by agents to join a labor pool or participate in a social program. Suppose that we are studying a solvency measure of a sample of insurance firms. If the firm becomes bankrupt or evolves into another type of financial distress, then we may not be able to examine financial statistics associated with the firm. Nonetheless, this is exactly the situation in which we would anticipate observing low values of the solvency measure. The response of interest is related to our opportunity to observe the subject, a type of selection bias. Chapter 7 discusses the attrition problem in greater detail.

1.3 Longitudinal Data Models

When examining the benefits and drawbacks of longitudinal data modeling, it is also useful to consider the types of inference that are based on longitudinal data models, as well as the variety of modeling approaches. The type of application under consideration influences the choice of inference and modeling approaches.

Types of Inference

For many longitudinal data applications, the primary motivation for the analysis is to learn about the effect that an (exogenous) explanatory variable has on a response, controlling for other variables, including omitted variables. Users are interested in whether estimators of parameter coefficients, contained in the vector β, differ in a statistically significant fashion from zero. This is also the primary motivation for most studies that involve regression analysis; this is not surprising given that many models of longitudinal data are special cases of regression models.

Because longitudinal data are collected over time, they also provide us with an ability to predict future values of a response for a specific subject. Chapter 4 considers this type of inference, known as *forecasting*.

The focus of Chapter 4 is on the "estimation" of random variables, known as *prediction*. Because future values of a response are, to the analyst, random variables, forecasting is a special case of prediction. Another special case involves situations where we would like to predict the expected value of a future response from a specific subject, conditional on *latent* (unobserved) characteristics associated with the subject. For example, this conditional expected value is known in insurance theory as a *credibility premium*, a quantity that is useful in pricing of insurance contracts.

Social Science Statistical Modeling

Statistical models are mathematical idealizations constructed to represent the behavior of data. When a statistical model is constructed (designed) to represent a data set with little regard to the underlying functional field from which the data emanate, we may think of the model as essentially data driven. For example, we might examine a data set of the form $(x_1, y_1), \ldots, (x_n, y_n)$ and posit a regression model to capture the association between x and y. We will call this type of model a *sampling-based model*, or, following the econometrics literature, we say that the model arises from the *data-generating process*.

In most cases, however, we will know something about the units of measurement of x and y and anticipate a type of relationship between x and y based on knowledge of the functional field from which these variables arise. To continue our example in a finance context, suppose that x represents a return from a market index and that y represents a stock return from an individual security. In this case, financial economics theory suggests a linear regression relationship of y on x. In the economics literature, Goldberger (1972E) defines a *structural model* to be a statistical model that represents causal relationships, as opposed to relationships that simply capture statistical associations. Chapter 6 further develops the idea of causal inference.

If a sampling-based model adequately represents statistical associations in our data, then why bother with an extra layer of theory when considering statistical models? In the context of binary dependent variables, Manski (1992E) offers three motivations: interpretation, precision, and extrapolation.

Interpretation is important because the primary purpose of many statistical analyses is to assess relationships generated by theory from a scientific field. A sampling-based model may not have sufficient structure to make this assessment, thus failing the primary motivation for the analysis.

Structural models utilize additional information from an underlying functional field. If this information is utilized correctly, then in some sense the structural model should provide a better representation than a model without this information. With a properly utilized structural model, we anticipate getting more precise estimates of model parameters and other characteristics. In practical terms, this improved precision can be measured in terms of smaller standard errors.

At least in the context of binary dependent variables, Manski (1992E) feels that extrapolation is the most compelling motivation for combining theory from a functional field with a sampling-based model. In a time-series context, extrapolation means forecasting; this is generally the main impetus for an analysis. In a regression context, extrapolation means inference about responses for sets of predictor variables "outside" of those realized in the sample. Particularly for public policy analysis, the goal of a statistical analysis is to infer the likely behavior of data outside of those realized.

Modeling Issues

This chapter has portrayed longitudinal data modeling as a special type of regression modeling. However, in the biometrics literature, longitudinal data models have their roots in multivariate analysis. Under this framework, we view the responses from an individual as a vector of responses; that is, $\mathbf{y}_i = (y_{i1}, y_{i2}, \ldots, y_{iT})'$. Within the biometrics framework, the first applications are referred to as *growth curve* models. These classic examples use the height of children as the response to examine the changes in height and growth, over time (see Chapter 5). Within the econometrics literature, Chamberlain (1982E, 1984E) exploited the multivariate structure. The multivariate analysis approach is most effective with balanced data at points equally spaced in time. However, compared to the regression approach, there are several limitations of the multivariate approach. These include the following:

- It is harder to analyze missing data, attrition, and different accrual patterns.
- Because there is no explicit allowance for time, it is harder to forecast and predict at time points between those collected (interpolation).

Even within the regression approach for longitudinal data modeling, there are still a number of issues that need to be resolved in choosing a model. We have already introduced the issue of modeling heterogeneity. Recall that there are two important types of models of heterogeneity, fixed- and random-effects models (the subjects of Chapters 2 and 3).

Another important issue is the structure for modeling the dynamics; this is the subject of Chapter 8. We have described imposing a serial correlation on the disturbance terms. Another approach, described in Section 8.2, involves using lagged (endogenous) responses to account for temporal patterns. These models are important in econometrics because they are more suitable for structural modeling where a greater tie exists between economic theory and statistical modeling than models that are based exclusively on features of the data. When the number of (time) observations per subject, T, is small, then simple correlation structures of the disturbance terms provide an adequate fit for many data sets. However, as T increases, we have greater opportunities to model the dynamic structure. The Kalman filter, described in Section 8.5, provides a computational technique that allows the analyst to handle a broad variety of complex dynamic patterns.

Many of the longitudinal data applications that appear in the literature are based on linear model theory. Hence, this text is predominantly (Chapters 1 through 8) devoted to developing linear longitudinal data models. However, nonlinear models represent an area of recent development where examples of their importance to statistical practice appear with greater frequency. The phrase "nonlinear models" in this context refers to instances where the distribution of the response cannot be reasonably approximated using a normal curve. Some examples of this occur when the response is binary or consists of other types of count data, such as the number of accidents in a state, and when the response is from a very heavy tailed distribution, such as with insurance claims. Chapters 9 through 11 introduce techniques from this budding literature to handle these types of nonlinear models.

Types of Applications

A statistical model is ultimately useful only if it provides an accurate approximation to real data. Table 1.1 outlines the data sets used in this text to underscore the importance of longitudinal data modeling.

1.4 Historical Notes

The term "panel study" was coined in a marketing context when Lazarsfeld and Fiske (1938O) considered the effect of radio advertising on product sales. Traditionally, hearing radio advertisements was thought to increase the likelihood of purchasing a product. Lazarsfeld and Fiske considered whether those that bought the product would be more likely to hear the advertisement, thus positing a reverse in the direction of causality. They proposed repeatedly interviewing a set of people (the "panel") to clarify the issue.

Baltes and Nesselroade (1979EP) trace the history of longitudinal data and methods with an emphasis on childhood development and psychology. They describe longitudinal research as consisting of "a variety of methods connected by the idea that the entity under investigation is observed repeatedly as it exists and evolves over time." Moreover, they trace the need for longitudinal research to at least as early as the nineteenth century.

Table 1.1. *Several illustrative longitudinal data sets*

Data title	Subject area	File name	Unit of analysis	Description
Bond maturity	Finance	Bondmat	Subjects are $n = 328$ firms over $T = 10$ years: 1980–1989. $N = 3{,}280$ observations.	Examine the maturity of debt structure in terms of corporate financial characteristics.
Capital structure	Finance	Capital	Subjects are $n = 361$ Japanese firms over $T = 15$ years: 1984–1998. $N = 5{,}415$ observations.	Examine changes in capital structure before and after the market crash for different types of cross-holding structures.
Charitable contributions	Accounting	Charity	Subjects are $n = 47$ taxpayers over $T = 10$ years: 1979–1988. $N = 470$ observations.	Examine characteristics of taxpayers to determine factors that influence the amount of charitable giving.
Divorce	Sociology	Divorce	Subjects are $n = 51$ states over $T = 4$ years: 1965, 1975, 1985, and 1995. $N = 204$ observations.	Assess socioeconomic variables that affect the divorce rate.
Group term life data	Insurance	Glife	Subjects are $n = 106$ credit unions over $T = 7$ years. $N = 742$ observations.	Forecast group term life insurance claims of Florida credit unions.
Housing prices	Real estate	Hprice	Subjects are $n = 36$ metropolitan statistical areas (MSAs) over $T = 9$ years: 1986–1994. $N = 324$ observations.	Examine annual housing prices in terms of MSA demographic and economic indices.
Lottery sales	Marketing	Lottery	Subjects are $n = 50$ postal code areas over $T = 40$ weeks.	Examine effects of area economic and demographic characteristics on lottery sales.
Medicare hospital costs	Social insurance	Medicare	Subjects are $n = 54$ states over $T = 6$ years: 1990–1995. $N = 324$ observations.	Forecast Medicare hospital costs by state based on utilization rates and past history.
Property and liability insurance	Insurance	Pdemand	Subjects are $n = 22$ countries over $T = 7$ years: 1987–1993. $N = 154$ observations.	Examine the demand for property and liability insurance in terms of national economic and risk-aversion characteristics.
Student achievement	Education	Student	Subjects are $n = 400$ students from 20 schools observed over $T = 4$ grades (3–6). $N = 1{,}012$ observations.	Examine student math achievement based on student and school demographic and socioeconomic characteristics.
Tax preparers	Accounting	Taxprep	Subjects are $n = 243$ taxpayers over $T = 5$ years: 1982–1984, 1986, 1987. $N = 1{,}215$ observations.	Examine characteristics of taxpayers to determine the demand for a professional tax preparer.
Tort filings	Insurance	Tfiling	Subjects are $n = 19$ states over $T = 6$ years: 1984–1989. $N = 114$ observations.	Examine demographic and legal characteristics of states that influence the number of tort filings.
Workers' compensation	Insurance	Workerc	Subjects are $n = 121$ occupation classes over $T = 7$ years. $N = 847$ observations.	Forecast workers' compensation claims by occupation class.

Toon (2000EP) cites Engel's 1857 budget survey, examining how the amount of money spent on food changes as a function of income, as perhaps the earliest example of a study involving repeated measurements from the same set of subjects.

As noted in Section 1.2, in early panel data studies, pooled cross-sectional data were analyzed by estimating cross-sectional parameters using regression and then using time-series methods to model the regression parameter estimates, treating the estimates as known with certainty. Dielman (1989O) discusses this approach in more detail and provides examples. Early applications in economics of the basic fixed-effects model include those by Kuh (1959E), Johnson (1960E), Mundlak (1961E) and Hoch (1962E). Chapter 2 introduces this and related models in detail.

Balestra and Nerlove (1966E) and Wallace and Hussain (1969E) introduced the (random-effects) error-components model, the model with $\{\alpha_i\}$ as random variables. Chapter 3 introduces this and related models in detail.

Wishart (1938B), Rao (1965B), and Potthoff and Roy (1964B) were among the first contributors in the biometrics literature to use multivariate analysis for analyzing growth curves. Specifically, they considered the problem of fitting polynomial growth curves of serial measurements from a group of subjects. Chapter 5 contains examples of growth curve analysis.

This approach to analyzing longitudinal data was extended by Grizzle and Allen (1969B), who introduced covariates, or explanatory variables, into the analysis. Laird and Ware (1982B) made the other important transition from multivariate analysis to regression modeling. They introduce the two-stage model that allows for both fixed and random effects. Chapter 3 considers this modeling approach.

2

Fixed-Effects Models

Abstract. This chapter introduces the analysis of longitudinal and panel data using the general linear model framework. Here, longitudinal data modeling is cast as a regression problem by using fixed parameters to represent the heterogeneity; nonrandom quantities that account for the heterogeneity are known as *fixed effects*. In this way, ideas of model representation and data exploration are introduced using regression analysis, a toolkit that is widely known. Analysis of covariance, from the general linear model, easily handles the many parameters needed to represent the heterogeneity.

Although longitudinal and panel data can be analyzed using regression techniques, it is also important to emphasize the special features of these data. Specifically, the chapter emphasizes the wide cross section and the short time series of many longitudinal and panel data sets, as well as the special model specification and diagnostic tools needed to handle these features.

2.1 Basic Fixed-Effects Model

Data

Suppose that we are interested in explaining hospital costs for each state in terms of measures of utilization, such as the number of discharged patients and the average hospital stay per discharge. Here, we consider the state to be the unit of observation, or *subject*. We differentiate among states with the index i, where i may range from 1 to n, and n is the number of subjects. Each state is observed T_i times and we use the index t to differentiate the observation times. With these indices, let y_{it} denote the response of the ith subject at the tth time point. Associated with each response y_{it} is a set of explanatory variables, or *covariates*. For example, for state hospital costs, these explanatory variables include the number of discharged patients and the average hospital stay per discharge. In general, we assume there are K explanatory variables $x_{it,1}, x_{it,2}, \ldots, x_{it,K}$ that may vary by subject i and time t. We achieve a more compact notational form by expressing the K explanatory variables as a $K \times 1$ column vector

$$\mathbf{x}_{it} = \begin{pmatrix} x_{it,1} \\ x_{it,2} \\ \vdots \\ x_{it,K} \end{pmatrix}.$$

To save space, it is customary to use the alternate expression $\mathbf{x}_{it} = (x_{it,1}, x_{it,2}, \ldots, x_{it,K})'$, where the prime means transpose. (You will find that some sources prefer to use a superscript "T" for transpose. Here, T will refer to the number of time replications.) Thus, the data for the ith subject consists of

$$\begin{gathered} \{x_{i1,1}, \ldots, x_{i1,K}, y_{i1}\} \\ \vdots \\ \{x_{iT_i,1}, \ldots, x_{iT_i,K}, y_{iT_i}\}, \end{gathered}$$

which can be expressed more compactly as

$$\begin{gathered} \{\mathbf{x}'_{i1}, y_{i1}\} \\ \vdots \\ \{\mathbf{x}'_{iT_i}, y_{iT_i}\}. \end{gathered}$$

Unless specified otherwise, we allow the number of responses to vary by subject, indicated with the notation T_i. This is known as the *unbalanced* case. We use the notation $T = \max\{T_1, T_2, \ldots, T_n\}$ to be the maximal number of responses for a subject. Recall from Section 1.1 that the case $T_i = T$ for each i is called the *balanced* case.

Basic Models

To analyze relationships among variables, the relationships between the response and the explanatory variables are summarized through the *regression function*

$$\mathrm{E}y_{it} = \alpha + \beta_1 x_{it,1} + \beta_2 x_{it,2} + \cdots + \beta_K x_{it,K}, \tag{2.1}$$

which is linear in the parameters $\alpha, \beta_1, \ldots, \beta_K$. For applications where the explanatory variables are nonrandom, the only restriction of Equation (2.1) is that we believe that the variables enter linearly. As we will see in Chapter 6, for applications where the explanatory variables are random, we may interpret the expectation in Equation (2.1) as conditional on the observed explanatory variables.

We focus attention on assumptions that concern the observable variables, $\{x_{it,1}, \ldots, x_{it,K}, y_{it}\}$.

Assumptions of the Observables Representation of the Linear Regression Model

F1. $\mathrm{E} y_{it} = \alpha + \beta_1 x_{it,1} + \beta_2 x_{it,2} + \cdots + \beta_K x_{it,K}$.
F2. $\{x_{it,1}, \ldots, x_{it,K}\}$ are nonstochastic variables.
F3. Var $y_{it} = \sigma^2$.
F4. $\{y_{it}\}$ are independent random variables.

The "observables representation" is based on the idea of conditional linear expectations (see Goldberger, 1991E, for additional background). One can motivate Assumption F1 by thinking of $(x_{it,1}, \ldots, x_{it,K}, y_{it})$ as a draw from a population, where the mean of the conditional distribution of y_{it} given $\{x_{it,1}, \ldots, x_{it,K}\}$ is linear in the explanatory variables. Inference about the distribution of y is conditional on the observed explanatory variables, so that we may treat $\{x_{it,1}, \ldots, x_{it,K}\}$ as nonstochastic variables. When considering types of sampling mechanisms for thinking of $(x_{it,1}, \ldots, x_{it,K}, y_{it})$ as a draw from a population, it is convenient to think of a stratified random sampling scheme, where values of $\{x_{it,1}, \ldots, x_{it,K}\}$ are treated as the strata. That is, for each value of $\{x_{it,1}, \ldots, x_{it,K}\}$, we draw a random sample of responses from a population. This sampling scheme also provides motivation for Assumption F4, the independence among responses. To illustrate, when drawing from a database of firms to understand stock return performance (y), one can choose large firms (measured by asset size), focus on an industry (measured by standard industrial classification), and so forth. You may not select firms with the largest stock return performance because this is stratifying based on the response, not the explanatory variables.

A fifth assumption that is often implicitly required in the linear regression model is the following:

F5 $\{y_{it}\}$ is normally distributed.

This assumption is not required for all statistical inference procedures because central limit theorems provide approximate normality for many statistics of interest. However, formal justification for some, such as t-statistics, do require this additional assumption.

In contrast to the observables representation, the classical formulation of the linear regression model focuses attention on the "errors" in the regression, defined as $\varepsilon_{it} = y_{it} - (\alpha + \beta_1 x_{it,1} + \beta_2 x_{it,2} + \cdots + \beta_K x_{it,K})$.

Assumptions of the Error Representation of the Linear Regression Model

E1. $y_{it} = \alpha + \beta_1 x_{it,1} + \beta_2 x_{it,2} + \cdots + \beta_K x_{it,K} + \varepsilon_{it}$, where $\mathrm{E}\,\varepsilon_{it} = 0$.
E2. $\{x_{it,1}, \ldots, x_{it,K}\}$ are nonstochastic variables.
E3. $\mathrm{Var}\,\varepsilon_{it} = \sigma^2$.
E4. $\{\varepsilon_{it}\}$ are independent random variables.

The "error representation" is based on the Gaussian theory of errors (see Stigler, 1986G, for a historical background). As already described, the linear regression function incorporates the additional knowledge from independent variables through the relation $\mathrm{E}y_{it} = \alpha + \beta_1 x_{it,1} + \beta_2 x_{it,2} + \cdots + \beta_K x_{it,K}$. Other unobserved variables that influence the measurement of y are encapsulated in the "error" term ε_{it}, which is also known as the "disturbance" term. The independence of errors, F4, can be motivated by assuming that $\{\varepsilon_{it}\}$ are realized through a simple random sample from an unknown population of errors.

Assumptions E1–E4 are equivalent to assumptions F1–F4. The error representation provides a useful springboard for motivating goodness-of-fit measures. However, a drawback of the error representation is that it draws the attention from the observable quantities $(x_{it,1}, \ldots, x_{it,K}, y_{it})$ to an unobservable quantity, $\{\varepsilon_{it}\}$. To illustrate, consider that the sampling basis, viewing $\{\varepsilon_{it}\}$ as a simple random sample, is not directly verifiable because one cannot directly observe the sample $\{\varepsilon_{it}\}$. Moreover, the assumption of additive errors in E1 will be troublesome when we consider nonlinear regression models in Chapters 9–11. Our treatment focuses on the observable representation in Assumptions F1–F4.

In Assumption F1, the slope parameters $\beta_1, \beta_2, \ldots, \beta_K$ are associated with the K explanatory variables. For a more compact expression, we summarize the parameters as a column vector of dimension $K \times 1$, denoted by

$$\boldsymbol{\beta} = \begin{pmatrix} \beta_1 \\ \vdots \\ \beta_K \end{pmatrix}.$$

With this notation, we may rewrite Assumption F1 as

$$\mathrm{E}y_{it} = \alpha + \mathbf{x}_{it}' \boldsymbol{\beta}, \tag{2.2}$$

because of the relation $\mathbf{x}_{it}'\beta = \beta_1 x_{it,1} + \beta_2 x_{it,2} + \cdots + \beta_K x_{it,K}$. We call the representation in Equation (2.2) cross sectional because, although it relates the explanatory variables to the response, it does not use the information in the repeated measurements on a subject. Because it also does not include (subject-specific) heterogeneous terms, we also refer to the Equation (2.2) representation as part of a *homogeneous model.*

Our first representation that uses the information in the repeated measurements on a subject is

$$\mathrm{E}y_{it} = \alpha_i + \mathbf{x}_{it}'\beta. \tag{2.3}$$

Equation (2.3) and Assumptions F2–F4 comprise the *basic fixed-effects model.* Unlike Equation (2.2), in Equation (2.3) the intercept terms, α_i, are allowed to vary by subject.

Parameters of Interest

The parameters $\{\beta_j\}$ are common to each subject and are called global, or *population*, parameters. The parameters $\{\alpha_i\}$ vary by subject and are known as individual, or *subject-specific*, parameters. In many applications, we will see that population parameters capture broad relationships of interest and hence are the parameters of interest. The subject-specific parameters account for the different features of subjects, not broad population patterns. Hence, they are often of secondary interest and are called *nuisance parameters.*

As we saw in Section 1.3, the subject-specific parameters represent our first device that helps control for the heterogeneity among subjects. We will see that estimators of these parameters use information in the repeated measurements on a subject. Conversely, the parameters $\{\alpha_i\}$ are nonestimable in cross-sectional regression models without repeated observations. That is, with $T_i = 1$, the model

$$y_{i1} = \alpha_i + \beta_1 x_{i1,1} + \beta_2 x_{i1,2} + \cdots + \beta_K x_{i1,K} + \varepsilon_{i1}$$

has more parameters $(n + K)$ than observations (n) and thus we cannot identify all the parameters. Typically, the disturbance term ε_{it} includes the information in α_i in cross-sectional regression models. An important advantage of longitudinal data models compared to cross-sectional regression models is the ability to separate the effects of $\{\alpha_i\}$ from the disturbance terms $\{\varepsilon_{it}\}$. By separating out subject-specific effects, our estimates of the variability become more precise and we achieve more accurate inferences.

Subject and Time Heterogeneity

We will argue that the subject-specific parameter α_i captures much of the time-constant information in the responses. However, the basic fixed-effects model assumes that $\{y_{it}\}$ are independent terms and, in particular, that

- no *serial correlation* (correlation over time) exists and
- no *contemporaneous correlation* (correlation across subjects) exists.

Thus, no special relationships between subjects and time periods are assumed. By interchanging the roles of i and t, we may consider the function

$$\mathrm{E}y_{it} = \lambda_t + \mathbf{x}_{it}'\boldsymbol{\beta}, \tag{2.4}$$

where the parameter λ_t is a time-specific variable that does not depend on subjects.

For most longitudinal data applications, the number of subjects, n, substantially exceeds the maximal number of time periods, T. Further, generally the heterogeneity among subjects explains a greater proportion of variability than the heterogeneity among time periods. Thus, we begin with the "basic" function $\mathrm{E}y_{it} = \alpha_i + \mathbf{x}_{it}'\boldsymbol{\beta}$. This model allows explicit parameterization of the subject-specific heterogeneity.

Both functions in Equations (2.3) and (2.4) are based on traditional one-way analysis of covariance models. For this reason, the basic fixed-effects model is also called the *one-way fixed-effects model*. By using binary (dummy) variables for the time dimension, we can incorporate time-specific parameters into the population parameters. In this way, it is straightforward to consider the function

$$\mathrm{E}y_{it} = \alpha_i + \lambda_t + \mathbf{x}_{it}'\boldsymbol{\beta}. \tag{2.5}$$

Equation (2.5) with Assumptions F2–F4 is called the *two-way fixed-effects model*.

Example 2.1: Urban Wages Glaeser and Maré (2001E) investigated the effects of determinants on wages, with the goal of understanding why workers in cities earn more than their nonurban counterparts. They examined two-way fixed-effects models using data from the National Longitudinal Survey of Youth (NLSY); they also used data from the PSID to assess the robustness of their results to another sample. For the NLSY data, they examined $n = 5{,}405$ male heads of households over the years 1983–1993, consisting of a total of $N = 40{,}194$ observations. The dependent variable was logarithmic hourly wage. The primary explanatory variable of interest was a three-level categorical variable that measures the city size in which workers reside. To capture this variable, two binary (dummy) variables were used: (1) a variable to indicate whether the worker resides in a large city (with more than one-half million residents), a "dense metropolitan area," and (2) a variable to indicate whether the worker resides in a metropolitan area that does not contain a large city, a "nondense metropolitan area." The omitted category is nonmetropolitan area. Several other control variables were included to capture effects of a worker's experience, occupation, education, and race. When including time dummy variables, there were $K = 30$ explanatory variables in the reported regressions.

2.2 Exploring Longitudinal Data

Why Explore?

The models that we use to represent reality are simplified approximations. As stated by George Box (1979G), "All models are wrong, but some are useful." The inferences that we draw by examining a model calibrated with a data set depend on the data characteristics; we expect a reasonable proximity between the model assumptions and the data. To assess this proximity, we explore the many important features of the data. By *data exploration*, we mean summarizing the data, either numerically or graphically, without reference to a model.

Data exploration provides hints of the appropriate model. To draw reliable inferences from the modeling procedure, it is important that the data be congruent with the model. Further, exploring the data also alerts us to any unusual observations or subjects. Because standard inference techniques are generally nonrobust to unusual features, it is important to identify these features early in the modeling process.

Data exploration also provides an important communication device. Because data exploration techniques are not model dependent, they may be better understood than model-dependent inference techniques. Thus, they can be used to communicate features of a data set, often supplementing model-based inferences.

Data Exploration Techniques

Longitudinal data analysis is closely linked to multivariate analysis and regression analysis. Thus, the data exploration techniques developed in these fields are applicable to longitudinal data and will not be developed here. The reader may consult Tukey (1977G) for the original source on exploratory data analysis. The following list summarizes commonly used data exploration techniques that will be demonstrated throughout this book:

- Examine graphically the distribution of y and each x through histograms, density estimates, boxplots, and so on.
- Examine numerically the distribution of y and each x through statistics such as means, medians, standard deviations, minimums, and maximums.
- Examine the relationship between y and each x through correlations and scatter plots.

In addition, summary statistics and graphs by time period may be useful for detecting temporal patterns.

Three data exploration techniques that are specific to longitudinal data are (1) multiple time-series plots, (2) scatter plots with symbols, and (3) added-variable plots. Because these techniques are specific to longitudinal data analysis, they are less widely known and are described in the following. Another way to examine data is through *diagnostic techniques*, described in Section 2.4. In contrast to data exploration techniques, diagnostic techniques are performed after the fit of a preliminary model.

Multiple Time-Series Plots

A *multiple time-series plot* is a plot of a variable, generally the response, y_{it}, versus time t. Within the context of longitudinal data, we serially (over time) connect observations over a common subject. This graph helps us to (1) detect patterns in the response, by subject and over time, (2) identify unusual observations and/or subjects, and (3) visualize the heterogeneity.

Scatter Plots with Symbols

In the context of regression, a plot of the response, y_{it}, versus an explanatory variable, x_{itj}, helps us to assess the relationship between these variables. In the context of longitudinal data, it is often useful to add a plotting symbol to the scatter plot to identify the subject. This allows us to see the relationship between the response and explanatory variable yet account for the varying intercepts. Further, if there is a separation in the explanatory variable, such as an increase over time, then we can serially connect the observations. In this case, we may not require a separate plotting symbol for each subject.

Basic Added-Variable Plot

A *basic added-variable* plot is a scatter plot of $\{y_{it} - \bar{y}_i\}$ versus $\{x_{itj} - \bar{x}_{ij}\}$, where $\bar{y}_i$ and $\bar{x}_{ij}$ are averages of $\{y_{it}\}$ and $\{x_{itj}\}$ over time. An added-variable plot is a standard regression diagnostic technique and is described in further detail in Section 2.4. Although the basic added-variable plot can be viewed as a special case of the more general diagnostic technique, it can also be motivated without reference to a model. That is, in many longitudinal data sets, the subject-specific parameters account for a large portion of the variability. This plot allows us to visualize the relationship between y and each x, without forcing our eye to adjust for the heterogeneity of the subject-specific intercepts.

Example 2.2: Medicare Hospital Costs We consider $T = 6$ years, 1990–1995, of data for inpatient hospital charges that are covered by the Medicare program. The data were obtained from the Center for Medicare and Medicaid Services. To illustrate, in 1995 the total covered charges were \$157.8 billions for 12 million discharges. For this analysis, we use state as the subject, or risk class. Thus, we consider $n = 54$ states, which include the 50 states in the Union, the District of Columbia, Virgin Islands, Puerto Rico, and an unspecified "other" category.

The response variable of interest is the severity component, covered claims per discharge, which we label as CCPD. The variable CCPD is of interest to actuaries because the Medicare program reimburses hospitals on a per-stay basis. Also, many managed care plans reimburse hospitals on a per-stay basis. Because CCPD varies over state and time, both the state and time (YEAR = 1, . . . , 6) are potentially important explanatory variables. We do not assume a priori that frequency is independent of severity. Thus, number of discharges, NUM_DSCHG, is another potential explanatory variable. We also investigate the importance of another component of hospital utilization, AVE_DAYS, defined to be the average hospital stay per discharge in days.

Table 2.1 summarizes these basic variables by year. Here, we see that both claims and number of discharges increase over time whereas the average hospital stay decreases. The standard deviations and extreme values indicate that there is substantial variability among states.

Figure 2.1 illustrates the multiple time-series plot. Here, we see that not only are overall claims increasing but also that claims increase for each state. Different levels of hospital costs among states are also apparent; we call this feature heterogeneity. Further, Figure 2.1 indicates that there is greater variability among states than over time.

Figure 2.2 illustrates the scatter plot with symbols. This plot of CCPD versus number of discharges, connecting observations over time, shows a positive overall relationship between CCPD and the number of discharges. Like CCPD, we see a substantial state variation of different numbers of discharges. Also like CCPD, the number of discharges increases over time, so that, for each state, there

Table 2.1. *Summary statistics of covered claims per discharge, number of discharges, and average hospital stay, by year*

Variable	Time period[b]	Mean	Median	Standard deviation	Minimum	Maximum
Covered claims	1990	8,503	7,992	2,467	3,229	16,485
per discharge	1991	9,473	9,113	2,712	2,966	17,637
(CCPD)[a]	1992	10,443	10,055	3,041	3,324	19,814
	1993	11,160	10,667	3,260	4,138	21,122
	1994	11,523	10,955	3,346	4,355	21,500
	1995	11,797	11,171	3,278	5,058	21,032
Total		10,483	10,072	3,231	2,966	21,500
Number of	1990	197.73	142.59	202.99	0.53	849.37
discharges	1991	203.14	142.69	210.38	0.52	885.92
(NUM_DSCHG)	1992	210.89	143.25	218.92	0.65	908.59
(in thousands)	1993	211.25	143.67	219.82	0.97	894.22
	1994	218.87	150.08	226.78	1.16	905.62
	1995	222.51	152.70	229.46	1.06	902.48
Total		210.73	144.28	216.72	0.52	908.59
Average	1990	9.05	8.53	2.08	6.33	17.48
hospital stay	1991	9.82	8.57	7.23	6.14	60.25
(AVE_DAYS)	1992	8.62	8.36	1.86	5.83	16.35
	1993	8.52	8.11	2.11	5.83	17.14
	1994	7.90	7.56	1.73	5.38	14.39
	1995	7.34	7.14	1.44	5.12	12.80
Total		8.54	8.07	3.47	5.12	60.25

[a] The variable CCPD is in dollars of claim per discharge.
[b] Each year summarizes $n = 54$ states. The total summarizes $(6 \times 54) = 324$ observations.
Source: Center for Medicare and Medicaid Services.

is a positive relationship between CCPD and number of discharges. The slope is higher for those states with fewer discharges. This plot also suggests that the number of discharges lagged by one year is an important predictor of CCPD.

Figure 2.3 is a scatter plot of CCPD versus average total days, connecting observations over time. This plot demonstrates the unusual nature of the second observation for the 54th state. We also see evidence of this point through the maximum statistic of the average hospital stay in Table 2.1. This point does not appear to follow the same pattern as the rest of our data and turns out to have a large impact on our fitted models.

Figure 2.4 illustrates the basic added-variable plot. This plot portrays CCPD versus year, after excluding the second observation for the 54th state. In

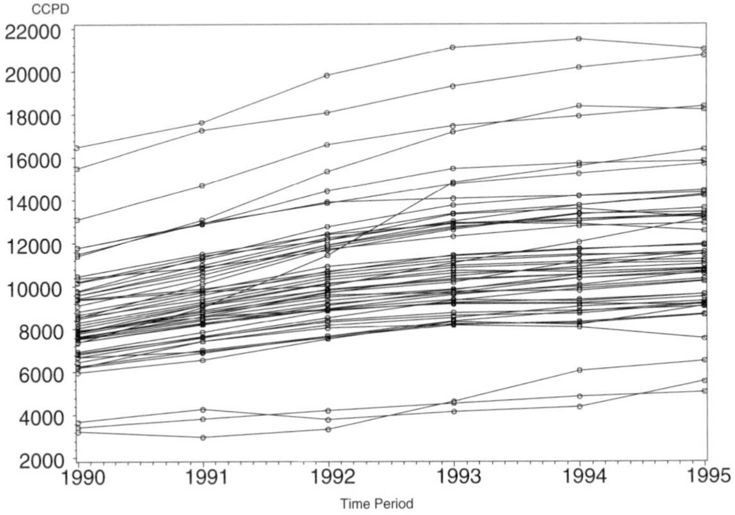

Figure 2.1. Multiple time-series plot of CCPD. Covered claims per discharge (CCPD) are plotted over $T = 6$ years, 1990–1995. The line segments connect states; thus, we see that CCPD increases for almost every state over time.

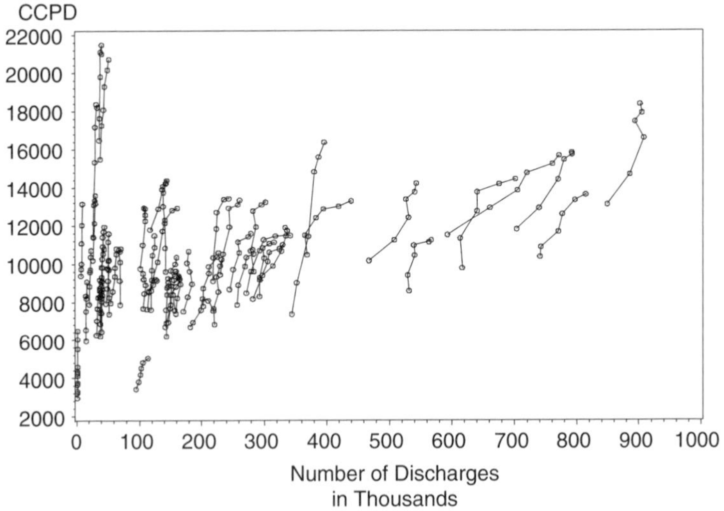

Figure 2.2. Scatter plot of CCPD versus number of discharges. The line segments connect observations within a state over 1990–1995. We see a substantial state variation of numbers of discharges. There is a positive relationship between CCPD and number of discharges for each state. Slopes are higher for those states with fewer discharges.

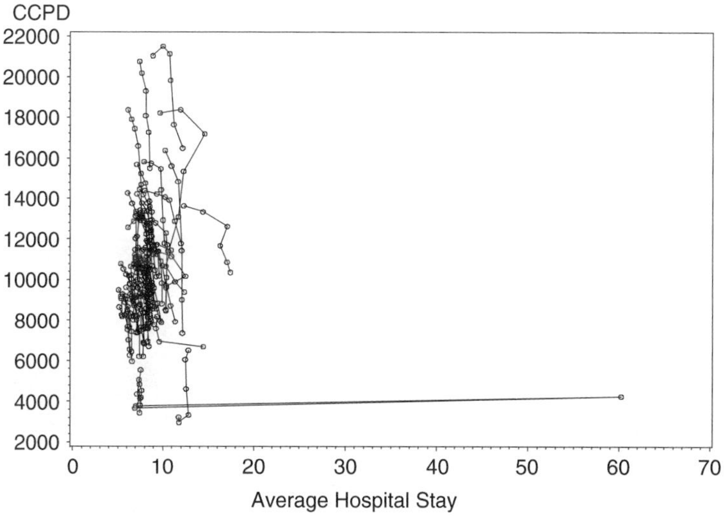

Figure 2.3. Scatter plot of CCPD versus average hospital stay. The line segments connect states over 1990–1995. This figure demonstrates that the second observation for the 54th state is unusual.

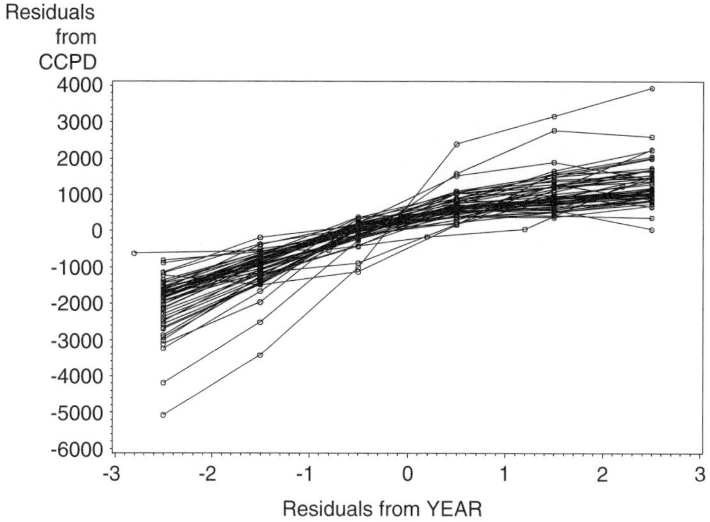

Figure 2.4. Added-variable plot of CCPD versus year. Here, we have controlled for the state factor. In this figure, the second observation for the 54th state has been excluded. We see that the rate of increase of CCPD over time is approximately consistent among states, yet there exist important variations. The rate of increase is substantially larger for the 31st state (New Jersey).

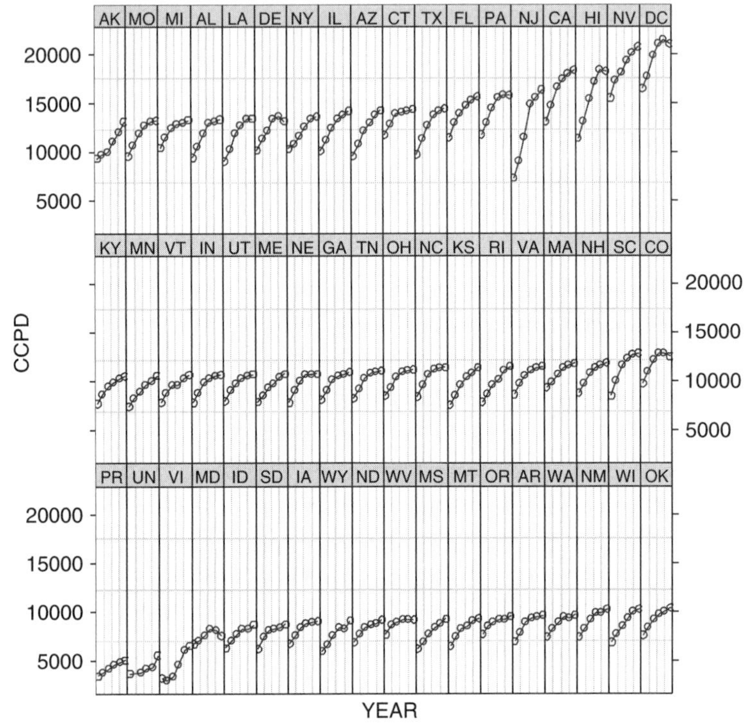

Figure 2.5. Trellis plot of CCPD versus year. Each of the fifty-four panels represents a plot of CCPD versus YEAR, from 1990 to 1995 (with the horizontal axis being suppressed).

Figure 2.4 we have controlled for the state factor, which we observed to be an important source of variation. Figure 2.4 shows that the rate of increase of CCPD over time is approximately consistent among states, yet there exist important variations. The rate of increase is substantially larger for the 31st state (New Jersey).

Trellis Plot

A technique for graphical display that has recently become popular in the statistical literature is a *trellis plot*. This graphical technique takes its name from a trellis, which is a structure of open latticework. When viewing a house or garden, one typically thinks of a trellis as being used to support creeping plants such as vines. We will use this lattice structure and refer to a trellis plot as consisting of one or more panels arranged in a rectangular array. Graphs that contain multiple versions of a basic graphical form, with each version portraying a variation of the basic theme, promote comparisons and assessments of change. By repeating a basic graphical form, we promote the process of communication. Trellis plots have been advocated by Cleveland (1993G), Becker, Cleveland, and Shyu (1996G), Venables and Ripley (1999G), and Pinherio and Bates (2000S).

Tufte (1997G) states that using small multiples in graphical displays achieves the same desirable effects as using parallel structure in writing. Parallel structure in writing is successful because it allows readers to identify a sentence relationship only once and then focus on the meaning of each individual sentence element, such as a word, phrase, or clause. Parallel structure helps achieve economy of expression and draws together related ideas for comparison and contrast. Similarly, small multiples in graphs allow us to visualize complex relationships across different groups and over time.

Figure 2.5 illustrates the use of small multiples. In each panel, the plot portrayed is identical except that it is based on a different state; this use of parallel structure allows us to demonstrate the increasing CCPD for each state. Moreover, by organizing the states by average CCPD, we can see the overall level of CCPD for each state as well as variations in the slope (rate of increase). This plot was produced using the statistical package R.

2.3 Estimation and Inference

Least-Squares Estimation

Returning to our model in Equation (2.3), we now consider estimation of the regression coefficients $\boldsymbol{\beta}$ and α_i and then the variance parameter σ^2. By the Gauss–Markov theorem, the best linear unbiased estimators of $\boldsymbol{\beta}$ and α_i are the ordinary least-squares (OLS) estimators. These are given by

$$\mathbf{b} = \left(\sum_{i=1}^{n} \sum_{t=1}^{T_i} (\mathbf{x}_{it} - \bar{\mathbf{x}}_i)(\mathbf{x}_{it} - \bar{\mathbf{x}}_i)' \right)^{-1} \left(\sum_{i=1}^{n} \sum_{t=1}^{T_i} (\mathbf{x}_{it} - \bar{\mathbf{x}}_i)(y_{it} - \bar{y}_i) \right), \quad (2.6)$$

where $\mathbf{b} = (b_1, b_2, \ldots, b_K)'$ and

$$a_i = \bar{y}_i - \bar{\mathbf{x}}_i' \mathbf{b}. \quad (2.7)$$

The derivations of these estimators are in Appendix 2A.1.[1]

[1] We now begin to use matrix notation extensively. You may wish to review this set of notation in Appendix A, focusing on the definitions and basic operations in A.1–A.3, before proceeding.

Statistical and econometric packages are widely available and thus users will rarely have to code the least-squares estimator expressions. Nonetheless, the expressions in Equations (2.6) and (2.7) offer several valuable insights.

First, we note that there are $n + K$ unknown regression coefficients in Equation (2.3), n for the $\{\alpha_i\}$ parameters and K for the $\boldsymbol{\beta}$ parameters. Using standard regression routines, this calls for the inversion of an $(n + K) \times (n + K)$ matrix. However, the calculation of the OLS estimators in Equation (2.6) requires inversion of only a $K \times K$ matrix. This technique of treating the subject identifier as a categorical explanatory variable known as a *factor* is a standard feature of analysis of covariance models.

Second, the OLS estimator of $\boldsymbol{\beta}$ can also be expressed as a weighted average of subject-specific estimators. Specifically, suppose that all parameters are subject-specific, so that the regression function is $\mathrm{E}y_{it} = \alpha_i + \mathbf{x}'_{it}\boldsymbol{\beta}_i$. Then, routine calculations show that the OLS estimator of $\boldsymbol{\beta}_i$ is

$$\mathbf{b}_i = \left(\sum_{t=1}^{T_i} (\mathbf{x}_{it} - \bar{\mathbf{x}}_i)(\mathbf{x}_{it} - \bar{\mathbf{x}}_i)' \right)^{-1} \left(\sum_{t=1}^{T_i} (\mathbf{x}_{it} - \bar{\mathbf{x}}_i)(y_{it} - \bar{y}_i) \right).$$

Now, define a weight matrix

$$\mathbf{W}_i = \sum_{t=1}^{T_i} (\mathbf{x}_{it} - \bar{\mathbf{x}}_i)(\mathbf{x}_{it} - \bar{\mathbf{x}}_i)',$$

so that a simpler expression for $\mathbf{b}_i$ is

$$\mathbf{b}_i = \mathbf{W}_i^{-1} \sum_{t=1}^{T_i} (\mathbf{x}_{it} - \bar{\mathbf{x}}_i)(y_{it} - \bar{y}_i).$$

With this weight, we can express the estimator of $\boldsymbol{\beta}$ as

$$\mathbf{b} = \left(\sum_{i=1}^{n} \mathbf{W}_i \right)^{-1} \sum_{i=1}^{n} \mathbf{W}_i \mathbf{b}_i, \tag{2.8}$$

a (matrix) weighted average of subject-specific parameter estimates. To help interpret Equation (2.8), consider Figure 2.2. Here, we see that the response (CCPD) is positively related to number of discharges *for each state*. Thus, because each subject-specific coefficient is positive, we expect the weighted average of coefficients to also be positive.

For a third insight from Equations (2.6) and (2.7), consider another weighting vector

$$\mathbf{W}_{it,1} = \left(\sum_{i=1}^{n} \mathbf{W}_i \right)^{-1} (\mathbf{x}_{it} - \bar{\mathbf{x}}_i).$$

With this vector, another expression for Equation (2.6) is

$$\mathbf{b} = \sum_{i=1}^{n} \sum_{t=1}^{T_i} \mathbf{W}_{it,1} y_{it}. \tag{2.9}$$

From this, we see that the regression coefficients in **b** are linear combinations of the responses. By the linearity, if the responses are normally distributed (Assumption F5), then so are the regression coefficients in **b**.

Fourth, regression coefficients associated with time-constant variables cannot be estimated using Equation (2.6). Specifically, suppose that the *j*th variable does not depend on time, so that $x_{it,j} = \bar{x}_{i,j}$. Then, elements in the *j*th row and column of

$$\sum_{i=1}^{n} \sum_{t=1}^{T_i} (\mathbf{x}_{it} - \bar{\mathbf{x}}_i)(\mathbf{x}_{it} - \bar{\mathbf{x}}_i)'$$

are identically zero and so the matrix is not invertible. Thus, regression coefficients cannot be calculated using Equation (2.6) and, in fact, are not estimable when one of the explanatory variables is constant in time.

Other Properties of Estimators

Both a_i and **b** have the usual (finite-sample) properties of OLS regression estimators. In particular, they are unbiased estimators. Further, by the Gauss–Markov theorem, they are minimum variance among the class of unbiased estimators. If the responses are normally distributed (Assumption F5), then so are a_i and **b**. Further, using Equation (2.9), it is easy to check that the variance of **b** turns out to be

$$\operatorname{Var} \mathbf{b} = \sigma^2 \left(\sum_{i=1}^{n} \mathbf{W}_i \right)^{-1}. \tag{2.10}$$

ANOVA Table and Standard Errors

The estimator of the variance parameter σ^2 follows from the customary regression setup. That is, it is convenient to first define residuals and the *analysis of variance* (ANOVA) table. From this, we get an estimator of σ^2, as well as standard errors for the regression coefficient estimators.

To this end, define the *residuals* as $e_{it} = y_{it} - (a_i + \mathbf{x}'_{it}\mathbf{b})$, the difference between the observed and fitted values. In ANOVA terminology, the sum of squared residuals is called the *error sum of squares*, denoted by *Error SS* $= \sum_{it} e_{it}^2$. The *mean square error* is our estimator of σ^2, denoted by

$$s^2 = \frac{\textit{Error SS}}{N - (n + K)} = \textit{Error MS}. \tag{2.11}$$

Table 2.2. *ANOVA table*

Source	Sum of squares	*df*	Mean square
Regression	*Regression SS*	$n - 1 + K$	*Regression MS*
Error	*Error SS*	$N - (n + K)$	*Error MS*
Total	*Total SS*	$N - 1$	

The corresponding positive square root is the *residual standard deviation*, denoted by s. Here, recall that $T_1 + T_2 + \cdots + T_n = N$ is the total number of observations. These calculations are summarized in Table 2.2. To complete the definitions of the expressions in Table 2.2, we have

$$\textit{Total SS} = \sum_{it} (y_{it} - \bar{y})^2$$

and

$$\textit{Regression SS} = \sum_{it} (a_i + \mathbf{x}'_{it}\mathbf{b} - \bar{y})^2.$$

Further, the mean-square quantities are the sum of square quantities divided by their respective degrees of freedom (*df*). The ANOVA table calculations are often reported through the goodness-of-fit statistic called the *coefficient of determination*,

$$R^2 = \frac{\textit{Regression SS}}{\textit{Total SS}},$$

or the version adjusted for degrees of freedom,

$$R_{\mathrm{a}}^2 = 1 - \frac{(\textit{Error SS}) \,/\, (N - (n + K))}{(\textit{Total SS}) \,/\, (N - 1)}.$$

An important function of the residual standard deviation is to estimate standard errors associated with parameter estimators. Using the ANOVA table and Equation (2.10), we obtain the estimated variance matrix of the vector of regression coefficients: $\widehat{\mathrm{Var}}\, \mathbf{b} = s^2 (\sum_{i=1}^{n} \mathbf{W}_i)^{-1}$. Thus, the *standard error* for the jth regression coefficient b_j is

$$se(b_j) = s \sqrt{j\text{th diagonal element of } \left(\sum_{i=1}^{n} \mathbf{W}_i \right)^{-1}}.$$

Standard errors are the basis for the t-ratios, arguably the most important (or at least most widely cited) statistics in applied regression analysis. To illustrate,

```
Display 2.1 SAS OUTPUT
General Linear Models Procedure
Dependent Variable: CCPD
                                      Sum of                  Mean
Source                  DF           Squares                Square   F Value     Pr > F
Model                   57      3258506185.0            57166775.2    203.94     0.0001
Error                  265        74284379.1              280318.4
Corrected Total        322      3332790564.1
                  R-Square              C.V.              Root MSE               CCPD Mean
                  0.977711          5.041266             529.45105               10502.344
                                     T for H0:           Pr > |T|    Std Error of
Parameter               Estimate    Parameter=0                        Estimate
YEAR                  710.884203          26.51          0.0001        26.8123882
AVE_DAYS              361.290071           6.23          0.0001        57.9789849
NUM_DCHG               10.754717           4.18          0.0001         2.5726119
YR_31                1262.456077           9.82          0.0001       128.6088909
```

consider the t-ratio for the jth regression coefficient b_j:

$$t(b_j) = \frac{b_j}{se(b_j)} = \frac{b_j}{s\sqrt{j\text{th diagonal element of } \left(\sum_{i=1}^{n} \mathbf{W}_i\right)^{-1}}}.$$

Assuming the responses are normally distributed, $t(b_j)$ has a t-distribution with $N-(n+K)$ degrees of freedom.

Example 2.2: Medicare Hospital Costs (continued) To illustrate, we return to the Medicare example. Figures 2.1–2.4 suggested that the state categorical variable is important. Further, NUM_DSCH, AVE_DAYS, and YEAR are also potentially important. From Figure 2.4, we noted that the increase in CCPD is higher for New Jersey than for other states. Thus, we also included a special interaction variable YEAR*(STATE = 31) that allowed us to represent the unusually large time slope for the 31st state, New Jersey. Thus, we estimate the function

$$\begin{aligned} \text{E CCPD}_{it} &= \alpha_i + \beta_1 \text{ YEAR}_t + \beta_2 \text{ AVE\_DAYS}_{it} + \beta_3 \text{ NUM\_DSCH}_{it} \\ &\quad + \beta_4 \text{YEAR}_t^{*}(\text{STATE} = 31). \end{aligned} \tag{2.12}$$

The fitted model appears in Display 2.1, using the statistical package SAS.

Example 2.1: Urban Wages (continued) Table 2.3 summarizes three regression models reported by Glaeser and Maré (2001E) in their investigation of determinants of hourly wages. The two homogeneous models do not include worker-specific intercepts, whereas these are included in the fixed-effects model. For the homogeneous model without controls, the only two explanatory variables are the binary variables for indicating whether a worker resides in a (dense or nondense) metropolitan area. The omitted category in this regression is nonmetropolitan area, so we interpret the 0.263 coefficient to mean that workers

Table 2.3. *Regression coefficient estimators of several hourly wage models*

Variable	Homogenous model without controls	Homogeneous model with controls[a]	Two-way fixed-effects model
Dense metropolitan premium	0.263	0.245 (0.01)	0.109 (0.01)
Nondense metropolitan premium	0.175	0.147 (0.01)	0.070 (0.01)
Coefficient of determination R^2	1.4	33.1	38.1
Adjusted R^2	1.4	33.0	28.4

[a] Standard errors are in parentheses.
Source: Glaeser and Maré (2001E).

in dense metropolitan areas on average earn 0.263 log dollars, or 26.3%, more than their non–metropolitan area counterparts. Similarly, those in nondense metropolitan areas earn 17.5% more than their non–metropolitan area counterparts.

Wages may also be influenced by a worker's experience, occupation, education, and race, and there is no guarantee that these characteristics are distributed uniformly over different city sizes. Thus, a regression model with these controls is also reported in Table 2.3. Table 2.3 shows that workers in cities, particularly dense cities, still receive more than workers in nonurban areas, even when controlling for a worker's experience, occupation, education, and race.

Glaeser and Maré offer additional explanations as to why workers in cities earn more than their nonurban counterparts, including higher cost of living and urban amenities (while also discounting these explanations as of less importance). They do posit an "omitted ability bias"; that is, they suggest that the ability of a worker is an important wage determinant that should be controlled for. They suggest that higher ability workers may flock to cities because if cities speed the flow of information, then this might be more valuable to gifted workers (those who have high "human capital"). Moreover, cities may be centers of consumption that cater to the rich. Ability is a difficult attribute to measure. (For example, they examine a proxy, the Armed Forces Qualification Test, and find that it is not useful.) However, if one treats ability as constant in time, then its effects on wages will be captured in the time-constant worker-specific intercept α_i. Table 2.3 reports on the fixed-effects regression that includes a worker-specific intercept. Here, we see that the parameter estimates for city premiums have been substantially reduced, although they are still statistically significantly. One interpretation is that a worker will receive a 10.9% (7%)

higher wage for working in a dense (nondense) city when compared to a non-metropolitan worker, even when controlling for a worker's experience, occupation, education, and race and omitting time-constant attributes such as "ability." Section 7.2 will present a much more detailed discussion of this omitted-variable interpretation.

Large-Sample Properties of Estimators

In typical regression situations, responses are at best only approximately normally distributed. Nonetheless, hypothesis tests and predictions are based on regression coefficients that are approximately normally distributed. This premise is reasonable because of central limit theorems, which roughly state that weighted sums of independent random variables are approximately normally distributed and that the approximation improves as the number of random variables increases. Regression coefficients can be expressed as weighted sums of responses. Thus, if sample sizes are large, then we may assume approximate normality of regression coefficient estimators.

In the longitudinal data context, "large samples" means that either the number of subjects, n, and/or the number of observations per subject, T, becomes large. In this chapter, we discuss the case where n becomes large while T remains fixed. Our motivation is that, in many data applications, the number of subjects is large relative to the number of time periods observed. Chapter 8 will discuss the problem of large T.

As n becomes large yet T remains fixed, most of the properties of $\mathbf{b}$ are retained from the standard regression situations. To illustrate, we have that $\mathbf{b}$ is a weakly consistent estimate of $\boldsymbol{\beta}$. Specifically, weak consistency means approaching (convergence) in probability. This is a direct result of the unbiasedness and an assumption that $\Sigma_i \mathbf{W}_i$ grows without bound. Further, under mild regularity conditions, we have a central limit theorem for the slope estimator. That is, $\mathbf{b}$ is approximately normally distributed even though the responses may not be.

The situation for the estimators of subject-specific intercepts α_i differs dramatically. For instance, the least-squares estimator of α_i is not consistent as n becomes large. Moreover, if the responses are not normally distributed, then a_i is not even approximately normal. Intuitively, this is because we assume that the number of observations per subject, T_i, is a bounded number.

As n grows, the number of parameters *grows*, a situation called "infinite dimensional nuisance parameters" in the literature (see, for example, Neyman and Scott, 1948E, for a classic treatment). When the number of parameters grows with sample size, the usual large-sample properties of estimators may not be valid. Section 7.1 and Chapter 9 will discuss this issue further.

2.4 Model Specification and Diagnostics

Inference based on a fitted statistical model often may be criticized because the features of the data are not in congruence with the model assumptions. *Diagnostic techniques* are procedures that examine this congruence. Because we sometimes use discoveries about model inadequacies to improve the model specification, this group of procedures is also called *model specification*, or *misspecification*, tests or procedures. The broad distinction between diagnostics and the Section 2.2 data exploration techniques is that the former are performed after a preliminary model fit whereas the latter are done before fitting models with data.

When an analyst fits a great number of models to a data set, this leads to difficulties known as *data snooping*. That is, with several explanatory variables, one can generate a large number of linear models and an infinite number of nonlinear models. By searching over many models, it is possible to "overfit" a model, so that standard errors are smaller than they should be and insignificant relationships appear significant.

There are widely different philosophies espoused in the literature for model specification. At one end of the spectrum are those who argue that data snooping is a problem endemic to all data analysis. Proponents of this philosophy argue that a model should be fully specified before data examination can begin; in this way, inferences drawn from the data are not mitigated from data snooping. At the other end of the spectrum are those who argue that inferences from a model are unreliable if the data are not in accordance with model assumptions. Proponents of this philosophy argue that a model summarizes important characteristics of the data and that the best model should be sought through a series of specification tests.

These distinctions are widely discussed in the applied statistical modeling literature. We present here several specification tests and procedures that can be used to describe how well the data fit the model. Results from the specification tests and procedures can then be used to either respecify the model or interpret model results, according to one's beliefs in model fitting.

2.4.1 Pooling Test

A *pooling test*, also known as a *test for heterogeneity*, examines whether or not the intercepts take on a common value, say α. An important advantage of longitudinal data models, compared to cross-sectional regression models, is that we can allow for heterogeneity among subjects, generally through subject-specific parameters. Thus, an important first procedure is to justify the need for subject-specific effects.

The null hypothesis of homogeneity can be expressed as $\mathrm{H}_0 : \alpha_1 = \alpha_2 = \cdots = \alpha_n = \alpha$. Testing this null hypothesis is simply a special case of the general linear hypothesis and can be handled directly as such. Here is one way to perform a partial F-(Chow) test.

Procedure for the Pooling Test

1. Run the "full model" with $\mathrm{E}y_{it} = \alpha_i + \mathbf{x}'_{it}\boldsymbol{\beta}$ to get *Error SS* and s^2.
2. Run the "reduced model" with $\mathrm{E}y_{it} = \alpha + \mathbf{x}'_{it}\boldsymbol{\beta}$ to get (*Error* SS)$_{\text{reduced}}$.
3. Compute the partial F-statistic,

$$F\text{-ratio} = \frac{(Error\ SS)_{\text{reduced}} - Error\ SS}{(n-1)s^2}.$$

4. Reject H_0 if the F-ratio exceeds a percentile from an F-distribution with numerator degrees of freedom $df_1 = n - 1$ and denominator degrees of freedom $df_2 = N - (n + K)$. The percentile is one minus the significance level of the test.

This is an exact test in the sense that it does not require large sample sizes yet does require normality of the responses (Assumption F5). Studies have shown that the F-test is not sensitive to departures from normality (see, for example, Layard, 1973G). Further, note that if the denominator degrees of freedom, df_2, is large, then we may approximate the distribution by a chi-square distribution with $n - 1$ degrees of freedom.

Example 2.2: Medicare Hospital Costs (continued) To test for heterogeneity in Medicare hospital cost, we test the null hypothesis $\mathrm{H}_0 : \alpha_1 = \alpha_2 = \cdots = \alpha_{51}$. From Display 2.1, we have *Error SS* = 74,284,379.1 and $s^2 =$ 280,318.4. Fitting the pooled cross-sectional regression function (with common effects α)

$$\begin{aligned}\mathrm{ECCPD}_{it} &= \alpha + \beta_1\ \mathrm{YEAR}_t + \beta_2\ \mathrm{AVE\_DAYS}_{it} + \beta_3\ \mathrm{NUM\_DSCH}_{it} \\ &\quad + \beta_4\ \mathrm{YEAR}^*_t(\mathrm{STATE} = 31)\end{aligned}$$

yields (*Error SS*)$_{\text{reduced}}$ = 2,373,115,932.9. Thus, the test statistic is

$$F\text{-ratio} = \frac{2{,}373{,}115{,}932.9 - 74{,}284{,}379.1}{(54-1)280{,}318.4} = 154.7.$$

For an F-distribution with $df_1 = 53$ and $df_2 = 323 - (54 + 4) = 265$ the associated p-value is less than 0.0001. This provides strong evidence for the case for retaining subject-specific parameters α_i in the model specification.

2.4.2 Added-Variable Plots

An *added-variable plot*, also known as a *partial regression plot*, is a standard graphical device used in regression analysis; see, for example, Cook and Weisberg (1982G). It allows one to view the relationship between a response and an explanatory variable, after controlling for the linear effects of other explanatory variables. Thus, added-variable plots allow analysts to visualize the relationship between y and each x, without forcing the eye to adjust for the differences induced by the other explanatory variables. The Section 2.2 *basic* added-variable plot is a special case of the following procedure that can be used for additional regression variables.

To produce an added-variable plot, one first selects an explanatory variable, say x_j, and then uses the following procedure.

Procedure for Producing an Added-Variable Plot

1. Run a regression of y on the other explanatory variables (omitting x_j) and calculate the residuals from this regression. Call these residuals e_1.
2. Run a regression of x_j on the other explanatory variables (omitting x_j) and calculate the residuals from this regression. Call these residuals e_2.
3. Produce a plot of e_1 versus e_2. This is an added-variable plot.

Correlations and Added-Variable Plots

To help interpret added-variable plots, use Equation (2.1) to express the disturbance term as

$$\varepsilon_{it} = y_{it} - (\alpha_i + \beta_1 x_{it,1} + \cdots + \beta_K x_{it,K}).$$

That is, we may think of the error as the response, after controlling for the linear effects of the explanatory variables. The residual e_1 is an approximation of the error, interpreted to be the response after controlling for the effects of explanatory variables. Similarly, we may interpret e_2 to be the jth explanatory variable, after controlling for the effects of other explanatory variables. Thus, we interpret the added-variable plot as a graph of the relationship between y and x_j, after controlling for the effects of other explanatory variables.

As with all scatter plots, the added-variable plot can be summarized numerically through a correlation coefficient that we will denote by corr(e_1, e_2). It is related to the t-statistic of x_j, $t(b_j)$, from the full regression equation (including x_j) through the expression

$$\operatorname{corr}(e_1, e_2) = \frac{t(b_j)}{\sqrt{t(b_j)^2 + N - (n+K)}},$$

where $n + K$ is the number of regression coefficients in the full regression equation and N is the number of observations. Thus, the t-statistic from the full regression equation can be used to determine the correlation coefficient of the added-variable plot without running the three-step procedure. However, unlike correlation coefficients, the added-variable plot allows us to visualize potential nonlinear relationships between y and x_j.

2.4.3 Influence Diagnostics

Traditional influence diagnostics are important because they allow an analyst to understand the impact of individual observations on the estimated model. That is, influence statistics allow analysts to perform a type of "sensitivity analysis," in which one can calibrate the effect of individual observations on regression coefficients. *Cook's distance* is a diagnostic statistic that is widely used in regression analysis and is reviewed in Appendix 2A.3. For the fixed-effects longitudinal data models of Chapter 2, observation-level diagnostic statistics are of less interest because the effect of unusual observations is absorbed by subject-specific parameters. Of greater interest is the impact that an entire subject has on the population parameters.

To assess the impact that a subject has on estimated regression coefficients, we use the statistic

$$B_i(\mathbf{b}) = \left(\mathbf{b} - \mathbf{b}_{(i)}\right)' \left(\sum_{i=1}^{n} \mathbf{W}_i\right) \left(\mathbf{b} - \mathbf{b}_{(i)}\right) / K,$$

where $\mathbf{b}_{(i)}$ is the OLS estimator $\mathbf{b}$ calculated with the ith subject omitted. Thus, $B_i(\mathbf{b})$ measures the distance between regression coefficients calculated with and without the ith subject. In this way, we can assess the effect of the ith subject. The longitudinal data influence diagnostic is similar to Cook's distance for regression. However, Cook's distance is calculated at the observation level whereas $B_i(\mathbf{b})$ is at the subject level.

Observations with a "large" value of $B_i(\mathbf{b})$ may be influential on the parameter estimates. Banerjee and Frees (1997S) showed that the statistic $B_i(\mathbf{b})$ has an approximate χ^2 (chi-square) distribution with K degrees of freedom. Thus, we may use quantiles of the χ^2 to quantify the adjective "large." Influential observations warrant further investigation; they may require correction for coding errors, additional variable specification to accommodate the patterns they emphasize, or deletion from the data set.

From the definition of $B_i(\mathbf{b})$, it appears that the calculation of the influence statistic is computationally intensive. This is because the definition requires $n+1$ regression computations, one for $\mathbf{b}$ and one for each $\mathbf{b}_{(i)}$. However, as with Cook's distance at the observation level, shortcut calculation procedures are available. The details are in Appendix 2A.3.

Example 2.2: Medicare Hospital Costs (continued) Figure 2.3 alerted us to the unusual value of AVE_DAYS that occurred in the $i = 54$ subject at the $t = 2$ time point. It turns out that this observation has a substantial impact on the fitted regression model. Fortunately, the graphical procedure in Figure 2.3 and the summary statistics in Table 2.1 were sufficient to detect this unusual point. Influence diagnostic statistics provide another tool for detecting unusual observations and subjects. Suppose that the model in Equation (2.12) was fit using the full data set of $N = 324$ observations. Cook's distance can be calculated as $D_{54,2} = 17.06$ for the $(i = 54, t = 2)$ point, strongly indicating an influential observation. The corresponding subject-level statistic was $B_{54} = 244.3$. Compared to a chi-square distribution with $K = 4$ degrees of freedom, this indicates that something about the 54th subject was unusual. For comparison, the diagnostic statistics were calculated under a fitted regression model after removing the $(i = 54, t = 2)$ point. The largest value of Cook's distance was 0.0907 and the largest value of the subject-level statistic was 0.495. Neither value indicates substantial influential behavior after the unusual $(i = 54, t = 2)$ point was removed.

2.4.4 Cross-Sectional Correlation

Our basic model relies on Assumption F4, the independence among observations. In traditional cross-sectional regression models, this assumption is untestable without a parametric assumption. However, with repeated measurements on a subject, it is possible to examine this assumption. As is traditional in the statistics literature, when testing for independence, we are really testing for zero correlation. That is, we are interested in the null hypothesis, $\text{H}_0 : \rho_{ij} = \text{Corr}(y_{it}, y_{jt}) = 0$ for $i \neq j$.

To understand how violations of this assumption may arise in practice, suppose that the true model is $y_{it} = \lambda_t + \mathbf{x}'_{it}\boldsymbol{\beta} + \varepsilon_{it}$, where we use λ_t for a random temporal effect that is common to all subjects. Because it is common, it induces correlation among subjects, as follows. We first note that the variance of a response is Var $y_{it} = \sigma_\lambda^2 + \sigma^2$, where Var $\varepsilon_{it} = \sigma^2$ and Var $\lambda_t = \sigma_\lambda^2$. From here, basic calculations show that the covariance between observations at the same time but from different subjects is $\mathrm{Cov}(y_{it}, y_{jt}) = \sigma_\lambda^2$, for $i \neq j$. Thus, the cross-sectional correlation is $\mathrm{Corr}(y_{it}, y_{jt}) = \sigma_\lambda^2/(\sigma_\lambda^2 + \sigma^2)$. Hence, a positive cross-sectional correlation may be due to unobserved temporal effects that are common among subjects.

Testing for Nonzero Cross-Sectional Correlation

To test $\mathrm{H}_0 : \rho_{ij} = 0$ for all $i \neq j$, we use a procedure developed in Frees (1995E) where balanced data were assumed so that $T_i = T$.

Procedure for Computing Cross-Sectional Correlation Statistics

1. Fit a regression model and calculate the model residuals, $\{e_{it}\}$.
2. For each subject i, calculate the ranks of each residual. That is, define $\{r_{i,1}, \ldots, r_{i,T}\}$ to be the ranks of $\{e_{i,1}, \ldots, e_{i,T}\}$. These ranks will vary from 1 to T, so that the average rank is $(T + 1)/2$.
3. For the ith and jth subject, calculate the rank correlation coefficient (Spearman's correlation)

$$sr_{ij} = \frac{\sum_{t=1}^{T} (r_{i,t} - (T+1)/2)(r_{j,t} - (T+1)/2)}{\sum_{t=1}^{T} (r_{i,t} - (T+1)/2)^2}.$$

4. Calculate the average Spearman's correlation and the average squared Spearman's correlation:

$$R_{\mathrm{AVE}} = \frac{1}{n(n-1)/2} \sum_{\{i<j\}} sr_{ij} \quad \text{and} \quad R_{\mathrm{AVE}}^2 = \frac{1}{n(n-1)/2} \sum_{\{i<j\}} (sr_{ij})^2,$$

where $\sum_{\{i<j\}}$ means sum over $j = 2, \ldots, n$ and $i = 1, \ldots, j-1$.

Calibration of Cross-Sectional Correlation Test Statistics

Large values of the statistics R_{AVE} and R_{AVE}^2 indicate the presence of nonzero cross-sectional correlations. In applications where either positive or negative cross-sectional correlations prevail, one should consider the R_{AVE} statistic. Friedman (1937G) showed that $FR = (T-1)((n-1)R_{\mathrm{AVE}} + 1)$ follows a chi-square distribution (with $T - 1$ degrees of freedom) asymptotically, as n becomes large. Friedman devised the test statistic FR to determine the equality of treatment means in a two-way analysis of variance. The statistic FR is also used in the problem of "n-rankings." In this context, n "judges" are asked to rank T

items and the data are arranged in a two-way analysis of variance layout. The statistic R_{AVE} is interpreted to be the average agreement among judges.

The statistic R^2_{AVE} is useful for detecting a broader range of alternatives than can be found with the statistic R_{AVE}. For hypothesis testing purposes, we compare R^2_{AVE} to a distribution that is a weighted sum of chi-square random variables. Specifically, define

$$Q = a(T)\left(\chi_1^2 - (T-1)\right) + b(T)\left(\chi_2^2 - T(T-3)/2\right),$$

where χ_1^2 and χ_2^2 are independent chi-square random variables with $T-1$ and $T(T-3)/2$ degrees of freedom, respectively, and the constants are

$$a(T) = \frac{4(T+2)}{5(T-1)^2(T+1)} \quad \text{and} \quad b(T) = \frac{2(5T+6)}{5T(T-1)(T+1)}.$$

Frees (1995E) showed that $n(R^2_{\text{AVE}} - (T-1)^{-1})$ follows a Q-distribution asymptotically, as n becomes large. Thus, one rejects H_0 if R^2_{AVE} exceeds $(T-1)^{-1} + Q_q/n$, where Q_q is an appropriate quantile from the Q distribution.

Because the Q-distribution is a weighted sum of chi-square random variables, computing quantiles may be tedious. For an approximation, it is much faster to compute the variance of Q and use a normal approximation. Exercise 2.13 illustrates the use of this approximation.

The statistics R_{AVE} and R^2_{AVE} are averages over $n(n-1)/2$ correlations, which may be computationally intensive for large values of n. Appendix 2A.4 describes some shortcut calculation procedures.

Example 2.2: Medicare Hospital Costs (continued) The main drawback of the R_{AVE} and R^2_{AVE} statistics is that the asymptotic distributions are only available for balanced data. To achieve balanced data for the Medicare hospital costs data, we omit the 54th state. The model in Equation (2.12) was fit to the remaining 53 states and the residuals were calculated. The values of the correlation statistics were calculated as $R_{\text{AVE}} = 0.281$ and $R^2_{\text{AVE}} = 0.388$. Both statistics are statistically significant with p-values less than 0.001. This result indicates substantial cross-sectional correlation, indicating some type of "comovement" among states over time that is not captured by our simple model.

For comparison, the model was refit using YEAR as a categorical variable in lieu of a continuous one. This is equivalent to including six indicator variables, one for each year. The values of the correlation statistics were found to be $R_{\text{AVE}} = 0.020$ and $R^2_{\text{AVE}} = 0.419$. Thus, we have captured some of the positive "comovement" among state Medicare hospital costs with time indicator variables.

2.4.5 Heteroscedasticity

When fitting regression models to data, an important assumption is that the variability is common among all observations. This assumption of common variability is called *homoscedasticity*, meaning "same scatter." Indeed, the least-squares procedure assumes that the expected variability of each observation is constant; it gives the same weight to each observation when minimizing the sum of squared deviations. When the scatter varies by observation, the data are said to be *heteroscedastic*. Heteroscedasticity affects the efficiency of the regression coefficient estimators, although these estimators remain unbiased even in the presence of heteroscedasticity.

In the longitudinal data context, the variability Var y_{it} may depend on the subject through i, or the time period through t, or both. Several techniques are available for handling heteroscedasticity. First, heteroscedasticity may be mitigated through a transformation of the response variable. See Carroll and Ruppert (1988G) for a broad treatment of this approach. Second, heteroscedasticity may be explained by weighting variables, denoted by $\mathbf{w}_{it}$. Third, the heteroscedasticity may be ignored in the estimation of the regression coefficients yet accounted for in the estimation of the standard errors. Section 2.5.3 expands on this approach. Further, as we will see in Chapter 3, the variability of y_{it} may vary over i and t through a "random-effects" specification.

One method for detecting heteroscedasticity is to perform a preliminary regression fit of the data and plot the residuals versus the fitted values. The preliminary regression fit removes many of the major patterns in the data and leaves the eye free to concentrate on other patterns that may influence the fit. We plot residuals versus fitted values because the fitted values are an approximation of the expected value of the response and, in many situations, the variability grows with the expected response.

More formal tests of heteroscedasticity are also available in the regression literature. For an overview, see Judge et al. (1985E) or Greene (2002E). To illustrate, let us consider a test due to Breusch and Pagan (1980E). Specifically, this test examines the alternative hypothesis $\mathrm{H_a}$: Var $y_{it} = \sigma^2 + \gamma' \mathbf{w}_{it}$, where $\mathbf{w}_{it}$ is a known vector of weighting variables and γ is a p-dimensional vector of parameters. Thus, the null hypothesis is $\mathrm{H_0}$: Var $y_{it} = \sigma^2$.

Procedure for Testing for Heteroscedasticity

1. Fit a regression model and calculate the model residuals, $\{e_{it}\}$.
2. Calculate squared standardized residuals, $e_{it}^{*2} = e_{it}^2 / (\mathit{Error\ SS}/N)$.
3. Fit a regression model of e_{it}^{*2} on $\mathbf{w}_{it}$.

4. The test statistic is $LM = (Regress\ SS_w)/2$, where $Regress\ SS_w$ is the regression sum of squares from the model fit in Step 3.
5. Reject the null hypothesis if LM exceeds a percentile from a chi-square distribution with p degrees of freedom. The percentile is one minus the significance level of the test.

Here, we use LM to denote the test statistic because Breusch and Pagan derived it as a Lagrange multiplier statistic; see Breusch and Pagan (1980E) for more details. Appendix C.7 reviews Lagrange multiplier statistics and related hypothesis tests.

A common approach for handling heteroscedasticity involves computing standard errors that are robust to the homoscedasticity specification. This is the topic of Section 2.5.3.

2.5 Model Extensions

To introduce extensions of the basic model, we first provide a more compact representation using matrix notation. A matrix form of the Equation (2.2) function is

$$\mathrm{E}\mathbf{y}_i = \alpha_i \mathbf{1}_i + \mathbf{X}_i \boldsymbol{\beta}, \tag{2.13}$$

where $\mathbf{y}_i$ is the $T_i \times 1$ vector of responses for the ith subject, $\mathbf{y}_i = (y_{i1}, y_{i2}, \ldots, y_{iT_i})'$, and $\mathbf{X}_i$ is a $T_i \times K$ matrix of explanatory variables,

$$\mathbf{X}_i = \begin{pmatrix} x_{i1,1} & x_{i1,2} & \cdots & x_{i1,K} \\ x_{i2,1} & x_{i2,2} & \cdots & x_{i2,K} \\ \vdots & \vdots & \ddots & \vdots \\ x_{iT_i,1} & x_{iT_i,2} & \cdots & x_{iT_i,K} \end{pmatrix} = \begin{pmatrix} \mathbf{x}'_{i1} \\ \mathbf{x}'_{i2} \\ \vdots \\ \mathbf{x}'_{iT_i} \end{pmatrix}. \tag{2.14}$$

This can be expressed more compactly as $\mathbf{X}_i = (\mathbf{x}_{i1}, \mathbf{x}_{i2}, \ldots, \mathbf{x}_{iT_i})'$. Finally, $\mathbf{1}_i$ is the $T_i \times 1$ vector of ones.

2.5.1 Serial Correlation

In longitudinal data, subjects are measured repeatedly over time and repeated measurements of a subject tend to be related to one another. As we have seen, one way to capture this relationship is through common subject-specific parameters. Alternatively, this relationship can be captured through the correlation among observations within a subject. Because this correlation is among observations taken over time, we refer to it as a *serial correlation*.

As with time-series analysis, it is useful to measure tendencies in time patterns through a correlation structure. Nonetheless, it is also important to note that time patterns can be handled through the use of time-varying explanatory variables. As a special case, temporal indicator (dummy) variables can be used for time patterns in the data. Although the distinction is difficult to isolate when examining data, time-varying explanatory variables account for time patterns in the mean response, whereas serial correlations are used to account for time patterns in the second moment of the response. In Chapters 6 and 8 we will explore other methods for modeling time patterns, such as using lagged dependent variables.

Timing of Observations

The actual times that observations are taken are important when examining serial correlations. This section assumes that observations are taken equally spaced in time, such as quarterly or annually. The only degree of imbalance that we explicitly allow for is the number of observations per subject, denoted by T_i. Chapter 7 introduces issues of missing data, attrition, and other forms of imbalance. Chapter 8 introduces techniques for handling data that are not equally spaced in time.

Temporal Covariance Matrix

For a full set of observations, we use $\mathbf{R}$ to denote the $T \times T$ temporal variance–covariance matrix. This is defined by $\mathbf{R} = \mathrm{Var}\,\mathbf{y}$, where $\mathbf{R}_{rs} = \mathrm{Cov}\,(y_r, y_s)$ is the element in the rth row and sth column of $\mathbf{R}$. There are at most $T(T+1)/2$ unknown elements of $\mathbf{R}$. We denote this dependence of $\mathbf{R}$ on parameters using the notation $\mathbf{R}(\boldsymbol{\tau})$, where $\boldsymbol{\tau}$ is a vector of parameters.

For less than a full set of observations, consider the ith subject that has T_i observations. Here, we define $\mathrm{Var}\,\mathbf{y}_i = \mathbf{R}_i(\boldsymbol{\tau})$, a $T_i \times T_i$ matrix. The matrix $\mathbf{R}_i(\boldsymbol{\tau})$ can be determined by removing certain rows and columns of the matrix $\mathbf{R}(\boldsymbol{\tau})$. We assume that $\mathbf{R}_i(\boldsymbol{\tau})$ is positive-definite and only depends on i through its dimension.

The matrix $\mathbf{R}(\boldsymbol{\tau})$ depends on $\boldsymbol{\tau}$, a vector of unknown parameters called *variance components*. Let us examine several important special cases of $\mathbf{R}$ as summarized in Table 2.4:

- $\mathbf{R} = \sigma^2\mathbf{I}$, where $\mathbf{I}$ is a $T \times T$ identity matrix. This is the case of no serial correlation, or independent case.
- $\mathbf{R} = \sigma^2((1-\rho)\mathbf{I} + \rho\mathbf{J})$, where $\mathbf{J}$ is a $T \times T$ matrix of ones. This is the *compound symmetry* model, also known as the *uniform correlation model*.
- $\mathbf{R}_{rs} = \sigma^2\rho^{|r-s|}$. This is the *autoregressive* of order 1 model, denoted by $AR(1)$.
- The no-structure case in which no additional assumptions on $\mathbf{R}$ are made.

Table 2.4. *Covariance structure examples*

Structure	Example	Variance comp ($\boldsymbol{\tau}$)	Structure	Example	Variance comp ($\boldsymbol{\tau}$)
Independent	$\mathbf{R} = \begin{pmatrix} \sigma^2 & 0 & 0 & 0 \\ 0 & \sigma^2 & 0 & 0 \\ 0 & 0 & \sigma^2 & 0 \\ 0 & 0 & 0 & \sigma^2 \end{pmatrix}$	σ^2	Autoregressive of order 1, $AR(1)$	$\mathbf{R} = \sigma^2 \begin{pmatrix} 1 & \rho & \rho^2 & \rho^3 \\ \rho & 1 & \rho & \rho^2 \\ \rho^2 & \rho & 1 & \rho \\ \rho^3 & \rho^2 & \rho & 1 \end{pmatrix}$	σ^2, ρ
Compound symmetry	$\mathbf{R} = \sigma^2 \begin{pmatrix} 1 & \rho & \rho & \rho \\ \rho & 1 & \rho & \rho \\ \rho & \rho & 1 & \rho \\ \rho & \rho & \rho & 1 \end{pmatrix}$	σ^2, ρ	No structure	$\mathbf{R} = \begin{pmatrix} \sigma_1^2 & \sigma_{12} & \sigma_{13} & \sigma_{14} \\ \sigma_{12} & \sigma_2^2 & \sigma_{23} & \sigma_{24} \\ \sigma_{13} & \sigma_{23} & \sigma_3^2 & \sigma_{34} \\ \sigma_{14} & \sigma_{24} & \sigma_{34} & \sigma_4^2 \end{pmatrix}$	σ_1^2, σ_2^2 $\sigma_3^2, \sigma_4^2,$ σ_{12}, σ_{13} σ_{14}, σ_{23} σ_{24}, σ_{34}

To see how the compound symmetry model may occur, consider the model $y_{it} = \alpha_i + \varepsilon_{it}$, where α_i is a random cross-sectional effect. This yields $\mathbf{R}_{tt} = \text{Var } y_{it} = \sigma_\alpha^2 + \sigma_\varepsilon^2 = \sigma^2$. Similarly, for $r \neq s$, we have $\mathbf{R}_{rs} = \text{Cov}(y_{ir}, y_{is}) = \sigma_\alpha^2$. To write this in terms of σ^2, note that the correlation is $\text{Corr}(y_{ir}, y_{is}) = \sigma_\alpha^2/(\sigma_\alpha^2 + \sigma_\varepsilon^2) = \rho$. Thus, $\mathbf{R}_{rs} = \sigma^2\rho$ and $\mathbf{R} = \sigma^2((1-\rho)\mathbf{I} + \rho\mathbf{J})$.

The autoregressive model is a standard representation used in time-series analysis. Suppose that u_{it} are independent errors and the disturbance terms are determined sequentially through the relation $\varepsilon_{it} = \rho\varepsilon_{i,t-1} + u_{it}$. This implies the relation $\mathbf{R}_{rs} = \sigma^2\rho^{|r-s|}$, with $\sigma^2 = (\text{Var } u_{it})/(1-\rho^2)$. This model may also be extended to the case where time observations are not equally spaced in time; Section 8.2 provides further details. For the unstructured model, there are $T(T+1)/2$ unknown elements of $\mathbf{R}$. It turns out that this choice is nonestimable for a fixed-effects model with individual-specific intercepts. Chapter 7 provides additional details.

2.5.2 Subject-Specific Slopes

In the Medicare hospital costs example, we found that a desirable model of covered claims per discharge, was of the form

$$\text{E CCPD}_{it} = \alpha_i + \beta_1 \text{ YEAR}_t + \mathbf{x}'_{it}\boldsymbol{\beta}.$$

Thus, one could interpret β_1 as the expected annual change in CCPD; this may be due, for example, to a medical inflation component. Suppose that the analyst anticipates that the medical inflation component will vary by state (as suggested by Figure 2.5). To address this concern, we consider instead the model

$$\text{E CCPD}_{it} = \alpha_{i1} + \alpha_{i2} \text{ YEAR}_t + \mathbf{x}'_{it}\boldsymbol{\beta}.$$

Here, subject-specific intercepts are denoted by $\{\alpha_{i1}\}$ and we allow for subject-specific slopes associated with year through the notation $\{\alpha_{i2}\}$.

Thus, in addition to letting intercepts vary by subject, it is also useful to let one or more slopes vary by subject. We will consider regression functions of the form

$$\text{E} y_{it} = \mathbf{z}'_{it}\boldsymbol{\alpha}_i + \mathbf{x}'_{it}\boldsymbol{\beta}. \tag{2.15}$$

Here, the subject-specific parameters are $\boldsymbol{\alpha}_i = (\alpha_{i1}, \ldots, \alpha_{iq})'$ and the q explanatory variables are $\mathbf{z}_{it} = (z_{it1}, z_{it2}, \ldots, z_{itq})'$; both column vectors are of dimension $q \times 1$. Equation (2.15) is shorthand notation for the function

$$\text{E} y_{it} = \alpha_{i1} z_{it1} + \alpha_{i2} z_{it2} + \cdots + \alpha_{iq} z_{itq} + \beta_1 x_{it1} + \beta_2 x_{it2} + \cdots + \beta_K x_{itK}.$$

To provide a more compact representation using matrix notation, we define $\mathbf{Z}_i = (\mathbf{z}_{i1}, \mathbf{z}_{i2}, \ldots, \mathbf{z}_{iT_i})'$, a $T_i \times q$ matrix of explanatory variables. With this notation, as in Equation (2.13), a matrix form of Equation (2.15) is

$$\mathrm{E}\mathbf{y}_i = \mathbf{Z}_i \boldsymbol{\alpha}_i + \mathbf{X}_i \boldsymbol{\beta}. \tag{2.16}$$

The responses between subjects are independent, yet we allow for temporal correlation and heteroscedasticity through the assumption that $\mathrm{Var}\, \mathbf{y}_i = \mathbf{R}_i(\boldsymbol{\tau}) = \mathbf{R}_i$.

Taken together, these assumptions comprise what we term the *fixed-effects linear longitudinal data model.*

Assumptions of the Fixed-Effects Linear Longitudinal Data Model

F1. $\mathrm{E}\mathbf{y}_i = \mathbf{Z}_i \boldsymbol{\alpha}_i + \mathbf{X}_i \boldsymbol{\beta}$.
F2. $\{x_{it,1}, \ldots, x_{it,K}\}$ and $\{z_{it,1}, \ldots, z_{it,q}\}$ are nonstochastic variables.
F3. $\mathrm{Var}\, \mathbf{y}_i = \mathbf{R}_i(\boldsymbol{\tau}) = \mathbf{R}_i$.
F4. $\{\mathbf{y}_i\}$ are independent random vectors.
F5. $\{y_{it}\}$ are normally distributed.

Note that we use the same letters, F1–F5, to denote the assumptions of the fixed-effects linear longitudinal data model and the linear regression model. This is because the models differ only through their mean and variance functions.

Sampling and Model Assumptions

We can use a model that represents how the sample is drawn from the population to motivate the assumptions of the fixed-effects linear longitudinal data model. Specifically, assume the data arise as a *stratified sample*, in which the subjects are the strata. For example, in the Section 2.2 example we would identify each state as a stratum. Under stratified sampling, one assumes independence among different subjects (Assumption F4). For observations within a stratum, unlike stratified sampling in a survey context, we allow for serial dependence to arise in a time-series pattern through Assumption F3. In general, selecting subjects based on exogenous characteristics suggests stratifying the population and using a fixed-effects model. For example, many panel data studies have analyzed selected large countries, firms, or important CEOs (chief executive officers). When a sample is selected based on exogenous explanatory variables and these explanatory variables are treated as fixed, yet variable, we treat the subject-specific terms as fixed, yet variable.

Least-Squares Estimators

The estimators are derived in Appendix 2A.2 assuming that the temporal correlation matrix $\mathbf{R}_i$ is known. Section 3.5 will address the problems of estimating the parameters that determine this matrix. Moreover, Section 7.1 will emphasize some of the special problems of estimating these parameters in the presence of fixed-effects heterogeneity. With known $\mathbf{R}_i$, the regression coefficient estimators are given by

$$\mathbf{b}_{\mathrm{FE}} = \left(\sum_{i=1}^{n} \mathbf{X}_i' \mathbf{R}_i^{-1/2} \mathbf{Q}_{Z,i} \mathbf{R}_i^{-1/2} \mathbf{X}_i \right)^{-1} \sum_{i=1}^{n} \mathbf{X}_i' \mathbf{R}_i^{-1/2} \mathbf{Q}_{Z,i} \mathbf{R}_i^{-1/2} \mathbf{y}_i \tag{2.17}$$

and

$$\mathbf{a}_{\mathrm{FE},i} = \left(\mathbf{Z}_i' \mathbf{R}_i^{-1} \mathbf{Z}_i \right)^{-1} \mathbf{Z}_i' \mathbf{R}_i^{-1} \left(\mathbf{y}_i - \mathbf{X}_i \mathbf{b}_{\mathrm{FE}} \right), \tag{2.18}$$

where $\mathbf{Q}_{Z,i} = \mathbf{I}_i - \mathbf{R}_i^{-1/2} \mathbf{Z}_i \left(\mathbf{Z}_i' \mathbf{R}_i^{-1} \mathbf{Z}_i \right)^{-1} \mathbf{Z}_i' \mathbf{R}_i^{-1/2}$.

2.5.3 Robust Estimation of Standard Errors

Equations (2.17) and (2.18) show that the regression estimators are linear combinations of the responses and thus it is straightforward to determine the variance of these estimations. To illustrate, we have

$$\operatorname{Var} \mathbf{b}_{\mathrm{FE}} = \left(\sum_{i=1}^{n} \mathbf{X}_i' \mathbf{R}_i^{-1/2} \mathbf{Q}_{Z,i} \mathbf{R}_i^{-1/2} \mathbf{X}_i \right)^{-1}.$$

Thus, standard errors for the components of $\mathbf{b}_{\mathrm{FE}}$ are readily determined by using estimates of $\mathbf{R}_i$ and taking square roots of diagonal elements of $\operatorname{Var} \mathbf{b}_{\mathrm{FE}}$.

It is common practice to ignore serial correlation and heteroscedasticity initially when estimating $\boldsymbol{\beta}$, so that one can assume $\mathbf{R}_i = \sigma^2 \mathbf{I}_i$. With this assumption, the least-squares estimator of $\boldsymbol{\beta}$ is

$$\mathbf{b} = \left(\sum_{i=1}^{n} \mathbf{X}_i' \mathbf{Q}_i \mathbf{X}_i \right)^{-1} \sum_{i=1}^{n} \mathbf{X}_i' \mathbf{Q}_i \mathbf{y}_i,$$

with $\mathbf{Q}_i = \mathbf{I}_i - \mathbf{Z}_i \left(\mathbf{Z}_i' \mathbf{Z}_i \right)^{-1} \mathbf{Z}_i'$. This is an unbiased and asymptotically normal estimator of $\boldsymbol{\beta}$, although it is less efficient than $\mathbf{b}_{\mathrm{FE}}$. Basic calculations show that it has variance

$$\operatorname{Var} \mathbf{b} = \left(\sum_{i=1}^{n} \mathbf{X}_i' \mathbf{Q}_i \mathbf{X}_i \right)^{-1} \left[\sum_{i=1}^{n} \mathbf{X}_i' \mathbf{Q}_i \mathbf{R}_i \mathbf{Q}_i \mathbf{X}_i \right] \left(\sum_{i=1}^{n} \mathbf{X}_i' \mathbf{Q}_i \mathbf{X}_i \right)^{-1}.$$

To estimate this, Huber (1967G), White (1980E), and Liang and Zeger (1986B) suggested replacing $\mathbf{R}_i$ by $\mathbf{e}_i\mathbf{e}_i'$, where $\mathbf{e}_i$ is the vector of residuals, to get an estimate that is robust to unsuspected serial correlation and heteroscedasticity. Thus, a robust standard error of b_j is

$$se(b_j) = \sqrt{j\text{th diagonal element of } \left(\sum_{i=1}^{n} \mathbf{X}_i'\mathbf{Q}_i\mathbf{X}_i\right)^{-1} \left[\sum_{i=1}^{n} \mathbf{X}_i'\mathbf{Q}_i\mathbf{e}_i\mathbf{e}_i'\mathbf{Q}_i\mathbf{X}_i\right] \left(\sum_{i=1}^{n} \mathbf{X}_i'\mathbf{Q}_i\mathbf{X}_i\right)^{-1}}. \tag{2.19}$$

For contrast, consider a pooled cross-sectional regression model (based on Equation (2.1)) so that $\mathbf{Q}_i = \mathbf{I}_i$ and assume no serial correlation. Then, the OLS estimator of $\boldsymbol{\beta}$ has variance

$$\mathrm{Var}\,\mathbf{b} = \left(\sum_{i=1}^{n} \mathbf{X}_i'\mathbf{X}_i\right)^{-1} \left[\sum_{i=1}^{n} \mathbf{X}_i'\mathbf{R}_i\mathbf{X}_i\right] \left(\sum_{i=1}^{n} \mathbf{X}_i'\mathbf{X}_i\right)^{-1},$$

where $\mathbf{R}_i = \sigma_i^2\mathbf{I}_i$ for heteroscedasticity. Further, using the estimator $s_i^2 = \mathbf{e}_i'\mathbf{e}_i/T_i$ for σ_i^2 yields the usual (White's) robust standard errors. By way of comparison, the robust standard error in Equation (2.19) accommodates heterogeneity (through the $\mathbf{Q}_i$ matrix) and also accounts for unsuspected serial correlation by using the $T_i \times T_i$ matrix $\mathbf{e}_i\mathbf{e}_i'$ in lieu of the scalar estimate $s_i^2 = \mathbf{e}_i'\mathbf{e}_i/T_i$.

Further Reading

Fixed-effects modeling can be best understood based on a solid foundation in regression analysis and analysis using the general linear model. The books by Draper and Smith (1981G) and Seber (1977G) are two classic references that introduce regression using matrix algebra. Treatments that emphasize categorical covariates in the general linear model context include those by Searle (1987G) and Hocking (1985G). Alternatively, most introductory graduate econometrics textbooks cover this material; see, for example, Greene (2002E) or Hayashi (2000E).

This book actively uses matrix algebra concepts to develop the subtleties and nuances of longitudinal data analysis. Appendix A provides a brief overview of the key results. Graybill (1969G) provides additional background.

Early applications of basic fixed-effects panel data models are by Kuh (1959E), Johnson (1960E), Mundlak (1961E), and Hoch (1962E).

Kiefer (1980E) discussed the basic fixed-effects model in the presence of an unstructured serial covariance matrix. He showed how to construct two-stage generalized least-squares (GLS) estimators of global parameters. He also gave a conditional maximum likelihood interpretation of the GLS estimator. Extensions of this idea and additional references are in the thesis by Kung (1996O); see also Section 7.1.

Empirical work on estimating subject-specific slope models has been limited in a fixed-effects context. An example is provided by Polachek and Kim (1994E); they used subject-specific slopes in fixed-effects models when examining gaps in earnings between males and females. Mundlak (1978bE) provided some basic motivation that will be described in Section 7.3.

Appendix 2A Least-Squares Estimation

2A.1 Basic Fixed-Effects Model: Ordinary Least-Squares Estimation

We first calculate the ordinary least-squares (OLS) estimators of the linear parameters in the function

$$\mathrm{E}y_{it} = \alpha_i + \mathbf{x}_{it}'\boldsymbol{\beta}, \quad i = 1, \ldots, n, \ t = 1, \ldots, T_i.$$

To this end, let $a_i^*, b_1^*, b_2^*, \ldots, b_k^*$ be "candidate" estimators of the parameters $\alpha_i, \beta_1, \beta_2, \ldots, \beta_K$. For these candidates, define the sum of squares

$$\mathrm{SS}(\mathbf{a}^*, \mathbf{b}^*) = \sum_{i=1}^{n} \sum_{t=1}^{T_i} (y_{it} - (a_i^* + \mathbf{x}_{it}'\mathbf{b}^*))^2,$$

where $\mathbf{a}^* = (a_1^*, \ldots, a_n^*)'$ and $\mathbf{b}^* = (b_1^*, \ldots, b_K^*)'$. Specifically, $\mathbf{a}^*$ and $\mathbf{b}^*$ are arguments of the sum of squares function *SS*. To minimize this quantity, first examine partial derivatives with respect to a_i^* to get

$$\frac{\partial}{\partial a_i^*}\mathrm{SS}(\mathbf{a}^*, \mathbf{b}^*) = (-2) \sum_{t=1}^{T_i} (y_{it} - (a_i^* + \mathbf{x}_{it}'\mathbf{b}^*)).$$

Setting these partial derivatives to zero yields the least-squares estimators of α_i, $a_i^*(\mathbf{b}^*) = \bar{y}_i - \bar{\mathbf{x}}_i'\mathbf{b}^*$, where $\bar{\mathbf{x}}_i = (\sum_{t=1}^{T_i} \mathbf{x}_{it})/T_i$. The sum of squares evaluated at this value of intercept is

$$\mathrm{SS}(\mathbf{a}^*(\mathbf{b}^*), \mathbf{b}^*) = \sum_{i=1}^{n} \sum_{t=1}^{T_i} (y_{it} - \bar{y}_i - (\mathbf{x}_{it} - \bar{\mathbf{x}}_i)'\mathbf{b}^*)^2.$$

To minimize this sum of squares, take a partial derivative with respect to each component of $\mathbf{b}^*$. For the jth component, we have

$$\frac{\partial}{\partial b_j^*} \mathrm{SS}(\mathbf{a}^*(\mathbf{b}^*), \mathbf{b}^*) = (-2) \sum_{i=1}^{n} \sum_{t=1}^{T_i} (x_{itj} - \bar{x}_{ij})(y_{it} - \bar{y}_i - (\mathbf{x}_{it} - \bar{\mathbf{x}}_i)'\mathbf{b}^*)).$$

Setting this equal to zero, for each component, yields the "normal equations"

$$\sum_{i=1}^{n} \sum_{t=1}^{T_i} (\mathbf{x}_{it} - \bar{\mathbf{x}}_i)(\mathbf{x}_{it} - \bar{\mathbf{x}}_i)' \mathbf{b}^* = \sum_{i=1}^{n} \sum_{t=1}^{T_i} (\mathbf{x}_{it} - \bar{\mathbf{x}}_i)(y_{it} - \bar{y}_i).$$

These normal equations yield the OLS estimators

$$\mathbf{b} = \left(\sum_{i=1}^{n} \sum_{t=1}^{T_i} (\mathbf{x}_{it} - \bar{\mathbf{x}}_i)(\mathbf{x}_{it} - \bar{\mathbf{x}}_i)' \right)^{-1} \sum_{i=1}^{n} \sum_{t=1}^{T_i} (\mathbf{x}_{it} - \bar{\mathbf{x}}_i)(y_{it} - \bar{y}_i)$$

and

$$a_i = \bar{y}_i - \bar{\mathbf{x}}_i' \mathbf{b}.$$

2A.2 Fixed-Effects Models: Generalized Least-Squares Estimation

We consider linear longitudinal data models with regression functions

$$\mathrm{E}\mathbf{y}_i = \mathbf{Z}_i \boldsymbol{\alpha}_i + \mathbf{X}_i \boldsymbol{\beta}, \quad i = 1, \ldots, n,$$

where the variance–covariance matrix of $\mathbf{y}_i$, $\mathbf{R}_i$, is assumed known.

The generalized least-squares sum of squares is

$$\mathrm{SS}(\mathbf{a}^*, \mathbf{b}^*) = \sum_{i=1}^{n} (\mathbf{y}_i - (\mathbf{Z}_i \mathbf{a}_i^* + \mathbf{X}_i \mathbf{b}^*))' \mathbf{R}_i^{-1} (\mathbf{y}_i - (\mathbf{Z}_i \mathbf{a}_i^* + \mathbf{X}_i \mathbf{b}^*)).$$

Here, $\mathbf{a}^*$ and $\mathbf{b}^*$ are "candidate" estimators of $\boldsymbol{\alpha} = (\boldsymbol{\alpha}_1, \ldots, \boldsymbol{\alpha}_n)'$ and $\boldsymbol{\beta} = (\beta_1, \ldots, \beta_K)'$; they are arguments of the function SS.

Following the same strategy as in the previous section, begin by taking partial derivatives of SS with respect to each subject-specific term. That is,

$$\frac{\partial}{\partial \mathbf{a}_i^*} \mathrm{SS}(\mathbf{a}^*, \mathbf{b}^*) = (-2) \mathbf{Z}_i' \mathbf{R}_i^{-1} (\mathbf{y}_i - (\mathbf{Z}_i \mathbf{a}_i^* + \mathbf{X}_i \mathbf{b}^*)).$$

Setting this equal to zero yields

$$\mathbf{a}_i(\mathbf{b}^*) = \left(\mathbf{Z}_i' \mathbf{R}_i^{-1} \mathbf{Z}_i \right)^{-1} \mathbf{Z}_i' \mathbf{R}_i^{-1} (\mathbf{y}_i - \mathbf{X}_i \mathbf{b}^*).$$

We work with the projection

$$\mathbf{P}_{Z,i} = \mathbf{R}_i^{-1/2}\mathbf{Z}_i\left(\mathbf{Z}_i'\mathbf{R}_i^{-1}\mathbf{Z}_i\right)^{-1}\mathbf{Z}_i'\mathbf{R}_i^{-1/2},$$

which is symmetric and idempotent ($\mathbf{P}_{Z,i}\mathbf{P}_{Z,i} = \mathbf{P}_{Z,i}$). With this notation, we have

$$\mathbf{R}_i^{-1/2}\mathbf{Z}_i\mathbf{a}_i(\mathbf{b}^*) = \mathbf{P}_{Z,i}\mathbf{R}_i^{-1/2}(\mathbf{y}_i - \mathbf{X}_i\mathbf{b}^*)$$

and

$$\begin{aligned}\mathbf{R}_i^{-1/2}(\mathbf{y}_i - (\mathbf{Z}_i\mathbf{a}_i^* + \mathbf{X}_i\mathbf{b}^*)) &= \mathbf{R}_i^{-1/2}(\mathbf{y}_i - \mathbf{X}_i\mathbf{b}^*) - \mathbf{P}_{Z,i}\mathbf{R}_i^{-1/2}(\mathbf{y}_i - \mathbf{X}_i\mathbf{b}^*)\\ &= (\mathbf{I} - \mathbf{P}_{Z,i})\mathbf{R}_i^{-1/2}(\mathbf{y}_i - \mathbf{X}_i\mathbf{b}^*).\end{aligned}$$

Now, define the projection, $\mathbf{Q}_{Z,i} = \mathbf{I} - \mathbf{P}_{Z,i}$, which is also symmetric and idempotent. With this notation, the sum of squares is

$$SS(\mathbf{a}^*, \mathbf{b}^*) = \sum_{i=1}^{n}(\mathbf{y}_i - \mathbf{X}_i\mathbf{b}^*)'\mathbf{R}_i^{-1/2}\mathbf{Q}_{Z,i}\mathbf{R}_i^{-1/2}(\mathbf{y}_i - \mathbf{X}_i\mathbf{b}^*).$$

To minimize the sum of squares, take a partial derivative with respect to $\mathbf{b}^*$. Setting this equal to zero yields the generalized least-squares estimators

$$\mathbf{b}_{\mathrm{FE}} = \left(\sum_{i=1}^{n}\mathbf{X}_i'\mathbf{R}_i^{-1/2}\mathbf{Q}_{Z,i}\mathbf{R}_i^{-1/2}\mathbf{X}_i\right)^{-1}\sum_{i=1}^{n}\mathbf{X}_i'\mathbf{R}_i^{-1/2}\mathbf{Q}_{Z,i}\mathbf{R}_i^{-1/2}\mathbf{y}_i$$

and

$$\mathbf{a}_{\mathrm{FE},i} = \left(\mathbf{Z}_i'\mathbf{R}_i^{-1}\mathbf{Z}_i\right)^{-1}\mathbf{Z}_i'\mathbf{R}_i^{-1}\left(\mathbf{y}_i - \mathbf{X}_i\mathbf{b}_{\mathrm{FE}}\right).$$

2A.3 Diagnostic Statistics

Observation-Level Diagnostic Statistic

We use Cook's distance to diagnose unusual observations. For brevity, we assume $\mathbf{R}_i = \sigma^2\mathbf{I}_i$. To define Cook's distance, first let $\mathbf{a}_{i,(it)}$ and $\mathbf{b}_{(it)}$ denote OLS fixed-effects estimators of α_i and β, calculated without the observation from the ith subject at the tth time point. Without this observation, the fitted value for the jth subject at the rth time point is

$$\hat{y}_{jr(it)} = \mathbf{z}_{jr}'\mathbf{a}_{j,(it)} + \mathbf{x}_{jr}'\mathbf{b}_{(it)}.$$

We define Cook's distance to be

$$D_{it} = \frac{\sum_{j=1}^{n}\sum_{r=1}^{T_i}\left(\hat{y}_{jr} - \hat{y}_{jr(it)}\right)^2}{(nq + K)\,s^2},$$

where the fitted value is calculated as $\hat{y}_{jr} = \mathbf{z}'_{jr}\mathbf{a}_{j,\text{FE}} + \mathbf{x}'_{jr}\mathbf{b}_{\text{FE}}$. We calibrate D_{it} using an F-distribution with numerator $df_1 = nq + K$ degrees of freedom and denominator $df_2 = N - (nq + K)$ degrees of freedom. The shortcut calculation form is as follows:

1. Calculate the leverage for the ith subject and tth time point as

$$h_{it} = \mathbf{z}'_{it}(\mathbf{Z}'_i\mathbf{Z}_i)^{-1}\mathbf{z}_{it} + \mathbf{x}'_{it}\left(\sum_{i=1}^{n}\mathbf{X}'_i\mathbf{X}_i\right)^{-1}\mathbf{x}_{it}.$$

2. Residuals are calculated as $e_{it} = y_{it} - (\mathbf{a}'_{i,\text{FE}}\mathbf{z}_{it} + \mathbf{b}'_{\text{FE}}\mathbf{x}_{it})$. The mean-square error is

$$s^2 = \frac{1}{N - (nq + K)}\sum_{i=1}^{n}\sum_{t=1}^{T_i} e_{it}^2.$$

3. Cook's distance is calculated as

$$D_{it} = \frac{e_{it}^2}{(1 - h_{it})^2}\frac{h_{it}}{(nq + K)s^2}.$$

Subject-Level Diagnostic Statistic

From Banerjee and Frees (1997S), the generalization of Section 2.4.3 to the fixed-effects longitudinal data model defined in Section 2.5 is

$$B_i(\mathbf{b}) = \left(\mathbf{b}_{\text{FE}} - \mathbf{b}_{\text{FE}(i)}\right)'\left(\sum_{i=1}^{n}\mathbf{X}'_i\mathbf{R}_i^{-1/2}\mathbf{Q}_{Z,i}\mathbf{R}_i^{-1/2}\mathbf{X}_i\right)\left(\mathbf{b}_{\text{FE}} - \mathbf{b}_{\text{FE}(i)}\right)/K,$$

where $\mathbf{Q}_{Z,i} = \mathbf{I}_i - \mathbf{R}_i^{-1/2}\mathbf{Z}_i(\mathbf{Z}'_i\mathbf{R}_i^{-1}\mathbf{Z}_i)^{-1}\mathbf{Z}'_i\mathbf{R}_i^{-1/2}$. The shortcut calculation form is

$$B_i(\mathbf{b}) = \mathbf{e}'_i\mathbf{Q}_{Z,i}\mathbf{R}_i^{-1/2}(\mathbf{I}_i - \mathbf{H}_i)^{-1}\mathbf{H}_i(\mathbf{I}_i - \mathbf{H}_i)^{-1}\mathbf{R}_i^{-1/2}\mathbf{Q}_{Z,i}\mathbf{e}_i/K$$

where

$$\mathbf{H}_i = \mathbf{R}_i^{-1/2}\mathbf{Q}_{Z,i}\mathbf{X}_i\left(\sum_{i=1}^{n}\mathbf{X}'_i\mathbf{R}_i^{-1/2}\mathbf{Q}_{Z,i}\mathbf{R}_i^{-1/2}\mathbf{X}_i\right)^{-1}\mathbf{X}'\mathbf{Q}_{Z,i}\mathbf{R}_i^{-1/2}$$

is the leverage matrix and $\mathbf{e}_i = \mathbf{y}_i - \mathbf{X}_i\mathbf{b}_{\text{FE}}$. We calibrate B_i using the chi-square distribution with K degrees of freedom.

2A.4 Cross-Sectional Correlation: Shortcut Calculations

The statistics R_{AVE} and R^2_{AVE} are averages over $n(n-1)/2$ correlations, which may be computationally intensive for large values of n. For a shortcut calculation for R_{AVE}, we compute Friedman's statistic directly,

$$FR = \frac{12}{nT(T+1)} \sum_{t=1}^{T} \left(\sum_{i=1}^{n} r_{i,t} \right)^2 - 3n(T+1),$$

and then use the relation $R_{\text{AVE}} = (FR - (T-1))/((n-1)(T-1))$.

For a shortcut calculation for R^2_{AVE}, first define the quantity

$$Z_{i,t,u} = \frac{1}{T^3 - T} 12(r_{i,t} - (T+1)/2)(r_{i,u} - (T+1)/2).$$

With this quantity, an alternative expression for R^2_{AVE} is

$$R^2_{\text{AVE}} = \frac{1}{n(n-1)} \sum_{\{t,u\}} \left(\left(\sum_i Z_{i,t,u} \right)^2 - \sum_i Z^2_{i,t,u} \right),$$

where $\Sigma_{\{t,u\}}$ means sum over $t = 1, \ldots, T$ and $u = 1, \ldots, T$. Although more complex in appearance, this is a much faster computational form for R^2_{AVE}.

Exercises and Extensions

Section 2.1

2.1 *Estimate longitudinal data models using regression routines* Consider a fictitious data set with $x_{it} = i \times t$, for $i = 1, 2, 3, 4$ and $t = 1, 2$. That is, we have the following set of values:

i	1	1	2	2	3	3	4	4
t	1	2	1	2	1	2	1	2
x_{it}	1	2	2	4	3	6	4	8

Consider the usual regression model of the form $\mathbf{y} = \mathbf{X}\boldsymbol{\beta} + \boldsymbol{\epsilon}$, where the matrix of explanatory variables is

$$\mathbf{X} = \begin{pmatrix} x_{11} & x_{12} & \cdots & x_{1K} \\ x_{21} & x_{22} & \cdots & x_{2K} \\ \vdots & \vdots & \ddots & \vdots \\ x_{n1} & x_{n2} & \cdots & x_{nK} \end{pmatrix}.$$

You wish to express your longitudinal data model in terms of the usual regression model.

a. Provide an expression for the matrix **X** for the regression model in Equation (2.1). Specify the dimension of the matrix as well as each entry of the matrix in terms of the data provided.
b. Consider the basic fixed-effects model in Equation (2.3). Express this in terms of the usual regression model by using binary (dummy) variables. Provide an expression for the matrix **X**.
c. Provide an expression for the matrix **X** for the fixed-effects model in Equation (2.4).
d. Provide an expression for the matrix **X** for the fixed-effects model in Equation (2.5).
e. Suppose now that you have $n = 400$ instead of 4 subjects and $T = 10$ observations per subject instead of 2. What is the dimension of your design matrices in parts (a)–(d)? What is the dimension of the matrix, $\mathbf{X}'\mathbf{X}$, that regression routines need to invert?

Section 2.3

2.2 Standard errors for regression coefficients Consider the basic fixed-effects model in Equation (2.3), with $\{\varepsilon_{it}\}$ identically and independently distributed (i.i.d.) with mean zero and variance σ^2.

a. Check Equation (2.10); that is, prove that

$$\mathrm{Var}\,\mathbf{b} = \sigma^2 \left(\sum_{i=1}^{n} \mathbf{W}_i \right)^{-1},$$

where

$$\mathbf{W}_i = \sum_{t=1}^{T_i} (\mathbf{x}_{it} - \bar{\mathbf{x}}_i)(\mathbf{x}_{it} - \bar{\mathbf{x}}_i)'.$$

b. Determine the variance of the ith intercept, $\mathrm{Var}\,a_i$.
c. Determine the covariance among intercepts; that is, determine $\mathrm{Cov}\,(a_i, a_j)$, for $i \neq j$.
d. Determine the covariance between an intercept and the slope estimator; that is, determine $\mathrm{Cov}(a_i, b)$.
e. Determine $\mathrm{Var}(a_i + \mathbf{x}^{*\prime}\mathbf{b})$, where $\mathbf{x}^*$ is a known vector of explanatory variables. For what value of $\mathbf{x}^*$ is this a minimum?

2.3 Least squares

a. Suppose that the regression function is $\mathrm{E}y_{it} = \alpha_i$. Determine the OLS estimator for α_i.
b. Suppose that the regression function is $\mathrm{E}y_{it} = \alpha_i x_{it}$, where x_{it} is a scalar. Determine the OLS estimator for α_i.

c. Suppose that the regression function is $\mathrm{E}y_{it} = \alpha_i x_{it} + \beta$. Determine the OLS estimator for α_i.

2.4 Two population slope interpretations Consider the basic fixed-effects model in Equation (2.3) and suppose that $K = 1$ and that x is a binary variable. Specifically, let $n_{1,i} = \sum_{t=1}^{T_i} x_{it}$ be the number of ones for the ith subject and let $n_{2,i} = T_i - n_{1,i}$ be the number of zeros. Further, define $\bar{y}_{1,i} = (\sum_{t=1}^{T_i} x_{it} y_{it})/n_{1,i}$ to be the average y when $x = 1$, for the ith subject, and similarly $\bar{y}_{2,i} = (\sum_{t=1}^{T_i} (1 - x_{it}) y_{it})/n_{2,i}$.

a. Show that we may write the fixed-effects slope, given in Equation (2.6), as

$$b = \frac{\sum_{i=1}^{n} w_i (\bar{y}_{1,i} - \bar{y}_{2,i})}{\sum_{i=1}^{n} w_i},$$

with weights $w_i = n_{1,i} n_{2,i} / T_i$.

b. Interpret this slope coefficient b.

c. Show that $\mathrm{Var}\, b = \sigma^2 / (\sum_{i=1}^{n} w_i)$.

d. Suppose that you would like to minimize Var b and that the set of observations numbers $\{T_1, \ldots, T_n\}$ is fixed. How could you design the binary variables x (and thus $n_{1,i}$ and $n_{1,i}$) to minimize Var b?

e. Suppose that $\bar{x}_i = 0$ for half the subjects and $\bar{x}_i = 1$ for the other half. What is Var b? Interpret this result.

f. Suppose that the ith subject is designed so that $\bar{x}_i = 0$. What is the contribution of this subject to $\sum_{i=1}^{n} w_i$?

2.5 Least-squares bias Suppose that the analyst should use the heterogeneous model in Equation (2.2) but instead decides to use a simpler, homogeneous model of the form $\mathrm{E}y_{it} = \alpha + \mathbf{x}_{it}'\boldsymbol{\beta}$.

a. Call the least-squares slope estimator $\mathbf{b}_{\mathrm{H}}$ (H for homogeneous). Show that the slope estimator is

$$\mathbf{b}_{\mathrm{H}} = \left(\sum_{i=1}^{n} \sum_{t=1}^{T_i} (\mathbf{x}_{it} - \bar{\mathbf{x}})(\mathbf{x}_{it} - \bar{\mathbf{x}})' \right)^{-1} \left(\sum_{i=1}^{n} \sum_{t=1}^{T_i} (\mathbf{x}_{it} - \bar{\mathbf{x}})(y_{it} - \bar{y}) \right).$$

b. Show that the deviation of $\mathbf{b}_{\mathrm{H}}$ from the slope $\boldsymbol{\beta}$ is

$$\mathbf{b}_{\mathrm{H}} - \boldsymbol{\beta} = \left(\sum_{i=1}^{n} \sum_{t=1}^{T_i} (\mathbf{x}_{it} - \bar{\mathbf{x}})(\mathbf{x}_{it} - \bar{\mathbf{x}})' \right)^{-1} \times \left(\sum_{i=1}^{n} \sum_{t=1}^{T_i} (\mathbf{x}_{it} - \bar{\mathbf{x}})(\alpha_i - \bar{\alpha} + \varepsilon_{it} - \bar{\varepsilon}) \right).$$

c. Assume that $K = 1$. Show that the bias in using $\mathbf{b}_{\mathrm{H}}$ can be expressed as

$$\mathrm{E}b_{\mathrm{H}} - \beta = \frac{1}{(N-1)s_x^2} \sum_{i=1}^{n} T_i \alpha_i \, (\bar{x}_i - \bar{x}),$$

where

$$s_x^2 = \frac{1}{N-1} \sum_{i=1}^{n} \sum_{t=1}^{T_i} (x_{it} - \bar{x})^2$$

is the sample variance of x.

2.6 Residuals Consider the basic fixed-effects model in Equation (2.3) and suppose that $K = 1$. Define the residuals of the OLS fit as $e_{it} = y_{it} - (a_i + x_{it} b)$.

a. Show that the average residual is zero; that is, show that $\bar{e} = 0$.
b. Show that the average residual for the ith subject is zero; that is, show that $\bar{e}_i = 0$.
c. Show that $\sum_{i=1}^{n} \sum_{t=1}^{T_i} e_{it} x_{it,j} = 0$.
d. Why does (c) imply that the estimated correlation between the residuals and the jth explanatory variable is zero?
e. Show that the estimated correlation between the residuals and the fitted values is zero.
f. Show that the estimated correlation between the residuals and the observed dependent variables is, in general, not equal to zero.
g. What are the implications of parts (e) and (f) for residual analysis?

2.7 Group interpretation Consider the basic fixed-effects model in Equation (2.3) and suppose that $K = 1$. Suppose that we are considering $n = 5$ groups. Each group was analyzed separately, with standard deviations and regression slope coefficients given as follows:

Group (i)	1	2	3	4	5
Observations per group (T_i)	11	9	11	9	11
Sample standard deviation ($s_{x,i}$)	1	3	5	8	4
Slope (b_j)	1	3	4	−3	0

For group i, the sample standard deviation of the explanatory variable is given by

$$s_{x,i}^2 = \frac{1}{T_i - 1} \sum_{t=1}^{T_i} (x_{it} - \bar{x}_i)^2.$$

a. Use Equation (2.8) to determine the overall slope estimator, b.
b. Discuss the influence of the group sample standard deviations and size on b.

2.8 Consistency Consider the basic fixed-effects model in Equation (2.3) and suppose that $K = 1$. A sufficient condition for (weak) consistency of b is the mean-square error tends to zero; that is, $\mathrm{E}(b - \beta)^2 \to 0$ as $n \to \infty$ (and T_i remains bounded).

a. Show that we require a sufficient amount of variability in the set of explanatory variables $\{x_{it}\}$ to ensure consistency of b. Explicitly, what does the phrase "a sufficient amount of variability" mean in this context?
b. Suppose that $x_{it} = (-2)^i$ for all i and t. Does this set of explanatory variables meet our sufficient condition to ensure consistency of b?
c. Suppose that $x_{it} = t(-2)^i$ for all i and t. Does this set of explanatory variables meet our sufficient condition to ensure consistency of b?
d. Suppose that $x_{it} = t(-1/2)^i$ for all i and t. Does this set of explanatory variables meet our sufficient condition to ensure consistency of b?

2.9 Least squares For the ith subject, consider the regression function $\mathrm{E}y_{it} = \alpha_i + \mathbf{x}'_{it}\boldsymbol{\beta}_i, t = 1, \ldots, T_i$.

a. Write this as a regression function of the form $\mathrm{E}\mathbf{y}_i = \mathbf{X}_i^* \boldsymbol{\beta}_i^*$ by giving appropriate definitions for $\mathbf{X}_i^*$ and $\boldsymbol{\beta}_i^*$.
b. Use a result on partitioned matrices (Equation (A.1) of Appendix A) to show that the least-squares estimator of $\boldsymbol{\beta}_i$ is

$$\mathbf{b}_i = \left(\sum_{t=1}^{T_i} (\mathbf{x}_{it} - \bar{\mathbf{x}}_i)(\mathbf{x}_{it} - \bar{\mathbf{x}}_i)' \right)^{-1} \left(\sum_{t=1}^{T_i} (\mathbf{x}_{it} - \bar{\mathbf{x}}_i)(y_{it} - \bar{y}_i) \right).$$

Section 2.4

2.10 Pooling test

a. Assume balanced data with $T = 5$ and $K = 5$. Use a statistical software package to show that the 95th percentile of the F-distribution with $df_1 = n - 1$ and $df_2 = N - (n + K) = 4n - 5$ behaves as follows:

n	10	15	20	25	50	100	250	500	1,000
95th percentile	2.1608	1.8760	1.7269	1.6325	1.4197	1.2847	1.1739	1.1209	1.0845

b. For the pooling test statistic defined in Section 2.4.1, show that F-ratio $\to 1$ as $n \to \infty$ (use weak or strong consistency). Interpret the results of (a) in terms of this result.

2.11 Added-variable plot Consider the basic fixed-effects model in Equation (2.3) and suppose that $K = 1$.

a. Begin with the model without any explanatory variables, $y_{it} = \alpha_i + \varepsilon_{it}$. Determine the residuals for this model, denoted by $e_{it,1}$.
b. Now consider the "model" $x_{it} = \alpha_i + \varepsilon_{it}$. Determine the residuals for this representation, denoted by $e_{it,2}$.
c. Explain why a plot of $\{e_{it,1}\}$ versus $\{e_{it,2}\}$ is a special case of added-variable plots.
d. Determine the sample correlation between $\{e_{it,1}\}$ and $\{e_{it,2}\}$. Denote this correlation as corr (e_1, e_2).
e. For the basic fixed-effects model in Equation (2.3) with $K = 1$, show that

$$b = \left(\sum_{i=1}^{n} \sum_{t=1}^{T_i} e_{it,2}^2 \right)^{-1} \left(\sum_{i=1}^{n} \sum_{t=1}^{T_i} e_{it,1} e_{it,2} \right).$$

f. For the basic fixed-effects model in Equation (2.3) with $K = 1$, show that

$$(N - (n+1))\, s^2 = \left(\sum_{i=1}^{n} \sum_{t=1}^{T_i} e_{it,1}^2 \right) - b \left(\sum_{i=1}^{n} \sum_{t=1}^{T_i} e_{it,1} e_{it,2} \right).$$

g. For the basic fixed-effects model in Equation (2.3) with $K = 1$, establish the relationship described in Section 2.4.2 between the partial correlation coefficient and the t-statistic. That is, use (d)–(f) to show that

$$\operatorname{corr}(e_1, e_2) = \frac{t(b)}{\sqrt{t(b)^2 + N - (n+1)}}.$$

2.12 Observation-level diagnostics We now establish a shortcut formula for calculating the usual Cook's distance in linear regression models. To this end, we consider the linear regression function $\mathrm{E}\mathbf{y} = \mathbf{X}\boldsymbol{\beta}$, which consists of N rows, and use the subscript "o" for a generic observation. Thus, let $\mathbf{x}_o'$ be the oth row, or observation. Further, define $\mathbf{X}_{(o)}$ to be the matrix of explanatory variables without the oth observation, and similarly for $\mathbf{y}_{(o)}$. The OLS estimator of $\boldsymbol{\beta}$ with all observations is $\mathbf{b} = (\mathbf{X}'\mathbf{X})^{-1}\mathbf{X}'\mathbf{y}$.

a. Use the equation just below Equation (A.3) in Appendix A to show that

$$\left(\mathbf{X}'_{(o)}\mathbf{X}_{(o)}\right)^{-1} = (\mathbf{X}'\mathbf{X} - \mathbf{x}_o\mathbf{x}'_o)^{-1} = (\mathbf{X}'\mathbf{X})^{-1} + \frac{(\mathbf{X}'\mathbf{X})^{-1}\mathbf{x}_o\mathbf{x}'_o(\mathbf{X}'\mathbf{X})^{-1}}{1 - h_{oo}},$$

where $h_{oo} = \mathbf{x}'_o(\mathbf{X}'\mathbf{X})^{-1}\mathbf{x}_o$ is the leverage for the oth observation.

b. The estimator of $\boldsymbol{\beta}$ without the oth observation is $\mathbf{b}_{(o)} = (\mathbf{X}'_{(o)}X_{(o)})^{-1}\mathbf{X}'_{(o)}\mathbf{y}_{(o)}$. Use part (a) to show that

$$\mathbf{b}_{(o)} = \mathbf{b} - \frac{(\mathbf{X}'\mathbf{X})^{-1}\mathbf{x}_o e_o}{1 - h_{oo}},$$

where $e_o = y_o - \mathbf{x}'_o\mathbf{b}$ is the oth residual.

c. Cook's distance is defined to be

$$D_o = \frac{\left(\hat{\mathbf{y}} - \hat{\mathbf{y}}_{(o)}\right)'\left(\hat{\mathbf{y}} - \hat{\mathbf{y}}_{(o)}\right)}{\text{ncol}(\mathbf{X})s^2},$$

where $\hat{\mathbf{y}} = \mathbf{Xb}$ is the vector of fitted values, ncol($\mathbf{X}$) is the number of columns of $\mathbf{X}$, and $\hat{\mathbf{y}}_{(o)} = \mathbf{Xb}_{(o)}$ is the vector of fitted values without the oth observation. Show that

$$D_o = \left(\frac{e_o}{s\sqrt{1 - h_{oo}}}\right)^2 \frac{h_{oo}}{\text{ncol}(\mathbf{X})(1 - h_{oo})}.$$

d. Use the expression in part (c) to verify the formula for Cook's distance given in Appendix 2A.3.

2.13 Cross-sectional correlation test statistic

a. Calculate the variance of the random variable Q in Section 2.4.4.
b. The following table provides the 95th percentile of the Q random variable as a function of T.

T	4	5	6	7	8	9	10	12	15
95th percentile	0.832	0.683	0.571	0.495	0.431	0.382	0.344	0.286	0.227

Compute the corresponding cutoffs using the normal approximation and your answer in part (a). Discuss the performance of the approximation as T increases.

Section 2.5

2.14 Serial correlations Consider the compound symmetry model, where the error variance–covariance matrix is given by $\mathbf{R} = \sigma^2((1 - \rho)\mathbf{I} + \rho\mathbf{J})$.

a. Check that the inverse of $\mathbf{R}$ is $\mathbf{R}^{-1} = \sigma^{-2}(1-\rho)^{-1}(\mathbf{I} - \frac{\rho}{\rho(\mathbf{T}-1)+1}\mathbf{J})$. Do this by showing that $\mathbf{R}^{-1} = \mathbf{I}$, the identity matrix.
b. Use this form of $\mathbf{R}$ in Equation (2.17) to show that the fixed-effects estimator of $\boldsymbol{\beta}$, $\mathbf{b}_{\text{FE}}$, equals the OLS estimator, $\mathbf{b}$, given in Section 2.5.3.

2.15 *Regression model* Consider the general fixed-effects longitudinal data model given in Equation (2.16). Write this model as a regression function in the form $\text{E}\mathbf{y} = \mathbf{X}^*\boldsymbol{\beta}^*$, being sure to do the following:

a. Describe specifically how to use the matrices of explanatory variables $\{\mathbf{Z}_i\}$ and $\{\mathbf{X}_i\}$ to form $\mathbf{X}^*$.
b. Describe specifically how to use the vectors of parameters $\{\boldsymbol{\alpha}_i\}$ and $\boldsymbol{\beta}$ to form $\boldsymbol{\beta}^*$.
c. Identify the dimensions of $\mathbf{y}$, $\mathbf{X}^*$, and $\boldsymbol{\beta}^*$.

2.16 *Interpreting the slope as a weighted average of subject-specific slopes* Consider the general fixed-effects longitudinal data model given in Equation (2.16) and define the weight matrix $\mathbf{W}_i = \mathbf{X}_i'\mathbf{R}_i^{-1/2}\mathbf{Q}_{Z,i}\mathbf{R}_i^{-1/2}\mathbf{X}_i$.

a. Based on data from only the ith subject, show that the least-squares slope estimator is

$$\mathbf{b}_{\text{FE},i} = \mathbf{W}_i^{-1}\mathbf{X}_i'\mathbf{R}_i^{-1/2}\mathbf{Q}_{Z,i}\mathbf{R}_i^{-1/2}\mathbf{y}_i.$$

(*Hint*: Consider Equation (2.17) with $n = 1$.)
b. Show that the fixed-effects estimator of $\boldsymbol{\beta}$ can be expressed as a matrix weighted average of the form

$$\mathbf{b}_{\text{FE}} = \left(\sum_{i=1}^{n}\mathbf{W}_i\right)^{-1}\sum_{i=1}^{n}\mathbf{W}_i\mathbf{b}_{\text{FE},i}.$$

2.17 *Fixed-effects linear longitudinal data estimators* Consider the regression coefficient estimators of the fixed-effects linear longitudinal data model in Equations (2.17) and (2.18). Show that if we assume no serial correlation, $q = 1$ and $z_{it,1} = 1$, then these expressions reduce to the estimators given in Equations (2.6) and (2.7).

2.18 *Ordinary least squares based on differenced data* Consider the basic fixed-effects model in Equation (2.3) and use ordinary least squares based on differenced data. That is, data will be differenced over time, so that the response is $\Delta y_{it} = y_{it} - y_{i,t-1}$ and the vector of covariates is $\Delta\mathbf{x}_{it} = \mathbf{x}_{it} - \mathbf{x}_{i,t-1}$.

a. Show that the OLS estimator of β based on differenced data is

$$\mathbf{b}_\Delta = \left(\sum_{i=1}^{n} \sum_{t=2}^{T_i} \Delta \mathbf{x}_{it} \Delta \mathbf{x}_{it}' \right)^{-1} \left(\sum_{i=1}^{n} \sum_{t=2}^{T_i} \Delta \mathbf{x}_{it} \Delta y_{it} \right).$$

b. Now compute the variance of this estimator. To this end, define the vector of differenced responses $\Delta \mathbf{y}_i = (\Delta y_{i2}, \ldots, \Delta y_{iT_i})'$ and the corresponding matrix of differenced covariates $\Delta \mathbf{X}_i = (\Delta \mathbf{x}_{i2}, \ldots, \Delta \mathbf{x}_{iT_i})'$. With this notation, show that

$$\operatorname{Var} \mathbf{b}_\Delta = \left(\sum_{i=1}^{n} \Delta \mathbf{X}_i' \Delta \mathbf{X}_i \right)^{-1} \left(\sum_{i=1}^{n} \Delta \mathbf{X}_i' \mathbf{R}_i^* \Delta \mathbf{X}_i \right) \left(\sum_{i=1}^{n} \Delta \mathbf{X}_i' \Delta \mathbf{X}_i \right)^{-1},$$

where

$$\mathbf{R}_i^* = \operatorname{Var} \Delta \mathbf{y}_i = \sigma^2 \begin{pmatrix} 2 & -1 & 0 & \cdots & 0 & 0 \\ -1 & 2 & -1 & \cdots & 0 & 0 \\ 0 & -1 & 2 & \cdots & 0 & 0 \\ \vdots & \vdots & \vdots & \ddots & \vdots & \vdots \\ 0 & 0 & 0 & \cdots & 2 & -1 \\ 0 & 0 & 0 & \cdots & -1 & 2 \end{pmatrix}.$$

c. Now assume balanced data so that $T_i = T$. Further assume that $\{\mathbf{x}_{it}\}$ are i.i.d. with mean $\mathrm{E}\mathbf{x}_{it} = \boldsymbol{\mu}_x$ and variance $\operatorname{Var} \mathbf{x}_{it} = \boldsymbol{\Sigma}_x$. Using Equation (2.10), show that

$$\lim_{n \to \infty} n \left(\operatorname{Var} (\mathbf{b} \mid \mathbf{X}_1, \ldots, \mathbf{X}_n) \right) = \sigma^2 \frac{1}{(T-1)} \boldsymbol{\Sigma}_x^{-1},$$

with probability one.

d. Use the assumptions of part (c). Using part (b), show that

$$\lim_{n \to \infty} n \left(\operatorname{Var} (\mathbf{b}_\Delta \mid \mathbf{X}_1, \ldots, \mathbf{X}_n) \right) = \sigma^2 \frac{3T-4}{2(T-1)^2} \boldsymbol{\Sigma}_x^{-1},$$

with probability one.

e. From the results of parts (c) and (d), argue that the Equation (2.6) fixed-effects estimator is more efficient than the least-squares estimator based on differenced data. What is the limiting (as $T \to \infty$) efficiency ratio?

Empirical Exercises

2.19 Charitable contributions We analyze individual income tax returns data from the 1979–1988 Statistics of Income (SOI) Panel of Individual Returns. The SOI Panel is a subset of the IRS Individual Tax Model File and represents

Table 2E.1. *Taxpayer characteristics*

Variable	Description
SUBJECT	Subject identifier, 1–47
TIME	Time identifier, 1–10
CHARITY	The sum of cash and other property contributions, excluding carry-overs from previous years
INCOME	Adjusted gross income
PRICE	One minus the marginal tax rate. Here, the marginal tax rate is defined on income prior to contributions
AGE	A binary variable that equals one if a taxpayer is over sixty-four-years old and equals zero otherwise
MS	A binary variable that equals one if a taxpayer is married and equals zero otherwise
DEPS	Number of dependents claimed on the taxpayer's form

a simple random sample of individual income tax returns filed each year. Based on the individual returns data, the goal is to investigate whether a taxpayer's marginal tax rate affects private charitable contributions, and secondly, if the tax revenue losses due to deductions of charitable contributions is less than the gain of charitable organizations. To address these issues, we consider a price and income model of charitable contributions, considered by Banerjee and Frees (1997S).

They define price as the complement of an individual's federal marginal tax rate, using taxable income prior to contributions. Income of an individual is defined as the adjusted gross income. The dependent variable is total charitable contributions, which is measured as the sum of cash and other property contributions, excluding carry-overs from previous years. Other covariates included in the model are age, marital status, and the number of dependents of an individual taxpayer. Age is a binary variable representing whether a taxpayer is over sixty-four years or not. Similarly, marital status represents whether an individual is married or single.

The population consists of all U.S. taxpayers who itemize their deductions. Specifically, these are the individuals who are likely to have and to record charitable contribution deductions in a given year. Among the 1,413 taxpayers in our subset of the SOI Panel, approximately 22% itemized their deductions each year during the period 1979–1988. A random sample of 47 individuals was selected from the latter group. These data are analyzed in Banerjee and Frees (1997S). Table 2E.1 lists the variables.

a. *Basic summary statistics.*
   i. Summarize each variable. For the binary variables, AGE and MS, provide only averages. For the other variables, CHARITY, INCOME,

PRICE, and DEPS, provide the mean, median, standard deviation, minimum, and maximum. Further, summarize the average response variable CHARITY over TIME.

ii. Create a multiple time-series plot of CHARITY versus TIME.

iii. Summarize the relationship among CHARITY, INCOME, PRICE, DEPS, and TIME. Do this by calculating correlations and scatter plots for each pair.

b. *Basic fixed-effects model.*

i. Run a one-way fixed-effects model of CHARITY on INCOME, PRICE, DEPS, AGE, and MS. State which variables are statistically significant and justify your conclusions.

ii. Produce an added-variable plot of CHARITY versus INCOME, controlling for the effects of PRICE, DEPS, AGE, and MS. Interpret this plot.

c. *Incorporating temporal effects.* Is there an important time pattern?

i. Rerun the model in (b)(i) and include TIME as an additional explanatory (continuous) variable.

ii. Rerun the model in (b)(i) and include TIME through dummy variables, one for each year.

iii. Rerun the model in (b)(i) and include an $AR(1)$ component for the error.

iv. Which of the three methods for incorporating temporal effects do you prefer? Be sure to justify your conclusion.

d. *Unusual observations.*

i. Rerun the model in (b)(i) and calculate Cook's distance to identify unusual observations.

ii. Rerun the model in (b)(i) and calculate the influence statistic for each subject. Identify the subject with the largest influence statistic. Rerun your model by omitting the subject that you have identified. Summarize the effects on the global parameter estimates.

2.20 Tort filings The response that we consider is FILINGS, the number of tort actions against insurance companies per 100,000 population (y). For each of six years, 1984–1989, the data were obtained from 19 states. Thus, there are $6 \times 19 = 114$ observations available. The issue is to try to understand how state legal, economic, and demographic characteristics affect FILINGS. Table 2E.2 describes these characteristics. More extensive motivation is provided in Section 10.2.

a. *Basic summary statistics.*

i. Provide a basic table of univariate summary statistics, including the mean, median, standard deviation, minimum, and maximum.

Table 2E.2. *State characteristics*

Dependent variable	Description
FILINGS	Number of filings of tort actions against insurance companies per 100,000 population
State legal characteristics	
JSLIAB	An indicator of joint and several liability reform
COLLRULE	An indicator of collateral source reform
CAPS	An indicator of caps on noneconomic reform
PUNITIVE	An indicator of limits of punitive damage
State economic and demographic characteristics	
POPLAWYR	The population per lawyer
VEHCMILE	Number of automobiles miles per mile of road, in thousands
GSTATEP	Percentage of gross state product from manufacturing and construction
POPDENSY	Number of people per ten square miles of land
WCMPMAX	Maximum workers' compensation weekly benefit
URBAN	Percentage of population living in urban areas
UNEMPLOY	State unemployment rate, in percentages

Source: Lee (1994O).

ii. Calculate the correlation between the dependent variable, FILINGS, and each of the explanatory variables. Note the correlation between WCMPMAX and FILINGS.
iii. Examine the mean of each variable over time. (You may also wish to explore the data in other ways, through looking at distributions and basic plots.)

b. Fit a pooled cross-sectional regression model (based on Equation (2.1)) using VEHCMILE, GSTATEP, POPDENSY, WCMPMAX, URBAN, UNEMPLOY, and JSLIAB as explanatory variables.
c. Fit a one-way fixed-effects model using state as the subject identifier and VEHCMILE, GSTATEP, POPDENSY, WCMPMAX, URBAN, UNEMPLOY, and JSLIAB as explanatory variables.
d. Fit a two-way fixed-effects model using state and time subject identifiers and VEHCMILE, GSTATEP, POPDENSY, WCMPMAX, URBAN, UNEMPLOY, and JSLIAB as explanatory variables.
e. Perform partial F-tests to see which of the models in parts (b), (c), and (d) you prefer.
f. The one-way fixed-effects model using state as the subject identifier produced a statistically significant, positive coefficient on WCMPMAX. However, this was not in congruence with the correlation coefficient. To explain this coefficient, produce an added-variable plot of FILINGS versus

WCMPMAX, controlling for state and the other explanatory variables, VEHCMILE, GSTATEP, POPDENSY, URBAN, UNEMPLOY, and JSLIAB.

g. For your model in part (c), calculate the influence statistic for each subject. Identify the subject with the largest influence. (You may also wish to rerun the part (c) model without this subject.)
h. Rerun the part (c) model, but assuming an autoregressive of order error structure. Is this $AR(1)$ coefficient statistically significant?

2.21 Housing prices In this problem, we will examine models of housing prices in U.S. metropolitan areas. Many studies have addressed the housing market (see, for example, Green and Malpezzi, 2003O for an introduction). The prices of houses are influenced by demand-side factors such as income and demographic variables. Supply-side factors, such as the regulatory environment of a metropolitan area, may also be important.

The data consist of annual observations from 36 metropolitan statistical areas (MSAs) over the nine-year period 1986–1994. The response variable is NARSP, an MSA's average sale price based on transactions reported through the Multiple Listing Service, National Association of Realtors. As part of a preliminary analysis, the response variable has been transformed using a natural logarithm. For this problem, the demand-side variables are time varying yet the supply-side variables do not vary with time. Table 2E.3 summarizes the variables.

a. *Basic summary statistics.*
   i. Begin by summarizing the time-constant supply-side explanatory variables. Provide means for the binary variables and the mean, median, standard deviation, minimum, and maximum for the other variables.
   ii. Assess the strength of the relationship among the supply-side explanatory variables by calculating correlations.
   iii. Summarize the additional variables. Provide mean, median, standard deviation, minimum, and maximum for these variables.
   iv. Examine trends over time producing the means over time for the non–supply side variables.
   v. Produce a multivariate time-series plot of NARSP.
   vi. Calculate correlations among NARSP, PERYPC, PERPOP, and YEAR.
   vii. Plot PERYPC versus YEAR. Comment on the unusual behavior in 1993 and 1994.
   viii. Produce an added-variable plot of NARSP versus PERYPC, controlling for the effects of PERPOP and YEAR. Interpret this plot.

Table 2E.3. *MSA characteristics*

Response variable	Description
NARSP	An MSA's average sale price, in logarithmic units, based on transactions reported through the Multiple Listing Service
Demand-side explanatory variables	
YPC	Annual per capita income, from the Bureau of Economic Analysis
POP	Population, from the Bureau of Economic Analysis
PERYPC	Annual percentage growth of per capita income
PERPOP	Annual percentage growth of population
Supply-side explanatory variables	
REGTEST	Regulatory index from Malpezzi (1996O)
RCDUM	Rent control dummy variable
SREG1	Sum of American Institute of Planners state regulatory questions regarding use of environmental planning and management
AJWTR	Indicates whether the MSA is adjacent to a coastline
AJPARK	Indicates whether the MSA is adjacent to one or more large parks, military bases, or reservations
Additional variables	
MSA	Subject (MSA) identifier, 1–36
TIME	Time identifier, 1–9

b. *Basic fixed-effects model.*

i. Fit a homogeneous model, as in Equation (2.1), using PERYPC, PERPOP, and YEAR as explanatory variables. Comment on the statistical significance of each variable and the overall model fit.

ii. Run a one-way fixed-effects model of NARSP on PERYPC, PERPOP, and YEAR. State which variables are statistically significant and justify your conclusions. Comment on the overall model fit.

iii. Compare the models in parts (b)(i) and (b)(ii) using a partial F-test. State which model you prefer based on this test.

iv. Rerun the step in (b)(ii) by excluding YEAR as an additional explanatory (continuous) variable yet including an $AR(1)$ component for the error. State whether or not YEAR should be included in the model.

v. For the model in (b)(ii) calculate the influence statistic for each MSA. Identify the MSA with the largest influence statistic. Rerun your model by omitting the MSA that you have identified. Summarize the effects on the global parameter estimates.

Table 2E.4. *Bond maturity*

Variable	Description
DEBTMAT	The book value–weighted average of the firm's debt (the response variable)
SIC	Standard Industrial Classification (SIC) of the firm
FIRMID	Subject (firm) identifier, 1–328
TIME	Time identifier, 1–10
MVBV	The market value of the firm (proxied by the sum of the book value of assets and the market value of equity less the book value of equity) scaled by the book value of assets
SIZE	The natural logarithm of the estimate of firm value measured in 1982 dollars using the Producer Price Index deflator
CHANGEEPS	The difference between next year's earnings per share and this year's earnings per share scaled by this year's common stock price per share
ASSETMAT	The book value–weighted average of the maturities of current assets and net property plant and equipment
VAR	Ratio of the standard deviation of the first difference in earnings before interest, depreciation, and taxes to the average of assets over the period 1980–1989
TERM	The difference between the long-term and short-term yields on government bonds
BONDRATE	The firm's S&P bond rating
TAXRATE	Ratio of income taxes paid to pretax income
LEVERAGE	Ratio of total debt to the market value of the firm

c. *Additional analyses.*
   i. We have not yet tried to fit any supply-side variables. Redo the model fit in part (b)(i), yet including supply-side variables REGTEST, RCDUM, SREG1, AJPARK, and AJWTR. Comment on the statistical significance of each variable and the overall model fit.
   ii. Rerun the model in part (c)(i) and include a dummy variable for each MSA, resulting in a one-way fixed-effects model. Comment on the difficulty of achieving unique parameter estimates with this procedure.

2.22 *Bond maturity (unstructured problem)* These data consist of observations of 328 nonregulated firms over the period 1980–1989. The goal is to assess the debt maturity structure of a firm.

For this exercise, develop a fixed-effects model using the variables in Table 2E.4. For one approach, see Stohs and Mauer (1996O).

3

Models with Random Effects

Abstract. This chapter considers the Chapter 2 data structure but here the heterogeneity is modeled using random quantities in lieu of fixed parameters; these random quantities are known as *random effects*. By introducing random quantities, the analysis of longitudinal and panel data can now be cast in the mixed linear model framework.

Although mixed linear models are an established part of statistical methodology, their use is not as widespread as regression. Thus, the chapter introduces this modeling framework, beginning with the special case of a single random intercept known as the *error-components* model and then focusing on the *linear mixed-effects* model, which is particularly important for longitudinal data. After introducing the models, this chapter describes estimation of regression coefficients and variance components, as well as hypothesis testing for regression coefficients.

3.1 Error-Components/Random-Intercepts Model

Sampling and Inference

Suppose that you are interested in studying the behavior of individuals who are randomly selected from a population. For example, in Section 3.2 we will study the effects that an individual's economic and demographic characteristics have on the amount of income tax paid. Here, the set of subjects that we will study is randomly selected from a larger database, which is itself a random sample of the U.S. taxpayers. In contrast, the Chapter 2 Medicare example dealt with a fixed set of subjects. That is, it is difficult to think of the 54 states as a subset from some "superpopulation" of states. For both situations, it is natural to use subject-specific parameters, $\{\alpha_i\}$, to represent the heterogeneity among subjects. Unlike Chapter 2, Chapter 3 discusses situations in which it is more reasonable to represent $\{\alpha_i\}$ as random variables instead of fixed, yet unknown, parameters. By arguing that $\{\alpha_i\}$ are draws from a distribution, we will have the ability to make inferences about subjects in a population that are not included in the sample.

Basic Model and Assumptions

The *error-components model* equation is

$$y_{it} = \alpha_i + \mathbf{x}'_{it}\boldsymbol{\beta} + \varepsilon_{it}. \tag{3.1}$$

This portion of the notation is the same as the error representation of the basic fixed-effects model. However, now the term α_i is assumed to be a random variable, not a fixed, unknown parameter. The term α_i is known as a *random effect*. *Mixed-effects* models are ones that include random as well as fixed effects. Because Equation (3.1) includes both random effects (α_i) and fixed effects ($\boldsymbol{\beta}$), the error-components model is a special case of the *mixed linear model*.

To complete the specification of the error-components model, we assume that $\{\alpha_i\}$ are identically and independently distributed (i.i.d.) with mean zero and variance σ_α^2. Further, we assume that $\{\alpha_i\}$ are independent of the error random variables, $\{\varepsilon_{it}\}$. For completeness, we still assume that $\boldsymbol{x}_{it}$ is a vector of covariates, or explanatory variables, and that $\boldsymbol{\beta}$ is a vector of fixed, yet unknown, population parameters. Note that because $\mathrm{E}\,\alpha_i = 0$, it is customary to include a constant within the vector $\boldsymbol{x}_{it}$. This was not true of the fixed-effects models in Chapter 2 where we did not center the subject-specific terms about 0.

Linear combinations of the form $\mathbf{x}'_{it}\boldsymbol{\beta}$ quantify the effect of known variables that may affect the response. Additional variables, which are either unimportant or unobservable, comprise the "error term." In the error-components model, we may think of a regression model $y_{it} = \mathbf{x}'_{it}\boldsymbol{\beta} + \eta_{it}$, where the error term η_{it} is decomposed into two components so that $\eta_{it} = \alpha_i + \varepsilon_{it}$. The term α_i represents the time-constant portion whereas ε_{it} represents the remaining portion. To identify the model parameters, we assume that the two terms are independent. In the biological sciences, the error-components model is known as the *random-intercepts model*; this descriptor is used because the intercept α_i is a random variable. We will use the descriptors "error components" and "random intercepts" interchangeably, although, for simplicity, we often use only the former term.

Traditional ANOVA Setup

In the error-components model, the terms $\{\alpha_i\}$ account for the heterogeneity among subjects. To help interpret this feature, consider the special case where

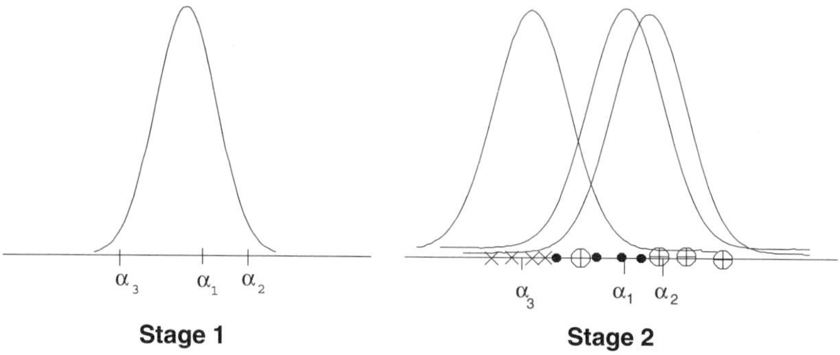

Figure 3.1. Two-stage-random-effects sampling. In the left panel, unobserved subject-specific components are drawn from an unobserved population. In the right panel, several observations are drawn for each subject. These observations are centered about the unobserved subject-specific components from the first stage. Different plotting symbols represent draws for different subjects.

$K = 1$ and $x_{it} = 1$ and denote $\mu = \beta_1$. In this case, Equation (3.1) contains no explanatory variables and reduces to

$$y_{it} = \mu + \alpha_i + \varepsilon_{it},$$

the traditional random-effects, one-way ANOVA model. Neter and Wasserman (1974G) describe this classic model. This model can be interpreted as arising from a two-stage sampling scheme:

Stage 1. Draw a random sample of n subjects from a population. The subject-specific parameter α_i is associated with the ith subject.

Stage 2. Conditional on α_i, draw realizations of $\{y_{it}\}$, for $t = 1, \ldots, T_i$, for the ith subject.

That is, in the first stage, we draw a sample from a population of subjects. In the second stage, we observe each subject over time. Because the first stage is considered a random draw from a population of subjects, we represent characteristics that do not depend on time through the random quantity α_i. Figure 3.1 illustrates the two-stage sampling scheme.

Within this traditional model, interest generally centers about the distribution of the population of subjects. For example, the parameter $\mathrm{Var}\,\alpha_i = \sigma_\alpha^2$ summarizes the heterogeneity among subjects. In Chapter 2 on fixed-effects models, we examined the heterogeneity issue through a test of the null hypothesis $\mathrm{H}_0 : \alpha_1 = \alpha_2 = \cdots = \alpha_n$. In contrast, under the random-effects model, we examine the null hypothesis $\mathrm{H}_0 : \sigma_\alpha^2 = 0$. Furthermore, estimates of σ_α^2 are of interest but require scaling to interpret. A more useful quantity to report is $\sigma_\alpha^2/(\sigma_\alpha^2+\sigma^2)$, the *intraclass correlation*. As we saw in Section 2.5.1, this quantity can be interpreted as the correlation between observations within a subject. The correlation is constrained to lie between 0 and 1 and does not depend on the units of measurement for the response. Further, it can also be interpreted as the proportion of variability of a response that is due to heterogeneity among subjects.

Sampling and Model Assumptions

The Section 2.1 basic fixed-effects and error-components models are similar in appearance yet, as will be discussed in Section 7.2, can lead to different substantive conclusions in the context of a specific application. As we have described, the choice between these two models is dictated primarily by the method in which the sample is drawn. On the one hand, selecting subjects based on a two-stage, or *cluster*, sample implies use of the random-effects model. On the other hand, selecting subjects based on exogenous characteristics suggests a stratified sample and thus use of a fixed-effects model.

The sampling basis allows us to restate the error-components model, as follows.

Error-Components Model Assumptions

R1. $\mathrm{E}(y_{it}\,|\,\alpha_i)=\alpha_i+\mathbf{x}_{it}'\boldsymbol{\beta}$.
R2. $\{x_{it,1},\ldots,x_{it,K}\}$ are nonstochastic variables.
R3. $\mathrm{Var}(y_{it}\,|\,\alpha_i)=\sigma^2$.
R4. $\{y_{it}\}$ are independent random variables, conditional on $\{\alpha_1,\ldots,\alpha_n\}$.
R5. y_{it} is normally distributed, conditional on $\{\alpha_1,\ldots,\alpha_n\}$.
R6. $\mathrm{E}\alpha_i=0$, $\mathrm{Var}\,\alpha_i=\sigma_\alpha^2$, and $\{\alpha_1,\ldots,\alpha_n\}$ are mutually independent.
R7. $\{\alpha_i\}$ is normally distributed.

Assumptions R1–R5 are similar to the fixed-effects models assumptions F1–F5; the main difference is that we now condition on random subject-specific terms, $\{\alpha_1,\ldots,\alpha_n\}$. Assumptions R6 and R7 summarize the sampling basis of the subject-specific terms. Taken together, these assumptions comprise our *error-components model*.

However, Assumptions R1–R7 do not provide an "observables" representation of the model because they are based on unobservable quantities, $\{\alpha_1,\ldots,\alpha_n\}$. We summarize the effects of Assumptions R1–R7 on the observable variables, $\{x_{it,1},\ldots,x_{it,K},y_{it}\}$.

Observables Representation of the Error-Components Model

RO1. $\mathrm{E} y_{it} = \mathbf{x}_{it}' \boldsymbol{\beta}$.
RO2. $\{x_{it,1}, \ldots, x_{it,K}\}$ are nonstochastic variables.
RO3. $\mathrm{Var}\, y_{it} = \sigma^2 + \sigma_\alpha^2$ and $\mathrm{Cov}(y_{ir}, y_{is}) = \sigma_\alpha^2$, for $r \neq s$.
RO4. $\{\mathbf{y}_i\}$ are independent random vectors.
RO5. $\{\mathbf{y}_i\}$ are normally distributed.

To reiterate, the properties RO1–RO5 are a consequence of R1–R7. As we progress into more complex situations, our strategy will consist of using sampling bases to suggest basic assumptions, such as R1–R7, and then converting them into testable properties such as RO1–RO5. Inference about the testable properties then provides information about the more basic assumptions. When considering nonlinear models beginning in Chapter 9, this conversion will not be as direct. In some instances, we will focus on the observable representation directly and refer to it as a *marginal* or *population-averaged model*. The marginal version emphasizes the assumption that observations are correlated within subjects (Assumption RO3), not the random-effects mechanism for inducing the correlation.

For more complex situations, it will be useful to describe these assumptions in matrix notation. As in Equation (2.13), the regression function can be expressed more compactly as

$$\mathrm{E}\,(\mathbf{y}_i \mid \alpha_i) = \alpha_i \mathbf{1}_i + \mathbf{X}_i \boldsymbol{\beta}$$

and thus,

$$\mathrm{E}\mathbf{y}_i = \mathbf{X}_i \boldsymbol{\beta}. \tag{3.2}$$

Recall that $\mathbf{1}_i$ is a $T_i \times 1$ vector of ones and, from Equation (2.14), that $\mathbf{X}_i$ is a $T_i \times K$ matrix of explanatory variables, $\mathbf{X}_i = (\mathbf{x}_{i1} \mathbf{x}_{i2} \cdots \mathbf{x}_{iT_i})'$. The expression for $\mathrm{E}(\mathbf{y}_i \mid \alpha_i)$ is a restatement of Assumption R1 in matrix notation. Equation (3.2) is a restatement of Assumption RO1. Alternatively, Equation (3.2) is due to the law of iterated expectations and Assumptions R1 and R6, because $\mathrm{E}\mathbf{y}_i = \mathrm{E}\,\mathrm{E}(\mathbf{y}_i \mid \alpha_i) = \mathrm{E}\alpha_i \mathbf{1}_i + \mathbf{X}_i \boldsymbol{\beta} = \mathbf{X}_i \boldsymbol{\beta}$. For Assumption RO3, we have

$$\mathrm{Var}\, \mathbf{y}_i = \mathbf{V}_i = \sigma_\alpha^2 \mathbf{J}_i + \sigma^2 \mathbf{I}_i. \tag{3.3}$$

Here, recall that $\mathbf{J}_i$ is a $T_i \times T_i$ matrix of ones, and $\mathbf{I}_i$ is a $T_i \times T_i$ identity matrix.

Structural Models

Model assumptions are often dictated by sampling procedures. However, we also wish to consider stochastic models that represent causal relationships suggested by a substantive field; these are known as *structural models*. Section 6.1 describes structural modeling in longitudinal and panel data analysis. For example, in models of economic applications, it is important to consider carefully what one means by the "population of interest." Specifically, when considering choices of economic entities, a standard defense for a probabilistic approach to analyzing economic decisions is that, although there may be a finite number of economic entities, there is an infinite range of economic decisions. For example, in the Chapter 2 Medicare hospital cost example, one may argue that each state faces a distribution of infinitely many economic outcomes and that this is the population of interest. This viewpoint argues that one should use an error-components model. Here, we interpret $\{\alpha_i\}$ to represent those aspects of the economic outcome that are unobservable yet constant over time. In contrast, in Chapter 2 we implicitly used the sampling-based model to interpret $\{\alpha_i\}$ as fixed effects.

This viewpoint is the standard rationale for studying stochastic economics. A quote from Haavelmo (1944E) helps illustrate this point:

> The class of populations we are dealing with does not consist of an infinity of different individuals, it consists of an infinity of possible decisions which might be taken with respect to the value of y.

This defense is well summarized by Nerlove and Balestra in a monograph edited by Mátyás and Sevestre (1996E, Chapter 1) in the context of panel data modeling.

Inference

When designing a longitudinal study and considering whether to use a fixed- or random-effects model, keep in mind the purposes of the study. If you would like to make statements about a population larger than the sample, then use the random-effects model. Conversely, if you are simply interested in controlling for subject-specific effects (treating them as nuisance parameters) or in making predictions for a specific subject, then use the fixed-effects model.

Time-Constant Variables

When designing a longitudinal study and considering whether to use a fixed- or random-effects model, also keep in mind the variables of interest. Often, the primary interest is in testing for the effect of a time-constant variable. To illustrate, in our taxpayer example, we may be interested in the effects that gender may have on an individual's tax liability; we assume that this variable does not change for an individual over the course of our study. Another important example of a time-constant variable is a variable that classifies subjects into groups. Often, we wish to compare the performance of different groups, for example, a "treatment group" and a "control group."

In Section 2.3, we saw that time-constant variables are perfectly collinear with subject-specific intercepts and hence are inestimable. In contrast, coefficients associated with time-constant variables are estimable in a random-effects model. Hence, if a time-constant variable such as gender or treatment group is the primary variable of interest, one should design the longitudinal study so that a random-effects model can be used.

Degrees of Freedom

When designing a longitudinal study and considering whether to use a fixed- or random-effects model, also keep in mind the size of the data set necessary for inference. In most longitudinal data studies, inference about the population parameters $\boldsymbol{\beta}$ is the primary goal, whereas the terms $\{\alpha_i\}$ are included to control for the heterogeneity. In the basic fixed-effects model, we have seen that there are $n + K$ linear regression parameters plus 1 variance parameter. This is compared to only $1 + K$ regression plus 2 variance parameters in the basic random-effects model. Particularly in studies where the time dimension is small (such as $T = 2$ or 3), a design suggesting a random-effects model may be preferable because fewer degrees of freedom are necessary to account for the subject-specific parameters.

Generalized Least-Squares Estimation

Equations (3.2) and (3.3) summarize the mean and variance of the vector of responses. To estimate regression coefficients, this chapter uses generalized least-squares (GLS) equations of the form

$$\left(\sum_{i=1}^{n} \mathbf{X}_i' \mathbf{V}_i^{-1} \mathbf{X}_i\right) \boldsymbol{\beta} = \sum_{i=1}^{n} \mathbf{X}_i' \mathbf{V}_i^{-1} \mathbf{y}_i.$$

The solution of these equations yields GLS estimators, which, in this context, we call the *error-components* estimator of $\boldsymbol{\beta}$. Additional algebra (Exercise 3.1) shows that this estimator can be expressed as

$$\mathbf{b}_{\mathrm{EC}} = \left(\sum_{i=1}^{n} \mathbf{X}_i' \left(\mathbf{I}_i - \frac{\zeta_i}{T_i}\mathbf{J}_i\right) \mathbf{X}_i\right)^{-1} \sum_{i=1}^{n} \mathbf{X}_i' \left(\mathbf{I}_i - \frac{\zeta_i}{T_i}\mathbf{J}_i\right) \mathbf{y}_i, \qquad (3.4)$$

where the quantity

$$\zeta_i = \frac{T_i \sigma_\alpha^2}{T_i \sigma_\alpha^2 + \sigma^2}$$

is a function of the variance components σ_α^2 and σ^2. In Chapter 4, we will refer to this quantity as the *credibility factor*. Further, the variance of the error-components estimator turns out to be

$$\mathrm{Var}\, \mathbf{b}_{\mathrm{EC}} = \sigma^2 \left(\sum_{i=1}^{n} \mathbf{X}_i' \left(\mathbf{I}_i - \frac{\zeta_i}{T_i} \mathbf{J}_i \right) \mathbf{X}_i \right)^{-1}.$$

To interpret $\mathbf{b}_{\mathrm{EC}}$, we give an alternative form for the corresponding Chapter 2 fixed-effects estimator. That is, from Equation (2.6) and some algebra, we have

$$\mathbf{b} = \left(\sum_{i=1}^{n} \mathbf{X}_i' \left(\mathbf{I}_i - T_i^{-1} \mathbf{J}_i \right) \mathbf{X}_i \right)^{-1} \sum_{i=1}^{n} \mathbf{X}_i' \left(\mathbf{I}_i - T_i^{-1} \mathbf{J}_i \right) \mathbf{y}_i.$$

Thus, we see that the random-effects $\mathbf{b}_{\mathrm{EC}}$ and fixed-effects $\mathbf{b}$ are approximately equal when the credibility factors are close to one. This occurs when σ_α^2 is large relative to σ^2. Intuitively, when there is substantial separation among the intercept terms, relative to the uncertainty in the observations, we anticipate that the fixed- and random-effects estimators will behave similarly. Conversely, Equation (3.4) shows that $\mathbf{b}_{\mathrm{EC}}$ is approximately equal to an ordinary least-squares estimator when σ^2 is large relative to σ_α^2 (so that the credibility factors are close to zero). Section 7.2 further develops the comparison among these alternative estimators.

Feasible GLS Estimator

The calculation of the GLS estimator in Equation (3.4) assumes that the variance components σ_α^2 and σ^2 are known.

Procedure for Computing a "Feasible" GLS Estimator

1. First run a regression assuming $\sigma_\alpha^2 = 0$, resulting in an OLS estimate of β.
2. Use the residuals from Step 1 to determine estimates of σ_α^2 and σ^2.
3. Using the estimates of σ_α^2 and σ^2 from Step 2, determine $\mathbf{b}_{\mathrm{EC}}$ using Equation (3.4).

For Step 2, there are many ways of estimating the variance components. Section 3.5 provides details. This procedure could be iterated. However, studies have shown that iterated versions do not improve the performance of the one-step estimators. See, for example, Carroll and Rupert (1988G).

To illustrate, we consider some simple moment-based estimators of σ_α^2 and σ^2 due to Baltagi and Chang (1994E). Define the residuals $e_{it} = y_{it} - (a_i + \mathbf{x}_{it}'\mathbf{b})$ using α_i and $\mathbf{b}$ according to the Chapter 2 fixed-effects estimators in Equations (2.6) and (2.7). Then, the estimator of σ^2 is s^2, as given in Equation (2.11). The estimator of σ_α^2 is

$$s_\alpha^2 = \frac{\sum_{i=1}^n T_i \, (a_i - \bar{a}_w)^2 - s^2 c_n}{N - \sum_{i=1}^n T_i^2/N},$$

where $\bar{a}_w = N^{-1} \sum_{i=1}^n T_i \, a_i$ and

$$c_n = n - 1 + \operatorname{trace}\left\{ \left(\sum_{i=1}^n \sum_{t=1}^{T_i} (\mathbf{x}_{it} - \bar{\mathbf{x}}_i)(\mathbf{x}_{it} - \bar{\mathbf{x}}_i)' \right)^{-1} \sum_{i=1}^n T_i (\bar{\mathbf{x}}_i - \bar{\mathbf{x}})(\bar{\mathbf{x}}_i - \bar{\mathbf{x}})' \right\}.$$

A potential drawback is that a particular realization of s_α^2 may be negative; this feature is undesirable for a variance estimator.

Pooling Test

As with the traditional random-effects ANOVA model, the *test for heterogeneity*, or *pooling test*, is written as a test of the null hypothesis $\mathrm{H}_0 : \sigma_\alpha^2 = 0$. That is, under the null hypothesis, we do not have to account for subject-specific effects. Although this is a difficult issue for the general case, in the special case of error components, desirable test procedures have been developed. We discuss here a test that extends a Lagrange multiplier test statistic, due to Breusch and Pagan (1980E), to the unbalanced data case. (See Appendix C for an introduction to Lagrange multiplier statistics.) This test is a simpler version of one developed by Baltagi and Li (1990E) for a more complex model (specifically, a two-way error-component model that we will introduce in Chapter 6).

Pooling Test Procedure

1. Run the pooled cross-sectional regression model $y_{it} = \mathbf{x}_{it}'\boldsymbol{\beta} + \varepsilon_{it}$ to get residuals e_{it}.
2. For each subject, compute an estimator of σ_α^2,

$$s_i = \frac{1}{T_i(T_i - 1)} \left(T_i^2 \bar{e}_i^2 - \sum_{t=1}^{T_i} e_{it}^2 \right),$$

where $\bar{e}_i = T_i^{-1} \sum_{t=1}^{T_i} e_{it}$.

3. Compute the test statistic,

$$TS = \frac{1}{2n}\left(\frac{\sum_{i=1}^{n} s_i \sqrt{T_i(T_i - 1)}}{N^{-1}\sum_{i=1}^{n}\sum_{t=1}^{T_i} e_{it}^2}\right)^2 .$$

4. Reject H_0 if TS exceeds a percentile from a chi-square distribution with one degree of freedom. The percentile is one minus the significance level of the test.

Note that the pooling test procedure uses estimators of σ_α^2, s_i, which may be negative with positive probability. Section 5.4 discusses alternative procedures where we restrict variance estimators to be nonnegative.

3.2 Example: Income Tax Payments

In this section, we study the effects that an individual's economic and demographic characteristics have on the amount of income tax paid. Specifically, the response of interest is LNTAX, defined as the natural logarithm of the liability on the tax return. Table 3.1 describes several taxpayer characteristics that may affect tax liability.

The data for this study are from the Statistics of Income (SOI) panel of Individual Returns, a part of the Ernst and Young/University of Michigan Tax Research Database. The SOI panel represents a simple random sample of unaudited individual income tax returns filed for tax years 1979–1990. The data are compiled from a stratified probability sample of unaudited individual income tax returns (Forms 1040, 1040A, and 1040EZ) filed by U.S. taxpayers. The estimates obtained from these data are intended to represent all returns filed for the income tax years under review. All returns processed are subjected to sampling except tentative and amended returns.

We examine a balanced panel from 1982–1984 and 1986–1987 taxpayers included in the SOI panel; a 4% sample of this comprises our sample of 258 taxpayers. These years are chosen because they contain the interesting information on paid-preparer usage. Specifically, these data include line-item tax return data plus a binary variable noting the presence of a paid tax preparer for years 1982–1984 and 1986–1987. These data are also analyzed in Frischmann and Frees (1999O).

The primary goal of this analysis is to determine whether tax preparers significantly affect tax liability. To motivate this question, we note that preparers have the opportunity to impact virtually every line item on a tax return. Our variables are selected because they appear consistently in prior research and are

Table 3.1. *Taxpayer characteristics*

Characteristic	Description
Demographic	
MS	a binary variable; one if the taxpayer is married and zero otherwise
HH	a binary variable; one if the taxpayer is the head of household and zero otherwise
DEPEND	the number of dependents claimed by the taxpayer
AGE	a binary variable; one if the taxpayer is age 65 or over and zero otherwise
Economic	
LNTPI	the natural logarithm of the sum of all positive income line items on the return, in 1983 dollars
MR	the marginal tax rate; computed on total personal income less exemptions and the standard deduction
EMP	a binary variable; one if Schedule C or F is present and zero otherwise; self-employed taxpayers have greater need for professional assistance to reduce the reporting risks of doing business.
PREP	a variable indicating the presence of a paid preparer
LNTAX	the natural logarithm of the tax liability, in 1983 dollars; this is the response variable of interest

Table 3.2. *Averages of binary variables* ($n = 258$)

YEAR	MS	HH	AGE	EMP	PREP
1982	0.597	0.081	0.085	0.140	0.450
1983	0.597	0.093	0.105	0.159	0.442
1984	0.624	0.085	0.112	0.155	0.484
1986	0.647	0.081	0.132	0.147	0.508
1987	0.647	0.093	0.147	0.147	0.516

largely outside the influence of tax preparers (that is, they are "exogenous"). Briefly, our explanatory variables are as follows: MS, HH, AGE, EMP, and PREP are binary variables coded for married, head of household, at least 65 years of age, self-employed, and paid preparer, respectively. Further, DEPEND is the number of dependents and MR is the marginal tax rate measure. Finally, LNTPI and LNTAX are the total positive income and tax liability as stated on the return in 1983 dollars, in logarithmic units.

Tables 3.2 and 3.3 describe the basic taxpayer characteristics used in our analysis. The binary variables in Table 3.2 indicate that over half the sample is married (MS) and approximately half the sample uses a paid preparer (PREP).

Table 3.3. *Summary statistics for other variables*

Variable	Mean	Median	Minimum	Maximum	Standard deviation
DEPEND	2.419	2.000	0.000	6.000	1.338
LNTPI	9.889	10.051	−0.128	13.222	1.165
MR	23.523	22.000	0.000	50.000	11.454
LNTAX	6.880	7.701	0.000	11.860	2.695

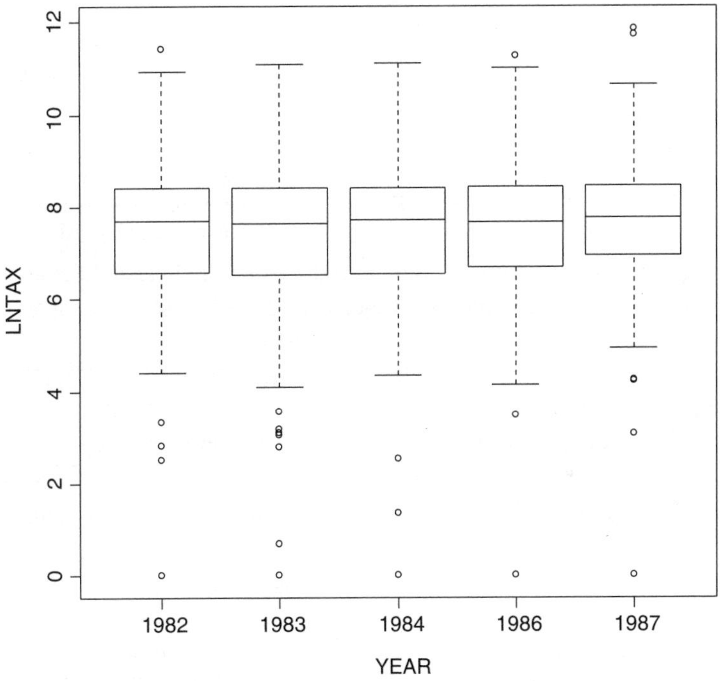

Figure 3.2. Boxplot of LNTAX versus YEAR. Logarithmic tax liability (in real dollars) is stable over the years 1982–1987.

Preparer use appears highest in 1986 and 1987, years straddling significant tax law change. Slightly less than 10% of the sample is 65 or older (AGE) in 1982. The presence of self-employment income (EMP) also varies over time.

The summary statistics for the other nonbinary variables are in Table 3.3. Further analyses indicate an increasing income trend, even after adjusting for inflation, as measured by total positive income (LNTPI). Moreover, both the mean and median marginal tax rates (MR) are decreasing, although mean and median tax liabilities (LNTAX) are stable (see Figure 3.2). These results are

Table 3.4. *Averages of logarithmic tax by level of explanatory variable*

Level of explanatory variable	Explanatory variable				
	MS	HH	AGE	EMP	PREP
0	5.973	7.013	6.939	6.983	6.624
1	7.430	5.480	6.431	6.297	7.158

Table 3.5. *Correlation coefficients*

	DEPEND	LNTPI	MR
LNTPI	0.278		
MR	0.128	0.796	
LNTAX	0.085	0.718	0.747

consistent with congressional efforts to reduce rates and expand the tax base through broadening the definition of income and eliminating deductions.

To explore the relationship between each indicator variable and logarithmic tax, Table 3.4 presents the average logarithmic tax liability by level of indicator variable. This table shows that married filers pay greater tax, head-of-household filers pay less tax, taxpayers 65 or over pay less, taxpayers with self-employed income pay less, and taxpayers who use a professional tax preparer pay more.

Table 3.5 summarizes basic relations among logarithmic tax and the other nonbinary explanatory variables. Both LNTPI and MR are strongly correlated with logarithmic tax whereas the relationship between DEPEND and logarithmic tax is positive, yet weaker. Table 3.5 also shows that LNTPI and MR are strongly positively correlated.

Although not presented in detail here, exploration of the data revealed several other interesting relationships among the variables. To illustrate, we show a basic added-variable plot in Figure 3.3. Notice the strong relation between logarithmic tax liability and total income, even after controlling for subject-specific time-constant effects.

The error-components model described in Section 3.1 was fit using the explanatory variables described in Table 3.1. The estimated model appears in Display 3.1, from a fit using the statistical package SAS. Display 3.1 shows that HH, EMP, LNTPI, and MR are statistically significant variables that affect LNTAX. Somewhat surprisingly, the PREP variable was not statistically significant.

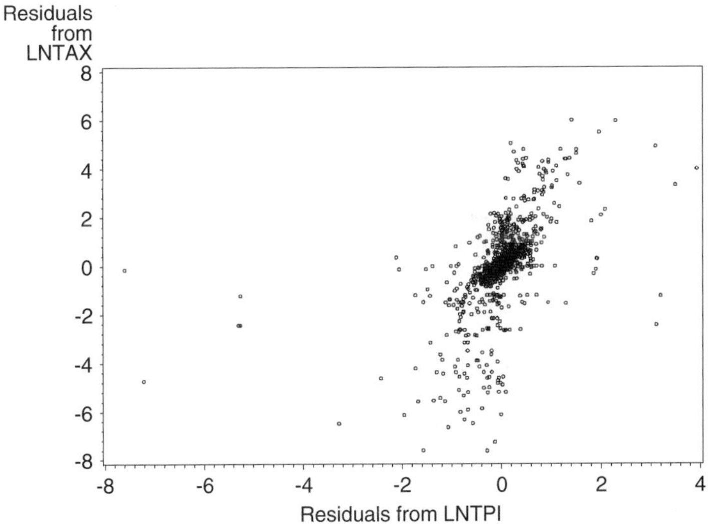

Figure 3.3. Added-variable plot of LNTAX versus LNTPI.

Display 3.1 Selected SAS Output

Iteration History

Iteration	Evaluations	-2 Log Like	Criterion
0	1	4984.68064143	
1	2	4791.25465804	0.00000001

Convergence criteria met.

Covariance Parameter Estimates

Cov Parm	Subject	Estimate
Intercept	SUBJECT	0.9217
Residual		1.8740

Fit Statistics

-2 Log Likelihood	4791.3
AIC (smaller is better)	4813.3
AICC (smaller is better)	4813.5
BIC (smaller is better)	4852.3

Solution for Fixed Effects

Effect	Estimate	Standard Error	DF	t Value	Pr > |t|
Intercept	-2.9604	0.5686	257	-5.21	<.0001
MS	0.03730	0.1818	1024	0.21	0.8375
HH	-0.6890	0.2312	1024	-2.98	0.0029
AGE	0.02074	0.1993	1024	0.10	0.9171
EMP	-0.5048	0.1674	1024	-3.02	0.0026
PREP	-0.02170	0.1171	1024	-0.19	0.8530
LNTPI	0.7604	0.06972	1024	10.91	<.0001
DEPEND	-0.1128	0.05907	1024	-1.91	0.0566
MR	0.1154	0.007288	1024	15.83	<.0001

To test for the importance of heterogeneity, the Section 3.1 pooling test was performed. A fit of the pooled cross-sectional model, with the same explanatory variables, produced residuals and an error sum of squares equal to *Error SS* = 3599.73. Thus, with $T = 5$ years and $n = 258$ subjects, the test statistic is $TS =$ 273.5. Comparing this test statistic to a chi-square distribution with one degree of freedom indicates that the null hypothesis of homogeneity is rejected. As we will see in Chapter 7, there are some unusual features of this data set that cause this test statistic to be large.

3.3 Mixed-Effects Models

Similar to the extensions for the fixed-effects model described in Section 2.5, we now extend the error-components model to allow for variable slopes, serial correlation, and heteroscedasticity.

3.3.1 Linear Mixed-Effects Model

We now consider conditional regression functions of the form

$$\mathrm{E}(y_{it} \mid \boldsymbol{\alpha}_i) = \mathbf{z}_{it}'\boldsymbol{\alpha}_i + \mathbf{x}_{it}'\boldsymbol{\beta}, \tag{3.5}$$

where the term $\mathbf{z}_{it}'\boldsymbol{\alpha}_i$ comprises the *random-effects* portion of the model and the term $\mathbf{x}_{it}'\boldsymbol{\beta}$ comprises the *fixed-effects* portion. As with Equation (2.15) for fixed effects, Equation (3.5) is shorthand notation for

$$\mathrm{E}(y_{it}|\boldsymbol{\alpha}_i) = \alpha_{i1}z_{it1} + \alpha_{i2}z_{it2} + \cdots + \alpha_{iq}z_{itq} + \beta_1 x_{it1} + \beta_2 x_{it2} + \cdots + \beta_K x_{itK}.$$

As in Equation (2.16), a matrix form of Equation (3.5) is

$$\mathrm{E}(\mathbf{y}_i \mid \boldsymbol{\alpha}_i) = \mathbf{Z}_i\boldsymbol{\alpha}_i + \mathbf{X}_i\boldsymbol{\beta}. \tag{3.6}$$

We also wish to allow for serial correlation and heteroscedasticity. Similarly to our method in Section 2.5.1 for fixed effects, we can incorporate these extensions through the notation $\mathrm{Var}\,(\mathbf{y}_i \mid \boldsymbol{\alpha}_i) = \mathbf{R}_i$. We maintain the assumption that the responses between subjects are independent.

Further, we assume that the subject-specific effects $\{\boldsymbol{\alpha}_i\}$ are independent with mean $\mathrm{E}\,\boldsymbol{\alpha}_i = \mathbf{0}$ and variance–covariance matrix $\mathrm{Var}\,\boldsymbol{\alpha}_i = \mathbf{D}$, a $q \times q$ positive definite matrix. By assumption, the random effects have mean zero; thus, any nonzero mean for a random effect must be expressed as part of the fixed-effects terms. The columns of $\mathbf{Z}_i$ are usually a subset of the columns of $\mathbf{X}_i$.

Taken together, these assumptions comprise what we term the *linear mixed-effects model.*

Linear Mixed-Effects Model Assumptions

R1. $\mathrm{E}(\mathbf{y}_i \mid \boldsymbol{\alpha}_i) = \mathbf{Z}_i \boldsymbol{\alpha}_i + \mathbf{X}_i \boldsymbol{\beta}$.
R2. $\{x_{it,1}, \ldots, x_{it,K}\}$ and $\{z_{it,1}, \ldots, z_{it,q}\}$ are nonstochastic variables.
R3. $\mathrm{Var}(\mathbf{y}_i \mid \boldsymbol{\alpha}_i) = \mathbf{R}_i$.
R4. $\{\mathbf{y}_i\}$ are independent random vectors, conditional on $\{\boldsymbol{\alpha}_1, \ldots, \boldsymbol{\alpha}_n\}$.
R5. $\{\mathbf{y}_i\}$ are normally distributed, conditional on $\{\boldsymbol{\alpha}_1, \ldots, \boldsymbol{\alpha}_n\}$.
R6. $\mathrm{E}\,\boldsymbol{\alpha}_i = \mathbf{0}$, $\mathrm{Var}\,\boldsymbol{\alpha}_i = \mathbf{D}$, and $\{\boldsymbol{\alpha}_1, \ldots, \boldsymbol{\alpha}_n\}$ are mutually independent.
R7. $\{\boldsymbol{\alpha}_i\}$ are normally distributed.

With Assumptions R3 and R6, the variance of each subject can be expressed as

$$\mathrm{Var}\,\mathbf{y}_i = \mathbf{Z}_i \mathbf{D} \mathbf{Z}_i' + \mathbf{R}_i = \mathbf{V}_i(\boldsymbol{\tau}) = \mathbf{V}_i. \tag{3.7}$$

The notation $\mathbf{V}_i(\boldsymbol{\tau})$ means that the variance–covariance matrix of $\mathbf{y}_i$ depends on variance components $\boldsymbol{\tau}$. Section 2.5.1 provided several examples that illustrate how $\mathbf{R}_i$ may depend on $\boldsymbol{\tau}$; we will give special cases to show how $\mathbf{V}_i$ may depend on $\boldsymbol{\tau}$.

With this, we may summarize the effects of Assumptions R1–R7 on the observables variables, $\{x_{it,1}, \ldots, x_{it,K}, z_{it,1}, \ldots, z_{it,q}, y_{it}\}$.

Observables Representation of the Linear Mixed-Effects Model

RO1. $\mathrm{E}\mathbf{y}_i = \mathbf{X}_i \boldsymbol{\beta}$.
RO2. $\{x_{it,1}, \ldots, x_{it,K}\}$ and $\{z_{it,1}, \ldots, z_{it,q}\}$ are nonstochastic variables.
RO3. $\mathrm{Var}\,\mathbf{y}_i = \mathbf{Z}_i \mathbf{D} \mathbf{Z}_i' + \mathbf{R}_i = \mathbf{V}_i(\boldsymbol{\tau}) = \mathbf{V}_i$.
RO4. $\{\mathbf{y}_i\}$ are independent random vectors.
RO5. $\{\mathbf{y}_i\}$ are normally distributed.

As in Chapter 2 and Section 3.1, the properties RO1–RO5 are a consequence of R1–R7. We focus on these properties because they are the basis for testing our specification of the model. The observable representation is also known as a *marginal* or *population-averaged model*.

Example 3.1: Trade Localization Feinberg, Keane, and Bognano (1998E) studied $n = 701$ U.S.-based multinational corporations over the period 1983–1992. Using firm-level data available from the Bureau of Economic Analysis of the U.S. Department of Commerce, they documented how large corporations' allocation of employment and durable assets (property, plant, and equipment) of Canadian affiliates changed in response to changes in Canadian and U.S. tariffs. Specifically, their model was

$$\begin{aligned}\ln y_{it} &= \beta_{1i}\,\mathrm{CT}_{it} + \beta_{2i}\,\mathrm{UT}_{it} + \beta_{3i}\,\mathrm{Trend}_t + \mathbf{x}_{it}^{*\prime}\boldsymbol{\beta}^* + \varepsilon_{it}\\ &= (\beta_1 + \alpha_{1i})\,\mathrm{CT}_{it} + (\beta_2 + \alpha_{2i})\,\mathrm{UT}_{it} + (\beta_3 + \alpha_{3i})\,\mathrm{Trend}_t + \mathbf{x}_{it}^{*\prime}\boldsymbol{\beta}^* + \varepsilon_{it}\\ &= \alpha_{1i}\,\mathrm{CT}_{it} + \alpha_{2i}\,\mathrm{UT}_{it} + \alpha_{3i}\,\mathrm{Trend}_t + \mathbf{x}_{it}^{\prime}\boldsymbol{\beta} + \varepsilon_{it}.\end{aligned}$$

Here, CT_{it} is the sum over all industry Canadian tariffs in which firm i belongs, and similarly for UT_{it}. The vector $\mathbf{x}_{it}$ includes CT_{it}, UT_{it}, and Trend_t (for the mean effects), as well as real U.S. and Canadian wages, gross domestic product, price–earnings ratio, real U.S. interest rates, and a measure of transportation costs. For the response, they used both Canadian employment and durable assets.

The first equation emphasizes that response to changes in Canadian and U.S. tariffs, as well as time trends, is firm-specific. The second equation provides the link to the third expression, which is in terms of the linear mixed-effects model form. Here, we have included CT_{it}, UT_{it}, and Trend_t in $\mathbf{x}_{it}^*$ to get $\mathbf{x}_{it}$. With this reformulation, the mean of each random slope is zero; that is, $\mathrm{E}\,\alpha_{1i} = \mathrm{E}\,\alpha_{2i} = \mathrm{E}\,\alpha_{3i} = 0$. In the first specification, the means are $\mathrm{E}\,\beta_{1i} = \beta_1$, $\mathrm{E}\,\beta_{2i} = \beta_2$, and $\mathrm{E}\,\beta_{3i} = \beta_3$. Feinberg, Keane, and Bognano found that a significant portion of the variation was due to firm-specific slopes; they attribute this variation to idiosyncratic firm differences such as technology and organization. They also allowed for heterogeneity in the time trend. This allows for unobserved time-varying factors (such as technology and demand) that affect individual firms differently.

A major finding of this paper is that Canadian tariff levels were negatively related to assets and employment in Canada; this finding contradicts the hypothesis that lower tariffs would undermine Canadian manufacturing.

Special Cases

To help interpret linear mixed-effects models, we consider several important special cases. We begin by emphasizing the case where $q = 1$ and $z_{it} = 1$. In this case, the linear mixed-effects model reduces to the error-components model, introduced in Section 3.1. For this model, we have only subject-specific intercepts, no subject-specific slopes, and no serial correlation.

Repeated Measures Design

Another classic model is the *repeated measures design*. Here, several measurements are collected on a subject over a relatively short period of time, under controlled experimental conditions. Each measurement is subject to a different treatment but the order of treatments is randomized so that no serial correlation is assumed.

Specifically, we consider $i = 1, \ldots, n$ subjects. A response for each subject is measured based on each of T treatments, where the order of treatments is randomized. The mathematical model is as follows:

$$\underbrace{\textit{response}}_{y_{it}} = \underbrace{\textit{random subject effect}}_{\alpha_i} + \underbrace{\textit{fixed treatment effect}}_{\beta_t} + \underbrace{\textit{error}}_{\varepsilon_{it}}$$

The main research question of interest is $\mathbf{H}_0 : \beta_1 = \beta_2 = \cdots = \beta_T$; that is, the null hypothesis is no treatment differences.

The repeated measures design is a special case of Equation (3.5), taking $q = 1$, $z_{it} = 1$, $T_i = T$, and $K = T$ and using the tth explanatory variable, $x_{it,t}$, to indicate whether the tth treatment has been applied to the response.

Random-Coefficients Model

We now return to the linear mixed-effects model and suppose that $q = K$ and $\mathbf{z}_{it} = \mathbf{x}_{it}$. In this case the linear mixed-effects model reduces to a *random-coefficients* model of the form

$$\mathrm{E}(y_{it} \mid \boldsymbol{\alpha}_i) = \mathbf{x}_{it}'(\boldsymbol{\alpha}_i + \boldsymbol{\beta}) = \mathbf{x}_{it}'\boldsymbol{\beta}_i, \tag{3.8}$$

where $\{\boldsymbol{\beta}_i\}$ are random vectors with mean $\boldsymbol{\beta}$. The random-coefficients model can be easily interpreted as a two-stage sampling model. In the first stage, one draws the ith subject from a population that yields a vector of parameters $\boldsymbol{\beta}_i$. From the population, this vector has mean $\mathrm{E}\boldsymbol{\beta}_i = \boldsymbol{\beta}$ and variance $\mathrm{Var}\,\boldsymbol{\beta}_i = \mathbf{D}$. At the second stage, one draws T_i observations for the ith observation, conditional on having observed $\boldsymbol{\beta}_i$. The mean and variance of the observations are $\mathrm{E}(\mathbf{y}_i \mid \boldsymbol{\beta}_i) = \mathbf{X}_i\boldsymbol{\beta}_i$ and $\mathrm{Var}(\mathbf{y}_i \mid \boldsymbol{\beta}_i) = \mathbf{R}_i$. Putting these two stages together yields

$$\mathrm{E}\mathbf{y}_i = \mathbf{X}_i \mathrm{E}\boldsymbol{\beta}_i = \mathbf{X}_i\boldsymbol{\beta}$$

and

$$\begin{aligned}\mathrm{Var}\,\mathbf{y}_i &= \mathrm{E}(\mathrm{Var}(\mathbf{y}_i \mid \boldsymbol{\beta}_i)) + \mathrm{Var}(\mathrm{E}(\mathbf{y}_i \mid \boldsymbol{\beta}_i)) \\ &= \mathbf{R}_i + \mathrm{Var}(\mathbf{X}_i\boldsymbol{\beta}_i) = \mathbf{R}_i + \mathbf{X}_i'\mathbf{D}\mathbf{X}_i = \mathbf{V}_i.\end{aligned}$$

Example: Taxpayer Study (continued) The random-coefficients model was fit using the taxpayer data with $K = 8$ variables. The model fitting was done using the statistical package SAS, with the MIVQUE(0) variance components estimation techniques, described in Section 3.5. The resulting fitting $\mathbf{D}$ matrix appears in Table 3.6. Display 3.2 provides additional details of the model fit.

Table 3.6. *Values of the estimated* $\mathbf{D}$ *matrix*

	INTERCEPT	MS	HH	AGE	EMP	PREP	LNTPI	MR	DEPEND
INTERCEPT	47.86								
MS	−0.40	20.64							
HH	−1.26	1.25	23.46						
AGE	18.48	2.61	−0.79	22.33					
EMP	−8.53	0.92	0.12	0.21	20.60				
PREP	4.22	−0.50	−1.85	−0.21	−0.50	21.35			
LNTPI	−4.54	−0.17	0.15	−2.38	1.18	−0.38	21.44		
MR	0.48	0.06	−0.03	0.14	−0.09	0.04	−0.09	20.68	
DEPEND	3.07	0.41	0.29	−0.60	−0.40	−0.35	−0.34	0.01	20.68

Display 3.2 Selected SAS Output for the Random Coefficients Model

```
                  Fit Statistics
      -2 Res Log Likelihood             7876.0
      AIC (smaller is better)           7968.0
      AICC (smaller is better)          7971.5
      BIC (smaller is better)           8131.4
            Solution for Fixed Effects
                     Standard
Effect        Estimate      Error     DF    t Value    Pr > |t|
Intercept      -9.5456     2.1475    253      -4.44      <.0001
MS             -0.3183     1.0664     41      -0.30      0.7668
HH             -1.0514     1.4418     16      -0.73      0.4764
AGE            -0.4027     1.1533     20      -0.35      0.7306
EMP            -0.1498     0.9019     31      -0.17      0.8691
PREP           -0.2156     0.6118     67      -0.35      0.7257
LNTPI           1.6118     0.3712    257       4.34      <.0001
DEPEND         -0.2814     0.4822     70      -0.58      0.5613
MR             0.09303     0.2853    250       0.33      0.7446
```

Variations of the Random-Coefficients Model

Certain variations of the two-stage interpretation of the random-coefficients models lead to other forms of the random-effects model in Equation (3.6). To illustrate, in Equation (3.6), we may take the columns of $\mathbf{Z}_i$ to be a strict subset of the columns of $\mathbf{X}_i$. This is equivalent to assuming that certain components of $\boldsymbol{\beta}_i$ associated with $\mathbf{Z}_i$ are stochastic whereas other components that are associated with $\mathbf{X}_i$ (but not $\mathbf{Z}_i$) are nonstochastic.

Note that the convention in Equation (3.6) is to assume that the mean of the random effects $\boldsymbol{\alpha}_i$ is known and equal to zero. Alternatively, we could assume that $\boldsymbol{\alpha}_i$ are unknown with mean, say, $\boldsymbol{\alpha}$, that is, $\mathrm{E}\boldsymbol{\alpha}_i = \boldsymbol{\alpha}$. However, this is equivalent to specifying additional fixed-effects terms $\mathbf{Z}_i\boldsymbol{\alpha}$ in Equation (3.6). By convention, we absorb these additional terms into the "$\mathbf{X}_i\boldsymbol{\beta}$" portion of the model. Thus, it is customary to include those explanatory variables in the $\mathbf{Z}_i$ design matrix as part of the $\mathbf{X}_i$ design matrix.

Another variation of the two-stage interpretation uses known variables $\mathbf{B}_i$ such that $\mathrm{E}\boldsymbol{\beta}_i = \mathbf{B}_i\boldsymbol{\beta}$. Then, we have $\mathrm{E}\mathbf{y}_i = \mathbf{X}_i\mathbf{B}_i\boldsymbol{\beta}$ and $\mathrm{Var}\,\mathbf{y}_i = \mathbf{R}_i + \mathbf{X}_i'\mathbf{D}\mathbf{X}_i$. This is equivalent to our Equation (3.6) model with $\mathbf{X}_i\mathbf{B}_i$ replacing $\mathbf{X}_i$ and $\mathbf{X}_i$ replacing $\mathbf{Z}_i$. Hsiao (1986E, Section 6.5) refers to this as a variable-coefficients model with coefficients that are functions of other exogenous variables. Chapter 5 describes this approach in greater detail.

Example 3.2: Lottery Sales Section 4.5 will describe a case in which we wish to predict lottery sales. The response variable y_{it} is logarithmic lottery sales in week t for a geographic unit i. No time-varying variables are available for these data, so a basic explanation of lottery sales is through the one-way random-effects ANOVA model of the form $y_{it} = \alpha_i^* + \varepsilon_{it}$. We can interpret α_i^* to be the (conditional) mean lottery sales for the ith geographic unit. In addition, we will have available several time-constant variables that describe the geographic unit, including population, median household income, and median home value. Denote this set of variables that describe the ith geographic unit as $\mathbf{B}_i$. With the two-stage interpretation, we could use these variables to explain the mean lottery sales with the representation $\alpha_i^* = \mathbf{B}_i'\boldsymbol{\beta} + \alpha_i$. Note that the variable α_i^* is unobservable, so this model is not estimable by itself. However, when combined with the ANOVA model, we have

$$y_{it} = \alpha_i + \mathbf{B}_i'\boldsymbol{\beta} + \varepsilon_{it},$$

our error-components model. This combined model is estimable.

Group Effects

In many applications of longitudinal data analysis, it is of interest to assess differences of responses from different groups. In this context, the term "group" refers to a category of the population. For example, in the Section 3.2 taxpayer example, we may be interested in studying the differences in tax liability due to gender or due to political party affiliation.

A typical model that includes group effects can be expressed as a special case of the linear mixed-effects model, using $q = 1$, $z_{it} = 1$, and the expression

$$\mathrm{E}(y_{git} \mid \alpha_{gi}) = \alpha_{gi} + \delta_g + \mathbf{x}_{git}'\boldsymbol{\beta}.$$

Here, the subscripts range over $g = 1, \ldots, G$ groups, $i = 1, \ldots, n_g$ subjects in each group, and $t = 1, \ldots, T_{gi}$ observations of each subject. The terms $\{\alpha_{gi}\}$ represent random, subject-specific effects and $\{\delta_g\}$ represent fixed differences among groups. An interesting aspect of the random-effects portion is that subjects need not change groups over time for the model to be estimable. To illustrate, if we were interested in gender differences in tax liability, we would not expect individuals to change gender over such a small sample. This is in contrast to the fixed-effects model, where group effects are not estimable owing to their collinearity with subject-specific effects.

Time-Constant Variables

The study of time-constant variables provides strong motivation for designing a panel, or longitudinal, study that can be analyzed as a linear mixed-effects model. Within a linear mixed-effects model, both the heterogeneity terms $\{\alpha_i\}$ and parameters associated with time-constant variables can be analyzed simultaneously. This was not the case for the fixed-effects models, where the heterogeneity terms and time-constant variables are perfectly collinear. The group effect just discussed is a specific type of time-constant variable. Of course, it is also possible to analyze group effects where individuals switch groups over time, such as with political party affiliation. This type of problem can be handled directly using binary variables to indicate the presence or absence of a group type, and represents no particular difficulties.

We may split the explanatory variables associated with the population parameters into those that vary by time and those that do not (time-constant parameters). Thus, we can write our linear mixed-effects conditional regression function as

$$\mathrm{E}(y_{it} \mid \boldsymbol{\alpha}_i) = \boldsymbol{\alpha}_i' \mathbf{z}_{it} + \mathbf{x}_{1i}' \boldsymbol{\beta}_1 + \mathbf{x}_{2it}' \boldsymbol{\beta}_2.$$

This model is a generalization of the group effects model.

3.3.2 Mixed Linear Models

In the Section 3.3.1 linear mixed-effects models, we assumed independence among subjects (Assumption RO4). This assumption is not tenable for all models of repeated observations on a subject over time, so it is of interest to introduce a generalization known as the *mixed linear model*. This model equation is given by

$$\mathbf{y} = \mathbf{Z}\boldsymbol{\alpha} + \mathbf{X}\boldsymbol{\beta} + \boldsymbol{\epsilon}. \tag{3.9}$$

Here, $\mathbf{y}$ is an $N \times 1$ vector of responses, $\boldsymbol{\epsilon}$ is a an $N \times 1$ vector of errors, $\mathbf{Z}$ and $\mathbf{X}$ are $N \times q$ and $N \times K$ matrices of explanatory variables, respectively, and $\boldsymbol{\alpha}$ and $\boldsymbol{\beta}$ are, respectively, $q \times 1$ and $K \times 1$ vectors of parameters.

For the mean structure, we assume $\mathrm{E}(\mathbf{y} \mid \boldsymbol{\alpha}) = \mathbf{Z}\boldsymbol{\alpha} + \mathbf{X}\boldsymbol{\beta}$ and $\mathrm{E}\,\boldsymbol{\alpha} = \mathbf{0}$, so that $\mathrm{E}\mathbf{y} = \mathbf{X}\boldsymbol{\beta}$. For the covariance structure, we assume $\mathrm{Var}(\mathbf{y} \mid \boldsymbol{\alpha}) = \mathbf{R}$, $\mathrm{Var}\,\boldsymbol{\alpha} = \mathbf{D}$, and $\mathrm{Cov}(\boldsymbol{\alpha}, \boldsymbol{\epsilon}) = \mathbf{0}$. This yields $\mathrm{Var}\,\mathbf{y} = \mathbf{ZDZ}' + \mathbf{R} = \mathbf{V}$.

Unlike the linear mixed-effects model in Section 3.3.1, the mixed linear model does not require independence between subjects. Further, the model is sufficiently flexible so that several complex hierarchical structures can be expressed as special cases of it. To see how the linear mixed-effects model is a special case of the mixed linear model, take $\mathbf{y} = (\mathbf{y}_1', \mathbf{y}_2', \ldots, \mathbf{y}_n')'$, $\boldsymbol{\epsilon} = (\boldsymbol{\epsilon}_1', \boldsymbol{\epsilon}_2', \ldots, \boldsymbol{\epsilon}_n')'$, $\boldsymbol{\alpha} = (\boldsymbol{\alpha}_1', \boldsymbol{\alpha}_2', \ldots, \boldsymbol{\alpha}_n')'$,

$$\mathbf{X} = \begin{pmatrix} \mathbf{X}_1 \\ \mathbf{X}_2 \\ \mathbf{X}_3 \\ \vdots \\ \mathbf{X}_n \end{pmatrix},$$

and

$$\mathbf{Z} = \begin{pmatrix} \mathbf{Z}_1 & \mathbf{0} & \mathbf{0} & \cdots & \mathbf{0} \\ \mathbf{0} & \mathbf{Z}_2 & \mathbf{0} & \cdots & \mathbf{0} \\ \mathbf{0} & \mathbf{0} & \mathbf{Z}_3 & \cdots & \mathbf{0} \\ \vdots & \vdots & \vdots & \ddots & \vdots \\ \mathbf{0} & \mathbf{0} & \mathbf{0} & \cdots & \mathbf{Z}_n \end{pmatrix}.$$

With these choices, the mixed linear model reduces to the linear mixed-effects model.

The *two-way error-components* model is an important panel data model that is not a specific type of linear mixed-effects model, although it is a special case of the mixed linear model. This model can be expressed as

$$y_{it} = \alpha_i + \lambda_t + \mathbf{x}_{it}'\boldsymbol{\beta} + \varepsilon_{it}. \tag{3.10}$$

This is similar to the error-components model but we have added a random time component, λ_t. We assume that $\{\lambda_t\}$, $\{\alpha_i\}$, and $\{\varepsilon_{it}\}$ are mutually independent. See Chapter 8 for additional details regarding this model.

In summary, the mixed linear model generalizes the linear mixed-effects model and includes other models that are of interest in longitudinal data analysis. Much of the estimation can be accomplished directly in terms of the mixed linear model. To illustrate, in this book many of the examples are analyzed using PROC MIXED, a procedure within the statistical package SAS specifically designed to analyze mixed linear models. The primary advantage of the linear mixed-effects model is that it provides a more intuitive platform for examining longitudinal data.

Example 3.3: Income Inequality Zhou (2000O) examined changes in income determinants for a sample of $n = 4{,}730$ urban Chinese residents over a period 1955–1994. Subjects were selected as a stratified random sample from twenty cities with the strata size being proportional to the city size. The income information was collected retrospectively at a single point in time for six time points before 1985 (1955, 1960, 1965, 1975, 1978, and 1984) and five time points after 1984 (1987, 1991, 1992, 1993, and 1994); the year 1985 marks the official beginning of an urban reform.

Specifically, the model was

$$\ln y_{c,i,t} = (1 - z_{1,t})\alpha_{c,1} + z_{1,t}\alpha_{c,2} + \lambda_t + \mathbf{x}'_{c,i,t}\boldsymbol{\beta} + z_{1,t}\mathbf{x}'_{c,i,t}\boldsymbol{\beta}_2 + \varepsilon_{c,i,t}.$$

Here, $y_{c,i,t}$ represents income for the ith subject in the cth city at time t. The vector $\mathbf{x}_{c,i,t}$ represents several control variables that include gender, age, age squared, education, occupation, and work organization (government, other public, and private firms). The variable $z_{1,t}$ is a binary variable defined to be one if $t \geq 1985$ and zero otherwise. Thus, the vector $\boldsymbol{\beta}$ represents parameter estimates for the explanatory variables before 1985 and $\boldsymbol{\beta}_2$ represents the differences after urban reform. The primary interest is in the change of the explanatory variable effects, $\boldsymbol{\beta}_2$.

For the other variables, the random effect λ_t is meant to control for undetected time effects. There are two city effects: $(1 - z_{1,t})\alpha_{c,1}$ is for cities before 1985 and $z_{1,t}\alpha_{c,2}$ is for after 1984. Note that these random effects are at the city level and not at the subject (i) level. Zhou used a combination of error components and autoregressive structure to model the serial relationships of the disturbance terms. Including these random effects accounted for clustering of responses within both cities and time periods, thus providing more accurate assessment of the regression coefficients $\boldsymbol{\beta}$ and $\boldsymbol{\beta}_2$.

Zhou found significant returns to education and these returns increased in the postreform era. Little change was found among organization effects, with the exception of significantly increased effects for private firms.

3.4 Inference for Regression Coefficients

Estimation of the linear mixed-effects model proceeds in two stages. In the first stage, we estimate the regression coefficients $\boldsymbol{\beta}$, assuming knowledge of the variance components $\boldsymbol{\tau}$. Then, in the second stage, the variance components $\boldsymbol{\tau}$ are estimated. Section 3.5 discusses variance component estimation. This section discusses regression coefficient inference, assuming that the variance components are known.

GLS Estimation

From Section 3.3, we have that the vector $\mathbf{y}_i$ has mean $\mathbf{X}_i\boldsymbol{\beta}$ and variance $\mathbf{Z}_i\mathbf{D}\mathbf{Z}_i' + \mathbf{R}_i = \mathbf{V}_i(\boldsymbol{\tau}) = \mathbf{V}_i$. Thus, direct calculations show that the GLS estimator of $\boldsymbol{\beta}$ is

$$\mathbf{b}_{\mathrm{GLS}} = \left(\sum_{i=1}^{n} \mathbf{X}_i'\mathbf{V}_i^{-1}\mathbf{X}_i\right)^{-1} \sum_{i=1}^{n} \mathbf{X}_i'\mathbf{V}_i^{-1}\mathbf{y}_i. \tag{3.11}$$

The GLS estimator of $\boldsymbol{\beta}$ takes the same form as in the error-components model estimator in Equation (3.4) yet with a more general variance covariance matrix $\mathbf{V}_i$. Furthermore, direct calculation shows that the variance is

$$\operatorname{Var} \mathbf{b}_{\mathrm{GLS}} = \left(\sum_{i=1}^{n} \mathbf{X}_i'\mathbf{V}_i^{-1}\mathbf{X}_i\right)^{-1}. \tag{3.12}$$

As with fixed-effects estimators, it is possible to express $\mathbf{b}_{\mathrm{GLS}}$ as a weighted average of subject-specific estimators. To this end, for the ith subject, define the GLS estimator $\mathbf{b}_{i,\mathrm{GLS}} = (\mathbf{X}_i'\mathbf{V}_i^{-1}\mathbf{X}_i)^{-1}\mathbf{X}_i'\mathbf{V}_i^{-1}\mathbf{y}_i$ and the weight $\mathbf{W}_{i,\mathrm{GLS}} = \mathbf{X}_i'\mathbf{V}_i^{-1}\mathbf{X}_i$. Then, we can write

$$\mathbf{b}_{\mathrm{GLS}} = \left(\sum_{i=1}^{n} \mathbf{W}_{i,\mathrm{GLS}}\right)^{-1} \sum_{i=1}^{n} \mathbf{W}_{i,\mathrm{GLS}}\mathbf{b}_{i,\mathrm{GLS}}.$$

Matrix Inversion Formula

To simplify the calculations and to provide better intuition for our expressions, we cite a formula for inverting $\mathbf{V}_i$. Note that the matrix $\mathbf{V}_i$ has dimension $T_i \times T_i$. From Appendix A.5, we have

$$\mathbf{V}_i^{-1} = (\mathbf{R}_i + \mathbf{Z}_i\mathbf{D}\mathbf{Z}_i')^{-1} = \mathbf{R}_i^{-1} - \mathbf{R}_i^{-1}\mathbf{Z}_i\left(\mathbf{D}^{-1} + \mathbf{Z}_i'\mathbf{R}_i^{-1}\mathbf{Z}_i\right)^{-1}\mathbf{Z}_i'\mathbf{R}_i^{-1}. \tag{3.13}$$

The expression on the right-hand side of Equation (3.13) is easier to compute than the left-hand side when the temporal covariance matrix $\mathbf{R}_i$ has an easily computable inverse and the dimension q is small relative to T_i. Moreover, because the matrix $\mathbf{D}^{-1} + \mathbf{Z}_i'\mathbf{R}_i^{-1}\mathbf{Z}_i$ is only a $q \times q$ matrix, it is easier to invert than $\mathbf{V}_i$, a $T_i \times T_i$ matrix.

Some special cases are of interest. First, note that, in the case of no serial correlation, we have $\mathbf{R}_i = \sigma^2\mathbf{I}_i$ and Equation (3.13) reduces to

$$\mathbf{V}_i^{-1} = \left(\sigma^2\mathbf{I}_i + \mathbf{Z}_i\mathbf{D}\mathbf{Z}_i'\right)^{-1} = \frac{1}{\sigma^2}\left(\mathbf{I}_i - \mathbf{Z}_i\left(\sigma^2\mathbf{D}^{-1} + \mathbf{Z}_i'\mathbf{Z}_i\right)^{-1}\mathbf{Z}_i'\right). \quad (3.14)$$

Further, in the error-components model considered in Section 3.1, we have $q = 1, \mathbf{D} = \sigma_\alpha^2$, and $\mathbf{Z}_i = \mathbf{1}_i$, so that Equation (3.13) reduces to

$$\mathbf{V}_i^{-1} = \left(\sigma^2\mathbf{I}_i + \sigma_\alpha^2\mathbf{Z}_i\mathbf{Z}_i'\right)^{-1} = \frac{1}{\sigma^2}\left(\mathbf{I}_i - \frac{\sigma_\alpha^2}{T_i\sigma_\alpha^2 + \sigma^2}\mathbf{J}_i\right) = \frac{1}{\sigma^2}\left(\mathbf{I}_i - \frac{\zeta_i}{T_i}\mathbf{J}_i\right), \quad (3.15)$$

where $\zeta_i = T_i\sigma_\alpha^2/(T_i\sigma_\alpha^2 + \sigma^2)$, as in Section 3.1. This demonstrates that Equation (3.4) is a special case of Equation (3.11).

For another special case, consider the random-coefficients model ($\mathbf{z}_{it} = \mathbf{x}_{it}$) with no serial correlation so that $\mathbf{R}_i = \sigma^2\mathbf{I}_i$. Here, the weight $\mathbf{W}_{i,\text{GLS}}$ takes on the simple form $\mathbf{W}_{i,\text{GLS}} = (\mathbf{D} + \sigma^2(\mathbf{X}_i'\mathbf{X}_i)^{-1})^{-1}$ (see Exercise 3.8). From this form, we see that subjects with "large" values of $\mathbf{X}_i'\mathbf{X}_i$ have a greater effect on $\mathbf{b}_{\text{GLS}}$ than subjects with smaller values.

Maximum Likelihood Estimation

With Assumption RO5, the log likelihood of a single subject is

$$l_i(\boldsymbol{\beta}, \boldsymbol{\tau}) = -\frac{1}{2}\left(T_i \ln(2\pi) + \ln\det\mathbf{V}_i(\boldsymbol{\tau}) + (\mathbf{y}_i - \mathbf{X}_i\boldsymbol{\beta})'\,\mathbf{V}_i(\boldsymbol{\tau})^{-1}\,(\mathbf{y}_i - \mathbf{X}_i\boldsymbol{\beta})\right). \quad (3.16)$$

With Equation (3.16), the log likelihood for the entire data set is

$$L(\boldsymbol{\beta}, \boldsymbol{\tau}) = \sum_{i=1}^{n} l_i(\boldsymbol{\beta}, \boldsymbol{\tau}).$$

The values of $\boldsymbol{\beta}$ and $\boldsymbol{\tau}$ that maximize $L(\boldsymbol{\beta}, \boldsymbol{\tau})$ are the maximum likelihood estimators (MLEs), which we denote as $\mathbf{b}_{\text{MLE}}$ and $\boldsymbol{\tau}_{\text{MLE}}$.[1]

The *score vector* is the vector of derivatives of the log-likelihood taken with respect to the parameters. We denote the vector of parameters by $\boldsymbol{\theta} = (\boldsymbol{\beta}', \boldsymbol{\tau}')'$. With this notation, the score vector is $\partial L(\boldsymbol{\theta})/\partial\boldsymbol{\theta}$. Typically, if this score has a

[1] We now begin to use likelihood inference extensively. You may wish to review Appendix B for additional background on joint normality and the related likelihood function. Appendix C reviews likelihood estimation in a general context.

root, then the root is a maximum likelihood estimator. To compute the score vector, first take derivatives with respect to $\boldsymbol{\beta}$ and find the root. That is,

$$\begin{aligned}\frac{\partial}{\partial \boldsymbol{\beta}} L(\boldsymbol{\beta}, \boldsymbol{\tau}) &= \sum_{i=1}^{n} \frac{\partial}{\partial \boldsymbol{\beta}} l_i(\boldsymbol{\beta}, \boldsymbol{\tau}) \\ &= -\frac{1}{2} \sum_{i=1}^{n} \frac{\partial}{\partial \boldsymbol{\beta}} \left((\mathbf{y}_i - \mathbf{X}_i \boldsymbol{\beta})' \mathbf{V}_i(\boldsymbol{\tau})^{-1} (\mathbf{y}_i - \mathbf{X}_i \boldsymbol{\beta}) \right) \\ &= \sum_{i=1}^{n} \mathbf{X}_i' \mathbf{V}_i(\boldsymbol{\tau})^{-1} (\mathbf{y}_i - \mathbf{X}_i \boldsymbol{\beta}).\end{aligned}$$

Setting the score vector equal to zero yields

$$\mathbf{b}_{\text{MLE}} = \left(\sum_{i=1}^{n} \mathbf{X}_i' \mathbf{V}_i(\boldsymbol{\tau})^{-1} \mathbf{X}_i \right)^{-1} \sum_{i=1}^{n} \mathbf{X}_i' \mathbf{V}_i(\boldsymbol{\tau})^{-1} \mathbf{y}_i = \mathbf{b}_{\text{GLS}}. \qquad (3.17)$$

Hence, for fixed covariance parameters $\boldsymbol{\tau}$, the maximum likelihood estimator and the generalized least-squares estimator are the same.

Robust Estimation of Standard Errors

Even without the assumption of normality, the maximum likelihood estimator $\mathbf{b}_{\text{MLE}}$ has desirable properties. It is unbiased, efficient, and asymptotically normal with covariance matrix given in Equation (3.12). However, the estimator does depend on knowledge of variance components. As an alternative, it can be useful to consider an alternative, weighted least-squares estimator

$$\mathbf{b}_{\text{W}} = \left(\sum_{i=1}^{n} \mathbf{X}_i' \mathbf{W}_{i,\text{RE}} \mathbf{X}_i \right)^{-1} \sum_{i=1}^{n} \mathbf{X}_i' \mathbf{W}_{i,\text{RE}} \mathbf{y}_i, \qquad (3.18)$$

where the weighting matrix $\mathbf{W}_{i,\text{RE}}$ depends on the application at hand. To illustrate, one could use the identity matrix so that $\mathbf{b}_{\text{W}}$ reduces to the ordinary least-squares estimator. Another choice is $\mathbf{Q}_i$ from Section 2.5.3, which yields fixed-effects estimators of $\boldsymbol{\beta}$. We explore this choice further in Section 7.2. The weighted least-squares estimator is an unbiased estimator of $\boldsymbol{\beta}$ and is asymptotically normal, although not efficient unless $\mathbf{W}_{i,\text{RE}} = \mathbf{V}_i^{-1}$. Basic calculations show that it has variance

$$\operatorname{Var} \mathbf{b}_{\text{W}} = \left(\sum_{i=1}^{n} \mathbf{X}_i' \mathbf{W}_{i,\text{RE}} \mathbf{X}_i \right)^{-1} \sum_{i=1}^{n} \mathbf{X}_i' \mathbf{W}_{i,\text{RE}} \mathbf{V}_i \mathbf{W}_{i,\text{RE}} \mathbf{X}_i \left(\sum_{i=1}^{n} \mathbf{X}_i' \mathbf{W}_{i,\text{RE}} \mathbf{X}_i \right)^{-1}.$$

As in Section 2.5.3, we may consider estimators that are robust to unsuspected serial correlation and heteroscedasticity. Specifically, following a suggestion made independently by Huber (1967G), White (1980E), and Liang and Zeger (1986B), we can replace $\mathbf{V}_i$ by $\mathbf{e}_i\mathbf{e}_i'$, where $\mathbf{e}_i = \mathbf{y}_i - \mathbf{X}_i\mathbf{b}_\mathrm{W}$ is the vector of residuals. Thus, a robust standard error of $b_{j,\mathrm{W}}$, the jth element of $\mathbf{b}_\mathrm{W}$, is

$$se(b_{j,\mathrm{W}})$$

$$= \sqrt{j\text{th diagonal element of} \left(\sum_{i=1}^{n} \mathbf{X}_i'\mathbf{W}_{i,\mathrm{RE}}\mathbf{X}_i\right)^{-1} \sum_{i=1}^{n} \mathbf{X}_i'\mathbf{W}_{i,\mathrm{RE}}\mathbf{e}_i\mathbf{e}_i'\mathbf{W}_{i,\mathrm{RE}}\mathbf{X}_i \left(\sum_{i=1}^{n} \mathbf{X}_i'\mathbf{W}_{i,\mathrm{RE}}\mathbf{X}_i\right)^{-1}}.$$

Testing Hypotheses

For many statistical analyses, testing the null hypothesis that a regression coefficient equals a specified value may be the main goal. That is, interest may lie in testing $\mathrm{H}_0 : \beta_j = \beta_{j,0}$, where the specified value $\beta_{j,0}$ is often (although not always) equal to 0. The customary procedure is to compute the relevant

$$t\text{-statistic} = \frac{b_{j,\mathrm{GLS}} - \beta_{j,0}}{se(b_{j,\mathrm{GLS}})},$$

where $b_{j,\mathrm{GLS}}$ is the jth component of $\mathbf{b}_\mathrm{GLS}$ from Equation (3.17) and $se(b_{j,\mathrm{GLS}})$ is the square root of the jth diagonal element of $(\sum_{i=1}^{n} \mathbf{X}_i'\mathbf{V}_i(\hat{\tau})^{-1}\mathbf{X}_i)^{-1}$ with $\hat{\tau}$ as the estimator of the variance component (to be described in Section 3.5). Then, one assesses H_0 by comparing the t-statistic to a standard normal distribution.

There are two widely used variants of this standard procedure. First, one can replace $se(b_{j,\mathrm{GLS}})$ by $se(b_{j,\mathrm{W}})$ to get "robust t-statistics." Second, one can replace the standard normal distribution with a t-distribution with the "appropriate" number of degrees of freedom. There are several methods for calculating the degrees of freedom and these depend on the data and the purpose of the analysis. To illustrate, in Display 3.2 you will see that the approximate degrees of freedom under the "DF" column is different for each variable. This is produced by the SAS default "containment method." For the applications in this text, we typically will have a large number of observations and will be more concerned with potential heteroscedasticity and serial correlation; thus, we will use robust t-statistics. For readers with smaller data sets interested in the second alternative, Littell et al. (1996S) describes the t-distribution approximation in detail.

For testing hypotheses concerning several regression coefficients simultaneously, the customary procedure is the likelihood ratio test. One may express the null hypothesis as $\mathrm{H}_0 : \mathbf{C}\boldsymbol{\beta} = \mathbf{d}$, where $\mathbf{C}$ is a $p \times K$ matrix with rank p, $\mathbf{d}$ is a $p \times 1$ vector (typically $\mathbf{0}$), and we recall that $\boldsymbol{\beta}$ is the $K \times 1$ vector of regression coefficients. Both $\mathbf{C}$ and $\mathbf{d}$ are user specified and depend on the application at hand. This null hypothesis is tested against the alternative $\mathrm{H}_0 : \mathbf{C}\boldsymbol{\beta} \neq \mathbf{d}$.

Likelihood Ratio Test Procedure

1. Using the unconstrained model, calculate maximum likelihood estimates and the corresponding likelihood, denoted as L_{MLE}.
2. For the model constrained using $\mathrm{H}_0 : \mathbf{C}\boldsymbol{\beta} = \mathbf{d}$, calculate maximum likelihood estimates and the corresponding likelihood, denoted as L_{Reduced}.
3. Compute the likelihood ratio test statistic, $LRT = 2(L_{\mathrm{MLE}} - L_{\mathrm{Reduced}})$.
4. Reject H_0 if LRT exceeds a percentile from a chi-square distribution with p degrees of freedom. The percentile is one minus the significance level of the test.

Of course, one may also use p-values to calibrate the significance of the test. See Appendix C.7 for more details on the likelihood ratio test.

The likelihood ratio test is the industry standard for assessing hypotheses concerning several regression coefficients. However, we note that better procedures may exist, particularly for small data sets. For example, Pinheiro and Bates (2000S) recommend the use of "conditional F-tests" when p is large relative to the sample size. As with testing individual regression coefficients, we shall be more concerned with potential heteroscedasticity for large data sets. In this case, a modification of the Wald test procedure is available.

The Wald procedure for testing $\mathrm{H}_0 : \mathbf{C}\boldsymbol{\beta} = \mathbf{d}$ is to compute the test statistic

$$(\mathbf{C}\mathbf{b}_{\mathrm{MLE}} - \mathbf{d})' \left(\mathbf{C} \left[\sum_{i=1}^{n} \mathbf{X}_i' \mathbf{V}_i(\boldsymbol{\tau}_{\mathrm{MLE}})^{-1} \mathbf{X}_i \right] \mathbf{C}' \right)^{-1} (\mathbf{C}\mathbf{b}_{\mathrm{MLE}} - \mathbf{d})$$

and compare this statistic to a chi-square distribution with p degrees of freedom. Compared to the likelihood ratio test, the advantage of the Wald procedure is that the statistic can be computed with just one evaluation of the likelihood, not two. However, the disadvantage is that, for general constraints such as $\mathbf{C}\boldsymbol{\beta} = \mathbf{d}$, specialized software is required.

An advantage of the Wald procedure is that it is straightforward to compute robust alternatives. For a robust alternative, we use the regression coefficient estimator defined in Equation (3.18) and compute

$$(\mathbf{Cb}_{\mathrm{W}}-\mathbf{d})' \left(\mathbf{C} \left[\left(\sum_{i=1}^{n} \mathbf{X}_i' \mathbf{W}_{i,\mathrm{RE}} \mathbf{X}_i \right)^{-1} \sum_{i=1}^{n} \mathbf{X}_i' \mathbf{W}_{i,\mathrm{RE}} \mathbf{e}_i \mathbf{e}_i' \mathbf{W}_{i,\mathrm{RE}} \mathbf{X}_i \left(\sum_{i=1}^{n} \mathbf{X}_i' \mathbf{W}_{i,\mathrm{RE}} \mathbf{X}_i \right)^{-1} \right] \mathbf{C}' \right)^{-1} (\mathbf{Cb}_{\mathrm{W}}-\mathbf{d}).$$

We compare this statistic to a chi-square distribution with p degrees of freedom.

3.5 Variance Components Estimation

In this section, we describe several methods for estimating the variance components. The two primary methods entail maximizing a likelihood function, in contrast to moment estimators. In statistical estimation theory (Lehmann, 1991G), there are well-known trade-offs when considering moment compared to likelihood estimation. Typically, likelihood functions are maximized by using iterative procedures that require starting values. At the end of this section, we describe how to obtain reasonable starting values for the iteration using moment estimators.

3.5.1 Maximum Likelihood Estimation

The log-likelihood was presented in Section 3.4. Substituting the expression for the GLS estimator in Equation (3.11) into the log likelihood in Equation (3.16) yields the *concentrated* or *profile log likelihood*

$$L(\mathbf{b}_{\mathrm{GLS}}, \boldsymbol{\tau}) = -\frac{1}{2} \sum_{i=1}^{n} \left(T_i \ln(2\pi) + \ln \det \mathbf{V}_i(\boldsymbol{\tau}) + (\mathit{Error\ SS})_i(\boldsymbol{\tau}) \right), \quad (3.19)$$

which is a function of $\boldsymbol{\tau}$. Here, the error sum of squares for the ith subject is

$$(\mathit{Error\ SS})_i(\boldsymbol{\tau}) = (\mathbf{y}_i - \mathbf{X}_i \mathbf{b}_{\mathrm{GLS}})' \mathbf{V}_i^{-1}(\boldsymbol{\tau}) (\mathbf{y}_i - \mathbf{X}_i \mathbf{b}_{\mathrm{GLS}}). \quad (3.20)$$

Thus, we now maximize the log likelihood as a function of $\boldsymbol{\tau}$ only. In only a few special cases can one obtain closed-form expressions for the maximizing variance components. Exercise 3.12 illustrates one such special case.

Special Case: Error-Components Model For this special case, the variance components are $\boldsymbol{\tau} = (\sigma^2, \sigma_\alpha^2)'$. Using Equation (A.5) in Appendix A.5, we have that $\ln \det \mathbf{V}_i = \ln \det(\sigma_\alpha^2 \mathbf{J}_i + \sigma^2 \mathbf{I}_i) = T_i \ln \sigma^2 + \ln(1 + T_i \sigma_\alpha^2 / \sigma^2)$. From this and Equation (3.19), we have that the concentrated likelihood is

$$
\begin{aligned}
L\left(\mathbf{b}_{\mathrm{GLS}}, \sigma_\alpha^2, \sigma^2\right) \\
= -\frac{1}{2} \sum_{i=1}^{n} \Bigg\{ T_i \ln(2\pi) + T_i \ln \sigma^2 + \ln\left(1 + T_i \frac{\sigma_\alpha^2}{\sigma^2}\right) \\
+ \frac{1}{\sigma^2} (\mathbf{y}_i - \mathbf{X}_i \mathbf{b}_{\mathrm{GLS}})' \left(\mathbf{I}_i - \frac{\sigma_\alpha^2}{T_i \sigma_\alpha^2 + \sigma^2} \mathbf{J}_i\right) (\mathbf{y}_i - \mathbf{X}_i \mathbf{b}_{\mathrm{GLS}}) \Bigg\},
\end{aligned}
$$

where $\mathbf{b}_{\mathrm{GLS}}$ is given in Equation (3.4). This likelihood can be maximized over $(\sigma^2, \sigma_\alpha^2)$ using iterative methods.

Iterative Estimation

In general, the variance components are estimated recursively. This can be done using either the Newton–Raphson or Fisher scoring method (see for example, Harville, 1977S, and Wolfinger et al., 1994S).

Newton–Raphson. Let $L = L(\mathbf{b}_{\mathrm{GLS}}(\boldsymbol{\tau}), \boldsymbol{\tau})$, and use the iterative method

$$
\boldsymbol{\tau}_{\mathrm{NEW}} = \boldsymbol{\tau}_{\mathrm{OLD}} - \left\{ \left(\frac{\partial^2 L}{\partial \boldsymbol{\tau} \partial \boldsymbol{\tau}'} \right)^{-1} \frac{\partial L}{\partial \boldsymbol{\tau}} \right\} \Bigg|_{\boldsymbol{\tau} = \boldsymbol{\tau}_{\mathrm{OLD}}}.
$$

Here, the matrix $-\partial^2 L / (\partial \boldsymbol{\tau} \partial \boldsymbol{\tau}')$ is called the *sample information matrix.*

Fisher scoring. Define the *expected information matrix*

$$
I(\boldsymbol{\tau}) = -\mathrm{E}\left(\frac{\partial^2 L}{\partial \boldsymbol{\tau} \partial \boldsymbol{\tau}'} \right)
$$

and use

$$
\boldsymbol{\tau}_{\mathrm{NEW}} = \boldsymbol{\tau}_{\mathrm{OLD}} + I\left(\boldsymbol{\tau}_{\mathrm{OLD}}\right)^{-1} \left\{ \frac{\partial L}{\partial \boldsymbol{\tau}} \right\} \Bigg|_{\boldsymbol{\tau} = \boldsymbol{\tau}_{\mathrm{OLD}}}.
$$

3.5.2 Restricted Maximum Likelihood

As the name suggests, *restricted maximum likelihood* (REML) is a likelihood-based estimation procedure. Thus, it shares many of the desirable properties of MLEs. Because it is based on likelihoods, it is not specific to a particular design matrix, as are analyses of variance estimators (Harville, 1977S). Thus, it can be readily applied to a wide variety of models. Like MLEs, REML estimators are translation invariant.

Maximum likelihood often produces biased estimators of the variance components $\boldsymbol{\tau}$. In contrast, estimation based on REML results in unbiased estimators of $\boldsymbol{\tau}$, at least for many balanced designs. Because maximum likelihood estimators are negatively biased, they often turn out to be negative, an intuitively undesirable situation for many users. Because of the unbiasedness of many REML estimators, there is less of a tendency to produce negative estimators (Corbeil and Searle, 1976a,S). As with MLEs, REML estimators can be defined to be parameter values for which the (restricted) likelihood achieves a maximum value over a constrained parameter space. Thus, as with maximum likelihood, it is straightforward to modify the method to produce nonnegative variance estimators.

The idea behind REML estimation is to consider the likelihood of linear combinations of the responses that do not depend on the mean parameters. To illustrate, consider the mixed linear model. We assume that the responses, denoted by the vector $\mathbf{y}$, are normally distributed, have mean $\mathrm{E}\mathbf{y} = \mathbf{X}\boldsymbol{\beta}$, and have variance–covariance matrix $\mathrm{Var}\,\mathbf{y} = \mathbf{V} = \mathbf{V}(\boldsymbol{\tau})$. The dimension of $\mathbf{y}$ is $N \times 1$, and the dimension of $\mathbf{X}$ is $N \times p$. With this notation, define the projection matrix $\mathbf{Q} = \mathbf{I} - \mathbf{X}(\mathbf{X}'\mathbf{X})^{-1}\mathbf{X}'$ and consider the linear combination of responses $\mathbf{Qy}$. Straightforward calculations show that $\mathbf{Qy}$ has mean $\mathbf{0}$ and variance–covariance matrix $\mathrm{Var}(\mathbf{Qy}) = \mathbf{QVQ}$. Because (i) $\mathbf{Qy}$ has a multivariate normal distribution and (ii) the mean and variance–covariance matrix do not depend on $\boldsymbol{\beta}$, the distribution of $\mathbf{Qy}$ does not depend on $\boldsymbol{\beta}$. Further, Appendix 3A.1 shows that $\mathbf{Qy}$ is independent of the GLS estimator $\mathbf{b}_{\mathrm{GLS}} = (\mathbf{X}'\mathbf{V}^{-1}\mathbf{X})^{-1}\mathbf{X}'\mathbf{V}^{-1}\mathbf{y}$.

The vector $\mathbf{Qy}$ is the residual vector from an OLS fit of the data. Hence, REML is also referred to as *residual* maximum likelihood estimation. Because the rank of $\mathbf{Q}$ is $N - p$, we lose some information by considering this transformation of the data; this motivates the use of the descriptor *restricted* maximum likelihood. (There is some information about $\boldsymbol{\tau}$ in the vector $\mathbf{b}_{\mathrm{GLS}}$ that we are not using for estimation.) Further, note that we could also use any linear transform of $\mathbf{Q}$, such as $\mathbf{AQ}$, in that $\mathbf{AQY}$ also has a multivariate normal distribution with a mean and variance–covariance matrix that do not depend on $\boldsymbol{\beta}$. Patterson and Thompson (1971S) and Harville (1974S, 1977S) showed that the likelihood does not depend on the choice of $\mathbf{A}$. They introduced the "restricted" log likelihood

$$L_{\mathrm{REML}}(\mathbf{b}_{\mathrm{GLS}}(\boldsymbol{\tau}), \boldsymbol{\tau}) \\ = -\frac{1}{2}[\ln \det(\mathbf{V}(\boldsymbol{\tau})) + \ln \det(\mathbf{X}'\mathbf{V}(\boldsymbol{\tau})^{-1}\mathbf{X}) + (\mathit{Error\ SS})(\boldsymbol{\tau})], \tag{3.21}$$

up to an additive constant. See Appendix 3A.2 for a derivation of this likelihood. Restricted maximum likelihood estimators $\boldsymbol{\tau}_{\mathrm{REML}}$ are defined to be maximizers of the function $L_{\mathrm{REML}}(\mathbf{b}_{\mathrm{GLS}}(\boldsymbol{\tau}), \boldsymbol{\tau})$. Here, the error sum of squares is

$$(Error\ SS)(\boldsymbol{\tau}) = (\mathbf{y} - \mathbf{X}\mathbf{b}_{\mathrm{GLS}}(\boldsymbol{\tau}))'\mathbf{V}(\boldsymbol{\tau})^{-1}(\mathbf{y} - \mathbf{X}\mathbf{b}_{\mathrm{GLS}}(\boldsymbol{\tau})). \tag{3.22}$$

Analogous to Equation (3.19), the usual log likelihood is

$$L(\mathbf{b}_{\mathrm{GLS}}(\boldsymbol{\tau}), \boldsymbol{\tau}) = -\frac{1}{2}\left[\ln \det(\mathbf{V}(\boldsymbol{\tau})) + (Error\ SS)(\boldsymbol{\tau})\right],$$

up to an additive constant. The only difference between the two likelihoods is the term $\ln \det(\mathbf{X}'\mathbf{V}(\boldsymbol{\tau})^{-1}\mathbf{X})$. Thus, iterative methods of maximization are the same – that is, using either Newton–Raphson or Fisher scoring. For linear mixed-effects models, this additional term is $\ln \det\left(\sum_{i=1}^{n} \mathbf{X}_i'\mathbf{V}_i(\boldsymbol{\tau})^{-1}\mathbf{X}_i\right)$.

For balanced analysis of variance data ($T_i = T$), Corbeil and Searle (1976a,S) established that the REML estimation reduces to standard analysis of variance estimators. Thus, REML estimators are unbiased for these designs. However, REML estimators and analysis of variance estimators differ for unbalanced data. The REML estimators achieve their unbiasedness by accounting for the degrees of freedom lost in estimating the fixed effects $\boldsymbol{\beta}$; in contrast, MLEs do not account for this loss of degrees of freedom. When p is large, the difference between REML estimators and MLEs is significant. Corbeil and Searle (1976b,S) showed that, in terms of mean-square errors, MLEs outperform REML estimators for small p (< 5), although the situation is reversed for large p with a sufficiently large sample.

Harville (1974S) gave a Bayesian interpretation of REML estimators. He pointed out that using only $\mathbf{Qy}$ to make inferences about $\boldsymbol{\tau}$ is equivalent to ignoring prior information about $\boldsymbol{\beta}$ and using all the data.

Some statistical packages present maximized values of restricted likelihoods, suggesting to users that these values can be used for inferential techniques, such as likelihood ratio tests. For likelihood ratio tests, one should use "ordinary" likelihoods, even when evaluated at REML estimators, not the "restricted" likelihoods used to determine REML estimators. Appendix 3A.3 illustrates the potentially disastrous consequences of using REML likelihoods for likelihood ratio tests.

Starting Values

Both the Newton–Raphson and Fisher scoring algorithms and the maximum likelihood and REML estimation methods involve recursive calculations that require starting values. We now describe two nonrecursive methods due to Swamy (1970E) and Rao (1970S), respectively. One can use the results of these nonrecursive methods as starting values in the Newton–Raphson and Fisher scoring algorithms.

Swamy's moment-based procedure appeared in the econometrics panel data literature. We consider a random-coefficients model, that is, Equation (3.8) with $\mathbf{x}_{it} = \mathbf{z}_{it}$ and $\mathbf{R}_i = \sigma_i^2 \mathbf{I}_i$.

Procedure for Computing Moment-Based Variance Component Estimators

1. Compute an OLS estimator of σ_i^2,

$$s_i^2 = \frac{1}{T_i - K} \mathbf{y}_i' \left(\mathbf{I}_i - \mathbf{X}_i (\mathbf{X}_i' \mathbf{X}_i)^{-1} \mathbf{X}_i' \right) \mathbf{y}_i.$$

   This is an OLS procedure in that it ignores $\mathbf{D}$.
2. Next, calculate $\mathbf{b}_{i,\mathrm{OLS}} = (\mathbf{X}_i' \mathbf{X}_i)^{-1} \mathbf{X}_i' \mathbf{y}_i$, a predictor of $\boldsymbol{\beta} + \boldsymbol{\alpha}_i$.
3. Finally, estimate $\mathbf{D}$ using

$$\mathbf{D}_{\mathrm{SWAMY}} = \frac{1}{n-1} \sum_{i=1}^{n} (\mathbf{b}_{i,\mathrm{OLS}} - \bar{\mathbf{b}})(\mathbf{b}_{i,\mathrm{OLS}} - \bar{\mathbf{b}})' - \frac{1}{n} \sum_{i=1}^{n} s_i^2 (\mathbf{X}_i' \mathbf{X}_i)^{-1},$$

   where

$$\bar{\mathbf{b}} = \frac{1}{n} \sum_{i=1}^{n} \mathbf{b}_{i,\mathrm{OLS}}.$$

The estimator of $\mathbf{D}$ can be motivated by examining the variance of $\mathbf{b}_{i,\mathrm{OLS}}$:

$$\begin{aligned}\mathrm{Var}(\mathbf{b}_{i,\mathrm{OLS}}) &= \mathrm{Var}\left((\mathbf{X}_i' \mathbf{X}_i)^{-1} \mathbf{X}_i' \left(\mathbf{X}_i (\boldsymbol{\beta} + \boldsymbol{\alpha}_i) + \boldsymbol{\epsilon}_i \right) \right) \\ &= \mathrm{Var}\left(\boldsymbol{\beta} + \boldsymbol{\alpha}_i + (\mathbf{X}_i' \mathbf{X}_i)^{-1} \mathbf{X}_i' \boldsymbol{\epsilon}_i \right) = \mathbf{D} + \sigma_i^2 (\mathbf{X}_i' \mathbf{X}_i)^{-1}.\end{aligned}$$

Using

$$\frac{1}{n-1} \sum_{i=1}^{n} (\mathbf{b}_{i,\mathrm{OLS}} - \bar{\mathbf{b}})(\mathbf{b}_{i,\mathrm{OLS}} - \bar{\mathbf{b}})'$$

and s_i^2 as estimators of $\mathrm{Var}(\mathbf{b}_{i,\mathrm{OLS}})$ and σ_i^2, respectively, yields $\mathbf{D}_{\mathrm{SWAMY}}$ as an estimator of $\mathbf{D}$.

Various modifications of this estimator are possible. One can iterate the procedure by using $\mathbf{D}_{\mathrm{SWAMY}}$ to improve the estimators s_i^2, and so on. Homoscedasticity of the $\boldsymbol{\epsilon}_i$ terms could also be assumed. Hsiao (1986E) recommends dropping the second term, $n^{-1} \sum_{i=1}^{n} s_i^2 (\mathbf{X}_i' \mathbf{X}_i)^{-1}$, to ensure that $\mathbf{D}_{\mathrm{SWAMY}}$ is nonnegative definite.

3.5.3 MIVQUEs

Another nonrecursive method is Rao's (1970S) *minimum-variance quadratic unbiased estimator* (MIVQUE). To describe this method, we return to the mixed linear model $\mathbf{y} = \mathbf{X}\boldsymbol{\beta} + \boldsymbol{\epsilon}$, in which $\mathrm{Var}\,\mathbf{y} = \mathbf{V} = \mathbf{V}(\boldsymbol{\tau})$. We wish to estimate the linear combination of variance components, $\sum_{k=1}^{r} c_k \tau_k$, where the c_k are specified constants and $\boldsymbol{\tau} = (\tau_1, \ldots, \tau_r)'$. We assume that $\mathbf{V}$ is linear in the sense that

$$\mathbf{V} = \sum_{k=1}^{r} \tau_k \frac{\partial}{\partial \tau_k} \mathbf{V}.$$

Thus, with this assumption, we have that the matrix of second derivatives (the Hessian) of $\mathbf{V}$ is zero (Graybill, 1969G). Although this assumption is generally viable, it is not satisfied by, for example, autoregressive models. It is not restrictive to assume that $\partial \mathbf{V}/\partial \tau_k$ is known even though the variance component τ_k is unknown. To illustrate, consider an error-components structure so that $\mathbf{V} = \sigma_\alpha^2 \mathbf{J} + \sigma^2 \mathbf{I}$. Then,

$$\frac{\partial}{\partial \sigma_\alpha^2} \mathbf{V} = \mathbf{J}$$

and

$$\frac{\partial}{\partial \sigma^2} \mathbf{V} = \mathbf{I}$$

are both known.

Quadratic estimators of $\sum_{k=1}^{r} c_k \tau_k$ are based on $\mathbf{y}'\mathbf{A}\mathbf{y}$, where $\mathbf{A}$ is a symmetric matrix to be specified. The variance of $\mathbf{y}'\mathbf{A}\mathbf{y}$, assuming normality, can easily be shown to be 2 trace($\mathbf{VAVA}$). We would like the estimator to be invariant to translation of $\boldsymbol{\beta}$. That is, we require

$$\mathbf{y}'\mathbf{A}\mathbf{y} = (\mathbf{y} - \mathbf{X}\mathbf{b}_0)'\mathbf{A}(\mathbf{y} - \mathbf{X}\mathbf{b}_0) \text{ for each } \mathbf{b}_0.$$

Thus, we restrict our choice of $\mathbf{A}$ to those that satisfy $\mathbf{AX} = \mathbf{0}$.

For unbiasedness, we would like $\sum_{k=1}^{r} c_k \tau_k = \mathrm{E}(\mathbf{y}'\mathbf{A}\mathbf{y})$. Using $\mathbf{AX} = \mathbf{0}$, we have

$$\begin{aligned}\mathrm{E}(\mathbf{y}'\mathbf{A}\mathbf{y}) &= \mathrm{E}(\boldsymbol{\epsilon}'\mathbf{A}\boldsymbol{\epsilon}) = \mathrm{trace}(\mathrm{E}(\boldsymbol{\epsilon}\boldsymbol{\epsilon}'\mathbf{A})) = \mathrm{trace}(\mathbf{VA}) \\ &= \sum_{k=1}^{r} \tau_k \,\mathrm{trace}\left(\left(\frac{\partial}{\partial \tau_k}\mathbf{V}\right)\mathbf{A}\right).\end{aligned}$$

Because this equality should be valid for all variance components τ_k, we require that **A** satisfy

$$c_k = \text{trace}\left(\left(\frac{\partial}{\partial \tau_k}\mathbf{V}\right)\mathbf{A}\right), \quad \text{for} \quad k = 1, \ldots, r. \tag{3.23}$$

Rao showed that the minimum value of trace(**VAVA**) satisfying $\mathbf{AX} = \mathbf{0}$ and the constraints in Equation (3.23) is attained at

$$\mathbf{A}_*(\mathbf{V}) = \sum_{k=1}^{r} \lambda_k \mathbf{V}^{-1}\mathbf{Q}\left(\frac{\partial}{\partial \tau_k}\mathbf{V}\right)\mathbf{V}^{-1}\mathbf{Q},$$

where $\mathbf{Q} = \mathbf{Q}(\mathbf{V}) = \mathbf{I} - \mathbf{X}(\mathbf{X}'\mathbf{V}^{-1}\mathbf{X})^{-1}\mathbf{X}'\mathbf{V}^{-1}$ and $(\lambda_1, \ldots, \lambda_r)$ is the solution of

$$\mathbf{S}(\lambda_1, \ldots, \lambda_r)' = (c_1, \ldots, c_r)'.$$

Here, the (i, j)th element of **S** is given by

$$\text{trace}\left(\mathbf{V}^{-1}\mathbf{Q}\left(\frac{\partial}{\partial \tau_i}\mathbf{V}\right)\mathbf{V}^{-1}\mathbf{Q}\left(\frac{\partial}{\partial \tau_j}\mathbf{V}\right)\right).$$

Thus, the MIVQUE of $\boldsymbol{\tau}$ is the solution of

$$\mathbf{S}\boldsymbol{\tau}_{\text{MIVQUE}} = \mathbf{G}, \tag{3.24}$$

where the kth element of **G** is given by $\mathbf{y}'\mathbf{V}^{-1}\mathbf{Q}(\partial\mathbf{V}/\partial\tau_k)\mathbf{V}^{-1}\mathbf{Q}\mathbf{y}$.

When comparing Rao's to Swamy's method, we note that the MIVQUEs are available for a larger class of models. For example, in the longitudinal data context, it is possible to handle serial correlation with the MIVQUEs. A drawback of the MIVQUE is that normality is assumed; this can be weakened to zero kurtosis for certain forms of **V** (Swallow and Searle, 1978S). Further, MIVQUEs require a prespecified estimate of **V**. A widely used specification is to use the identity matrix for **V** in Equation (3.24). This specification produces estimators called MIVQUE(0), an option in widely available statistical packages. It is the default option in PROC MIXED of the statistical package SAS.

Further Reading

When compared to regression and linear models, there are fewer textbook introductions to mixed linear models, although more are becoming available. Searle, Casella, and McCulloch (1992S) give an early technical treatment. A slightly less technical one is by Longford (1993EP). McCulloch, and Searle (2001G) give an excellent recent technical treatment. Other recent contributions integrate statistical software into their exposition. Little et al. (1996S) and Verbeke and Molenberghs (2000S) introduce mixed linear models using the SAS statistical package. Pinheiro and Bates (2000S) provide an introduction using the S and S-Plus statistical packages.

Random effects in ANOVA and regression models have been part of the standard statistical literature for quite some time; see, for example, Henderson (1953B), Scheffé (1959G), Searle (1971G), or Neter and Wasserman (1974G). Balestra and Nerlove (1966E) introduced the error-components model to the econometrics literature. The random-coefficients model was described early on by Hildreth and Houck (1968S).

As described in Section 3.5, most of the development of variance components estimators occurred in the 1970s. More recently, Baltagi and Chang (1994E) compared the relative performance of several variance components estimators for the error-components model.

Appendix 3A REML Calculations

3A.1 Independence of Residuals and Least-Squares Estimators

Assume that $\mathbf{y}$ has a multivariate normal distribution with mean $\mathbf{X}\boldsymbol{\beta}$ and variance–covariance matrix $\mathbf{V}$, where $\mathbf{X}$ has dimension $N \times p$ with rank p. Recall that the matrix $\mathbf{V}$ depends on the parameters $\boldsymbol{\tau}$.

We use the matrix $\mathbf{Q} = \mathbf{I} - \mathbf{X}(\mathbf{X}'\mathbf{X})^{-1}\mathbf{X}'$. Because $\mathbf{Q}$ is idempotent and has rank $N - p$, we can find an $N \times (N - p)$ matrix $\mathbf{A}$ such that

$$\mathbf{A}\mathbf{A}' = \mathbf{Q}, \qquad \mathbf{A}'\mathbf{A} = \mathbf{I}_N.$$

We also need $\mathbf{G} = \mathbf{V}^{-1}\mathbf{X}(\mathbf{X}'\mathbf{V}^{-1}\mathbf{X})^{-1}$, an $N \times p$ matrix. Note that $\mathbf{G}'\mathbf{y} = \mathbf{b}_{\mathrm{GLS}}$, the GLS estimator of $\boldsymbol{\beta}$.

With these two matrices, define the transformation matrix $\mathbf{H} = (\mathbf{A} : \mathbf{G})$, an $N \times N$ matrix. Consider the transformed variables

$$\mathbf{H}'\mathbf{y} = \begin{bmatrix} \mathbf{A}'\mathbf{y} \\ \mathbf{G}'\mathbf{y} \end{bmatrix} = \begin{bmatrix} \mathbf{A}'\mathbf{y} \\ \mathbf{b}_{\mathrm{GLS}} \end{bmatrix}.$$

Basic calculations show that

$$\mathbf{A}'\mathbf{y} \sim N(\mathbf{0}, \mathbf{A}'\mathbf{V}\mathbf{A})$$

and

$$\mathbf{G}'\mathbf{y} = \mathbf{b}_{\mathrm{GLS}} \sim N(\boldsymbol{\beta}, (\mathbf{X}'\mathbf{V}^{-1}\mathbf{X})^{-1}),$$

in which $\mathbf{z} \sim N(\boldsymbol{\mu}, \mathbf{V})$ denotes that a random vector $\mathbf{z}$ has a multivariate normal distribution with mean $\boldsymbol{\mu}$ and variance $\mathbf{V}$. Further, we have that $\mathbf{A}'\mathbf{y}$ and $\mathbf{b}_{\mathrm{GLS}}$ are independent. This is due to normality and the zero covariance matrix:

$$\mathrm{Cov}(\mathbf{A}'\mathbf{y}, \mathbf{b}_{\mathrm{GLS}}) = \mathrm{E}(\mathbf{A}'\mathbf{y}\mathbf{y}'\mathbf{G}) = \mathbf{A}'\mathbf{V}\mathbf{G} = \mathbf{A}'\mathbf{X}(\mathbf{X}'\mathbf{V}^{-1}\mathbf{X})^{-1} = \mathbf{0}.$$

We have $\mathbf{A}'\mathbf{X} = \mathbf{0}$ because $\mathbf{A}'\mathbf{X} = (\mathbf{A}'\mathbf{A})\mathbf{A}'\mathbf{X} = \mathbf{A}'\mathbf{Q}\mathbf{X}$ and $\mathbf{Q}\mathbf{X} = \mathbf{0}$. Zero covariance, together with normality, implies independence.

3A.2 Restricted Likelihoods

To develop the restricted likelihood, we first check the rank of the transformation matrix $\mathbf{H}$. Thus, with $\mathbf{H}$ as in Appendix 3A.1 and Equation (A.2) of Appendix A.5, we have

$$\begin{aligned}\det(\mathbf{H}^2) = \det(\mathbf{H}'\mathbf{H}) &= \det\left[\begin{bmatrix}\mathbf{A}' \\ \mathbf{G}'\end{bmatrix}\begin{bmatrix}\mathbf{A} & \mathbf{G}\end{bmatrix}\right] = \det\begin{bmatrix}\mathbf{A}'\mathbf{A} & \mathbf{A}'\mathbf{G} \\ \mathbf{G}'\mathbf{A} & \mathbf{G}'\mathbf{G}\end{bmatrix} \\ &= \det(\mathbf{A}'\mathbf{A})\det\left(\mathbf{G}'\mathbf{G} - \mathbf{G}'\mathbf{A}(\mathbf{A}'\mathbf{A})^{-1}\mathbf{A}'\mathbf{G}\right) \\ &= \det(\mathbf{G}'\mathbf{G} - \mathbf{G}'\mathbf{Q}\mathbf{G}) = \det\left(\mathbf{G}'\mathbf{X}(\mathbf{X}'\mathbf{X})^{-1}\mathbf{X}'\mathbf{G}\right) = \det\left((\mathbf{X}'\mathbf{X})^{-1}\right),\end{aligned}$$

using $\mathbf{G}'\mathbf{X} = \mathbf{I}$. Thus, the transformation $\mathbf{H}$ is nonsingular if and only if $\mathbf{X}'\mathbf{X}$ is nonsingular. In this case, no information is lost by considering the transformation $\mathbf{H}'\mathbf{y}$.

We now develop the restricted likelihood based on the probability density function of $\mathbf{A}'\mathbf{y}$. We first note a relationship used by Harville (1974S), concerning the probability density function of $\mathbf{G}'\mathbf{y}$. We write $f_{\mathbf{G}'\mathbf{y}}(\mathbf{z}, \boldsymbol{\beta})$ to denote the probability density function of the random vector $\mathbf{G}'\mathbf{y}$, evaluated at the (vector) point $\mathbf{z}$ with mean (vector) parameter $\boldsymbol{\beta}$. Because probability density functions integrate to 1, we have the relation

$$\begin{aligned}1 &= \int f_{\mathbf{G}'\mathbf{y}}(\mathbf{z}, \boldsymbol{\beta})d\mathbf{z} \\ &= \int \frac{1}{(2\pi)^{p/2}\det(\mathbf{X}'\mathbf{V}^{-1}\mathbf{X})^{-1/2}} \exp\left(-\frac{1}{2}(\mathbf{z} - \boldsymbol{\beta})'\mathbf{X}'\mathbf{V}^{-1}\mathbf{X}(\mathbf{z} - \boldsymbol{\beta})\right) d\mathbf{z} \\ &= \int \frac{1}{(2\pi)^{p/2}\det(\mathbf{X}'\mathbf{V}^{-1}\mathbf{X})^{-1/2}} \exp\left(-\frac{1}{2}(\mathbf{z} - \boldsymbol{\beta})'\mathbf{X}'\mathbf{V}^{-1}\mathbf{X}(\mathbf{z} - \boldsymbol{\beta})\right) d\boldsymbol{\beta} \\ &= \int f_{\mathbf{G}'\mathbf{y}}(\mathbf{z}, \boldsymbol{\beta})d\boldsymbol{\beta}, \quad \text{for each } \mathbf{z},\end{aligned}$$

with a change of variables.

Because of the independence of $\mathbf{A}'\mathbf{y}$ and $\mathbf{G}'\mathbf{y} = \mathbf{b}_{\mathrm{GLS}}$, we have $f_{\mathbf{H}'\mathbf{y}} = f_{\mathbf{A}'\mathbf{y}}f_{\mathbf{G}'\mathbf{y}}$, where $f_{\mathbf{H}'\mathbf{y}}$, $f_{\mathbf{A}'\mathbf{y}}$, and $f_{\mathbf{G}'\mathbf{y}}$ are the density functions of the random vectors $\mathbf{H}'\mathbf{y}$, $\mathbf{A}'\mathbf{y}$, and $\mathbf{G}'\mathbf{y}$, respectively. For notation, let $\mathbf{y}^*$ be a potential realization of the random vector $\mathbf{y}$. Thus, the probability density function of $\mathbf{A}'\mathbf{y}$ is

$$f_{\mathbf{A}'\mathbf{y}}(\mathbf{A}'\mathbf{y}^*) = \int f_{\mathbf{A}'\mathbf{y}}(\mathbf{A}'\mathbf{y}^*) f_{\mathbf{G}'\mathbf{y}}(\mathbf{G}'\mathbf{y}^*, \beta) d\beta$$
$$= \int f_{\mathbf{H}'\mathbf{y}}(\mathbf{H}'\mathbf{y}^*, \beta) d\beta = \int \det(\mathbf{H})^{-1} f_{\mathbf{y}}(\mathbf{y}^*, \beta) d\beta,$$

using a change of variables. Now, let $\mathbf{b}^*_{\mathrm{GLS}}$ be the realization of $\mathbf{b}_{\mathrm{GLS}}$ using y^*. Then, from a standard equality from analysis of variance,

$$(\mathbf{y}^* - \mathbf{X}\beta)'\mathbf{V}^{-1}(\mathbf{y}^* - \mathbf{X}\beta) = (\mathbf{y}^* - \mathbf{X}\mathbf{b}^*_{\mathrm{GLS}})'\mathbf{V}^{-1}(\mathbf{y}^* - \mathbf{X}\mathbf{b}^*_{\mathrm{GLS}})$$
$$+ (\mathbf{b}^*_{\mathrm{GLS}} - \beta)'\mathbf{X}'\mathbf{V}^{-1}\mathbf{X}(\mathbf{b}^*_{\mathrm{GLS}} - \beta).$$

With this equality, the probability density function $f_{\mathbf{y}}$ can be expressed as

$$f_{\mathbf{y}}(\mathbf{y}^*, \beta) = \frac{1}{(2\pi)^{N/2}\det(\mathbf{V})^{1/2}} \exp\left(-\frac{1}{2}(\mathbf{y}^* - \mathbf{X}\beta)'\mathbf{V}^{-1}(\mathbf{y}^* - \mathbf{X}\beta)\right)$$
$$= \frac{1}{(2\pi)^{N/2}\det(\mathbf{V})^{1/2}} \exp\left(-\frac{1}{2}(\mathbf{y}^* - \mathbf{X}\mathbf{b}^*_{\mathrm{GLS}})'\mathbf{V}^{-1}(\mathbf{y}^* - \mathbf{X}\mathbf{b}^*_{\mathrm{GLS}})\right)$$
$$\times \exp\left(-\frac{1}{2}(\mathbf{b}^*_{\mathrm{GLS}} - \beta)'\mathbf{X}'\mathbf{V}^{-1}\mathbf{X}(\mathbf{b}^*_{\mathrm{GLS}} - \beta)\right)$$
$$= \frac{(2\pi)^{p/2}\det(\mathbf{X}'\mathbf{V}^{-1}\mathbf{X})^{-1/2}}{(2\pi)^{N/2}\det(\mathbf{V})^{1/2}}$$
$$\times \exp\left(-\frac{1}{2}(\mathbf{y}^* - \mathbf{X}\mathbf{b}^*_{\mathrm{GLS}})'\mathbf{V}^{-1}(\mathbf{y}^* - \mathbf{X}\mathbf{b}^*_{\mathrm{GLS}})\right) f_{\mathbf{G}'\mathbf{y}}(\mathbf{b}^*_{\mathrm{GLS}}, \beta).$$

Thus,

$$f_{\mathbf{A}'\mathbf{y}}(\mathbf{A}^t\mathbf{y}^*) = \frac{(2\pi)^{p/2}\det(\mathbf{X}'\mathbf{V}^{-1}\mathbf{X})^{-1/2}}{(2\pi)^{N/2}\det(\mathbf{V})^{1/2}} \det(\mathbf{H})^{-1}$$
$$\times \exp\left(-\frac{1}{2}(\mathbf{y}^* - \mathbf{X}\mathbf{b}^*_{\mathrm{GLS}})'\mathbf{V}^{-1}(\mathbf{y}^* - \mathbf{X}\mathbf{b}^*_{\mathrm{GLS}})\right)$$
$$\times \int f_{\mathbf{G}'\mathbf{y}}(\mathbf{b}^*_{\mathrm{GLS}}, \beta) d\beta$$
$$= (2\pi)^{-(N-p)/2}\det(\mathbf{V})^{-1/2}\det(\mathbf{X}'\mathbf{X})^{1/2}\det(\mathbf{X}'\mathbf{V}^{-1}\mathbf{X})^{-1/2}$$
$$\times \exp\left(-\frac{1}{2}(\mathbf{y}^* - \mathbf{X}\mathbf{b}^*_{\mathrm{GLS}})'\mathbf{V}^{-1}(\mathbf{y}^* - \mathbf{X}\mathbf{b}^*_{\mathrm{GLS}})\right).$$

This yields the REML likelihood in Section 3.5, after taking logarithms and dropping constants that do not involve $\boldsymbol{\tau}$.

3A.3 Likelihood Ratio Tests and REML

Recall the likelihood ratio statistic, $LRT = 2(L(\boldsymbol{\theta}_{\text{MLE}}) - L(\boldsymbol{\theta}_{\text{Reduced}}))$. This is evaluated using the "concentrated" or "profile" log-likelihood given in Equations (3.19) and (3.20). For comparison, from Equation (3.21), the "restricted" log likelihood is

$$L_{\text{REML}}(\mathbf{b}_{\text{GLS}}, \boldsymbol{\tau}) = -\frac{1}{2}\sum_{i=1}^{n}\left(T_i \ln(2\pi) + \ln\det \mathbf{V}_i(\boldsymbol{\tau}) + (Error\ SS)_i\ (\boldsymbol{\tau})\right)$$
$$-\frac{1}{2}\ln\det\left(\sum_{i=1}^{n}\mathbf{X}_i'\mathbf{V}_i(\boldsymbol{\tau})^{-1}\mathbf{X}_i\right). \tag{3A.1}$$

To see why a REML likelihood does not work for likelihood ratio tests, consider the following example.

Special Case: Testing the Importance of a Subset of Regression Coefficients

For simplicity, we assume that $\mathbf{V}_i = \sigma^2\mathbf{I}_i$, so that there is no serial correlation. For this special case, we have the finite, and asymptotic, distribution of the partial F-test (Chow test). Because the asymptotic distribution is well known, we can easily judge whether or not REML likelihoods are appropriate.

Write $\boldsymbol{\beta} = (\boldsymbol{\beta}_1', \boldsymbol{\beta}_2')'$ and suppose that we wish to use the null hypothesis $\text{H}_0 : \boldsymbol{\beta}_2 = \mathbf{0}$. Assuming no serial correlation, the GLS estimator of $\boldsymbol{\beta}$ reduces to the OLS estimator; that is, $\mathbf{b}_{\text{GLS}} = \mathbf{b}_{\text{OLS}} = (\sum_{i=1}^{n}\mathbf{X}_i'\mathbf{X}_i)^{-1}\sum_{i=1}^{n}\mathbf{X}_i'\mathbf{y}_i$. Thus, from Equation (3.19), the concentrated likelihood is

$$L\left(\mathbf{b}_{\text{OLS}}, \sigma^2\right)$$
$$= -\frac{1}{2}\sum_{i=1}^{n}\left(T_i \ln(2\pi) + T_i \ln\sigma^2 + \frac{1}{\sigma^2}(\mathbf{y}_i - \mathbf{X}_i\mathbf{b}_{\text{OLS}})'(\mathbf{y}_i - \mathbf{X}_i\mathbf{b}_{\text{OLS}})\right)$$
$$= -\frac{1}{2}\left(N\ln(2\pi) + N\ln\sigma^2 + \frac{1}{\sigma^2}(Error\ SS)_{\text{Full}}\right),$$

where $(Error\ SS)_{\text{Full}} = \sum_{i=1}^{n}(\mathbf{y}_i - \mathbf{X}_i\mathbf{b}_{\text{OLS}})'(\mathbf{y}_i - \mathbf{X}_i\mathbf{b}_{\text{OLS}})$. The maximum likelihood estimator of σ^2 is $\sigma^2_{\text{MLE}} = (Error\ SS)_{\text{Full}}/N$, so the maximum likelihood is

$$L\left(\mathbf{b}_{\text{OLS}}, \sigma^2_{\text{MLE}}\right) = -\frac{1}{2}\left(N \ln(2\pi) + N \ln(\textit{Error SS})_{\text{Full}} - N \ln N + N\right).$$

Now, write $\mathbf{X}_i = (\mathbf{X}_{1i} \quad \mathbf{X}_{2i})$, where $\mathbf{X}_{1i}$ has dimension $T_i \times (K-r)$ and $\mathbf{X}_{2i}$ has dimension $T_i \times r$. Under H_0, the estimator of $\boldsymbol{\beta}_1$ is

$$\mathbf{b}_{\text{OLS, Reduced}} = \left(\sum_{i=1}^{n} \mathbf{X}_{1i}'\mathbf{X}_{1i}\right)^{-1} \sum_{i=1}^{n} \mathbf{X}_{1i}'\mathbf{y}_i.$$

Thus, under H_0, the log-likelihood is

$$\begin{aligned} &L\left(\mathbf{b}_{\text{OLS, Reduced}}, \sigma^2_{\text{MLE, Reduced}}\right) \\ &\quad = -\frac{1}{2}\left(N \ln(2\pi) + N \ln(\textit{Error SS})_{\text{Reduced}} - N \ln N + N\right), \end{aligned}$$

where

$$(\textit{Error SS})_{\text{Reduced}} = \sum_{i=1}^{n} (\mathbf{y}_i - \mathbf{X}_{1i}\mathbf{b}_{\text{OLS, Reduced}})'(\mathbf{y}_i - \mathbf{X}_{1i}\mathbf{b}_{\text{OLS, Reduced}}).$$

Thus, the likelihood ratio test statistic is

$$\begin{aligned} \text{LRT}_{\text{MLE}} &= 2\left(L\left(\mathbf{b}_{\text{OLS}}, \sigma^2_{\text{MLE}}\right) - L\left(\mathbf{b}_{\text{OLS, Reduced}}, \sigma^2_{\text{MLE, Reduced}}\right)\right) \\ &= N(\ln(\textit{Error SS})_{\text{Reduced}} - \ln(\textit{Error SS})_{\text{Full}}). \end{aligned}$$

From a Taylor-series approximation, we have

$$\ln y = \ln x + \frac{(y-x)}{x} - \frac{1}{2}\frac{(y-x)^2}{x^2} + \cdots.$$

Thus, we have

$$\text{LRT}_{\text{MLE}} = N\left(\frac{(\textit{Error SS})_{\text{Reduced}} - (\textit{Error SS})_{\text{Full}}}{(\textit{Error SS})_{\text{Full}}}\right) + \cdots,$$

which has an approximate chi-square distribution with r degrees of freedom.

For comparison, from Equation (3A.1), the "restricted" log likelihood is

$$\begin{aligned} L_{\text{REML}}\left(\mathbf{b}_{\text{OLS}}, \sigma^2\right) = &-\frac{1}{2}\left(N \ln(2\pi) + (N-K)\ln \sigma^2 + \frac{1}{\sigma^2}(\textit{Error SS})_{\text{Full}}\right) \\ &-\frac{1}{2}\ln \det\left(\sum_{i=1}^{n} \mathbf{X}_i'\mathbf{X}_i\right). \end{aligned}$$

The restricted maximum likelihood estimator of σ^2 is

$$\sigma^2_{\text{REML}} = (\textit{Error SS})_{\text{Full}}/(N-K).$$

Thus, the restricted maximum likelihood is

$$\begin{aligned}
&L_{\text{REML}}\left(\mathbf{b}_{\text{OLS}}, \sigma^2_{\text{REML}}\right) \\
&\quad = -\frac{1}{2}(N\ln(2\pi) + (N-K)\ln(\textit{Error SS})_{\text{Full}}) \\
&\qquad -\frac{1}{2}\ln\det\left(\sum_{i=1}^{n}\mathbf{X}_i'\mathbf{X}_i\right) + \frac{1}{2}((N-K)\ln(N-K) - (N-K)).
\end{aligned}$$

Under H_0, the restricted log likelihood is

$$\begin{aligned}
&L_{\text{REML}}\left(\mathbf{b}_{\text{OLS, Reduced}}, \sigma^2_{\text{REML, Reduced}}\right) \\
&\quad = -\frac{1}{2}(N\ln(2\pi) + (N-(K-q))\ln(\textit{Error SS})_{\text{Reduced}}) \\
&\qquad -\frac{1}{2}\ln\det\left(\sum_{i=1}^{n}\mathbf{X}_{1i}'\mathbf{X}_{1i}\right) \\
&\qquad +\frac{1}{2}((N-(K-q))\ln(N-(K-q)) - (N-(K-q))).
\end{aligned}$$

Thus, the likelihood ratio test statistic using a restricted likelihood is

$$\begin{aligned}
\text{LRT}_{\text{REML}} &= 2\left(L_{\text{REML}}\left(\mathbf{b}_{\text{OLS}}, \sigma^2_{\text{REML}}\right) - L_{\text{REML}}\left(\mathbf{b}_{\text{OLS, Reduced}}, \sigma^2_{\text{REML, Reduced}}\right)\right) \\
&= \frac{(N-K)}{N}LRT_{\text{MLE}} + \ln\det\left(\sum_{i=1}^{n}\mathbf{X}_{1i}'\mathbf{X}_{1i}\right) \\
&\quad - \ln\det\left(\sum_{i=1}^{n}\mathbf{X}_i'\mathbf{X}_i\right) + q\ln\left(\frac{(\textit{Error SS})_{\text{Reduced}}}{N-(K-q)} - 1\right) \\
&\quad + (N-K)\ln\left(1 - \frac{q}{N-(K-q)}\right).
\end{aligned}$$

The first term is asymptotically equivalent to the likelihood ratio test, using "ordinary" maximized likelihoods. The third and fourth terms tend to constants. The second term, $\ln\det\left(\sum_{i=1}^{n}\mathbf{X}_{1i}'\mathbf{X}_{1i}\right) - \ln\det\left(\sum_{i=1}^{n}\mathbf{X}_i'\mathbf{X}_i\right)$, may tend to plus or minus infinity, depending on the values of the explanatory variables. For example, in the special case that $\mathbf{X}_{1i}'\mathbf{X}_{2i} = \mathbf{0}$, we have

$$\ln\det\left(\sum_{i=1}^{n}\mathbf{X}_{1i}'\mathbf{X}_{1i}\right) - \ln\det\left(\sum_{i=1}^{n}\mathbf{X}_i'\mathbf{X}_i\right) = (-1)^r \ln\det\left(\sum_{i=1}^{n}\mathbf{X}_{2i}'\mathbf{X}_{2i}\right).$$

Thus, this term will tend to plus or minus infinity for most explanatory variable designs.

Exercises and Extensions

Section 3.1

3.1 GLS estimators For the error-components model, the variance of the vector of responses is given as

$$\mathbf{V}_i = \sigma_\alpha^2\mathbf{J}_i + \sigma^2\mathbf{I}_i.$$

a. By multiplying $\mathbf{V}_i$ by $\mathbf{V}_i^{-1}$, check that

$$\mathbf{V}_i^{-1} = \frac{1}{\sigma^2}\left(\mathbf{I}_i - \frac{\zeta_i}{T_i}\mathbf{J}_i\right).$$

b. Use this form of $\mathbf{V}_i^{-1}$ and the expression for a GLS estimator,

$$\mathbf{b}_{\text{EC}} = \left(\sum_{i=1}^{n}\mathbf{X}_i'\mathbf{V}_i^{-1}\mathbf{X}_i\right)^{-1}\sum_{i=1}^{n}\mathbf{X}_i'\mathbf{V}_i^{-1}\mathbf{y}_i,$$

to establish the formula in Equation (3.4).

c. Use Equation (3.4) to show that the basic random-effects estimator can be expressed as

$$\mathbf{b}_{\text{EC}} = \left(\sum_{i=1}^{n}\left\{\left(\sum_{t=1}^{T_i}\mathbf{x}_{it}\mathbf{x}_{it}'\right) - \zeta_i T_i \bar{\mathbf{x}}_i\bar{\mathbf{x}}_i'\right\}\right)^{-1}\sum_{i=1}^{n}\left\{\left(\sum_{t=1}^{T_i}\mathbf{x}_{it}y_{it}\right) - \zeta_i T_i\bar{\mathbf{x}}_i\bar{y}_i\right\}.$$

d. Show that

$$\mathbf{b} = \left(\sum_{i=1}^{n}\mathbf{X}_i'\left(\mathbf{I}_i - T_i^{-1}\mathbf{J}_i\right)\mathbf{X}_i\right)^{-1}\sum_{i=1}^{n}\mathbf{X}_i'\left(\mathbf{I}_i - T_i^{-1}\mathbf{J}_i\right)\mathbf{y}_i$$

is an alternative expression for the basic fixed-effects estimator given in Equation (2.6).

e. Suppose that σ_α^2 is large relative to σ^2 so that we assume that $\sigma_\alpha^2/\sigma^2 \to \infty$. Give an expression and interpretation for $\mathbf{b}_{\text{EC}}$.

f. Suppose that σ_α^2 is small relative to σ^2 so that we assume that $\sigma_\alpha^2/\sigma^2 \to 0$. Give an expression and interpretation for $\mathbf{b}_{\text{EC}}$.

3.2 GLS estimator as a weighted average Consider the error-components model and suppose that $K = 1$ and that $x_{it} = 1$. Show that

$$b_{\mathrm{EC}} = \frac{\sum_{i=1}^{n} \zeta_i \bar{y}_i}{\sum_{i=1}^{n} \zeta_i}.$$

3.3 Error-components model with one explanatory variable Consider the error-components model, $y_{it} = \alpha_i + \beta_0 + \beta_1 x_{it} + \varepsilon_{it}$. That is, consider the model in Equation 3.2 with $K = 2$ and $\mathbf{x}_{it} = (1 \; x_{it})'$.

a. Show that

$$\zeta_i = \frac{\sigma^2}{\sigma_\alpha^2} T_i \left(1 - \zeta_i\right).$$

b. Show that we may write the GLS estimators of β_0 and β_1 as

$$b_{1,\mathrm{EC}} = \frac{\sum_{i,t} x_{it} y_{it} - \sum_i T_i \zeta_i \bar{x}_i \bar{y}_i - \left(\sum_i (1 - \zeta_i) T_i\right) \bar{x}_{\mathrm{w}} \bar{y}_{\mathrm{w}}}{\sum_{i,t} x_{it}^2 - \sum_i T_i \zeta_i \bar{x}_i^2 - \left(\sum_i (1 - \zeta_i) T_i\right) \bar{x}_{\mathrm{w}}^2}$$

and

$$b_{0,\mathrm{EC}} = \bar{y}_{\mathrm{w}} - \bar{x}_{\mathrm{w}} b_{1,\mathrm{EC}},$$

where

$$\bar{x}_{\mathrm{w}} = \frac{\sum_i \zeta_i \bar{x}_i}{\sum_i \zeta_i} \quad \text{and} \quad \bar{y}_{\mathrm{w}} = \frac{\sum_i \zeta_i \bar{y}_i}{\sum_i \zeta_i}.$$

(*Hint*: Use the expression of $b_{1,\mathrm{EC}}$ in Exercise 3.1(c).)

c. Suppose that σ_α^2 is large relative to σ^2 so that we assume that $\sigma_\alpha^2/\sigma^2 \to \infty$. Give an expression and interpretation for $b_{1,\mathrm{EC}}$.

d. Suppose that σ_α^2 is small relative to σ^2 so that we assume that $\sigma_\alpha^2/\sigma^2 \to 0$. Give an expression and interpretation for $b_{1,\mathrm{EC}}$.

3.4 Two-population slope interpretation Consider the error-components model and suppose that $K = 1$ and that x is binary variable. Suppose further that x takes on the value of one for those from population 1 and minus one for those from population 2. Analogously to Exercise 2.4, let $n_{1,i}$ and $n_{2,i}$ be the number of ones and minus ones for the ith subject, respectively. Further, let $\bar{y}_{1,i}$ and $\bar{y}_{2,i}$ be the average response when x is one and minus one, for the ith subject, respectively.

Show that we may write the error-components estimator as

$$b_{1,\mathrm{EC}} = \frac{\sum_{i=1}^{n} (w_{1,i}\bar{y}_{1,i} - w_{2,i}\bar{y}_{2,i})}{\sum_{i=1}^{n} (w_{1,i} + w_{2,i})},$$

with weights $w_{1,i} = n_{1,i}(1 + \zeta_i - 2\zeta_i n_{1,i}/T_i)$ and $w_{2,i} = n_{2,i}(1 + \zeta_i - 2\zeta_i n_{2,i}/T_i)$.

(*Hint*: Use the expression of b_{EC} in Exercise 3.1(c).)

3.5 Unbiased variance estimators Perform the following steps to check that the variance estimators given by Baltagi and Chang (1994E) are unbiased variance estimators for the unbalanced error-components model introduced in Section 3.1. For notational simplicity, assume the model follows the form $y_{it} = \mu_\alpha + \alpha_i + \mathbf{x}'_{it}\beta + \varepsilon_{it}$, where μ_α is a fixed parameter representing the model intercept. As described in Section 3.1, we will use the residuals $e_{it} = y_{it} - (a_i + \mathbf{x}'_{it}\mathbf{b})$ with a_i and $\mathbf{b}$ according to the Chapter 2 fixed-effects estimators in Equations (2.6) and (2.7).

a. Show that response deviations can be expressed as

$$y_{it} - \bar{y}_i = (\mathbf{x}_{it} - \bar{\mathbf{x}}_i)'\beta + \varepsilon_{it} - \bar{\varepsilon}_i.$$

b. Show that the fixed-effects slope estimator can be expressed as

$$\mathbf{b} = \beta + \left(\sum_{i=1}^{n} \sum_{t=1}^{T_i} (\mathbf{x}_{it} - \bar{\mathbf{x}}_i)(\mathbf{x}_{it} - \bar{\mathbf{x}}_i)' \right)^{-1} \sum_{i=1}^{n} \sum_{t=1}^{T_i} (\mathbf{x}_{it} - \bar{\mathbf{x}}_i)\varepsilon_{it}.$$

c. Show that the residual can be expressed as

$$e_{it} = (\mathbf{x}_{it} - \bar{\mathbf{x}}_i)'(\beta - \mathbf{b}) + \varepsilon_{it} - \bar{\varepsilon}_i.$$

d. Show that the mean-square error defined in Equation (2.11) is an unbiased estimator for this model. That is, show that

$$\mathrm{E}s^2 = \mathrm{E}\frac{1}{N - (n + K)} \sum_{i=1}^{n} \sum_{t=1}^{T_i} e_{it}^2 = \sigma^2.$$

e. Show that $a_i - \bar{a}_{\mathrm{w}} = \alpha_i - \bar{\alpha}_{\mathrm{w}} + (\bar{\mathbf{x}}_i - \bar{\mathbf{x}})'(\beta - \mathbf{b}) + \bar{\varepsilon}_i - \bar{\varepsilon}$, where $\bar{\alpha}_{\mathrm{w}} = N^{-1}\sum_{i=1}^{n} T_i \alpha_i$.

f. Show that $\mathrm{E}(\alpha_i - \bar{\alpha}_{\mathrm{w}})^2 = \sigma_\alpha^2(1 + N^{-2}\sum_i T_i^2 - 2T_i/N)$.

g. Show that $Es_\alpha^2 = \sigma_\alpha^2$.

3.6 OLS estimator Perform the following steps to check that the OLS estimator of the slope coefficient still performs well when the error-components model is true.

a. Show that the OLS estimator for the model $y_{it} = \mathbf{x}'_{it}\boldsymbol{\beta} + \varepsilon_{it}$ can be expressed as

$$\mathbf{b}_{\text{OLS}} = \left(\sum_{i=1}^{n} \sum_{t=1}^{T_i} \mathbf{x}_{it}\mathbf{x}'_{it} \right)^{-1} \left(\sum_{i=1}^{n} \sum_{t=1}^{T_i} \mathbf{x}_{it} y_{it} \right).$$

b. Assuming the error-components model, $y_{it} = \alpha_i + \mathbf{x}'_{it}\boldsymbol{\beta} + \varepsilon_{it}$, show that the difference between the part (a) estimator and the vector of parameters is

$$\mathbf{b}_{\text{OLS}} - \boldsymbol{\beta} = \left(\sum_{i=1}^{n} \sum_{t=1}^{T_i} \mathbf{x}_{it}\mathbf{x}'_{it} \right)^{-1} \left(\sum_{i=1}^{n} \sum_{t=1}^{T_i} \mathbf{x}_{it} (\alpha_i + \varepsilon_{it}) \right).$$

c. Use part (b) to argue that the estimator given in part (a) is unbiased.
d. Calculate the variance of $\mathbf{b}_{\text{OLS}}$.
e. For $K = 1$, show that the variance calculated in part (d) is larger than the variance of the error-components estimator, $\text{Var}\, \mathbf{b}_{\text{EC}}$, given in Section 3.1.

3.7 Pooling test Perform the following steps to check that the test statistic for the pooling test given in Section 3.1 has an approximate chi-square distribution under the null hypothesis of a homogeneous model of the form $y_{it} = \mathbf{x}'_{it}\boldsymbol{\beta} + \varepsilon_{it}$.

a. Check that the residuals can be expressed as $e_{it} = \varepsilon_{it} + \mathbf{x}'_{it}(\boldsymbol{\beta} - \mathbf{b}_{\text{OLS}})$, where $\mathbf{b}_{\text{OLS}}$ is the OLS estimator of $\boldsymbol{\beta}$ in Exercise 3.6.
b. Check that

$$\text{E} \frac{1}{N-K} \sum_{i=1}^{n} \sum_{t=1}^{T_i} e_{it}^2 = \sigma^2.$$

c. Check that $T_i(T_i - 1)s_i = T_i^2 \bar{e}_i^2 - \sum_{t=1}^{T_i} e_{it}^2 = \sum_{r \neq s} e_{ir} e_{is}$, where the latter sum is over $\{(r, s)$ such that $s \neq r$ and $r = 1, \ldots, T_i, s = 1, \ldots, T_i\}$.
d. Check, for $s \neq r$, that $\text{E} e_{ir} e_{is} = -h_{ir,is}\sigma^2$, where

$$h_{ir,is} = \mathbf{x}'_{ir} \left(\sum_{i=1}^{n} \sum_{i=1}^{T_i} \mathbf{x}_{it}\mathbf{x}'_{it} \right)^{-1} \mathbf{x}_{is}$$

is an element of the hat matrix.

e. Establish conditions so that the bias is negligible. That is, check that

$$n^{-1/2} \sum_{i=1}^{n} \sum_{r \neq s} h_{ir,\, is} \Big/ \sqrt{T_i(T_i - 1)} \rightarrow 0.$$

f. Determine the approximate variance of s_i by showing that

$$\mathrm{E}\left(\frac{1}{T_i(T_i - 1)} \sum_{r \neq s} \varepsilon_{ir} \varepsilon_{is} \right)^2 = \frac{2\sigma^4}{T_i(T_i - 1)}.$$

g. Outline an argument to show that

$$\frac{1}{\sigma^2 \sqrt{2n}} \sum_{i=1}^{n} s_i \sqrt{T_i(T_i - 1)}$$

is approximately standard normal, thus, completing the argument for the behavior of the pooling test statistic under the null hypothesis.

Section 3.3

3.8 Nested models Let $y_{i,j,t}$ be the output of the jth firm in the ith industry for the tth time period. Assume that the error structure is given by

$$y_{i,j,t} = \mathrm{E} y_{i,j,t} + \delta_{i,j,t},$$

where $\delta_{i,j,t} = \alpha_i + \nu_{i,j} + \varepsilon_{i,j,t}$. Here, assume that each of $\{\alpha_i\}$, $\{\nu_{i,j}\}$, and $\{\varepsilon_{i,j,t}\}$ are independently and identically distributed and independent of one another.

a. Let $\mathbf{y}_i$ be the vector of responses for the ith industry. Write $\mathbf{y}_i$ as a function of $\{y_{i,j,t}\}$.
b. Use σ_α^2, σ_ν^2, and σ_ε^2 to denote the variance of each error component, respectively. Give an expression for $\mathrm{Var}\, \mathbf{y}_i$ in terms of these variance components.
c. Consider the linear mixed-effects model, $\mathbf{y}_i = \mathbf{Z}_i \boldsymbol{\alpha}_i + \mathbf{X}_i \boldsymbol{\beta} + \boldsymbol{\epsilon}_i$. Show how to write the quantity $\mathbf{Z}_i \boldsymbol{\alpha}_i$ in terms of the error components α_i and $\nu_{i,j}$ and the appropriate explanatory variables.

Section 3.4

3.9 GLS estimator as a weighted average of subject-specific GLS estimators Consider the random-coefficients model and the weighted average expression for the GLS estimator

$$\mathbf{b}_{\mathrm{GLS}} = \left(\sum_{i=1}^{n} \mathbf{W}_{i,\mathrm{GLS}} \right)^{-1} \sum_{i=1}^{n} \mathbf{W}_{i,\mathrm{GLS}}\, \mathbf{b}_{i,\mathrm{GLS}}.$$

a. Show that the weights can be expressed as $\mathbf{W}_{i,\mathrm{GLS}} = (\mathbf{D} + \sigma^2(\mathbf{X}_i'\mathbf{X}_i)^{-1})^{-1}$.
b. Show that $\mathrm{Var}\, \mathbf{b}_{i,\mathrm{GLS}} = \mathbf{D} + \sigma^2(\mathbf{X}_i'\mathbf{X}_i)^{-1}$.

3.10 Matched-pairs design Consider a pair of observations that have been "matched" in some way. The pair may consist of siblings, firms with similar characteristics but from different industries, or the same entity observed before and after some event of interest. Assume that there is reason to believe that the pair of observations are dependent in some fashion. Let (y_{i1}, y_{i2}) be the set of responses and $(\mathbf{x}_{i1}, \mathbf{x}_{i2})$ be the corresponding set of covariates. Because of the sampling design, the assumption of independence between y_{i1} and y_{i2} is not acceptable.

a. One alternative is to analyze the difference between the responses. Thus, let $y_i = y_{i1} - y_{i2}$. Assuming perfect matching, we might assume that $\mathbf{x}_{i1} = \mathbf{x}_{i2} = \mathbf{x}_i$, say, and use the model $y_i = \mathbf{x}_i'\boldsymbol{\gamma} + \varepsilon_i$. Without perfect matching, one could use the model $y_i = \mathbf{x}_{i1}'\boldsymbol{\beta}_1 - \mathbf{x}_{i2}'\boldsymbol{\beta}_2 + \eta_i$. Calculate the OLS estimator $\boldsymbol{\beta} = (\boldsymbol{\beta}_1', \boldsymbol{\beta}_2')'$, and call this estimator $\mathbf{b}_{\mathrm{PD}}$ because the responses are paired differences.
b. As another alternative, form the vectors $\mathbf{y}_i = (y_{i1}, y_{i2})'$ and $\boldsymbol{\epsilon}_i = (\varepsilon_{i1}, \varepsilon_{i2})'$, as well as the matrix

$$\mathbf{X}_i = \begin{pmatrix} \mathbf{x}_{i1}' & \mathbf{0} \\ \mathbf{0} & \mathbf{x}_{i2}' \end{pmatrix}.$$

Now, consider the linear mixed-effects model $\mathbf{y}_i = \mathbf{X}_i\boldsymbol{\beta} + \boldsymbol{\epsilon}_i$, where the dependence between responses is induced by the variance $\mathbf{R} = \mathrm{Var}\, \boldsymbol{\epsilon}_i$.
   i. Under this model specification, show that $\mathbf{b}_{\mathrm{PD}}$ is unbiased.
   ii. Compute the variance of $\mathbf{b}_{\mathrm{PD}}$.
   iii. Calculate the GLS estimator of $\boldsymbol{\beta}$, say $\mathbf{b}_{\mathrm{GLS}}$.
c. For yet another alternative, assume that the dependence is induced by a common latent random variable α_i. Specifically, consider the error-components model $y_{ij} = \alpha_i + \mathbf{x}_{ij}'\boldsymbol{\beta}_j + \varepsilon_{ij}$.
   i. Under this model specification, show that $\mathbf{b}_{\mathrm{PD}}$ is unbiased.
   ii. Calculate the variance of $\mathbf{b}_{\mathrm{PD}}$.
   iii. Let $\mathbf{b}_{\mathrm{EC}}$ be the GLS estimator under this model. Calculate the variance.
   iv. Show that if $\sigma_\alpha^2 \to \infty$, then the variance of $\mathbf{b}_{\mathrm{EC}}$ tends to the variance of $\mathbf{b}_{\mathrm{PD}}$. Thus, for highly correlated data, the two estimators have the same efficiency.
d. Continue with the model in part (c). We know that $\mathrm{Var}\, \mathbf{b}_{\mathrm{EC}}$ is "smaller" than $\mathrm{Var}\, \mathbf{b}_{\mathrm{PD}}$, because $\mathbf{b}_{\mathrm{EC}}$ is the GLS estimator of $\boldsymbol{\beta}$. To quantify this in a special case, assume asymptotically equivalent matching so that

$n^{-1}\sum_{i=1}^{n}\mathbf{x}_{ij}\mathbf{x}_{ij}' \rightarrow \Sigma_x$, $j = 1, 2$. Moreover, let $n^{-1}\sum_{i=1}^{n}\mathbf{x}_{i1}\mathbf{x}_{i2}' \rightarrow \Sigma_{12}$ and assume that Σ_{12} is symmetric. Suppose that we are interested in differences of the respondents, so that the (vector of) parameters of interest are $\beta_1 - \beta_2$. Let $\mathbf{b}_{\mathrm{EC}} = (\mathbf{b}_{1,\mathrm{EC}}', \mathbf{b}_{2,\mathrm{EC}}')'$ and $\mathbf{b}_{\mathrm{PD}} = (\mathbf{b}_{1,\mathrm{PD}}', \mathbf{b}_{2,\mathrm{PD}}')'$.

i. Show that

$$\frac{n}{\sigma^2}\,\mathrm{Var}(\mathbf{b}_{1,\mathrm{EC}} - \mathbf{b}_{2,\mathrm{EC}}) = \frac{2}{(1-\zeta/2)}\left(\Sigma_x + z\Sigma_{12}\right)^{-1},$$

where

$$z = \frac{\zeta/2}{(1-\zeta/2)} \quad \text{and} \quad \zeta = \frac{2\sigma_\alpha^2}{2\sigma_\alpha^2 + \sigma^2}.$$

ii. Show that

$$\frac{n}{\sigma^2}\,\mathrm{Var}(\mathbf{b}_{1,\mathrm{PD}} - \mathbf{b}_{2,\mathrm{PD}}) = 4\left(\Sigma_x + z\Sigma_{12}\right)^{-1}.$$

iii. Use parts (d)(i) and (d)(ii) to quantify the relative variances. For example, if $\Sigma_{12} = \mathbf{0}$, then the relative variances (efficiency) is $1/(2-\zeta)$, which is between 0.5 and 1.0.

3.11 *Robust standard errors* To estimate the linear mixed-effects model, consider the weighted least-squares estimator, given in Equation (3.18). The variance of this estimator, Var $\mathbf{b}_{\mathrm{W}}$, is also given in Section 3.3, along with the corresponding robust standard error of the jth component of $\mathbf{b}_{\mathrm{W}}$, denoted as $se(b_{\mathrm{W},j})$. Let us rewrite this as

$$se(b_{j,\mathrm{W}}) = \sqrt{j\text{th diagonal element of} \left(\sum_{i=1}^{n}\mathbf{X}_i'\mathbf{W}_{i,\mathrm{RE}}\mathbf{X}_i\right)^{-1} \sum_{i=1}^{n}\mathbf{X}_i'\mathbf{W}_{i,\mathrm{RE}}\hat{\mathbf{V}}_i\mathbf{W}_{i,\mathrm{RE}}\mathbf{X}_i \left(\sum_{i=1}^{n}\mathbf{X}_i'\mathbf{W}_{i,\mathrm{RE}}\mathbf{X}_i\right)^{-1}},$$

where $\hat{\mathbf{V}}_i = \mathbf{e}_i\mathbf{e}_i'$ is an estimator of $\mathbf{V}_i$. In particular, explain the following:

a. How does one go from a variance–covariance matrix to a standard error?
b. What about this standard error makes it "robust"?
c. Let's derive a new robust standard error. For simplicity, drop the i subscripts and define the "hat matrix"

$$\mathbf{H}_{\mathrm{W}} = \mathrm{W}_{\mathrm{RE}}^{1/2}\mathbf{X}(\mathbf{X}'\mathrm{W}_{\mathrm{RE}}\mathbf{X})^{-1}\mathbf{X}'\mathrm{W}_{\mathrm{RE}}^{1/2}.$$

i. Show that the weighted residuals can be expressed as a linear combination of weighted errors. Specifically, show that

$$\mathbf{W}_{\mathrm{RE}}^{1/2}\mathbf{e} = (\mathbf{I} - \mathbf{H}_{\mathrm{W}})\mathrm{W}_{\mathrm{RE}}^{1/2}\epsilon.$$

ii. Show that

$$\mathrm{E}(\mathbf{W}_{\mathrm{RE}}^{1/2}\mathbf{ee}'\mathbf{W}_{\mathrm{RE}}^{1/2}) = (\mathbf{I} - \mathbf{H}_{\mathrm{W}})\mathbf{W}_{\mathrm{RE}}^{1/2}\mathbf{V}\mathbf{W}_{\mathrm{RE}}^{1/2}(\mathbf{I} - \mathbf{H}_{\mathrm{W}}).$$

iii. Show that $\mathbf{ee}'$ is an unbiased estimator of a linear transform of $\mathbf{V}$. Specifically, show that

$$\mathrm{E}(\mathbf{ee}') = (\mathbf{I} - \mathbf{H}_{\mathrm{W}}^{*})\mathbf{V}(\mathbf{I} - \mathbf{H}_{\mathrm{W}}^{*}),$$

where $\mathbf{H}_{\mathrm{W}}^{*} = \mathbf{W}_{\mathrm{RE}}^{-1/2}\mathbf{H}_{\mathrm{W}}\mathbf{W}_{\mathrm{RE}}^{1/2}$.

d. Explain how the result in (c)(iii) suggests defining an alternative estimator of $\mathbf{V}_i$,

$$\hat{\mathbf{V}}_{1,i} = (\mathbf{I} - \mathbf{H}_{\mathrm{W},ii}^{*})^{-1}\mathbf{e}_i\mathbf{e}_i'(\mathbf{I} - \mathbf{H}_{\mathrm{W},ii}^{*})^{-1}.$$

Use this alternative estimator to suggest a new robust estimator of the standard error of $b_{\mathrm{W},j}$. (See Frees and Jin, 2004S, if you would like more details about the properties of this estimator.)

Section 3.5

3.12 *Bias of MLE and REML variance components estimators* Consider the error-components model and suppose that $T_i = T$, $K = 1$, and $x_{it} = 1$. Further, do not impose boundary conditions so that the estimators may be negative.

a. Show that the MLE of σ^2 may be expressed as

$$\hat{\sigma}_{\mathrm{ML}}^2 = \frac{1}{n(T-1)}\sum_{i=1}^{n}\sum_{t=1}^{T}(y_{it} - \bar{y}_i)^2.$$

b. Show that $\hat{\sigma}_{\mathrm{ML}}^2$ is an unbiased estimator of σ^2.

c. Show that the MLE of σ_α^2 may be expressed as

$$\hat{\sigma}_{\alpha,\mathrm{ML}}^2 = \frac{1}{n}\sum_{i=1}^{n}(\bar{y}_i - \bar{y})^2 - \frac{1}{T}\hat{\sigma}_{\mathrm{ML}}^2.$$

d. Show that $\hat{\sigma}_{\alpha,\mathrm{ML}}^2$ is a biased estimator of σ_α^2 and determine the bias.

e. Show that the REML estimator of σ^2 equals the corresponding MLE; that is, show $\hat{\sigma}_{\mathrm{REML}}^2 = \hat{\sigma}_{\mathrm{ML}}^2$.

f. Show that the REML estimator of σ_α^2 may be expressed as

$$\hat{\sigma}_{\alpha,\mathrm{REML}}^2 = \frac{1}{n-1}\sum_{i=1}^{n}(\bar{y}_i - \bar{y})^2 - \frac{1}{T}\hat{\sigma}_{\mathrm{ML}}^2.$$

g. Show that $\hat{\sigma}_{\alpha,\mathrm{REML}}^2$ is an unbiased estimator of σ_α^2.

Empirical Exercises

3.13 Charitable contributions[2]

a. Run an error-components model of CHARITY on INCOME, PRICE, DEPS, AGE, and MS. State which variables are statistically significant and justify your conclusions.
b. Rerun the step in part (a) by including the supply-side measures as additional explanatory variables. State whether or not these variables should be included in the model. Explain your reasoning.
c. Is there an important time pattern? For the model in part (a)(i), perform the following:
    i. Rerun it excluding YEAR as an explanatory variable yet including an $AR(1)$ serial component for the error.
    ii. Rerun it including YEAR as an explanatory variable and including an $AR(1)$ serial component for the error.
    iii. Rerun it including YEAR as an explanatory variable and including an unstructured serial component for the error. (This step may be difficult to achieve convergence of the algorithm!)
d. Which of the models from part (c) do you prefer, (i), (ii), or (iii)? Justify your choice. In your justification, discuss the nonstationarity of errors.
e. Now let's look at some *variable-slope models.*
    i. Rerun the model in part (a) including a variable slope for INCOME. State which of the two models is preferred and state your reason.
    ii. Rerun the model in part (a) including a variable slope for PRICE. State which of the two models is preferred and state your reason.
f. Which model do you think is best? Do not confine yourself to the options that you tested in the preceding parts. Justify your choice.

3.14 Tort Filings[3]

a. Run an error-components model using STATE as the subject identifier and VEHCMILE, GSTATEP, POPDENSY, WCMPMAX, URBAN, UNEMPLOY, and JSLIAB as explanatory variables.
b. Rerun the error-components model in part (a) and include the additional explanatory variables COLLRULE, CAPS, and PUNITIVE. Test whether these additional variables are statistically significant using the likelihood ratio test. State your null and alternative hypotheses, your test statistic, and decision-making rule.

[2] See Exercise 2.19 for the problem description.
[3] See Exercise 2.20 for the problem description.

c. Notwithstanding your answer in part (b), rerun the model in part (a) but also include variable random coefficients associated with WCMPMAX. Which model do you prefer, the model in part (a) or this one?
d. Just for fun, rerun the model in part (b) including variable random coefficients associated with WCMPMAX.
e. Rerun the error-components model in part (a) but include an autoregressive error of order 1. Test for the significance of this term.
f. Run the model in part (a) but with fixed effects. Compare this model to the random-effects version.

3.15 Housing prices[4]

a. *Basic summary statistics.*
   i. Produce a multiple time-series plot of NARSP.
   ii. Produce a multiple time-series plot of YPC.
   iii. Produce a scatter plot of NARSP versus YPC.
   iv. Produce an added-variable plot of NARSP versus YPC, controlling for the effects of MSA.
   v. Produce a scatter plot of NARSP versus YPC.
b. *Error-components model.*
   i. Run a one-way error-components model of NARSP on YPC and YEAR. State which variables are statistically significant.
   ii. Rerun the model in (b)(i) by including the supply-side measures as additional explanatory variables. State whether or not these variables should be included in the model. Explain your reasoning.
c. *Incorporating temporal effects.* Is there an important time pattern?
   i. Run a one-way error-components model of NARSP on YPC. Calculate residuals from this model. Produce a multiple time-series plot of residuals.
   ii. Rerun the model in part (c)(i) and include an $AR(1)$ serial component for the error. Discuss the stationarity of errors based on the output of this model fit and your analysis in part (c)(i).
d. *Variable-slope models.*
   i. Rerun the model in part (c)(i), including a variable slope for YPC. Assume that the two random effects (intercepts and slopes) are independent.
   ii. Rerun the model in part (d)(i) but allow for dependence between the two random effects. State which of the two models you prefer and why.

[4] See Exercise 2.21 for the problem description.

Table 3E.1. *Corporate structure*

Variable	Description
TANG	Tangibility: Net total fixed assets as a proportion of the book value of total assets.
MTB	Market-to-Book: Ratio of total assets at market value to total assets at book value.
LS	Logarithmic Sales: The natural logarithm of the amount of sales of goods and services to third parties, relating to the normal activities of the company.
PROF	Profitability: Earnings before interest and taxes plus depreciation, all divided by the book value of total assets.
STD	Volatility: The standard deviation of weekly, unlevered stock returns during the year. This variable proxies for the business risk facing the firm.
LVB	Total Leverage (Book): Total debt as a proportion of total debt plus book equity. Total debt is the sum of short-term and long-term debt. This is the dependent variable.

Source: Paker (2000O).

iii. Rerun the model in part (d)(ii) but incorporate the time-constant supply-side variables. Estimate the standard errors using robust standard errors. State which variables are statistically significant; justify your statement.

iv. Given the discussion of nonstationarity in part (c), describe why robust variance estimators are preferred when compared to the model-based standard errors.

3.16 Capital Structure (unstructured problem) During the 1980s, Japan's real economy was exhibiting a healthy rate of growth. The onset of the crash in the stock and real estate markets began at the end of December 1989, and the financial crisis soon spread to Japan's banking system. After more than ten years, the banking system remained weak and the economy struggles.

These data provide information on 361 industrial Japanese firms before and after the crash. Of the 361 firms in the sample, 355 are from the First Section of the Tokyo Stock Exchange; the remaining 6 are from the Second Section. Together, they constitute about 33% of the market capitalization of the Tokyo Stock Exchange.

The sample firms are classified as keiretsu or nonkeiretsu. That is, the main bank system is often part of a broader, cross-shareholding structure that includes corporate groups called "keiretsu." A useful function of such corporate groupings is to mitigate some of the informational and incentive problems in Japan's financial markets. The study helped to identify changes in financial structure before and after the crash and helped show how these changes are affected by whether or not a firm is classified as "keiretsu."

For this exercise, develop a model with random effects using the variables in Table 3E.1. Assess whether or not the keiretsu structure is an important determinant of leverage. For one approach, see Paker (2000O).

4

Prediction and Bayesian Inference

Abstract. Chapters 2 and 3 introduced models with fixed and random effects, focusing on estimation and hypothesis testing. This chapter discusses the third basic type of statistical inference, *prediction*. Prediction is particularly important in mixed-effects models where one often needs to summarize a random-effects component. In many cases the predictor can be interpreted as a *shrinkage estimator*, that is, a linear combination of local and global effects. In addition to showing how to predict the random-effects component, this chapter also shows how to predict disturbance terms, which are important for diagnostic work, as well as future responses, the main focus in forecasting applications.

The predictors are optimal in the sense that they are derived as minimum, mean-square (best) linear unbiased predictors. As an alternative to this classic frequentist setting, Bayesian predictors are also defined. Moreover, Bayesian predictors with a diffuse prior are equivalent to the minimum, mean-square linear unbiased predictors, thus providing additional motivation for these predictors.

4.1 Estimators versus Predictors

In the linear mixed-effects model, $y_{it} = \mathbf{z}'_{it}\boldsymbol{\alpha}_i + \mathbf{x}'_{it}\boldsymbol{\beta} + \varepsilon_{it}$, the random variables $\{\boldsymbol{\alpha}_i\}$ describe effects that are specific to a subject. Given the data $\{y_{it}, \mathbf{z}_{it}, \mathbf{x}_{it}\}$, in some problems we want to "summarize" subject effects. Chapters 2 and 3 discussed how to *estimate* fixed, unknown parameters. This chapter discusses applications where it is of interest to summarize random subject-specific effects. In general, we use the term *predictor* for an "estimator" of a random variable. Like estimators, a predictor is said to be *linear* if it is formed from a linear combination of the responses.

When would an analyst be interested in estimating a random variable? In some applications, the interest is in predicting the future values of a response, known as *forecasting*. To illustrate, Section 4.5 demonstrates forecasting in the context of Wisconsin lottery sales. This chapter also introduces tools and techniques for prediction in contexts other than forecasting. For example, in animal and plant breeding, one wishes to predict the production of milk for cows based on their lineage (random) and herds (fixed). In insurance, one wishes to predict expected claims for a policyholder given exposure to several risk factors (known as *credibility theory*). In sample surveys, one wishes to predict the size of a specific age–sex–race cohort within a small geographical area (known as *small-area estimation*). In a survey article, Robinson (1991S) also cites (1) ore reserve estimation in geological surveys, (2) measuring quality of a production plan, and (3) ranking baseball players' abilities.

4.2 Predictions for One-Way ANOVA Models

To begin, recall a special case of linear mixed-effects models, the traditional one-way random-effects ANOVA (analysis of variance) model,

$$y_{it} = \mu + \alpha_i + \varepsilon_{it}. \tag{4.1}$$

As described in Section 3.1, we assume that both α_i and ε_{it} are mean-zero, independent quantities. Suppose now that we are interested in summarizing the (conditional) mean effect of the ith subject, $\mu + \alpha_i$.

For contrast, recall the corresponding fixed-effects model. In this case, we did not explicitly express the overall mean but used the notation $y_{it} = \alpha_i + \varepsilon_{it}$. With this notation, α_i represents the mean of the ith subject. We saw that $\bar{y}_i$ is the "best" (Gauss–Markov) estimator of α_i. This estimator is unbiased; that is, $\mathrm{E}\bar{y}_i = \alpha_i$. Further, it is minimum variance ("best") among all linear unbiased estimators.

Shrinkage Estimator

For the model in Equation (4.1), it seems intuitively plausible that $\bar{y}$ is a desirable estimator of μ and that $\bar{y}_i - \bar{y}$ is a desirable "estimator" of α_i. Thus, $\bar{y}_i$ is a desirable predictor of $\mu + \alpha_i$. More generally, consider predictors of $\mu + \alpha_i$ that are linear combinations of $\bar{y}_i$ and $\bar{y}$, that is, $c_1 \bar{y}_i + c_2 \bar{y}$, for constants c_1 and c_2. To retain the unbiasedness, we use $c_2 = 1 - c_1$. Some basic calculations (see Exercise 4.1) show that the best value of c_1 that minimizes $\mathrm{E}\left(c_1 \bar{y}_i + (1 - c_1)\bar{y} - (\mu + \alpha_i)\right)^2$ is

$$c_1 = \frac{T_i^* \sigma_\alpha^2}{\sigma^2 + T_i^* \sigma_\alpha^2},$$

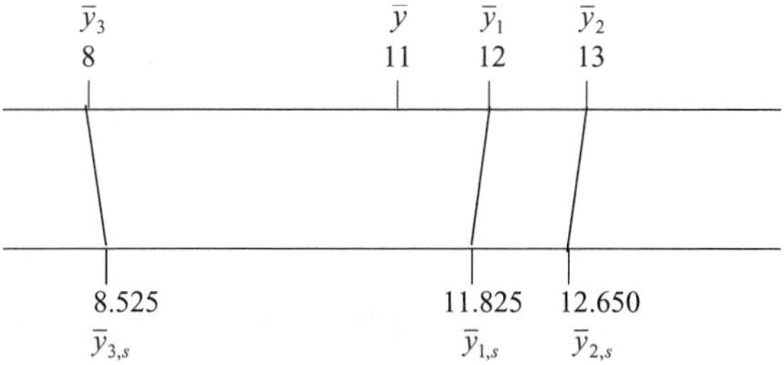

Figure 4.1. Comparison of subject-specific means to shrinkage estimates. For an illustrative data set, subject-specific and overall means are graphed on the upper scale. The corresponding shrinkage estimates are graphed on the lower scale. This figure shows the shrinkage aspect of models with random effects.

where

$$T_i^* = \frac{1 - \frac{2T_i}{N} + \frac{1}{N^2}\sum_{j=1}^{n} T_j^2}{T_i^{-1} - N^{-1}}.$$

Here, we use the notation σ_α^2 and σ^2 for the variance of α and ε, respectively. For interpretation, it is helpful to consider the case where the number of subjects, n, tends to infinity. This yields the *shrinkage estimator*, or predictor, of $\mu + \alpha_i$, defined as

$$\bar{y}_{i,s} = \zeta_i \bar{y}_i + (1 - \zeta_i)\bar{y}, \tag{4.2}$$

where

$$\zeta_i = \frac{T_i \sigma_\alpha^2}{T_i \sigma_\alpha^2 + \sigma^2} = \frac{T_i}{T_i + \sigma^2/\sigma_\alpha^2}$$

is the ith *credibility factor.*

Example 4.1: Visualizing Shrinkage Consider the following illustrative data: $\mathbf{y}_1 = (14, 12, 10, 12)'$, $\mathbf{y}_2 = (9, 16, 15, 12)'$, and $\mathbf{y}_3 = (8, 10, 7, 7)'$. That is, we have $n = 3$ subjects, each of which has $T = 4$ observations. The sample mean is $\bar{y} = 11$; the subject-specific sample means are $\bar{y}_1 = 12$, $\bar{y}_2 = 13$, and $\bar{y}_3 = 8$. We now fit the one-way random-effects ANOVA model in Equation (4.1). From the variance estimation procedures described in Section 3.5, we have that the REML estimates of σ^2 and σ_α^2 are 4.889 and 5.778, respectively. It follows that the estimated ζ_i weight is 0.825, and the corresponding predictions for the subjects are 11.825, 12.650, and 8.525, respectively.

Figure 4.1 compares subject-specific means to the corresponding predictions. Here, we see less spread in the predictions compared to the subject-specific means; each subject's estimate is "shrunk" to the overall mean, $\bar{y}$. These are the best predictors assuming α_i are random. In contrast, the subject-specific means are the best predictors assuming α_i are deterministic. Thus, this "shrinkage effect" is a consequence of the random-effects specification.

Under the random-effects ANOVA model, we have that $\bar{y}_i$ is an unbiased predictor of $\mu + \alpha_i$ in the sense that $\mathrm{E}(\bar{y}_i - (\mu + \alpha_i)) = 0$. However, $\bar{y}_i$ is inefficient in the sense that the shrinkage estimator $\bar{y}_{i,\mathrm{s}}$ has a smaller mean-square error than $\bar{y}_{i,\mathrm{s}}$. Intuitively, because $\bar{y}_{i,\mathrm{s}}$ is a linear combination of $\bar{y}_i$ and $\bar{y}$, we say that $\bar{y}_i$ has been "shrunk" toward the estimator $\bar{y}$. Further, because of the additional information in $\bar{y}$, it is customary to interpret a shrinkage estimator as "borrowing strength" from the estimator of the overall mean.

Note that the shrinkage estimator reduces to the fixed-effects estimator $\bar{y}_i$ when the credibility factor ζ_i becomes 1. It is easy to see that $\zeta_i \to 1$ as either (i) $T_i \to \infty$ or (ii) $\sigma_\alpha^2/\sigma^2 \to \infty$. That is, the best predictor approaches the subject mean as either (i) the number of observations per subject becomes large or (ii) the variability among subjects becomes large relative to the response variability. In actuarial language, either case supports the idea that the information from the ith subject is becoming more "credible."

Best Predictors

When the number of observations per subject varies, the shrinkage estimator defined in Equation (4.2) can be improved. This is because $\bar{y}$ is not the optimal estimator of μ. Using techniques described in Section 3.1, it is easy to check that (see Exercise 3.2) the GLS estimator of μ is

$$m_{\alpha,\mathrm{GLS}} = \frac{\sum_{i=1}^{n} \zeta_i \bar{y}_i}{\sum_{i=1}^{n} \zeta_i}. \tag{4.3}$$

In Section 4.3, we will see that the linear predictor of $\mu + \alpha_i$ that has minimum variance is

$$\bar{y}_{i,\mathrm{BLUP}} = \zeta_i\, \bar{y}_i + (1 - \zeta_i)\, m_{\alpha,\mathrm{GLS}}. \tag{4.4}$$

The acronym BLUP stands for *best linear unbiased predictor.*

Types of Predictors

This chapter focuses on predictors for the following three types of random variables:

1. Linear combinations of regression parameters and subject-specific effects. The statistic $\bar{y}_{i,\mathrm{BLUP}}$ provides an optimal predictor of $\mu + \alpha_i$. Thus, $\bar{y}_{i,\mathrm{BLUP}}$ is an example of a predictor of a linear combination of a global parameter and a subject-specific effect.

2. Residuals. Here, we wish to "predict" ε_{it}. The BLUP is then

$$e_{it,\mathrm{BLUP}} = y_{it} - \bar{y}_{i,\mathrm{BLUP}}. \tag{4.5}$$

These quantities, called BLUP *residuals*, are useful diagnostic statistics for developing a model. Further, unusual residuals help us understand if an observation is in line with others in the data set. For example, if the response is a salary or a stock return, unusual residuals may help us detect unusual salaries or stock returns.

3. Forecasts. Here, we wish to predict, for L lead time units into the future,

$$y_{i,T_i+L} = \mu + \alpha_i + \varepsilon_{i,T_i+L}. \tag{4.6}$$

Forecasting is similar to predicting a linear combination of global parameters and subject-specific effects, with an additional future error term. In the absence of serial correlation, we will see that the predictor is the same as the predictor of $\mu + \alpha_i$, although the mean-square error turns out to be larger. Serial correlation will lead to a different forecasting formula.

In this section, we have developed BLUPs using minimum-variance unbiased prediction. One can also develop BLUPs using normal distribution theory. That is, consider the case where α_i and $\{y_{i,1}, \ldots, y_{i,T_i}\}$ have a joint multivariate normal distribution. Then, it can be shown that

$$\mathrm{E}(\mu + \alpha_i \mid y_{i,1}, \ldots, y_{i,T_i}) = \zeta_i \, \bar{y}_i + (1 - \zeta_i)\mu.$$

This calculation is of interest because, if one were interested in estimating the unobservable α_i based on the observed responses $\{y_{i,1}, \ldots, y_{i,T_i}\}$, then normal theory suggests that the expectation is an optimal estimator. That is, consider asking the following question: What realization of $\mu + \alpha_i$ could be associated with $\{y_{i,1}, \ldots, y_{i,T_i}\}$? The expectation! The BLUP is the best linear unbiased estimator (BLUE) of $\mathrm{E}(\mu + \alpha_i \mid \{y_{i,1}, \ldots, y_{i,T_i}\})$; specifically, we need only replace μ by $m_{\alpha,\mathrm{GLS}}$. Section 4.6 will discuss these ideas more formally in a Bayesian context.

4.3 Best Linear Unbiased Predictors

This section develops BLUPs in the context of mixed-linear models. Section 4.4 then specializes the consideration to linear mixed-effects models. Section 8.4 will consider another specialization, to time-varying coefficient models. As described in Section 4.2, we develop BLUPs by examining the minimum mean-square-error predictor of a random variable w. This development is due to Harville (1976S) and also appears in his discussion of Robinson (1991S). However, the argument is originally due to Goldberger (1962E), who coined the phrase *best linear unbiased predictor*. The acronym BLUP was first used by Henderson (1973B).

Recall the mixed linear model presented in Section 3.3.2. That is, suppose that we observe an $N \times 1$ random vector $\mathbf{y}$ with mean $\mathrm{E}\mathbf{y} = \mathbf{X}\boldsymbol{\beta}$ and variance $\mathrm{Var}\,\mathbf{y} = \mathbf{V}$. The generic goal is to predict a random variable w such that

$$\mathrm{E}\,w = \boldsymbol{\lambda}'\boldsymbol{\beta}$$

and

$$\mathrm{Var}\,w = \sigma_w^2.$$

Denote the covariance between w and $\mathbf{y}$ as the $1 \times N$ vector $\mathrm{Cov}(w, \mathbf{y}) = \mathrm{E}\{(w - \mathrm{E}w)(\mathbf{y} - \mathrm{E}\mathbf{y})'\}$. The choice of w, and thus λ and σ_w^2, will depend on the application at hand; several examples will be given in Section 4.4.

Begin by assuming that the global regression parameters $\boldsymbol{\beta}$ are known. Then, Appendix 4A.1 shows that the best linear (in $\mathbf{y}$) predictor of w is

$$w^* = \mathrm{E}w + \mathrm{Cov}(w, \mathbf{y})\mathbf{V}^{-1}(\mathbf{y} - \mathrm{E}\mathbf{y}) = \boldsymbol{\lambda}'\boldsymbol{\beta} + \mathrm{Cov}(w, \mathbf{y})\mathbf{V}^{-1}(\mathbf{y} - \mathbf{X}\boldsymbol{\beta}).$$

As we will see in the Bayesian context in Section 4.6, if w, $\mathbf{y}$ have a multivariate joint normal distribution, then w^* equals $\mathrm{E}(w \mid \mathbf{y})$, so that w^* is a minimum mean-square predictor of w. Appendix 4A.2 shows that the predictor w^* is also a minimum mean-square predictor of w without the assumption of normality.

BLUPs as Predictors

Next, we assume that the global regression parameters $\boldsymbol{\beta}$ are not known. As in Section 3.4.2, we use $\mathbf{b}_{\mathrm{GLS}} = (\mathbf{X}'\mathbf{V}^{-1}\mathbf{X})^{-1}\mathbf{X}'\mathbf{V}^{-1}\mathbf{y}$ to be the GLS estimator of $\boldsymbol{\beta}$. This is the BLUE of $\boldsymbol{\beta}$. Replacing $\boldsymbol{\beta}$ by $\mathbf{b}_{\mathrm{GLS}}$ in the definition of w^*, we arrive at an expression for the BLUP of w,

$$\begin{aligned} w_{\mathrm{BLUP}} &= \boldsymbol{\lambda}'\mathbf{b}_{\mathrm{GLS}} + \mathrm{Cov}(w, \mathbf{y})\mathbf{V}^{-1}(\mathbf{y} - \mathbf{X}\mathbf{b}_{\mathrm{GLS}}) \\ &= (\boldsymbol{\lambda}' - \mathrm{Cov}(w, \mathbf{y})\mathbf{V}^{-1}\mathbf{X})\mathbf{b}_{\mathrm{GLS}} + \mathrm{Cov}(w, \mathbf{y})\mathbf{V}^{-1}\mathbf{y}. \qquad (4.7) \end{aligned}$$

Appendix 4A.2 establishes that w_{BLUP} is the BLUP of w in the sense that it is the best linear combination of responses that is unbiased and has the smallest-mean square error over all linear, unbiased predictors. From Appendix 4A.3, we also have the forms for the mean-square error and variance:

$$\begin{aligned}\operatorname{Var}(w_{\text{BLUP}} - w) \\ &= (\boldsymbol{\lambda}' - \operatorname{Cov}(w, \mathbf{y})\mathbf{V}^{-1}\mathbf{X})(\mathbf{X}'\mathbf{V}^{-1}\mathbf{X})^{-1}(\boldsymbol{\lambda}' - \operatorname{Cov}(w, \mathbf{y})\mathbf{V}^{-1}\mathbf{X})' \\ &\quad - \operatorname{Cov}(w, \mathbf{y})\mathbf{V}^{-1}\operatorname{Cov}(w, \mathbf{y})' + \sigma_w^2 \end{aligned} \tag{4.8}$$

and

$$\begin{aligned}\operatorname{Var} w_{\text{BLUP}} &= \operatorname{Cov}(w, \mathbf{y})\mathbf{V}^{-1}\operatorname{Cov}(w, \mathbf{y})' - (\boldsymbol{\lambda}' - \operatorname{Cov}(w, \mathbf{y})\mathbf{V}^{-1}\mathbf{X}) \\ &\quad \times (\mathbf{X}'\mathbf{V}^{-1}\mathbf{X})^{-1}(\boldsymbol{\lambda}' - \operatorname{Cov}(w, \mathbf{y})\mathbf{V}^{-1}\mathbf{X})'. \end{aligned} \tag{4.9}$$

From Equations (4.8) and (4.9), we see that

$$\sigma_w^2 = \operatorname{Var} w = \operatorname{Var}(w - w_{\text{BLUP}}) + \operatorname{Var} w_{\text{BLUP}}.$$

Hence, the prediction error, $w_{\text{BLUP}} - w$, and the predictor, w_{BLUP}, are uncorrelated. This fact will simplify calculations in subsequent examples.

The BLUPs are optimal, assuming the variance components implicit in $\mathbf{V}$ and $\operatorname{Cov}(w, \mathbf{y})$ are known. Applications of BLUPs typically require that the variance components be estimated, as described in Section 3.5. Best linear unbiased predictors with estimated variance components are known as *empirical* BLUPs, or *EBLUPs*. The formulas in Equations (4.8) and (4.9) do not account for the uncertainty in variance component estimation. Inflation factors that account for this additional uncertainty have been proposed (Kackar and Harville, 1984S), but they tend to be small, at least for data sets commonly encountered in practice. McCulloch and Searle (2001G) and Kenward and Roger (1997B) provide further discussions.

Special Case: One-Way Random-Effects ANOVA Model We now establish the one-way random-effects model BLUPs that were described in Equations (4.4)–(4.6) of Section 4.2. To do this, we first write the one-way random-effects ANOVA model as a special case of the mixed linear model. We then establish the predictions as special cases of w_{BLUP} given in Equation (4.7).

To express Equation (4.1) in terms of a mixed linear model, recall the error-components formulation in Section 3.1. Thus, we write (4.1) in vector form as

$$\mathbf{y}_i = \mu \mathbf{1}_i + \alpha_i \mathbf{1}_i + \boldsymbol{\varepsilon}_i,$$

where $\mathrm{Var}\,\mathbf{y}_i = \mathbf{V}_i = \sigma_\alpha^2 \mathbf{J}_i + \sigma^2 \mathbf{I}_i$ and

$$\mathbf{V}_i^{-1} = \frac{1}{\sigma^2}\left(\mathbf{I}_i - \frac{\zeta_i}{T_i}\mathbf{J}_i\right).$$

Thus, we have $\mathbf{y} = (\mathbf{y}_1', \ldots, \mathbf{y}_n')'$, $\mathbf{X} = (\mathbf{1}_1', \ldots, \mathbf{1}_n')'$, and $\mathbf{V} =$ block diagonal $(\mathbf{V}_1, \ldots, \mathbf{V}_n)$.

To develop expressions for the BLUPs, we begin with the GLS estimator, given in Equation (4.3). Thus, from Equation (4.7) and the block diagonal nature of $\mathbf{V}$, we have

$$\begin{aligned} w_{\mathrm{BLUP}} &= \lambda m_{\alpha,\mathrm{GLS}} + \mathrm{Cov}(w, \mathbf{y})\mathbf{V}^{-1}(\mathbf{y} - \mathbf{X} m_{\alpha,\mathrm{GLS}}) \\ &= \lambda m_{\alpha,\mathrm{GLS}} + \sum_{i=1}^{n} \mathrm{Cov}(w, \mathbf{y}_i)\mathbf{V}_i^{-1}(\mathbf{y}_i - \mathbf{1}_i m_{\alpha,\mathrm{GLS}}). \end{aligned}$$

Now, we have the relation

$$\begin{aligned} \mathbf{V}_i^{-1}(\mathbf{y}_i - \mathbf{1}_i m_{\alpha,\mathrm{GLS}}) &= \frac{1}{\sigma^2}\left(\mathbf{I}_i - \frac{\zeta_i}{T_i}\mathbf{J}_i\right)(\mathbf{y}_i - \mathbf{1}_i m_{\alpha,\mathrm{GLS}}) \\ &= \frac{1}{\sigma^2}((\mathbf{y}_i - \mathbf{1}_i m_{\alpha,\mathrm{GLS}}) - \zeta_i \mathbf{1}_i(\bar{y}_i - m_{\alpha,\mathrm{GLS}})). \end{aligned}$$

This yields

$$\begin{aligned} w_{\mathrm{BLUP}} &= \lambda m_{\alpha,\mathrm{GLS}} \\ &\quad + \frac{1}{\sigma^2}\sum_{i=1}^{n} \mathrm{Cov}(w, \mathbf{y}_i)((\mathbf{y}_i - \mathbf{1}_i m_{\alpha,\mathrm{GLS}}) - \zeta_i \mathbf{1}_i(\bar{y}_i - m_{\alpha,\mathrm{GLS}})). \end{aligned}$$

Now, suppose that we wish to predict $w = \mu + \alpha_i$. Then, we have $\lambda = 1$ and $\mathrm{Cov}(w, \mathbf{y}_i) = \mathbf{1}_i' \sigma_\alpha^2$ for the ith subject, and $\mathrm{Cov}(w, \mathbf{y}_i) = \mathbf{0}$ for all other subjects. Thus, we have

$$\begin{aligned} w_{\mathrm{BLUP}} &= m_{\alpha,\mathrm{GLS}} + \frac{\sigma_\alpha^2}{\sigma^2}\mathbf{1}_i'((\mathbf{y}_i - \mathbf{1}_i m_{\alpha,\mathrm{GLS}}) - \zeta_i \mathbf{1}_i(\bar{y}_i - m_{\alpha,\mathrm{GLS}})) \\ &= m_{\alpha,\mathrm{GLS}} + \frac{\sigma_\alpha^2}{\sigma^2} T_i((\bar{y}_i - m_{\alpha,\mathrm{GLS}}) - \zeta_i(\bar{y}_i - m_{\alpha,\mathrm{GLS}})) \\ &= m_{\alpha,\mathrm{GLS}} + \frac{\sigma_\alpha^2}{\sigma^2} T_i(1 - \zeta_i)(\bar{y}_i - m_{\alpha,\mathrm{GLS}}) = m_{\alpha,\mathrm{GLS}} + \zeta_i(\bar{y}_i - m_{\alpha,\mathrm{GLS}}), \end{aligned}$$

which confirms Equation (4.4).

For predicting residuals, we assume that $w = \varepsilon_{it}$. Thus, we have $\lambda = 0$ and $\mathrm{Cov}(\varepsilon_{it}, \mathbf{y}_i) = \sigma_\varepsilon^2 \mathbf{1}_{it}$ for the ith subject, and $\mathrm{Cov}\,(w, \mathbf{y}_i) = \mathbf{0}$ for all other subjects. Here, $\mathbf{1}_{it}$ denotes a $T_i \times 1$ vector with a one in the tth row and zeros otherwise. Thus, we have

$$w_{\text{BLUP}} = \sigma^2 \mathbf{1}_{\text{it}}' \mathbf{V}_i^{-1}(\mathbf{y}_i - \mathbf{X}_i \mathbf{b}_{\text{GLS}}) = y_{it} - \bar{y}_{i,\text{BLUP}},$$

which confirms Equation (4.5).

For forecasting, we use Equation (4.6) and choose $w = \mu + \alpha_i + \varepsilon_{i,T_i+L}$. Thus, we have $\lambda = 1$ and $\text{Cov}(w, \mathbf{y}_i) = \mathbf{1}_i \sigma_\alpha^2$ for the ith subject, and $\text{Cov}(w, \mathbf{y}_i) = \mathbf{0}$ for all other subjects. With this, our expression for w_{BLUP} is the same as that for the case when predicting $w = \mu + \alpha_i$.

4.4 Mixed-Model Predictors

Best linear unbiased predictors for mixed linear models were presented in Equation (4.7) with corresponding mean-square errors and variances in Equations (4.8) and (4.9), respectively. This section uses these results by presenting three broad classes of predictors that are useful for linear mixed-effects models, together with a host of special cases that provide additional interpretation. In some of the special cases, we point out that these results also pertain to the following:

- cross-sectional models, by choosing $\mathbf{D}$ to be a zero matrix, and
- fixed-effects models, by choosing $\mathbf{D}$ to be a zero matrix and incorporating $\mathbf{Z}_i \boldsymbol{\alpha}_i$ as fixed effects into the expected value of $\mathbf{y}_i$.

The three broad classes of predictors are (1) linear combinations of global parameters $\boldsymbol{\beta}$ and subject-specific effects $\boldsymbol{\alpha}_i$, (2) residuals, and (3) forecasts.

4.4.1 Linear Mixed-Effects Model

To see how the general Section 4.3 results apply to the linear mixed-effects model, recall from Equation (3.6) our matrix notation of the model: $\mathbf{y}_i = \mathbf{Z}_i \boldsymbol{\alpha}_i + \mathbf{X}_i \boldsymbol{\beta} + \boldsymbol{\epsilon}_i$. As described in Equation (3.9), we stack these equations to get $\mathbf{y} = \mathbf{Z}\boldsymbol{\alpha} + \mathbf{X}\boldsymbol{\beta} + \boldsymbol{\epsilon}$. Thus, $\text{E}\mathbf{y} = \mathbf{X}\boldsymbol{\beta}$ with $\mathbf{X} = (\mathbf{X}_1', \mathbf{X}_2', \ldots, \mathbf{X}_n')'$. Further, we have $\text{Var}\,\mathbf{y} = \mathbf{V}$, where the variance–covariance matrix is block diagonal of the form $\mathbf{V} = \text{block diagonal}\,(\mathbf{V}_1, \mathbf{V}_2, \ldots, \mathbf{V}_n)$, where $\mathbf{V}_i = \mathbf{Z}_i \mathbf{D} \mathbf{Z}_i' + \mathbf{R}_i$. Thus, from Equation (4.7), we have

$$\begin{aligned} w_{\text{BLUP}} &= \boldsymbol{\lambda}' \mathbf{b}_{\text{GLS}} + \text{Cov}(w, \mathbf{y}) \mathbf{V}^{-1}(\mathbf{y} - \mathbf{X}\mathbf{b}_{\text{GLS}}) \\ &= \boldsymbol{\lambda}' \mathbf{b}_{\text{GLS}} + \sum_{i=1}^{n} \text{Cov}(w, \mathbf{y}_i) \mathbf{V}_i^{-1}(\mathbf{y}_i - \mathbf{X}_i \mathbf{b}_{\text{GLS}}). \end{aligned} \tag{4.10}$$

Exercise 4.9 provides expressions for the BLUP mean-square error and variance.

4.4.2 Linear Combinations of Global Parameters and Subject-Specific Effects

Consider predicting linear combinations of the form $w = \mathbf{c}_1'\boldsymbol{\alpha}_i + \mathbf{c}_2'\boldsymbol{\beta}$, where $\mathbf{c}_1$ and $\mathbf{c}_2$ are known vectors of constants that are user-specified. Then, with this choice of w, straightforward calculations show that $\mathrm{E}w = \mathbf{c}_2'\,\boldsymbol{\beta}$, so that $\boldsymbol{\lambda} = \mathbf{c}_2$. Further, we have

$$\mathrm{Cov}(w, \mathbf{y}_j) = \begin{cases} \mathbf{c}_1'\mathbf{D}\mathbf{Z}_i' & \text{for} \quad j = i, \\ \mathbf{0} & \text{for} \quad j \neq i. \end{cases}$$

Putting this in Equation (4.10) yields

$$w_{\mathrm{BLUP}} = \mathbf{c}_1'\mathbf{D}\mathbf{Z}_i'\mathbf{V}_i^{-1}(\mathbf{y}_i - \mathbf{X}_i\,\mathbf{b}_{\mathrm{GLS}}) + \mathbf{c}_2'\,\mathbf{b}_{\mathrm{GLS}}.$$

To simplify this expression, we take $\mathbf{c}_2 = \mathbf{0}$ and use Wald's device. This yields the BLUP of $\boldsymbol{\alpha}_i$,

$$\mathbf{a}_{i,\mathrm{BLUP}} = \mathbf{D}\mathbf{Z}_i'\mathbf{V}_i^{-1}(\mathbf{y}_i - \mathbf{X}_i\mathbf{b}_{\mathrm{GLS}}). \tag{4.11}$$

With this notation, our BLUP of $w = \mathbf{c}_1'\boldsymbol{\alpha}_i + \mathbf{c}_2'\boldsymbol{\beta}$ is

$$w_{\mathrm{BLUP}} = \mathbf{c}_1'\mathbf{a}_{i,\mathrm{BLUP}} + \mathbf{c}_2'\mathbf{b}_{\mathrm{GLS}}. \tag{4.12}$$

Some additional special cases are of interest. For the random-coefficients model introduced in Section 3.3.1, with Equation (4.12) it is easy to check that the BLUP of $\boldsymbol{\beta} + \boldsymbol{\alpha}_i$ is

$$w_{\mathrm{BLUP}} = \boldsymbol{\zeta}_i\mathbf{b}_i + (1 - \boldsymbol{\zeta}_i)\mathbf{b}_{\mathrm{GLS}},$$

where $\mathbf{b}_i = (\mathbf{X}_i'\mathbf{V}_i^{-1}\mathbf{X}_i)^{-1}\mathbf{X}_i'\mathbf{V}_i^{-1}\mathbf{y}_i$ is the subject-specific GLS estimator and $\boldsymbol{\zeta}_i = \mathbf{D}\mathbf{X}_i'\mathbf{V}_i^{-1}\mathbf{X}_i$ is a weight matrix. This result generalizes the one-way random-effects predictors presented in Section 4.2.

In the case of the error-components model described in Section 3.1, we have $q = 1$ and $\mathbf{z}_{it} = 1$. Using Equation (4.11), the BLUP of α_i reduces to

$$a_{i,\mathrm{BLUP}} = \zeta_i(\bar{y}_i - \bar{\mathbf{x}}_i'\mathbf{b}_{\mathrm{GLS}}).$$

For comparison, recall from Chapter 2 that the fixed-effects parameter estimate is $a_i = \bar{y}_i - \bar{\mathbf{x}}_i'\mathbf{b}$. The other portion, $1 - \zeta_i$, is "borrowing strength" from zero, the mean of α_i.

Section 4.7 describes further examples from insurance credibility.

Example 3.1: Trade Localization (continued) Feinberg, Keane, and Bognano (1998E) used firm-level data to investigate U.S.-based multinational corporations employment and capital allocation decisions. From Chapter 3, their model can be written as

$$
\begin{aligned}
\ln y_{it} &= \beta_{1i}\mathrm{CT}_{it} + \beta_{2i}\mathrm{UT}_{it} + \beta_{3i}\mathrm{Trend}_t + \mathbf{x}_{it}^{*\prime}\boldsymbol{\beta}^{*} + \varepsilon_{it} \\
&= (\beta_1 + \alpha_{1i})\mathrm{CT}_{it} + (\beta_2 + \alpha_{2i})\,\mathrm{UT}_{it} + (\beta_3 + \alpha_{3i})\,\mathrm{Trend}_t + \mathbf{x}_{it}^{*\prime}\boldsymbol{\beta}^{*} + \varepsilon_{it} \\
&= \alpha_{1i}\mathrm{CT}_{it} + \alpha_{2i}\,\mathrm{UT}_{it} + \alpha_{3i}\,\mathrm{Trend}_t + \mathbf{x}_{it}^{\prime}\boldsymbol{\beta} + \varepsilon_{it},
\end{aligned}
$$

where $\mathrm{CT}_{it}(\mathrm{UT}_{it})$ is a measure of Canadian (U.S.) tariffs for firm i, and the response y is either employment or durable assets for the Canadian affiliate. Feinberg, et al. presented predictors of β_{1i} and β_{2i} using Bayesian methods (see the Section 4.6 discussion). In our notation, they predicted the linear combinations $\beta_1 + \alpha_{1i}$ and $\beta_2 + \alpha_{2i}$. The trend term was not of primary scientific interest and was included as a control variable. One major finding was that predictors for β_{1i} were negative for each firm, indicating that employment and assets in Canadian affiliates increased as Canadian tariffs decreased.

4.4.3 BLUP Residuals

For the second broad class, consider predicting a linear combination of residuals, $w = \mathbf{c}_{\varepsilon}^{\prime}\boldsymbol{\epsilon}_i$, where $\mathbf{c}_{\varepsilon}$ is a vector of constants. With this choice, we have $\mathrm{E}w = 0$; it follows that $\boldsymbol{\lambda} = \mathbf{0}$. Straightforward calculations show that

$$
\mathrm{Cov}(w, \mathbf{y}_j) = \begin{cases} \mathbf{c}_{\varepsilon}^{\prime}\mathbf{R}_i & \text{for} \quad j = i, \\ \mathbf{0} & \text{for} \quad j \neq i. \end{cases}
$$

Thus, from Equation (4.10) and Wald's device, we have the vector of BLUP residuals

$$
\mathbf{e}_{i,\mathrm{BLUP}} = \mathbf{R}_i \mathbf{V}_i^{-1}(\mathbf{y}_i - \mathbf{X}_i \mathbf{b}_{\mathrm{GLS}}), \tag{4.13a}
$$

which can also be expressed as

$$
\mathbf{e}_{i,\mathrm{BLUP}} = \mathbf{y}_i - (\mathbf{Z}_i \mathbf{a}_{i,\mathrm{BLUP}} + \mathbf{X}_i \mathbf{b}_{\mathrm{GLS}}). \tag{4.13b}
$$

Equation (4.13a) is appealing because it allows for direct computation of BLUP residuals; Equation (4.13b) is appealing because it is in the traditional "observed minus expected" form for residuals. We remark that the BLUP residual equals the GLS residual in the case that $\mathbf{D} = \mathbf{0}$; in this case, $\mathbf{e}_{i,\mathrm{BLUP}} = \mathbf{y}_i - \mathbf{X}_i \mathbf{b}_{\mathrm{GLS}} = \mathbf{e}_{i,\mathrm{GLS}}$. Also, recall the symbol $\mathbf{1}_{it}$, which denotes a $T_i \times 1$ vector that has a "one" in the tth position and is zero otherwise. Thus, we may define the BLUP residual as

$$
e_{it,\mathrm{BLUP}} = \mathbf{1}_{it}^{\prime}\mathbf{e}_{i,\mathrm{BLUP}} = \mathbf{1}_{it}^{\prime}\mathbf{R}_i \mathbf{V}_i^{-1}(\mathbf{y}_i - \mathbf{X}_i \mathbf{b}_{\mathrm{GLS}}).
$$

Equations (4.13a) and (4.13b) provide a generalization of the BLUP residual for the one-way random-effects model described in Equation (4.5). Further, using Equation (4.9), one can show that the BLUP residual has variance

$$\operatorname{Var} e_{it,\mathrm{BLUP}} = \mathbf{1}'_{it}\mathbf{R}_i\mathbf{V}_i^{-1}\left(\mathbf{V}_i - \mathbf{X}_i\left(\sum_{i=1}^{n}\mathbf{X}'_i\mathbf{V}_i^{-1}\mathbf{X}_i\right)^{-1}\mathbf{X}'_i\right)\mathbf{V}_i^{-1}\mathbf{R}_i\mathbf{1}_{it}.$$

Taking the square root of $\operatorname{Var} e_{it,\mathrm{BLUP}}$ with an estimated variance yields a standard error; this in conjunction with the BLUP residual is useful for diagnostic checking of the fitted model.

4.4.4 Predicting Future Observations

For the third broad class, suppose that the ith subject is included in the data set and we wish to predict

$$w = y_{i,T_i+L} = \mathbf{z}'_{i,T_i+L}\boldsymbol{\alpha}_i + \mathbf{x}'_{i,T_i+L}\boldsymbol{\beta} + \varepsilon_{i,T_i+L},$$

for L lead time units in the future. Assume that $\mathbf{z}_{i,T_i+L}$ and $\mathbf{x}_{i,T_i+L}$ are known. With this choice of w, it follows that $\boldsymbol{\lambda} = \mathbf{x}_{i,T_i+L}$. Further, we have

$$\operatorname{Cov}(w, \mathbf{y}_j) = \begin{cases} \mathbf{z}'_{i,T_i+L}\mathbf{D}\mathbf{Z}'_i + \operatorname{Cov}(\varepsilon_{i,T_i+L}, \boldsymbol{\epsilon}_i) & \text{for} \quad j = i, \\ \mathbf{0} & \text{for} \quad j \neq i. \end{cases}$$

Thus, using equations (4.10), (4.12), and (4.13), we have

$$\begin{aligned} \hat{y}_{i,T_i+L} &= w_{\mathrm{BLUP}} \\ &= (\mathbf{z}'_{i,T_i+L}\mathbf{D}\mathbf{Z}'_i + \operatorname{Cov}(\varepsilon_{i,T_i+L}, \boldsymbol{\epsilon}_i))\mathbf{V}_i^{-1}(\mathbf{y}_i - \mathbf{X}_i\mathbf{b}_{\mathrm{GLS}}) + \mathbf{x}'_{i,T_i+L}\mathbf{b}_{\mathrm{GLS}} \\ &= \mathbf{x}'_{i,T_i+L}\mathbf{b}_{\mathrm{GLS}} + \mathbf{z}'_{i,T_i+L}\mathbf{a}_{i,\mathrm{BLUP}} + \operatorname{Cov}(\varepsilon_{i,T_i+L}, \boldsymbol{\epsilon}_i)\mathbf{R}_i^{-1}\mathbf{e}_{i,\mathrm{BLUP}}. \end{aligned} \tag{4.14}$$

Thus, the BLUP forecast is the estimate of the conditional mean plus the serial correlation correction factor $\operatorname{Cov}(\varepsilon_{i,T_i+L}, \boldsymbol{\epsilon}_i)\mathbf{R}_i^{-1}\mathbf{e}_{i,\mathrm{BLUP}}$.

Using Equation (4.8), one can show that the variance of the forecast error is

$$\begin{aligned} &\operatorname{Var}(\hat{y}_{i,T_i+L} - y_{i,T_i+L}) \\ &= \left(\mathbf{x}'_{i,T_i+L} - (\mathbf{z}'_{i,T_i+L}\mathbf{D}\mathbf{Z}'_i + \operatorname{Cov}(\varepsilon_{i,T_i+L}, \boldsymbol{\epsilon}_i))\mathbf{V}_i^{-1}\mathbf{X}_i\right)\left(\sum_{i=1}^{n}\mathbf{X}'_i\mathbf{V}_i^{-1}\mathbf{X}_i\right) \\ &\quad \times \left(\mathbf{x}'_{i,T_i+L} - (\mathbf{z}'_{i,T_i+L}\mathbf{D}\mathbf{Z}'_i + \operatorname{Cov}(\varepsilon_{i,T_i+L}, \boldsymbol{\epsilon}_i))\mathbf{V}_i^{-1}\mathbf{X}_i\right)' \\ &\quad - \mathbf{z}'_{i,T_i+L}\mathbf{D}\mathbf{Z}'_i\mathbf{V}_i^{-1}\mathbf{Z}_i\mathbf{D}\mathbf{z}_{i,T_i+L} + \mathbf{z}'_{i,T_i+L}\mathbf{D}\mathbf{z}_{i,T_i+L} + \operatorname{Var}\varepsilon_{i,T_i+L}. \end{aligned} \tag{4.15}$$

Special Case: Autoregressive Serial Correlation To illustrate, consider the special case where we have autoregressive of order 1 ($AR(1)$), serially correlated errors. For stationary $AR(1)$ errors, the lag j autocorrelation coefficient ρ_j can be expressed as ρ^j. Thus, with an $AR(1)$ specification, we have

$$\mathbf{R} = \sigma^2 \begin{pmatrix} 1 & \rho & \rho^2 & \dots & \rho^{T-1} \\ \rho & 1 & \rho & \dots & \rho^{T-2} \\ \rho^2 & \rho & 1 & \dots & \rho^{T-3} \\ \vdots & \vdots & \vdots & \ddots & \vdots \\ \rho^{T-1} & \rho^{T-2} & \rho^{T-3} & \dots & 1 \end{pmatrix},$$

where we have omitted the i subscript. Straightforward matrix algebra results show that

$$\mathbf{R}^{-1} = \frac{1}{\sigma^2(1-\rho^2)} \begin{pmatrix} 1 & -\rho & 0 & \dots & 0 & 0 \\ -\rho & 1+\rho^2 & -\rho & \dots & 0 & 0 \\ 0 & -\rho & 1+\rho^2 & \dots & 0 & 0 \\ \vdots & \vdots & \vdots & \ddots & \vdots & \vdots \\ 0 & 0 & 0 & \dots & 1+\rho^2 & -\rho \\ 0 & 0 & 0 & \dots & -\rho & 1 \end{pmatrix}.$$

Further, omitting the i subscript on T, we have

$$\mathrm{Cov}(\varepsilon_{i,T_i+L}, \epsilon_i) = \sigma^2 \left(\rho^{T+L-1} \quad \rho^{T+L-2} \quad \rho^{T+L-3} \quad \dots \quad \rho^{L+1} \quad \rho^{L}\right).$$

Thus,

$$\begin{aligned} &\mathrm{Cov}(\varepsilon_{i,T_i+L}, \epsilon_i)\mathbf{R}_i^{-1} \\ &= \frac{1}{(1-\rho^2)} \left(\rho^{T+L-1} \quad \rho^{T+L-2} \quad \rho^{T+L-3} \quad \dots \quad \rho^{L+1} \quad \rho^{L}\right) \\ &\quad \times \begin{pmatrix} 1 & -\rho & 0 & \dots & 0 & 0 \\ -\rho & 1+\rho^2 & -\rho & \dots & 0 & 0 \\ 0 & -\rho & 1+\rho^2 & \dots & 0 & 0 \\ \vdots & \vdots & \vdots & \ddots & \vdots & \vdots \\ 0 & 0 & 0 & \dots & 1+\rho^2 & -\rho \\ 0 & 0 & 0 & \dots & -\rho & 1 \end{pmatrix} \\ &= \frac{1}{(1-\rho^2)} \left(0 \quad 0 \quad 0 \quad \cdots \quad 0 \quad -\rho^{L+2}+\rho^{L}\right) \\ &= \left(0 \quad 0 \quad 0 \quad \cdots \quad 0 \quad \rho^{L}\right). \end{aligned}$$

In summary, the L-step forecast is

$$\hat{y}_{i,T_i+L} = \mathbf{x}'_{i,T_i+L}\mathbf{b}_{\text{GLS}} + \mathbf{z}'_{i,T_i+L}\mathbf{a}_{i,\text{BLUP}} + \rho^L e_{iT_i,\text{BLUP}}.$$

That is, the L-step forecast equals the estimate of the conditional mean plus a correction factor of ρ^L times the most recent BLUP residual, $e_{iT_i,\text{BLUP}}$. This result was originally given by Goldberger (1962E) in the context of ordinary regression without random effects (that is, assuming $\mathbf{D} = \mathbf{0}$).

4.5 Example: Forecasting Wisconsin Lottery Sales

In this section, we forecast the sale of state lottery tickets from 50 postal (ZIP) codes in Wisconsin. Lottery sales are an important component of state revenues. Accurate forecasting helps in the budget-planning process. A model is useful in assessing the important determinants of lottery sales, and understanding the determinants of lottery sales is useful for improving the design of the lottery sales system. Additional details of this study are in Frees and Miller (2003O).

4.5.1 Sources and Characteristics of Data

State of Wisconsin lottery administrators provided weekly lottery sales data. We consider online lottery tickets that are sold by selected retail establishments in Wisconsin. These tickets are generally priced at \$1.00, so the number of tickets sold equals the lottery revenue. We analyze lottery sales (OLSALES) over a forty-week period (April 1998 through January 1999) from fifty randomly selected ZIP codes within the state of Wisconsin. We also consider the number of retailers within a ZIP code for each time (NRETAIL).

A budding body of literature, such as the work of Ashley, Liu, and Chang (1999O), suggests variables that influence lottery sales. Table 4.1 lists economic and demographic characteristics that we consider in this analysis. Much of the empirical literature on lotteries is based on annual data that examine the state as the unit of analysis. In contrast, we examine much finer economic units – the ZIP-code level – and weekly lottery sales. The economic and demographic characteristics were abstracted from the U.S. census. These variables summarize characteristics of individuals within ZIP codes at a single point in time and thus are not time-varying.

Table 4.2 summarizes the economic and demographic characteristics of fifty Wisconsin ZIP codes. For example, for the population variable (POPULATN), we see that the smallest ZIP code contained 280 people whereas the largest contained 39,098. The average, over fifty ZIP codes, was 9,311.04. Table 4.2

Table 4.1. *Lottery, economic, and demographic characteristics of fifty Wisconsin ZIP codes*

Characteristic	Description
Lottery	
OLSALES	Online lottery sales to individual consumers
NRETAIL	Number of listed retailers
Economic and demographic	
PERPERHH	Persons per household
MEDSCHYR	Median years of schooling
MEDHVL	Median home value in $1,000s for owner-occupied homes
PRCRENT	Percent of housing that is renter occupied
PRC55P	Percent of population that is 55 or older
HHMEDAGE	Household median age
MEDINC	Estimated median household income, in $1,000s
POPULATN	Population, in thousands

Source: Frees and Miller (2003O).

Table 4.2. *Summary statistics of lottery, economic, and demographic characteristics of fifty Wisconsin ZIP codes*

Variable	Mean	Median	Standard deviation	Minimum	Maximum
Average OLSALES	6,494.83	2,426.41	8,103.01	189	33,181
Average NRETAIL	11.94	6.36	13.29	1	68.625
PERPERHH	2.71	2.7	0.21	2.2	3.2
MEDSCHYR	12.70	12.6	0.55	12.2	15.9
MEDHVL	57.09	53.90	18.37	34.50	120
PRCRENT	24.68	24	9.34	6	62
PRC55P	39.70	40	7.51	25	56
HHMEDAGE	48.76	48	4.14	41	59
MEDINC	45.12	43.10	9.78	27.90	70.70
POPULATN	9.311	4.405	11.098	0.280	39.098

also summarizes average online sales and average number of retailers. Here, these are averages over forty weeks. To illustrate, we see that the forty-week average of online sales was as low as $189 and as high as $33,181.

It is possible to examine cross-sectional relationships between sales and economic demographic characteristics. For example, Figure 4.2 shows a positive relationship between average online sales and population. Further, the ZIP code

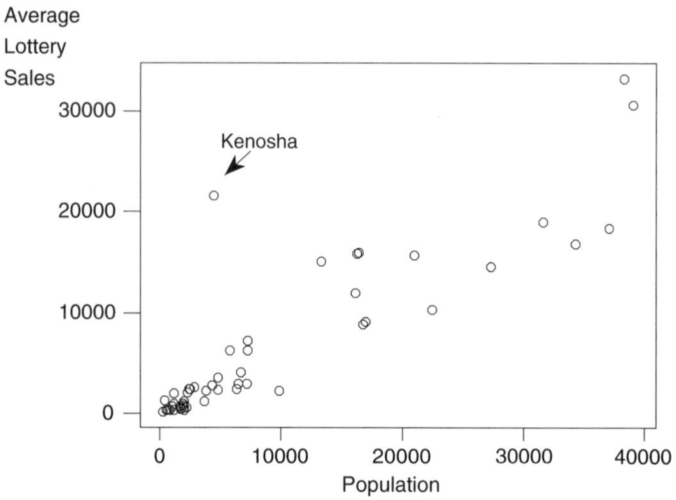

Figure 4.2. Scatter plot of average lottery sales versus population size. Note that sales for Kenosha are unusually large for its population size.

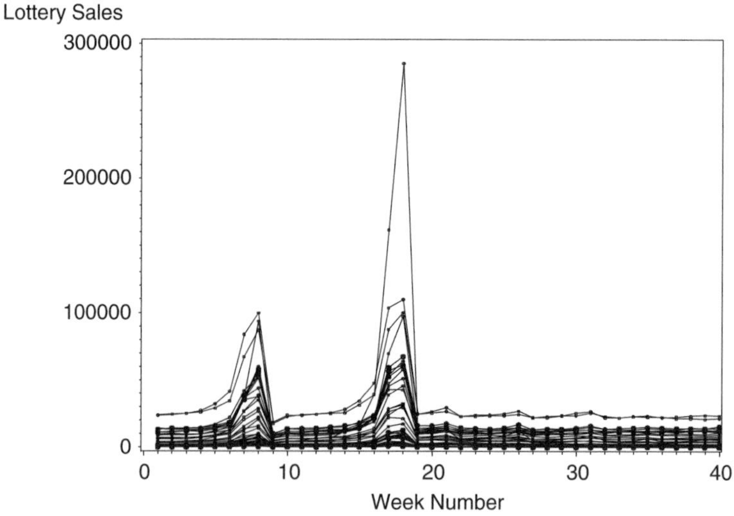

Figure 4.3. Multiple time-series plot of lottery sales. Note that sales at and around weeks 8 and 18 are unusually large owing to large PowerBall jackpots.

corresponding to the city of Kenosha, Wisconsin, has unusually large average sales for its population size.

However, cross-sectional relationships, such as correlations and plots similar to Figure 4.2, hide dynamic patterns of sales. Figure 4.3 presents a multiple

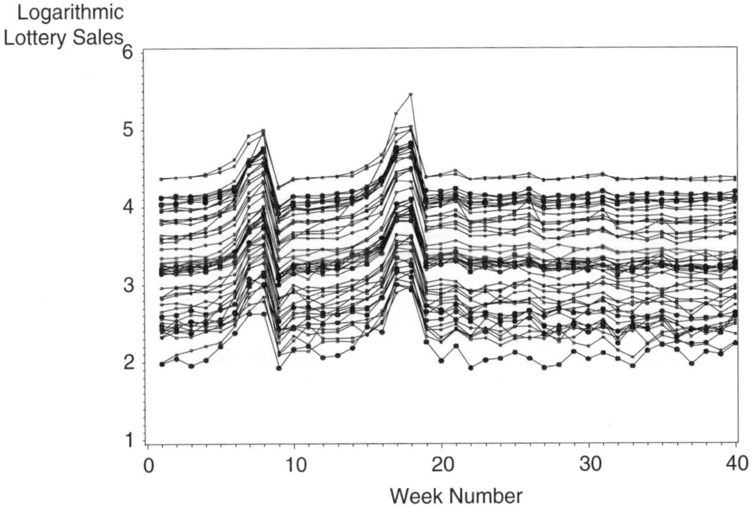

Figure 4.4. Multiple time-series plot of logarithmic (base 10) lottery sales.

time-series plot of (weekly) sales over time. Here, each line traces the sales patterns for a particular ZIP code. This figure shows the dramatic increase in sales for most ZIP codes, at approximately weeks 8 and 18. For both time points, the jackpot prize of one online game, PowerBall, grew to an amount in excess of $100 million. Interest in lotteries, and sales, increases dramatically when jackpot prizes reach such large amounts.

Figure 4.4 shows the same information as in Figure 4.3 but on a common (base 10) logarithmic scale. Here, we still see the effects of the PowerBall jackpots on sales. However, Figure 4.4 suggests a dynamic pattern that is common to all ZIP codes. Specifically, logarithmic sales for each ZIP code are relatively stable with the same approximate level of variability. Further, logarithmic sales for each ZIP code peak at the same time, corresponding to large PowerBall jackpots.

Another form of the response variable to consider is the proportional, or percentage, change. Specifically, define the percentage change to be

$$pchange_{it} = 100\left(\frac{sales_{it}}{sales_{i,t-1}} - 1\right). \tag{4.16}$$

A multiple times-series plot of the percentage change (not displayed here) shows autocorrelated serial patterns. We consider models of this transformed series in the following subsection on model selection.

Table 4.3. *Lottery model coefficient estimates based on in-sample data of $n = 50$ ZIP codes and $T = 35$ weeks*[a]

	Pooled cross-sectional model		Error-components model		Error-components model with $AR(1)$ term	
Variable	Parameter estimate	t-statistic	Parameter estimate	t-statistic	Parameter estimate	t-statistic
Intercept	13.821	10.32	18.096	2.47	15.255	2.18
PERPERHH	−1.085	−6.77	−1.287	−1.45	−1.149	−1.36
MEDSCHYR	−0.821	−11.90	−1.078	−2.87	−0.911	−2.53
MEDHVL	0.014	5.19	0.007	0.50	0.011	0.81
PRCRENT	0.032	8.51	0.026	1.27	0.030	1.53
PRC55P	−0.070	−5.19	−0.073	−0.98	−0.071	−1.01
HHMEDAGE	0.118	5.64	0.119	1.02	0.120	1.09
MEDINC	0.043	8.18	0.046	1.55	0.044	1.58
POPULATN	0.057	9.41	0.121	4.43	0.080	2.73
NRETAIL	0.021	5.22	−0.027	−1.56	0.004	0.20
Var α (σ_α^2)	—	—	0.607	—	0.528	—
Var ε (σ^2)	0.700	—	0.263	—	0.279	—
$AR(1)$ corr (ρ)	—	—	—	—	0.555	25.88
AIC	4353.25		2862.74		2270.97	

[a] The response is natural logarithmic sales.

4.5.2 In-Sample Model Specification

We now consider the specification of a model, a necessary component prior to forecasting. We decompose model specification criteria into two components, in-sample and out-of-sample criteria. To this end, we partition our data into two subsamples; we use the first thirty-five weeks to develop alternative fitted models and use the last five weeks to "predict" our held-out sample. The choice of five weeks for the out-of-sample validation is somewhat arbitrary; it was made with the rationale that lottery officials consider it reasonable to try to predict five weeks of sales based on thirty-five weeks of historical sales data.

Our first forecasting model is the pooled cross-sectional model. The model fits the data well; the coefficient of determination is $R^2 = 69.6\%$. The estimated regression coefficients appear in Table 4.3. From the corresponding t-statistics, we see that each variable is statistically significant.

Our second forecasting model is an error-components model. Table 4.3 provides parameter estimates and the corresponding t-statistics, as well as estimates of the variance components σ_α^2 and σ^2. As we have seen in other examples, allowing intercepts to vary by subject can result in regression coefficients for other variables becoming statistically insignificant.

When comparing this model to the pooled cross-sectional model, we may use the Lagrange multiplier test described in Section 3.1. The test statistic turns out to be $TS = 11{,}395.5$, indicating that the error-components model is strongly preferred to the pooled cross-sectional model. Another piece of evidence is *Akaike's Information Criterion (AIC)*. This criterion is defined as

$$AIC = -2 \times \ln(\textit{maximized likelihood}) + 2 \times (\textit{number of model parameters}).$$

The smaller this criterion is, the more preferred is the model. Appendix C.9 describes this criterion in further detail. Table 4.3 shows again that the error-components model is preferred compared to the pooled cross-sectional model based on the smaller value of the AIC statistic.

To assess further the adequacy of the error-components model, residuals from the fitted model were calculated. Several diagnostic tests and graphs were made using these residuals to improve the model fit. Figure 4.5 represents one such diagnostic graph, a plot of residuals versus lagged residuals. This figure shows a strong relationship between residuals, and lagged residuals, which we can represent using an autocorrelation structure for the error terms. To accommodate this pattern, we also consider an error-components model with an $AR(1)$ term; the fitted model appears in Table 4.3.

Figure 4.5 also shows a strong pattern of clustering, corresponding to weeks with large PowerBall jackpots. A variable that captures information about the

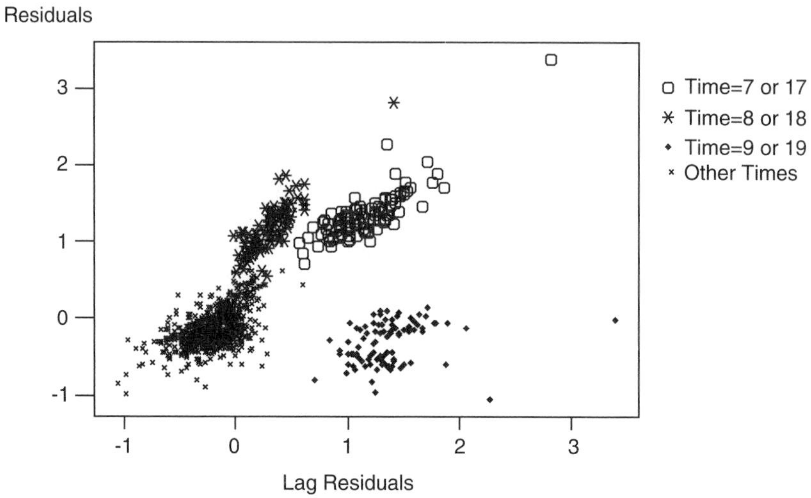

Figure 4.5. Scatter plot of residuals versus lagged residuals from an error-components model. The plot shows the strong autocorrelation tendency among residuals. Different plotting symbols indicate the clustering according to time. The four time symbols correspond to immediate prior to a jackpot (time = 7 or 17), the week during a jackpot (time = 8 or 18), the week following a jackpot (time = 9 or 19), and other weeks (time = 1–6, 10–16 or 20–35).

size of PowerBall jackpots would help in developing a model of lottery sales. However, for forecasting purposes, we require one or more variables that *anticipates* large PowerBall jackpots. That is, because the size of PowerBall jackpots is not known in advance, variables that proxy the event of large jackpots are not suitable for forecasting models. These variables could be developed through a separate forecasting model of PowerBall jackpots.

Other types of random-effects models for forecasting lottery sales could also be considered. To illustrate, we also fit a more parsimonious version of the $AR(1)$ version of the error-components model; specifically, we refit this model, deleting those variables with insignificant t-statistics. This fitted model did not perform substantially better in terms of overall model fit statistics such as AIC. Next, we explore alternative transforms of the response when examining a held-out sample.

4.5.3 Out-of-Sample Model Specification

We now compare the ability of several competing models to forecast values outside of the sample used for model parameter estimation. As in Section 4.5.2, we use the first thirty-five weeks of data to estimate model parameters. The remaining five weeks are used to assess the validity of model forecasts. For each model, we compute forecasts of lottery sales for weeks 36 through 40, by ZIP code level, based on the first thirty-five weeks. Denote these forecast values as $\mathrm{OL\hat{S}ALES}_{i,35+L}$, for $L = 1$ to 5. We summarize the accuracy of the forecasts through two statistics, the *mean absolute error*

$$MAE = \frac{1}{5n} \sum_{i=1}^{n} \sum_{L=1}^{5} \left| \mathrm{OL\hat{S}ALES}_{i,35+L} - \mathrm{OLSALES}_{i,35+L} \right| \qquad (4.17)$$

and the *mean absolute percentage error*

$$MAPE = \frac{100}{5n} \sum_{i=1}^{n} \sum_{L=1}^{5} \frac{\left| \mathrm{OL\hat{S}ALES}_{i,35+L} - \mathrm{OLSALES}_{i,35+L} \right|}{\mathrm{OLSALES}_{i,35+L}}. \qquad (4.18)$$

The several competing models include the three models of logarithmic sales summarized in Table 4.3. Because the autocorrelation term appears to be highly statistically significant in Table 4.3, we also fit a pooled cross-sectional model with an $AR(1)$ term. Further, we fit two modifications of the error-components model with the $AR(1)$ term. In the first case we use lottery sales as the response (not the logarithmic version) and in the second case we use percentage change of lottery sales, defined in Equation (4.16), as the response. Finally, the seventh model we consider is a basic fixed-effects model,

$$y_{it} = \alpha_i + \varepsilon_{it},$$

with an $AR(1)$ error structure. Recall that, for the fixed-effects models, the term α_i is treated as a fixed, not random, parameter. Because this parameter is time-invariant, it is not possible to include our time-invariant demographic and economic characteristics as part of the fixed-effects model.

Table 4.4 presents the model forecast criteria in Equations (4.17) and (4.18) for each of these seven models. We first note that Table 4.4 reconfirms the point that the $AR(1)$ term improves each model. Specifically, for both the pooled cross-sectional and the error-components model, the version with an $AR(1)$ term outperforms the analogous model without this term. Table 4.4 also shows that the error-components model dominates the pooled cross-sectional model. This was also anticipated by our pooling test, an in-sample test procedure.

Table 4.4 confirms that the error-components model with an $AR(1)$ term with logarithmic sales as the response is the preferred model, based on either the MAE or $MAPE$ criterion. The next best model was the corresponding fixed-effects model. We note that the models with sales as the response outperformed

Table 4.4. *Out-of-sample forecast comparison of six alternative models*

		Model forecast criteria	
Model	Model response	*MAE*	*MAPE*
Pooled cross-sectional model	logarithmic sales	3,012.68	83.41
Pooled cross-sectional model with $AR(1)$ term	logarithmic sales	680.64	21.19
Error-components model	logarithmic sales	1,318.05	33.85
Error-components model with $AR(1)$ term	logarithmic sales	571.14	18.79
Error-components model with $AR(1)$ term	sales	1,409.61	140.25
Error-components model with $AR(1)$ term	percentage change	1,557.82	48.70
Fixed-effects model with $AR(1)$ term	logarithmic sales	584.55	19.07

the model with percentage change as the response based on the *MAE* criterion, although the reverse is true based on the *MAPE* criterion.

4.5.4 Forecasts

We now forecast using the model that provides the best fit to the data, the error-components model with an $AR(1)$ term. The forecasts and variance of forecast errors for this model are special cases of the results for the linear mixed-effects model, given in Equations (4.14) and (4.15), respectively. Forecast intervals are calculated, using a normal curve approximation, as the point forecast plus or minus 1.96 times the square root of the estimated variance of the forecast error.

Figure 4.6 displays the BLUP forecasts and forecast intervals. Here, we use $T = 40$ weeks of data to estimate parameters and provide forecasts for $L = 5$ weeks. Calculation of the parameter estimates, point forecasts, and forecast intervals were done using logarithmic sales as the response. Then, point forecasts and forecast intervals were converted to dollars to display the ultimate impact of the model forecasting strategy.

Figure 4.6 shows the forecasts and forecast intervals for two selected postal codes. The lower forecast represents a postal code from Dane County, whereas the upper represents a postal code from Milwaukee. For each postal code, the middle line represents the point forecast and the upper and lower lines represent the bounds on a 95% forecast interval. Compared to the Dane County code, the

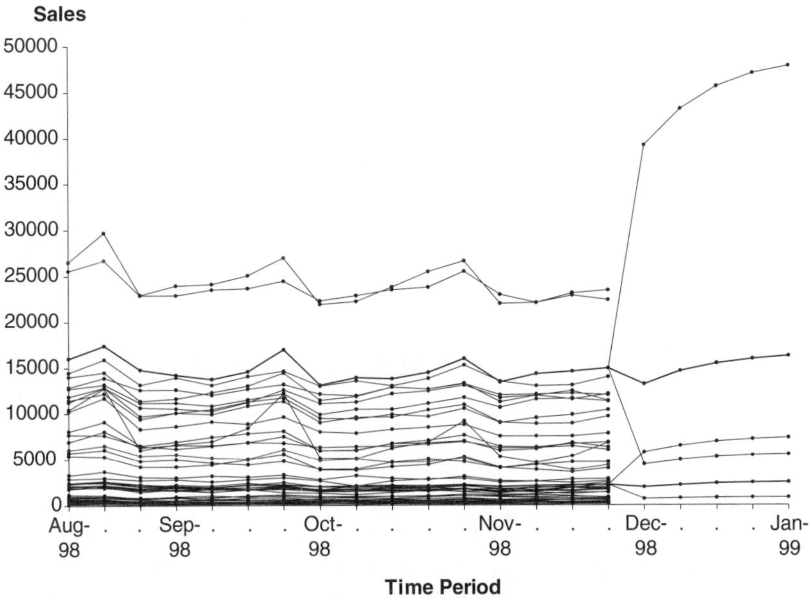

Figure 4.6. Forecast intervals for two selected postal codes. For each postal code, the middle line corresponds to point forecasts for five weeks. The upper and lower lines correspond to endpoints of 95% prediction intervals.

Milwaukee postal code has higher forecast sales. Thus, although standard errors on a logarithmic scale are about the same as those for Dane County, this higher point forecast leads to a larger interval when rescaled to dollars.

4.6 Bayesian Inference

With Bayesian statistical models, one views both the model parameters and the data as random variables. In this section, we use a specific type of Bayesian model, the *normal linear hierarchical model* discussed by, for example, Gelman et al. (2004G). As with the two-stage sampling scheme described in Section 3.3.1, the normal hierarchical linear model is one that is specified in stages. Specifically, we consider the following two-level hierarchy:

1. Given the parameters $\boldsymbol{\beta}$ and $\boldsymbol{\alpha}$, the response model is $\mathbf{y} = \mathbf{Z}\boldsymbol{\alpha} + \mathbf{X}\boldsymbol{\beta} + \boldsymbol{\epsilon}$. This level is an ordinary (fixed) linear model that was introduced in Section 3.2.2. Specifically, we assume that the responses $\mathbf{y}$ conditional on $\boldsymbol{\alpha}$ and $\boldsymbol{\beta}$ are normally distributed and that $\mathrm{E}(\mathbf{y} \mid \boldsymbol{\alpha}, \boldsymbol{\beta}) = \mathbf{Z}\boldsymbol{\alpha} + \mathbf{X}\boldsymbol{\beta}$ and $\mathrm{Var}(\mathbf{y} \mid \boldsymbol{\alpha}, \boldsymbol{\beta}) = \mathbf{R}$.

2. Assume that $\boldsymbol{\alpha}$ is distributed normally with mean $\boldsymbol{\mu}_\alpha$ and variance $\mathbf{D}$ and that $\boldsymbol{\beta}$ is distributed normally with mean $\boldsymbol{\mu}_\beta$ and variance $\boldsymbol{\Sigma}_\beta$, each independent of the other.

The technical differences between the mixed linear model and the normal hierarchical linear model are as follows:

- In the mixed linear model, $\boldsymbol{\beta}$ is an unknown, fixed parameter, whereas in the normal hierarchical linear model, $\boldsymbol{\beta}$ is a random vector, and
- the mixed linear model is distribution free, whereas distributional assumptions are made in each stage of the normal hierarchical linear model.

Moreover, there are important differences in interpretation. To illustrate, suppose that $\boldsymbol{\beta} = \mathbf{0}$ with probability one. In the classic non-Bayesian, also known as the *frequentist*, interpretation, we think of the distribution of $\{\boldsymbol{\alpha}\}$ as representing the likelihood of drawing a realization of $\boldsymbol{\alpha}_i$. The likelihood interpretation is most suitable when we have a population of firms or people and each realization is a draw from that population. In contrast, in the Bayesian case, one interprets the distribution of $\{\boldsymbol{\alpha}\}$ as representing the knowledge that one has of this parameter. This distribution may be subjective and allows the analyst a formal mechanism to inject his or her assessments into the model. In this sense the frequentist interpretation may be regarded as a special case of the Bayesian framework.

The joint distribution of $(\boldsymbol{\alpha}', \boldsymbol{\beta}')'$ is known as the *prior* distribution. In summary, the joint distribution of $(\boldsymbol{\alpha}', \boldsymbol{\beta}', \mathbf{y}')'$ is

$$\begin{pmatrix} \boldsymbol{\alpha} \\ \boldsymbol{\beta} \\ \mathbf{y} \end{pmatrix} \sim N \left(\begin{pmatrix} \boldsymbol{\mu}_\alpha \\ \boldsymbol{\mu}_\beta \\ \mathbf{Z}\boldsymbol{\mu}_\alpha + \mathbf{X}\boldsymbol{\mu}_\beta \end{pmatrix}, \begin{pmatrix} \mathbf{D} & \mathbf{0} & \mathbf{DZ}' \\ \mathbf{0} & \boldsymbol{\Sigma}_\beta & \boldsymbol{\Sigma}_\beta \mathbf{X}' \\ \mathbf{ZD} & \mathbf{X}\boldsymbol{\Sigma}_\beta & \mathbf{V} + \mathbf{X}\boldsymbol{\Sigma}_\beta\mathbf{X}' \end{pmatrix} \right), \tag{4.19}$$

where $\mathbf{V} = \mathbf{R} + \mathbf{ZDZ}'$.

The distribution of parameters given the data is known as the *posterior* distribution (see Appendix 9A). To calculate this conditional distribution, we use standard results from multivariate analysis (see Appendix B). Specifically, the posterior distribution of $(\boldsymbol{\alpha}', \boldsymbol{\beta}')'$ given $\mathbf{y}$ is normal. The conditional moments are

$$E\begin{pmatrix} \boldsymbol{\alpha} \\ \boldsymbol{\beta} \end{pmatrix} \Bigg| \mathbf{y} = \begin{pmatrix} \boldsymbol{\mu}_\alpha + \mathbf{DZ}'(\mathbf{V} + \mathbf{X}\boldsymbol{\Sigma}_\beta\mathbf{X}')^{-1}(\mathbf{y} - \mathbf{Z}\boldsymbol{\mu}_\alpha - \mathbf{X}\boldsymbol{\mu}_\beta) \\ \boldsymbol{\mu}_\beta + \boldsymbol{\Sigma}_\beta\mathbf{X}'(\mathbf{V} + \mathbf{X}\boldsymbol{\Sigma}_\beta\mathbf{X}')^{-1}(\mathbf{y} - \mathbf{Z}\boldsymbol{\mu}_\alpha - \mathbf{X}\boldsymbol{\mu}_\beta) \end{pmatrix} \tag{4.20}$$

and

$$\mathrm{Var}\begin{pmatrix}\boldsymbol{\alpha}\\ \boldsymbol{\beta}\end{pmatrix}\Bigg|\,\mathbf{y} = \begin{pmatrix}\mathbf{D} & \mathbf{0}\\ \mathbf{0} & \boldsymbol{\Sigma}_\beta\end{pmatrix} - \begin{pmatrix}\mathbf{DZ}'\\ \boldsymbol{\Sigma}_\beta\mathbf{X}'\end{pmatrix}(\mathbf{V}+\mathbf{X}\boldsymbol{\Sigma}_\beta\mathbf{X}')^{-1}(\mathbf{ZD} \quad \mathbf{X}\boldsymbol{\Sigma}_\beta). \quad (4.21)$$

Up to this point, the treatment of parameters $\boldsymbol{\alpha}$ and $\boldsymbol{\beta}$ has been symmetric. In longitudinal data applications, one typically has more information about the global parameters $\boldsymbol{\beta}$ than subject-specific parameters $\boldsymbol{\alpha}$. To see how the posterior distribution changes depending on the amount of information available, we consider two extreme cases.

First, consider the case $\boldsymbol{\Sigma}_\beta = \mathbf{0}$, so that $\boldsymbol{\beta} = \boldsymbol{\mu}_\beta$ with probability one. Intuitively, this means that $\boldsymbol{\beta}$ is precisely known, generally from collateral information. Then, from Equations (4.20) and (4.21), we have

$$\mathrm{E}(\boldsymbol{\alpha} \mid \mathbf{y}) = \boldsymbol{\mu}_\alpha + \mathbf{DZ}'\mathbf{V}^{-1}(\mathbf{y} - \mathbf{Z}\boldsymbol{\mu}_\alpha - \mathbf{X}\boldsymbol{\beta})$$

and

$$\mathrm{Var}(\boldsymbol{\alpha} \mid \mathbf{y}) = \mathbf{D} - \mathbf{DZ}'\mathbf{V}^{-1}\mathbf{ZD}.$$

Assuming that $\boldsymbol{\mu}_\alpha = \mathbf{0}$, the BLUE of $\mathrm{E}(\boldsymbol{\alpha}|\mathbf{y})$ is

$$\mathbf{a}_{\mathrm{BLUP}} = \mathbf{DZ}'\mathbf{V}^{-1}(\mathbf{y} - \mathbf{X}\,\mathbf{b}_{\mathrm{GLS}}).$$

Recall from Equation (4.11) that $\mathbf{a}_{\mathrm{BLUP}}$ is also the BLUP in the frequentist (non-Bayesian) model framework.

Second, consider the case where $\boldsymbol{\Sigma}_\beta^{-1} = \mathbf{0}$. In this case, prior information about the parameter $\boldsymbol{\beta}$ is vague; this is known as using a *diffuse* prior. To analyze the impact of this assumption, use Equation (A.4) of Appendix A.5 to get

$$\begin{aligned}(\mathbf{V}+\mathbf{X}\boldsymbol{\Sigma}_\beta\mathbf{X}')^{-1} &= \mathbf{V}^{-1} - \mathbf{V}^{-1}\mathbf{X}\big(\mathbf{X}'\mathbf{V}^{-1}\mathbf{X} + \boldsymbol{\Sigma}_\beta^{-1}\big)^{-1}\mathbf{X}'\mathbf{V}^{-1}\\ &\to \mathbf{V}^{-1} - \mathbf{V}^{-1}\mathbf{X}(\mathbf{X}'\mathbf{V}^{-1}\mathbf{X})^{-1}\mathbf{X}'\mathbf{V}^{-1} = \mathbf{Q}_{\mathrm{V}},\end{aligned}$$

as $\boldsymbol{\Sigma}_\beta^{-1} \to \mathbf{0}$. Note that $\mathbf{Q}_{\mathrm{V}}\mathbf{X} = \mathbf{0}$. Thus, with $\boldsymbol{\Sigma}_\beta^{-1} = \mathbf{0}$ and $\boldsymbol{\mu}_\alpha = \mathbf{0}$ we have $\boldsymbol{\alpha} \mid \mathbf{y} \sim \mathrm{N}$ with mean

$$\mathrm{E}(\boldsymbol{\alpha} \mid \mathbf{y}) = \mathbf{DZ}'\,\mathbf{Q}_{\mathrm{V}}\mathbf{y}$$

and variance

$$\mathrm{Var}(\boldsymbol{\alpha} \mid \mathbf{y}) = \mathbf{D} - \mathbf{DZ}'\mathbf{Q}_{\mathrm{V}}\,\mathbf{ZD}.$$

This summarizes the posterior distribution of $\boldsymbol{\alpha}$ given $\mathbf{y}$. Interestingly, from the expression for $\mathbf{Q}_V$, we have

$$\begin{aligned}\mathrm{E}(\alpha \mid \mathbf{y}) &= \mathbf{DZ}'(\mathbf{V}^{-1} - \mathbf{V}^{-1}\mathbf{X}(\mathbf{X}'\mathbf{V}^{-1}\mathbf{X})^{-1}\mathbf{X}'\mathbf{V}^{-1})\mathbf{y} \\ &= \mathbf{DZ}'\mathbf{V}^{-1}\mathbf{y} - \mathbf{DZ}'\mathbf{V}^{-1}\mathbf{X}\,\mathbf{b}_{\mathrm{GLS}} \\ &= \mathbf{a}_{\mathrm{BLUP}}.\end{aligned}$$

Similarly, one can check that $\mathrm{E}(\boldsymbol{\beta} \mid \mathbf{y}) \rightarrow \mathbf{b}_{\mathrm{GLS}}$ as $\boldsymbol{\Sigma}_{\beta}^{-1} \rightarrow \mathbf{0}$.

Thus, in both extreme cases, we arrive at the statistic $\mathbf{a}_{\mathrm{BLUP}}$ as a predictor of $\boldsymbol{\alpha}$. This analysis assumes $\mathbf{D}$ and $\mathbf{R}$ are matrices of fixed parameters. It is also possible to assume distributions for these parameters; typically, independent Wishart distributions are used for $\mathbf{D}^{-1}$ and $\mathbf{R}^{-1}$ as these are conjugate priors. (Appendix 9A introduces conjugate priors.) Alternatively, one can estimate $\mathbf{D}$ and $\mathbf{R}$ using methods described in Section 3.5. The general strategy of substituting point estimates for certain parameters in a posterior distribution is called *empirical Bayes estimation.*

To examine intermediate cases, we look to the following special case. Generalizations may be found in Luo, Young, and Frees (2001O).

Special Case: One-Way Random-Effects ANOVA Model We return to the model considered in Section 4.2 and, for simplicity, assume balanced data so that $T_i = T$. The goal is to determine the posterior distributions of the parameters. For illustrative purposes, we derive the posterior means and leave the derivation of posterior variances as an exercise for the reader. Thus, with Equation (4.1), the model is

$$y_{it} = \beta + \alpha_i + \varepsilon_{it},$$

where we use the random $\beta \sim N(\mu_\beta, \sigma_\beta^2)$ in lieu of the fixed mean μ. The prior distribution of α_i is independent with $\alpha_i \sim N(0, \sigma_\alpha^2)$.

Using Equation (4.20), we obtain the posterior mean of β,

$$\begin{aligned}\hat{\beta} = \mathrm{E}(\beta \mid \mathbf{y}) &= \mu_\beta + \boldsymbol{\Sigma}_\beta \mathbf{X}'(\mathbf{V} + \mathbf{X}\boldsymbol{\Sigma}_\beta\mathbf{X}')^{-1}(\mathbf{y} - \mathbf{X}\mu_\beta) \\ &= \left(\frac{1}{\sigma_\beta^2} + \frac{nT}{\sigma_\varepsilon^2 + T\sigma_\alpha^2}\right)^{-1}\left(\frac{nT}{\sigma_\varepsilon^2 + T\sigma_\alpha^2}\bar{y} + \frac{\mu_\beta}{\sigma_\beta^2}\right),\end{aligned}$$

after some algebra. Thus, $\hat{\beta}$ is a weighted average of the sample mean $\bar{y}$ and the prior mean μ_β. It is easy to see that $\hat{\beta}$ approaches the sample mean $\bar{y}$ as $\sigma_\beta^2 \rightarrow \infty$, that is, as prior information about β becomes "vague." Conversely, $\hat{\beta}$ approaches the prior mean μ_β as $\sigma_\beta^2 \rightarrow 0$, that is, as information about β becomes "precise."

Similarly, using Equation (4.20), the posterior mean of α_i is

$$\hat{\alpha}_i = \mathrm{E}(\alpha_i \mid \mathbf{y}) = \zeta((\bar{y}_i - \mu_\beta) - \zeta_\beta(\bar{y} - \mu_\beta))$$

where we recall that

$$\zeta = \frac{T\sigma_\alpha^2}{\sigma_\varepsilon^2 + T\sigma_\alpha^2}$$

and define

$$\zeta_\beta = \frac{nT\sigma_\beta^2}{\sigma_\varepsilon^2 + T\sigma_\alpha^2 + nT\sigma_\beta^2}.$$

Note that ζ_β measures the precision of knowledge about β. Specifically, we see that ζ_β approaches one as $\sigma_\beta^2 \to \infty$, and it approaches zero as $\sigma_\beta^2 \to 0$.

Combining these two results, we have that

$$\hat{\alpha}_i + \hat{\beta} = (1 - \zeta_\beta)((1 - \zeta)\mu_\beta + \zeta\bar{y}_i) + \zeta_\beta((1 - \zeta)\bar{y} + \zeta\bar{y}_i).$$

Thus, if our knowledge of the distribution of β is vague, then $\zeta_\beta = 1$ and the predictor reduces to the expression in Equation (4.4) (for balanced data). Conversely, if our knowledge of the distribution of β is precise, then $\zeta_\beta = 0$ and the predictor reduces to the expression given at the end of Section 4.2. With the Bayesian formulation, we may entertain situations in which knowledge is available although imprecise.

In summary, there are several advantages of the Bayesian approach. First, one can describe the entire distribution of parameters conditional on the data, such as through Equations (4.20) and (4.21). This allows one, for example, to provide probability statements regarding the likelihood of parameters. Second, this approach allows analysts to blend information known from other sources with the data in a coherent manner. In our development, we assumed that information may be known through the vector of $\boldsymbol{\beta}$ parameters, with their reliability controlled through the dispersion matrix $\boldsymbol{\Sigma}_\beta$. Values of $\boldsymbol{\Sigma}_\beta = \mathbf{0}$ indicate complete faith in values of $\boldsymbol{\mu}_\beta$, whereas values of $\boldsymbol{\Sigma}_\beta^{-1} = \mathbf{0}$ indicate complete reliance on the data in lieu of prior knowledge.

Third, the Bayesian approach provides a unified approach for estimating $(\boldsymbol{\alpha}, \boldsymbol{\beta})$. Chapter 3 on non-Bayesian methods required a separate section on variance components estimation. In contrast, in Bayesian methods, all parameters can be treated in a similar fashion. This is convenient for explaining results to consumers of the data analysis. Fourth, Bayesian analysis is particularly useful for forecasting future responses; we develop this aspect in Chapter 10.

4.7 Credibility Theory

Credibility is a technique for pricing insurance coverages that is widely used by health, group term life, and property and casualty actuaries. In the United States, the practical standards of application are described under the Actuarial Standard of Practice Number 25 published by the Actuarial Standards Board of the American Academy of Actuaries (http://www.actuary.org/). In addition, several insurance laws and regulation require the use of credibility.

The theory of credibility has been called a cornerstone of the field of actuarial science (Hickman and Heacox, 1999O). The basic idea is to use claims experience and additional information to develop a pricing formula through the relation

$$\textit{New Premium} = \zeta \times \textit{Claims Experience} + (1 - \zeta) \times \textit{Old Premium}, \quad (4.22)$$

where ζ, known as the "credibility factor," generally lies between zero and one. The case $\zeta = 1$ is known as "full credibility," where claims experience is used solely to determine the premium. The case $\zeta = 0$ can be thought of as "no credibility," where claims experience is ignored and external information is used as the sole basis for pricing.

Credibility has long found use in practice, with applications dating back to Mowbray (1914O). See Hickman and Heacox (1999O) and Venter (1996O) for historical accounts. The modern theory of credibility began with the work of Bühlmann (1967O), who showed how to express Equation (4.22) in what we now call a random-effects framework, thus removing the seemingly ad hoc nature of this procedure. Bühlmann expressed traditional credibility insurance prices as conditional expectations, where the conditioning is based on an unobserved risk type that he called a "structure variable."

Applications of credibility theory are considerably enhanced by accounting for known risk factors such as trends through (continuous) explanatory variables, different risk classes through categorical explanatory variables, and dynamic behavior through evolving distributions. These types of applications can be handled under the framework of mixed linear models; see Norberg (1986O) and Frees, Young, and Luo (1999O). This section shows that this class of models contains the standard credibility models as a special case.

By demonstrating that many important credibility models can be viewed in a longitudinal data framework, we restrict our consideration to certain types of credibility models. Specifically, the longitudinal data models accommodate only unobserved risks that are additive. Thus, we do not address models of nonlinear random effects, which have been investigated in the actuarial literature;

see, for example, Taylor (1977O) and Norberg (1980). Taylor (1977O) allowed insurance claims to be possibly infinite dimensional using Hilbert space theory and established credibility formulas in this general context. Norberg (1980O) considered the more concrete context, yet still general, of multivariate claims and established the relationship between credibility and statistical empirical Bayes estimation.

By expressing credibility rate-making applications in the framework of longitudinal data models, actuaries can realize several benefits:

- Longitudinal data models provide a wide variety of models from which to choose.
- Standard statistical software makes analyzing data relatively easy.
- Actuaries have another method for explaining the rate-making process.
- Actuaries can use graphical and diagnostic tools to select a model and assess its usefulness.

4.7.1 Credibility Theory Models

In this subsection, we demonstrate that commonly used credibility techniques are special cases of best linear unbiased prediction applied to the panel data model. For additional examples, see Frees, Young, and Luo (2001O).

Special Case: Basic Credibility Model of Bühlmann (1967O) Bühlmann (1967O) considered the one-way random-effects ANOVA model, which we now write as $y_{it} = \beta + \alpha_i + \varepsilon_{it}$ (see Section 4.2). If y_{it} represents the claims of the ith subject in period t, then β is the grand mean of the claims over the collection of subjects (policyholders, geographical regions, occupational classes, etc.), and α_i is the deviation of the ith subject's hypothetical mean from the overall mean β. Here, the hypothetical mean is the conditional expected value $\mathrm{E}(y_{it}|\alpha_i) = \beta + \alpha_i$. The disturbance term ε_{it} is the deviation of y_{it} from its hypothetical mean. One calls σ_α^2 the variance of the hypothetical means and σ^2 the process variance.

Special Case: Heteroscedastic Model of Bühlmann–Straub (1970O) Continue with the basic Bühlmann model and change only the variance–covariance matrix of the errors to $\mathbf{R}_i = \mathrm{Var}(\boldsymbol{\epsilon}_i) = \sigma^2 \text{ diagonal}(1/w_{i1}, \ldots, 1/w_{iT_i})$. By this change, we allow each observation to have a different exposure weight (Bühlmann and Straub, 1970O). For example, if a subject is a policyholder, then w_{it} measures the size of the ith policyholder's exposure during the tth period, possibly via payroll as for workers' compensation insurance.

Special Case: Regression Model of Hachemeister (1975O) Now assume a random-coefficients model, so that $\mathbf{x}_{it} = \mathbf{z}_{it}$. Then, with $\mathbf{R}$ as in the Bühlmann–Straub model, we have the regression model of Hachemeister (1975O). Hachemeister focused on the linear trend model for which $K = q = 2$, $\mathbf{x}_{it} = \mathbf{z}_{it} = (1 : t)'$.

Special Case: Nested Classification Model of Jewell (1975O) Suppose $y_{ijt} = \beta + \mu_i + \gamma_{ij} + \varepsilon_{ijt}$, a sum of uncorrelated components, in which β is the overall expected claims, μ_i is the deviation of the conditional expected claims of the ith sector from β, $i = 1, 2, \ldots, n$, γ_{ij} is the deviation of the conditional expected claims of the jth subject in the ith sector from the sector expectation of $\beta + \mu_i$, $j = 1, 2, \ldots, n_i$, and ε_{ijt} is the deviation of the observation y_{ijt} from $\beta + \mu_i + \gamma_{ij}$, $t = 1, 2, \ldots, T_{ij}$ (Jewell, 1975O). The conditional expected claims from the ith sector is $\mathrm{E}(y_{ijt} \mid \mu_i) = \beta + \mu_i$ and the conditional expected claims of subject j within the ith sector is $\mathrm{E}(y_{ijt} \mid \mu_i, \gamma_{ij}) = \beta + \mu_i + \gamma_{ij}$. If one were to apply this model to private passenger automobile insurance, for example, then the sector might be age of the insured and the subject geographical region of the insured. Note that one assumes with this model that the claims in region j for different ages are uncorrelated. If one believes this to be an unreasonable assumption, then a cross-classification model might be appropriate; see Dannenburg, Kaas, and Goovaerts (1996O). As an example for which this nested model might be more appropriate, one could let the sector be geographical region and the subject be the policyholder.

4.7.2 Credibility Rate-Making

In credibility rate-making, one is interested in predicting the expected claims, conditional on the random risk parameters of the ith subject, for time period $T_i + 1$. In our notation, the credibility rate-making problem is to "estimate" $\mathrm{E}(y_{i,T_i+1} \mid \boldsymbol{\alpha}_i) = \mathbf{x}'_{i,T_i+1}\boldsymbol{\beta} + \mathbf{z}'_{i,T_i+1}\boldsymbol{\alpha}_i$. From Section 4.4.2, the BLUP of claims is $\mathbf{x}'_{i,T_i+1}\mathbf{b}_{\mathrm{GLS}} + \mathbf{z}'_{i,T_i+1}\mathbf{a}_{i,\mathrm{BLUP}}$.

In Section 4.2, we saw how to predict claims for the Bühlmann model. We now illustrate this prediction for other familiar credibility models. A summary is in Table 4.5.

Special Case: Heteroscedastic Model of Bühlmann–Straub (1970O) (continued) In the Bühlmann–Straub case, we have $x_{it} = z_{it} = 1$, so that the predictor of claims is $b_{\mathrm{GLS}} + a_{i,\mathrm{BLUP}}$. Straightforward calculations, similar to the

Table 4.5. *Credibility factors and prediction of claims*

Notation	Credibility factors	Prediction of claims
Bühlmann		
$\bar{y}_i = \frac{1}{T_i}\sum_{t=1}^{T_i} y_{it}, \quad m_{\alpha,\mathrm{GLS}} = \frac{\sum_{i=1}^{n} \zeta_i \bar{y}_i}{\sum_{i=1}^{n} \zeta_i}$	$\zeta_i = \frac{T_i}{T_i + \sigma^2/\sigma_\alpha^2}$	For subject i, $(1-\zeta_i)m_{\alpha,\mathrm{GLS}} + \zeta_i \bar{y}_i$.
Bühlmann–Straub		
$\bar{y}_{i,w} = \frac{\sum_{t=1}^{T_i} w_{it} y_{it}}{\sum_{t=1}^{T_i} w_{it}}, \quad m_{\alpha,\mathrm{GLS}} = \frac{\sum_{i=1}^{n} \zeta_i \bar{y}_{i,w}}{\sum_{i=1}^{n} \zeta_i}$	$\zeta_i = \frac{\sum_{t=1}^{T_i} w_{it}}{\sum_{t=1}^{T_i} w_{it} + \sigma^2/\sigma_\alpha^2}$	For subject i, $(1-\zeta_i)m_{\alpha,\mathrm{GLS}} + \zeta_i \bar{y}_{i,w}$.
Hachemeister (linear trend)		
$\mathbf{W}_i = \begin{bmatrix} \sum_{t=1}^{T_i} w_{it} & \sum_{t=1}^{T_i} t w_{it} \\ \sum_{t=1}^{T_i} t w_{it} & \sum_{t=1}^{T_i} t^2 w_{it} \end{bmatrix}$	$\boldsymbol{\zeta}_i = \frac{\det(\mathbf{DW}_i)\mathbf{I}_2 + \sigma^2 \mathbf{DW}_i}{\det(\mathbf{DW}_i) + \sigma^2 \mathrm{trace}(\mathbf{DW}_i) + \sigma^4}$	For period $T_i + 1$, $(1 \quad T_i + 1)(\mathbf{b}_{\mathrm{GLS}} + \mathbf{a}_{i,\mathrm{BLUP}})$, where $\mathbf{b}_{\mathrm{GLS}} + \mathbf{a}_{i,\mathrm{BLUP}} = (1 - \boldsymbol{\zeta}_i)\mathbf{b}_{\mathrm{GLS}} + \boldsymbol{\zeta}_i \mathbf{W}_i^{-1} \begin{bmatrix} \sum_{t=1}^{T_i} w_{it} y_{it} \\ \sum_{t=1}^{T_i} t w_{it} y_{it} \end{bmatrix}$.
Jewell		
$A_i = \sum_{j=1}^{n_i} \frac{\sigma_\gamma^2 \sum_{t=1}^{T_{ij}} w_{ijt}}{\sigma_\gamma^2 \sum_{t=1}^{T_{ij}} w_{ijt} + \sigma^2}$	$\zeta_i = \frac{\sigma_\mu^2 A_i}{\sigma_\mu^2 A_i + \sigma_\gamma^2}$,	For sector i, $(1-\zeta_i)m_{\alpha,\mathrm{GLS}} + \zeta_i \bar{y}_{i,w}$. For subject j in sector i, $(1-\zeta_i)(1-\zeta_{ij})m_{\alpha,\mathrm{GLS}} + \zeta_i(1-\zeta_{ij})\bar{y}_{i,w} + \zeta_{ij}\bar{y}_{ij,w}$.
$\bar{y}_{ij,w} = \frac{\sum_{t=1}^{T_{ij}} w_{ijt} y_{ijt}}{\sum_{t=1}^{T_{ij}} w_{ijt}}$	$\zeta_{ij} = \frac{\sigma_\gamma^2 \sum_{t=1}^{T_{ij}} w_{ijt}}{\sigma_\gamma^2 \sum_{t=1}^{T_{ij}} w_{ijt} + \sigma^2}$	
$m_{\alpha,\mathrm{GLS}} = \frac{\sum_{i=1}^{n} \zeta_i \left(\frac{\sum_{j=1}^{n_i} \zeta_{ij} \bar{y}_{ij,w}}{\sum_{j=1}^{T_i} \zeta_{ij}} \right)}{\sum_{i=1}^{n} \zeta_i}$		

Bühlmann case, show that the predictor of claims is

$$(1-\zeta_i)m_{\alpha,\mathrm{GLS}}+\zeta_i\bar{y}_{i,w},$$

where now the credibility factor is

$$\zeta_i=\frac{\sum_{t=1}^{T_i}w_{it}}{\sum_{t=1}^{T_i}w_{it}+\sigma^2/\sigma_\alpha^2}$$

(see Table 4.5).

Special Case: Regression Model of Hachemeister (1975O) (continued) In the Hachemeister case, we have $\mathbf{x}_{it}=\mathbf{z}_{it}$. Define $\mathbf{b}_i=(\mathbf{X}_i'\mathbf{V}_i^{-1}\mathbf{X}_i)^{-1}\mathbf{X}_i'\mathbf{V}_i^{-1}\mathbf{y}_i$ to be the GLS estimator of $\boldsymbol{\alpha}_i+\boldsymbol{\beta}$ based only on the ith subject. In Exercise 4.3, we ask the reader to show that the BLUP (credibility estimator) of $\boldsymbol{\alpha}_i+\boldsymbol{\beta}$ is

$$\mathbf{a}_{i,\mathrm{BLUP}}+\mathbf{b}_{\mathrm{GLS}}=(\mathbf{I}-\boldsymbol{\zeta}_i)\mathbf{b}_{\mathrm{GLS}}+\boldsymbol{\zeta}_i\mathbf{b}_i,$$

in which $\boldsymbol{\zeta}_i=\mathbf{D}\mathbf{X}_i'\mathbf{V}_i^{-1}\mathbf{X}_i$ is the credibility factor. As in the Bühlmann case, again we see that the credibility estimator is a weighted average of a subject-specific statistic and a statistic that summarizes information from all subjects. This example is prominent in credibility theory because one can further express the GLS estimator of β as a weighted average of the $\mathbf{b}_i$ using the credibility factors as weights: $\mathbf{b}_{\mathrm{GLS}}=(\sum_{i=1}^n\boldsymbol{\zeta}_i)^{-1}\sum_{i=1}^n\boldsymbol{\zeta}_i\mathbf{b}_i$.

In Table 4.5, we show how to predict expected claims in the other examples that we considered in Section 4.7.1. In each case, the predicted claims for the ith subject is a weighted average of that subject's experience with the $\mathbf{b}_{\mathrm{GLS}}$, using the ith credibility factor as a weight.

Further Reading

For readers who would like more background in small-area estimation, please refer to Ghosh and Rao (1994S).

For readers who would like more background in credibility theory, please refer to Dannenburg, Kaas, and Goovaerts (1996O), Klugman, Panjer, and Willmot (1998O), and Venter (1996O). The Section 4.7 introduction to credibility theory does not include the important connection to Bayesian inference first pointed out by Bailey (1950O). See, for example, Klugman (1992O) and Pinquet (1997O). For connections with credibility and the Chapter 8 Kalman filter, see Klugman (1992O) and Ledolter, Klugman, and Lee (1991O). Bayesian inference is further described in Chapter 10.

Appendix 4A Linear Unbiased Prediction

4A.1 Minimum Mean-Square Predictor

Let c_1 be an arbitrary constant and let $\mathbf{c}_2$ be a vector of constants. For this choice of c_1 and $\mathbf{c}_2$, the mean-square error in using $c_1 + \mathbf{c}_2'\mathbf{y}$ to predict w is

$$\begin{aligned}\mathrm{MSE}(c_1, \mathbf{c}_2) &= \mathrm{E}(c_1 + c_2'\mathbf{y} - w)^2 \\ &= \mathrm{Var}(c_1 + c_2'\mathbf{y} - w) + \mathrm{E}(c_1 + c_2'\mathrm{E}\mathbf{y} - \mathrm{E}w)^2.\end{aligned}$$

Using $\mathrm{E}\mathbf{y} = \mathbf{X}\boldsymbol{\beta}$ and $\mathrm{E}w = \boldsymbol{\lambda}'\boldsymbol{\beta}$, we have

$$\frac{\partial}{\partial c_1}\mathrm{MSE}(c_1, \mathbf{c}_2) = \frac{\partial}{\partial c_1}(c_1 + (\mathbf{c}_2'\mathbf{X} - \boldsymbol{\lambda}')\boldsymbol{\beta})^2 = 2(c_1 + (\mathbf{c}_2'\mathbf{X} - \boldsymbol{\lambda}')\boldsymbol{\beta}).$$

Equating this to zero yields $c_1^* = c_1(\mathbf{c}_2) = (\boldsymbol{\lambda}' - \mathbf{c}_2'\mathbf{X})\boldsymbol{\beta}$. For this choice of c_1, we have

$$\begin{aligned}\mathrm{MSE}(c_1(\mathbf{c}_2), \mathbf{c}_2) &= \mathrm{E}(\mathbf{c}_2'(\mathbf{y} - \mathrm{E}\mathbf{y}) - (w - \mathrm{E}w))^2 \\ &= \mathrm{Var}(\mathbf{c}_2'\mathbf{y} - w) = \mathbf{c}_2'\mathbf{V}\mathbf{c}_2 + \sigma_w^2 - 2\mathrm{Cov}(w, \mathbf{y})\mathbf{c}_2.\end{aligned}$$

To find the best choice of $\mathbf{c}_2$, we have

$$\frac{\partial}{\partial \mathbf{c}_2}\mathrm{MSE}(c_1(\mathbf{c}_2), \mathbf{c}_2) = 2\mathbf{V}\mathbf{c}_2 - 2\mathrm{Cov}(w, \mathbf{y})'.$$

Setting this equal to zero yields $\mathbf{c}_2^* = \mathbf{V}^{-1}\mathrm{Cov}(w, \mathbf{y})'$. Thus, the minimum mean-square predictor is

$$c_1^* + \mathbf{c}_2^{*\prime}\mathbf{y} = (\boldsymbol{\lambda}' - \mathrm{Cov}(w, \mathbf{y})\mathbf{V}^{-1}\mathbf{X})\boldsymbol{\beta} + \mathrm{Cov}(w, \mathbf{y})\mathbf{V}^{-1}\mathbf{y}$$

as required.

4A.2 Best Linear Unbiased Predictor

To check that w_{BLUP} in Equation (4.7) is the best linear unbiased predictor, consider all other unbiased linear estimators of the form $w_{\mathrm{BLUP}} + \mathbf{c}'\mathbf{y}$, where $\mathbf{c}$ is a vector of constants. By the unbiasedness, we have that

$$\mathrm{E}\mathbf{c}'\mathbf{y} = \mathrm{E}w - \mathrm{E}w_{\mathrm{BLUP}} = 0.$$

Thus, $\mathbf{c}'\mathbf{y}$ is an unbiased estimator of 0. Following Harville (1976S), we require this of all possible distributions, so that a necessary and sufficient condition for $\mathrm{E}\mathbf{c}'\mathbf{y} = 0$ is $\mathbf{c}'\mathbf{X} = \mathbf{0}$.

We wish to minimize the mean-square prediction error over all choices of $\mathbf{c}$, so consider $\mathrm{E}(w_{\mathrm{BLUP}} + \mathbf{c}'\mathbf{y} - w)^2$. Now,

$$\begin{aligned}\mathrm{Cov}(w_{\mathrm{BLUP}} - w, \mathbf{c}'\mathbf{y}) &= \mathrm{Cov}(w_{\mathrm{BLUP}}, \mathbf{y})\mathbf{c} - \mathrm{Cov}(w, \mathbf{y})\mathbf{c} \\ &= \mathrm{Cov}(w, \mathbf{y})\mathbf{V}^{-1}\mathrm{Cov}(\mathbf{y}, \mathbf{y})\mathbf{c} + (\boldsymbol{\lambda}' - \mathrm{Cov}(w, \mathbf{y})\mathbf{V}^{-1}\mathbf{X}) \\ &\quad \times \mathrm{Cov}(\mathbf{b}_{\mathrm{GLS}}, \mathbf{y})\mathbf{c} - \mathrm{Cov}(w, \mathbf{y})\mathbf{c} \\ &= \mathrm{Cov}(w, \mathbf{y})\mathbf{c} + (\boldsymbol{\lambda}' - \mathrm{Cov}(w, \mathbf{y})\mathbf{V}^{-1}\mathbf{X}) \\ &\quad \times (\mathbf{X}'\mathbf{V}^{-1}\mathbf{X})^{-1}\mathbf{X}'\mathbf{V}^{-1}\mathrm{Cov}(\mathbf{y}, \mathbf{y})\mathbf{c} - \mathrm{Cov}(w, \mathbf{y})\mathbf{c} \\ &= (\boldsymbol{\lambda}' - \mathrm{Cov}(w, \mathbf{y})\mathbf{V}^{-1}\mathbf{X}))(\mathbf{X}'\mathbf{V}^{-1}\mathbf{X})^{-1}\mathbf{X}'\mathbf{c} \\ &= 0. \end{aligned} \tag{4A.1}$$

The last equality follows from $\mathbf{c}'\mathbf{X} = \mathbf{0}$. Thus, we have

$$\mathrm{E}(w_{\mathrm{BLUP}} + \mathbf{c}'\mathbf{y} - w)^2 = \mathrm{Var}(w_{\mathrm{BLUP}} - w) + \mathrm{Var}(\mathbf{c}'\mathbf{y}),$$

which can be minimized by choosing $\mathbf{c} = \mathbf{0}$.

4A.3 BLUP Variance

First note that

$$\mathrm{Cov}(\mathbf{y}, \mathrm{Cov}(w, \mathbf{y})\mathbf{V}^{-1}\mathbf{y} - w) = \mathrm{Cov}(\mathbf{y}, \mathbf{y})\mathbf{V}^{-1}\mathrm{Cov}(w, \mathbf{y})' - \mathrm{Cov}(\mathbf{y}, w) = \mathbf{0}.$$

Then, we have

$$\begin{aligned}\mathrm{Var}(w_{\mathrm{BLUP}} - w) &= \mathrm{Var}((\boldsymbol{\lambda}' - \mathrm{Cov}(w, \mathbf{y})\mathbf{V}^{-1}\mathbf{X})\mathbf{b}_{\mathrm{GLS}} + \mathrm{Cov}(w, \mathbf{y})\mathbf{V}^{-1}\mathbf{y} - w) \\ &= \mathrm{Var}((\boldsymbol{\lambda}' - \mathrm{Cov}(w, \mathbf{y})\mathbf{V}^{-1}\mathbf{X})\mathbf{b}_{\mathrm{GLS}}) \\ &\quad + \mathrm{Var}(\mathrm{Cov}(w, \mathbf{y})\mathbf{V}^{-1}\mathbf{y} - w).\end{aligned}$$

Also, we have

$$\begin{aligned}\mathrm{Var}(\mathrm{Cov}(w, \mathbf{y})\mathbf{V}^{-1}\mathbf{y} - w) &= \mathrm{Var}(\mathrm{Cov}(w, \mathbf{y})\mathbf{V}^{-1}\mathbf{y}) + \mathrm{Var}(w) \\ &\quad - 2\mathrm{Cov}(\mathrm{Cov}(w, \mathbf{y})\mathbf{V}^{-1}\mathbf{y}, w) \\ &= \mathrm{Cov}(w, \mathbf{y})\mathbf{V}^{-1}\mathrm{Cov}(w, \mathbf{y})' + \mathrm{Var}(w) \\ &\quad - 2\mathrm{Cov}(w, \mathbf{y})\mathbf{V}^{-1}\mathrm{Cov}(\mathbf{y}, w) \\ &= \sigma_w^2 - \mathrm{Cov}(w, \mathbf{y})\mathbf{V}^{-1}\mathrm{Cov}(w, \mathbf{y})'.\end{aligned}$$

Thus,

$$\begin{aligned}\mathrm{Var}(w_{\mathrm{BLUP}} - w) &= (\boldsymbol{\lambda}' - \mathrm{Cov}(w, \mathbf{y})\mathbf{V}^{-1}\mathbf{X})(\mathbf{X}'\mathbf{V}^{-1}\mathbf{X})^{-1} \\ &\quad \times (\boldsymbol{\lambda}' - \mathrm{Cov}(w, \mathbf{y})\mathbf{V}^{-1}\mathbf{X})' \\ &\quad - \mathrm{Cov}(w, \mathbf{y})\mathbf{V}^{-1}\mathrm{Cov}(w, \mathbf{y})' + \sigma_w^2,\end{aligned}$$

as in Equation (4.8).

From Equation (4A.1), we have $\mathrm{Cov}(w_{\mathrm{BLUP}} - w, w_{\mathrm{BLUP}}) = 0$. Thus,

$$\sigma_w^2 = \mathrm{Var}(w - w_{\mathrm{BLUP}} + w_{\mathrm{BLUP}}) = \mathrm{Var}(w - w_{\mathrm{BLUP}}) + \mathrm{Var}\, w_{\mathrm{BLUP}}.$$

With Equation (4.8), we have

$$\begin{aligned}\mathrm{Var}\, w_{\mathrm{BLUP}} = \sigma_w^2 - \mathrm{Var}(w - w_{\mathrm{BLUP}}) &= \mathrm{Cov}(w, \mathbf{y})\mathbf{V}^{-1}\mathrm{Cov}(w, \mathbf{y})' \\ &- (\boldsymbol{\lambda}' - \mathrm{Cov}(w, \mathbf{y})\mathbf{V}^{-1}\mathbf{X})(\mathbf{X}'\mathbf{V}^{-1}\mathbf{X})^{-1}(\boldsymbol{\lambda}' - \mathrm{Cov}(w, \mathbf{y})\mathbf{V}^{-1}\mathbf{X})',\end{aligned}$$

which is Equation (4.9).

Exercises and Extensions

Section 4.2

4.1 Shrinkage estimator Consider the Section 4.2 one-way random-effects model with $K = 1$, so that $y_{it} = \mu + \alpha_i + \varepsilon_{it}$.

a. Show that $\mathrm{E}(c_2(\bar{y} - \bar{y}_i) + \bar{y}_i - \alpha_i)^2$ is minimized over choices of c_2 at

$$c_2 = -\frac{\mathrm{Cov}(\bar{y} - \bar{y}_i, \bar{y}_i - \alpha_i)}{\mathrm{Var}(\bar{y} - \bar{y}_i)}.$$

b. Show that $\mathrm{Var}\, \bar{y}_i = \sigma_\alpha^2 + \sigma^2/T_i$, $\mathrm{Cov}(\bar{y}_i, \alpha_i) = \sigma_\alpha^2$, $\mathrm{Cov}(\bar{y}, \alpha_i) = (T_i/N)\sigma_\alpha^2$, $\mathrm{Cov}(\bar{y}, \bar{y}_i) = (\sigma^2 + T_i\sigma_\alpha^2)/N$, and $\mathrm{Var}\, \bar{y} = \sigma^2/N + (\sigma_\alpha^2/N^2)\sum_{j=1}^n T_j^2$.
c. Use part (b) to show that $\mathrm{Cov}(\bar{y} - \bar{y}_i, \bar{y}_i - \alpha_i) = \sigma^2(1/N - 1/T_i)$.
d. Use part (b) to show that

$$\mathrm{Var}(\bar{y} - \bar{y}_i) = \sigma^2\left(\frac{1}{T_i} - \frac{1}{N}\right) + \sigma_\alpha^2\left(1 - \frac{2T_i}{N} + \frac{1}{N^2}\sum_{j=1}^n T_j^2\right).$$

e. Use parts (a), (c), and (d) to show that the optimal choice of c_2 is

$$c_2 = \frac{\sigma^2}{\sigma^2 + T_i^*\sigma_\alpha^2}$$

and

$$c_1 = 1 - c_2 = \frac{T_i^*\sigma_\alpha^2}{\sigma^2 + T_i^*\sigma_\alpha^2},$$

where

$$T_i^* = \frac{1 - \frac{2T_i}{N} + \frac{1}{N^2}\sum_{j=1}^n T_j^2}{\frac{1}{T_i} - \frac{1}{N}}.$$

f. Use part (e) to show that, for balanced data with $T_i = T$, we have $T_i^* = T$.
g. Use part (e) to show that, as $N \to \infty$, we have $T_i^* = T_i$.

Section 4.4

4.2 BLUP predictor of random effects – error-components model Consider the Section 4.2 one-way random ANOVA model. Use Equation (4.11) to show that the BLUP of α_i is $a_{i,\mathrm{BLUP}} = \zeta_i(\bar{y}_i - \bar{\mathbf{x}}_i'\mathbf{b}_{\mathrm{GLS}})$.

4.3 Best linear unbiased prediction – random-coefficients model Consider the random-coefficients model with $K = q$ and $\mathbf{z}_{it} = \mathbf{x}_{it}$. Use Equation (4.12) to show that the BLUP of $\boldsymbol{\beta} + \boldsymbol{\alpha}_i$ is $w_{\mathrm{BLUP}} = \boldsymbol{\zeta}_i\mathbf{b}_i + (1 - \boldsymbol{\zeta}_i)\mathbf{b}_{\mathrm{GLS}}$, where $\mathbf{b}_i = (\mathbf{X}_i'\mathbf{V}_i^{-1}\mathbf{X}_i)^{-1}\mathbf{X}_i'\mathbf{V}_i^{-1}\mathbf{y}_i$ is the subject-specific GLS estimator and $\boldsymbol{\zeta}_i = \mathbf{D}\mathbf{X}_i'\mathbf{V}_i^{-1}\mathbf{X}_i$.

4.4 BLUP residuals Use Equations (4.11) and (4.13) to show $\mathbf{a}_{i,\mathrm{BLUP}} = \mathbf{D}\mathbf{Z}_i'\mathbf{R}_i^{-1}\,\mathbf{e}_{i,\mathrm{BLUP}}$.

4.5 BLUP subject-specific effects Use Equations (4.11) and (A.4) of Appendix A.5 to show

$$\mathbf{a}_{i,\mathrm{BLUP}} = \left(\mathbf{D}^{-1} + \mathbf{Z}_i'\mathbf{R}_i^{-1}\mathbf{Z}_i\right)^{-1}\mathbf{Z}_i'\mathbf{R}_i^{-1}(\mathbf{y}_i - \mathbf{X}_i\mathbf{b}_{\mathrm{GLS}}).$$

(For this alternative expression, one needs to only invert $\mathbf{R}_i$ and $q \times q$ matrices, not a $T_i \times T_i$ matrix.)

4.6 BLUP residuals Use Equation (4.11) to show that the BLUP residual in Equation (4.13a) can be expressed as Equation (4.13b), that is, as

$$\mathbf{e}_{i,\mathrm{BLUP}} = \mathbf{y}_i - (\mathbf{Z}_i\mathbf{a}_{i,\mathrm{BLUP}} + \mathbf{X}_i\mathbf{b}_{\mathrm{GLS}}).$$

4.7 Covariance Use Equation (4.11) to show that

$$\mathrm{Cov}(\mathbf{a}_{i,\mathrm{BLUP}}, \mathbf{b}_{\mathrm{GLS}}) = \mathbf{0}.$$

4.8 BLUP forecasts – random walk Assume a random-walk serial error correlation structure (see Section 8.2.2). Specifically, suppose that the subject-level dynamics are specified through $\varepsilon_{it} = \varepsilon_{i,t-1} + \eta_{it}$. Here, $\{\eta_{it}\}$ is an i.i.d. sequence with $\mathrm{Var}\,\eta_{it} = \sigma_\eta^2$ and assume that $\{\varepsilon_{i0}\}$ are unobserved constants. Then, we have $\mathrm{Var}\,\varepsilon_{it} = t\sigma_\eta^2$ and $\mathrm{Cov}\,(\varepsilon_{ir}, \varepsilon_{is}) = \mathrm{Var}\,\varepsilon_{ir} = r\sigma_\eta^2$, for $r < s$. This yields $\mathbf{R} = \sigma_\eta^2 \mathbf{R}_{\mathrm{RW}}$, where

$$\mathbf{R}_{\mathrm{RW}} = \begin{pmatrix} 1 & 1 & 1 & \ldots & 1 \\ 1 & 2 & 2 & \ldots & 2 \\ 1 & 2 & 3 & \ldots & 3 \\ \vdots & \vdots & \vdots & \ddots & \vdots \\ 1 & 2 & 3 & \ldots & T \end{pmatrix}.$$

a. Show that

$$\mathbf{R}_{\mathrm{RW}}^{-1} = \begin{pmatrix} 2 & -1 & 0 & \ldots & 0 & 0 \\ -1 & 2 & -1 & \ldots & 0 & 0 \\ 0 & -1 & 2 & \cdots & 0 & 0 \\ \vdots & \vdots & \vdots & \ddots & \vdots & \vdots \\ 0 & 0 & 0 & \ldots & 2 & -1 \\ 0 & 0 & 0 & \ldots & -1 & 1 \end{pmatrix}.$$

b. Show that

$$\mathrm{Cov}(\varepsilon_{i,T_i+L}, \boldsymbol{\epsilon}_i)' = \sigma_\eta^2 \begin{pmatrix} 1 & 2 & \ldots & T_i \end{pmatrix}.$$

c. Determine the one-step forecast; that is, determine the BLUP of y_{i,T_i+1}.
d. Determine the L-step forecast; that is, determine the BLUP of y_{i,T_i+L}.

4.9 BLUP mean-square errors – linear mixed-effects model Consider the linear mixed-effects model introduced in Section 4.4.

a. Use the general expression for the BLUP mean-square error to show that the mean-square error for the linear mixed-effects model can be expressed as

$$\begin{aligned} \mathrm{Var}(w_{\mathrm{BLUP}} - w) = {} & \left(\boldsymbol{\lambda}' - \sum_{i=1}^{n} \mathrm{Cov}(w, \mathbf{y}_i) \mathbf{V}_i^{-1} \mathbf{X}_i\right) \left(\sum_{i=1}^{n} \mathbf{X}_i' \mathbf{V}_i^{-1} \mathbf{X}_i\right)^{-1} \\ & \times \left(\boldsymbol{\lambda}' - \sum_{i=1}^{n} \mathrm{Cov}(w, \mathbf{y}_i) \mathbf{V}_i^{-1} \mathbf{X}_i\right)' \\ & - \sum_{i=1}^{n} \mathrm{Cov}(w, \mathbf{y}_i) \mathbf{V}_i^{-1} \mathrm{Cov}(w, \mathbf{y}_i)' + \sigma_w^2. \end{aligned}$$

b. Use the general expression for the BLUP variance to show that the variance for the linear mixed-effects model can be expressed as

$$\begin{aligned}\operatorname{Var} w_{\text{BLUP}} &= \sum_{i=1}^{n} \operatorname{Cov}(w, \mathbf{y}_i)\mathbf{V}_i^{-1}\operatorname{Cov}(w, \mathbf{y}_i)' \\ &\quad - \left(\boldsymbol{\lambda}' - \sum_{i=1}^{n} \operatorname{Cov}(w, \mathbf{y}_i)\mathbf{V}_i^{-1}\mathbf{X}_i\right)\left(\sum_{i=1}^{n} \mathbf{X}_i'\mathbf{V}_i^{-1}\mathbf{X}_i\right)^{-1} \\ &\quad \times \left(\boldsymbol{\lambda}' - \sum_{i=1}^{n} \operatorname{Cov}(w, \mathbf{y}_i)\mathbf{V}_i^{-1}\mathbf{X}_i\right)'.\end{aligned}$$

c. Now suppose that the BLUP of interest is a linear combination of global parameters and subject-specific effects of the form $w = \mathbf{c}_1'\boldsymbol{\alpha}_i + \mathbf{c}_2'\boldsymbol{\beta}$. Use part (a) to show that the mean-square error is

$$\begin{aligned}\operatorname{Var}(w_{\text{BLUP}} - w) &= (\mathbf{c}_2' - \mathbf{c}_1'\mathbf{D}\mathbf{Z}_i'\mathbf{V}_i^{-1}\mathbf{X}_i)\left(\sum_{i=1}^{n} \mathbf{X}_i'\mathbf{V}_i^{-1}\mathbf{X}_i\right)^{-1} \\ &\quad \times (\mathbf{c}_2' - \mathbf{c}_1'\mathbf{D}\mathbf{Z}_i'\mathbf{V}_i^{-1}\mathbf{X}_i)' - \mathbf{c}_1'\mathbf{D}\mathbf{Z}_i'\mathbf{V}_i^{-1}\mathbf{Z}_i\mathbf{D}\mathbf{c}_1 + \mathbf{c}_1'\mathbf{D}\mathbf{c}_1.\end{aligned}$$

d. Use direct calculations to show that the variance of the BLUP of $\boldsymbol{\alpha}_i$ is

$$\operatorname{Var} \mathbf{a}_{i,\text{BLUP}} = \mathbf{D}\mathbf{Z}_i'\mathbf{V}_i^{-1}\left(\mathbf{I}_i - \mathbf{X}_i\left(\mathbf{X}_i'\mathbf{V}_i^{-1}\mathbf{X}_i\right)^{-1}\mathbf{X}_i'\right)\mathbf{V}_i^{-1}\mathbf{Z}_i\mathbf{D}.$$

e. Use part (b) to establish the form of the variance of the BLUP residual in Section 4.4.3.

f. Use part (a) to establish the variance of the forecast error in Equation (4.15).

4.10 Henderson's mixed linear model justification of BLUPs Consider the model in Equation (3.9),

$$\mathbf{y} = \mathbf{Z}\boldsymbol{\alpha} + \mathbf{X}\boldsymbol{\beta} + \boldsymbol{\epsilon}.$$

In addition, assume that $\boldsymbol{\alpha}$, $\boldsymbol{\epsilon}$ are jointly multivariate normally distributed such that

$$\mathbf{y}|\boldsymbol{\alpha} \sim N(\mathbf{Z}\boldsymbol{\alpha} + \mathbf{X}\boldsymbol{\beta}, \mathbf{R})$$

and

$$\boldsymbol{\alpha} \sim N(\mathbf{0}, \mathbf{D}).$$

a. Show that the joint logarithmic probability density function of $\mathbf{y}, \boldsymbol{\alpha}$ is

$$l(\mathbf{y}, \boldsymbol{\alpha}) = -\frac{1}{2}(N \ln(2\pi) + \ln \det \mathbf{R} + (\mathbf{y} - (\mathbf{Z}\boldsymbol{\alpha} + \mathbf{X}\boldsymbol{\beta}))' \mathbf{R}^{-1} (\mathbf{y} - (\mathbf{Z}\boldsymbol{\alpha} + \mathbf{X}\boldsymbol{\beta}))) - \frac{1}{2}(q \ln(2\pi) + \ln \det \mathbf{D} + \boldsymbol{\alpha}'\mathbf{D}^{-1}\boldsymbol{\alpha}).$$

b. Treat this as a function of $\boldsymbol{\alpha}$ and $\boldsymbol{\beta}$. Take partial derivatives with respect to $\boldsymbol{\alpha}, \boldsymbol{\beta}$ to yield Henderson's (1984B) "mixed-model equations"

$$\mathbf{X}'\mathbf{R}^{-1}\mathbf{X}\boldsymbol{\beta} + \mathbf{X}'\mathbf{R}^{-1}\mathbf{Z}\boldsymbol{\alpha} = \mathbf{X}'\mathbf{R}^{-1}\mathbf{y},$$
$$\mathbf{Z}'\mathbf{R}^{-1}\mathbf{X}\boldsymbol{\beta} + (\mathbf{Z}'\mathbf{R}^{-1}\mathbf{Z} + \mathbf{D}^{-1})\boldsymbol{\alpha} = \mathbf{Z}'\mathbf{R}^{-1}\mathbf{y}.$$

c. Show that solving Henderson's mixed-model equations for unknowns $\boldsymbol{\alpha}, \boldsymbol{\beta}$ yields

$$\mathbf{b}_{\text{GLS}} = (\mathbf{X}'\mathbf{V}^{-1}\mathbf{X})^{-1}\mathbf{X}'\mathbf{V}^{-1}\mathbf{y},$$
$$\mathbf{a}_{\text{BLUP}} = \mathbf{D}\mathbf{Z}'\mathbf{V}^{-1}(\mathbf{y} - \mathbf{X}\mathbf{b}_{\text{GLS}}).$$

(*Hint*: Use Equation (A.4) of Appendix A.5.)

Empirical Exercises

4.11 Housing Prices[1] Here, we will calculate 95% prediction intervals for Chicago, the eleventh metropolitan area. Table 4E.1 provides the nine annual values of NARSP, PERPYC and PERPOP for Chicago.

a. Assume that you have fit a one-way fixed-effects model:

$$\text{NARSP}_{it} = \alpha_i + \beta_1 \text{PERPYC}_{it} + \beta_2 \text{PERPOP}_{it} + \beta_3 \text{YEAR}_t + \varepsilon_{it}.$$

You have fit the model using least squares and arrived at the estimates $b_1 = -0.008565$, $b_2 = -0.004347$, $b_3 = 0.036750$, $a_{11} = 0.285$, and $s = 0.0738$. Assume that next year's (1995) values for the explanatory variables are $\text{PERPYC}_{11,10} = 3.0$ and $\text{PERPOP}_{11,10} = 0.20$. Calculate a 95% prediction interval for Chicago's 1995 average sales price. When expressing your final answer, convert it from logarithmic dollars to dollars.

b. Assume that you have fit an error-components model that you have estimated using generalized least squares. You have fit the model using generalized least squares and arrived at the estimates $b_1 = -0.01$, $b_2 = -0.004$, $b_3 = 0.0367$, $\hat{\sigma}_\alpha = 0.10$, and $s = 0.005$. Assume that next

[1] See Exercise 2.21 for the problem description.

Table 4E.1. *Chicago housing prices, income and population growth, 1986–1994*

YEAR	NARSP	PERYPC	PERPOP
1	4.45551	5.83817	0.19823
2	4.50866	5.59691	0.32472
3	4.48864	7.80832	0.13056
4	4.67283	7.17689	0.33683
5	4.76046	5.90655	0.47377
6	4.87596	2.02724	0.99697
7	4.91852	−0.27135	−0.77503
8	4.95583	3.80041	0.19762
9	4.96564	3.66127	0.19723

year's (1995) values for the explanatory variables are $\text{PERPYC}_{11,10} = 3.0$ and $\text{PERPOP}_{11,10} = 0.20$. Calculate a 95% prediction interval for Chicago's 1995 average sales price. When expressing your final answer, convert it from logarithmic dollars to dollars.

c. Assume that you have fit an error-components model with an $AR(1)$ autocorrelation structure that you have estimated using generalized least squares. You have fit the model using generalized least squares and arrived at the estimates $b_1 = -0.01$, $b_2 = -0.004$, $b_3 = 0.0367$, $\hat{\rho} = 0.1$, $\hat{\sigma}_\alpha = 0.10$, and $s = 0.005$. Assume that next year's (1995) values for the explanatory variables are $\text{PERPYC}_{11,10} = 3.0$ and $\text{PERPOP}_{11,10} = 0.20$. Calculate a 95% prediction interval for Chicago's 1995 average sales price. When expressing your final answer, convert it from logarithmic dollars to dollars.

4.12 Workers' compensation (unstructured problem) Consider an example from workers' compensation insurance, examining losses due to permanent partial-disability claims. The data are from Klugman (1992O), who considers Bayesian model representations, and are originally from the National Council on Compensation Insurance. We consider $n = 121$ occupation, or risk, classes, over $T = 7$ years. To protect the data sources, further information on the occupation classes and years are not available.

The response variable of interest is the pure premium (PP), defined to be losses due to permanent partial disability per dollar of PAYROLL. The variable PP is of interest to actuaries because worker compensation rates are determined and quoted per unit of payroll. The exposure measure, PAYROLL, is one of the potential explanatory variables. Other explanatory variables are YEAR ($= 1, \ldots, 7$) and occupation class.

For this exercise, develop a random-effects model. Use this model to provide forecasts of the conditional mean of pure premium (known as credibility estimates in the actuarial literature). For one approach, see Frees et al. (2001O).

4.13 Group term life insurance (unstructured problem) We now consider claims data provided by a Wisconsin-based credit insurer. The data contain claims and exposure information for eighty-eight Florida credit unions. These are "life-savings" claims from a contract between the credit union and its members that provide a death benefit based on the members' savings deposited in the credit union. The dependent variable is $\mathrm{LN\_LSTC} = \ln(1 + \mathrm{LSTC}/1{,}000)$, where LSTC is the annual total claims from the life-savings contract. The exposure measure is $\mathrm{LN\_LSCV} = \ln(1 + \mathrm{LSCV}/1{,}000{,}000)$, where LSCV is the annual coverage for the life-savings contract. Also available is the contract YEAR.

For this exercise, develop a random-effects model. Use this model to provide forecasts of the conditional mean of pure premium. For one approach, see Frees et al. (2001O).

5
Multilevel Models

Abstract. This chapter describes a conditional modeling framework that takes into account hierarchical and clustered data structures. The data and models, known as *multilevel*, are used extensively in educational science and related disciplines in the social and behavioral sciences. We show that a multilevel model can be viewed as a linear mixed-effects model, and hence the statistical inference techniques introduced in Chapter 3 are readily applicable. By considering multilevel data and models as a separate unit, we expand the breadth of applications that linear mixed-effects models enjoy.

5.1 Cross-Sectional Multilevel Models

Educational systems are often described by structures in which the units of observation at one level are grouped within units at a higher level of structure. To illustrate, suppose that we are interested in assessing student performance based on an achievement test. Students are grouped into classes, classes are grouped into schools, and schools are grouped into districts. At each level, there are variables that may affect responses from a student. For example, at the class level, education of the teacher may be important, at the school level, the school size may be important, and at the district level, funding may be important. Further, each level of grouping may be of scientific interest. Finally, there may be not only relationships among variables within each group but also across groups that should be considered.

The term *multilevel* is used for this nested data structure. In our situation, we consider students to be the basic unit of observation; they are known as the "level-1" units of observation. The next level up is called "level 2" (classes in this example), and so forth.

We can imagine multilevel data being collected by a cluster sampling scheme. A random sample of districts is identified. For each district selected, a random sample of schools is chosen. From each school, a random sample of classes is taken, and from each class selected, a random sample of students. Mechanisms other than random sampling may be used, and this will influence the model selected to represent the data. Multilevel models are specified through conditional relationships, where the relationships described at one level are conditional on (generally unobserved) random coefficients of upper levels. Because of this conditional modeling framework, multilevel data and models are also known as *hierarchical*.

5.1.1 Two-Level Models

To illustrate the important features of the model, we initially consider only two levels. Suppose that we have a sample of n schools and, for the ith school, we randomly select n_i students (omitting class for the moment). For the jth student in the ith school, we assess the student's performance on an achievement test, y_{ij}, and information on the student's socioeconomic status, for example, the total family income z_{ij}. To assess achievement in terms of socioeconomic status, we could begin with a simple model of the form

$$y_{ij} = \beta_{0i} + \beta_{1i} z_{ij} + \varepsilon_{ij}. \tag{5.1}$$

Equation (5.1) describes a linear relation between socioeconomic status and expected performance, although we allow the linear relationship to vary by school through the notation β_{0i} and β_{1i} for school-specific intercepts and slopes. Equation (5.1) summarizes the level-1 model that concerns student performance as the unit of observation.

If we have identified a set of schools that are of interest, then we may simply think of the quantities $\{\beta_{0i}, \beta_{1i}\}$ as fixed parameters of interest. However, in educational research, it is customary to consider these schools to be a sample from a larger population; the interest is in making statements about this larger population. Thinking of the schools as a random sample, we model $\{\beta_{0i}, \beta_{1i}\}$ as random quantities. A simple representation for these quantities is

$$\beta_{0i} = \beta_0 + \alpha_{0i} \quad \text{and} \quad \beta_{1i} = \beta_1 + \alpha_{1i}, \tag{5.2}$$

where α_{0i}, α_{1i} are mean-zero random variables. Equations (5.2) represent a relationship about the schools and summarize the level-2 model.

Equations (5.1) and (5.2) describe models at two levels. For estimation, we combine (5.1) and (5.2) to yield

$$\begin{aligned} y_{ij} &= (\beta_0 + \alpha_{0i}) + (\beta_1 + \alpha_{1i}) z_{ij} + \varepsilon_{ij} \\ &= \alpha_{0i} + \alpha_{1i} z_{ij} + \beta_0 + \beta_1 z_{ij} + \varepsilon_{ij}. \end{aligned} \tag{5.3}$$

Equation (5.3) shows that the two-level model may be written as a single linear mixed-effects model. Specifically, we define $\boldsymbol{\alpha}_i = (\alpha_{0i}, \alpha_{1i})'$, $\mathbf{z}_{ij} = (1, z_{ij})'$, $\boldsymbol{\beta} = (\beta_0, \beta_1)'$, and $\mathbf{x}_{ij} = \mathbf{z}_{ij}$ to write

$$y_{ij} = \mathbf{z}'_{ij} \boldsymbol{\alpha}_i + \mathbf{x}'_{ij} \boldsymbol{\beta} + \varepsilon_{ij},$$

similar to Equation (3.5). Because we can write the combined multilevel model as a linear mixed-effects model, we can use the Chapter 3 techniques to estimate the model parameters. Note that we are now using the subscript "j" to denote replications within a stratum such as a school. This is because we interpret the replication to have no time ordering; generally we will assume no correlation among replications (conditional on the subject). In Section 5.2 we will reintroduce the "t" subscript when we consider time-ordered repeated measurements.

One desirable aspect of the multilevel model formulation is that we may modify conditional relationships at each level of the model, depending on the research interests of the study. To illustrate, we may wish to understand how characteristics of the school affect student performance. For example, Raudenbush and Bryk (2002EP) discussed an example where x_i indicates whether the school was a Catholic-based or a public school. A simple way to introduce this information is to modify the level-2 model in Equations (5.2) to

$$\beta_{0i} = \beta_0 + \beta_{01} x_i + \alpha_{0i} \quad \text{and} \quad \beta_{1i} = \beta_1 + \beta_{11} x_i + \alpha_{1i}. \tag{5.2a}$$

There are two level-2 regression models in (5.2a); analysts find it intuitively appealing to specify regression relationships that capture additional model variability. Note, however, that for each model the left-hand-side quantities are not observed. To emphasize this, Raudenbush and Bryk (2002EP) call these models "intercepts-as-outcomes" and "slopes-as-outcomes." In Section 5.3, we will learn how to predict these quantities.

Combining Equations (5.2a) with the level-1 model in Equation (5.1), we have

$$\begin{aligned} y_{ij} &= (\beta_0 + \beta_{01} x_i + \alpha_{0i}) + (\beta_1 + \beta_{11} x_i + \alpha_{1i}) z_{ij} + \varepsilon_{ij} \\ &= \alpha_{0i} + \alpha_{1i} z_{ij} + \beta_0 + \beta_{01} x_i + \beta_1 z_{ij} + \beta_{11} x_i z_{ij} + \varepsilon_{ij}. \end{aligned} \tag{5.4}$$

By defining $\boldsymbol{\alpha}_i = (\alpha_{0i}, \alpha_{1i})'$, $\mathbf{z}_{ij} = (1, z_{ij})'$, $\boldsymbol{\beta} = (\beta_0, \beta_{01}, \beta_1, \beta_{11})'$, and $\mathbf{x}_{ij} = (1, x_i, z_{ij}, x_i z_{ij})'$, we may again express this multilevel model as a single linear mixed-effects model.

The term $\beta_{11} x_i z_{ij}$, interacting between the level-1 variable z_{ij} and the level-2 variable x_i, is known as a *cross-level* interaction. For this example, suppose that we use $x = 1$ for Catholic schools and $x = 0$ for public schools. Then, β_{11} represents the difference between the marginal change in achievement scores, per unit of family income, between Catholic and public schools. Many researchers (see, for example, Raudenbush and Bryk, 2002EP) argue that understanding cross-level interactions is a major motivation for analyzing multilevel data.

Centering of Variables

It is customary in educational science to "center" explanatory variables to enhance the interpretability of model coefficients. To illustrate, consider the hierarchical models in (5.1), (5.2a), and (5.4). Using the "natural" metric for z_{ij}, we interpret β_{0i} to be the mean (conditional on the ith subject) response when $z = 0$. In many applications such as where z represents total income or test scores, a value of zero falls outside a meaningful range of values.

One possibility is to center level-1 explanatory variables about their overall mean and use $z_{ij} - \bar{z}_i$ as an explanatory variable in Equation (5.1). In this case, we may interpret the intercept $\beta_{0,i}$ to be the expected response for an individual with a score equal to the grand mean. This can be interpreted as an adjusted mean for the ith group.

Another possibility is to center each level-1 explanatory variable about its level-2 mean and use $z_{ij} - \bar{z}_i$ as an explanatory variable in Equation (5.1). In this case, we may interpret the intercept $\beta_{0,i}$ to be the expected response for an individual with a score equal to the mean of the ith group.

For longitudinal applications, you may wish to center the level-1 explanatory variables so that the intercept equals the expected random coefficient at a specific point in time, for example, at the start of a training program (see, for example, Kreft and deLeeuw, 1998EP).

Extended Two-Level Models

To consider many explanatory variables, we extend Equations (5.1) and (5.2). Consider a level-1 model of the form

$$y_{ij} = \mathbf{z}'_{1,ij}\boldsymbol{\beta}_i + \mathbf{x}'_{1,ij}\boldsymbol{\beta}_1 + \varepsilon_{ij}, \tag{5.5}$$

where $\mathbf{z}_{1,ij}$ and $\mathbf{x}_{1,ij}$ represent the set of level-1 variables associated with varying (over level 1) and fixed coefficients, respectively. The level-2 model is of the form

$$\boldsymbol{\beta}_i = \mathbf{X}_{2,i}\boldsymbol{\beta}_2 + \boldsymbol{\alpha}_i, \tag{5.6}$$

where $\mathrm{E}\boldsymbol{\alpha}_i = \mathbf{0}$. With this notation, the term $\mathbf{X}_{2,i}\boldsymbol{\beta}_2$ forms another set of effects with parameters to be estimated. Alternatively, we could write Equation (5.5) without explicitly recognizing the fixed coefficients $\boldsymbol{\beta}_1$ by including them in the random-coefficients Equation (5.6) but with zero variance. However, we prefer to recognize their presence explicitly because this helps in translating Equations (5.5) and (5.6) into computer statistical routines for implementation. Combining Equations (5.5) and (5.6) yields

$$\begin{aligned} y_{ij} &= \mathbf{z}'_{1,ij}(\mathbf{X}_{2,i}\boldsymbol{\beta}_2 + \boldsymbol{\alpha}_i) + \mathbf{x}'_{1,ij}\boldsymbol{\beta}_1 + \varepsilon_{ij} \\ &= \mathbf{z}'_{ij}\boldsymbol{\alpha}_i + \mathbf{x}'_{ij}\boldsymbol{\beta} + \varepsilon_{ij}, \end{aligned} \tag{5.7}$$

with the notation $\mathbf{x}'_{ij} = (\mathbf{x}'_{1,ij}\ \mathbf{z}'_{1,ij}\mathbf{X}_{2,i})$, $\mathbf{z}_{ij} = \mathbf{z}_{1,ij}$, and $\boldsymbol{\beta} = (\boldsymbol{\beta}'_1\ \boldsymbol{\beta}'_2)'$. Again, Equation (5.7) expresses this multilevel model in our usual linear mixed-effects model form.

It will be helpful to consider a number of special cases of Equations (5.5)–(5.7). To begin, suppose that $\boldsymbol{\beta}_i$ is a scalar and that $\mathbf{z}_{1,ij} = 1$. Then, the model in Equation (5.7) reduces to the error-components model introduced in Section 3.1. Raudenbush and Bryk (2002EP) further discuss the special case, where Equation (5.5) does not contain the fixed-effects $\mathbf{x}'_{1,ij}\boldsymbol{\beta}_1$ portion. In this case, Equation (5.7) reduces to

$$y_{ij} = \alpha_i + \mathbf{X}_{2,i}\boldsymbol{\beta}_2 + \varepsilon_{ij},$$

which Raudenbush and Bryk refer to as the "means-as-outcomes" model. This model, with only level-2 explanatory variables available, can be used to predict the means, or expected values, of each group *i*. We will study this prediction problem formally in Section 5.3.

Another special case of Equations (5.5)–(5.7) is the *random-coefficients* model. Here, we omit the level-1 fixed-effects portion $\mathbf{x}'_{1,ij}\boldsymbol{\beta}_1$ and use the identity matrix for $\mathbf{X}_{2,i}$. Then, Equation (5.7) reduces to

$$y_{ij} = \mathbf{z}'_{ij}(\boldsymbol{\beta}_2 + \boldsymbol{\alpha}_i) + \varepsilon_{ij}.$$

Example 5.1: High School Size and Student Achievement As reported in Lee (2000EP), Lee and Smith (1997EP) studied 9,812 grade-12 students in 1992 who attended 789 public, Catholic, and elite private high schools, drawn from a nationally representative sample from the National Education Longitudinal Study. The responses were achievement gains in reading and mathematics over four years of high school. The main variable of interest was a school-level variable: size of the high school. Educational research had emphasized that larger schools enjoy economies of scale and are able to offer a broader curriculum whereas smaller schools offer more positive social environments, as well as a more homogenous curriculum. Lee and Smith sought to investigate the optimal school size. To control for additional student-level effects, level-1 explanatory variables included gender, minority status, ability, and socioeconomic status. To control for additional school-level characteristics, level-2 explanatory variables included school-average minority concentration, school-average socioeconomic status, and type of school (Catholic, public, or elite private). Lee and Smith found that a medium school size, of approximately 600–900 students, produced the best achievement results.

Motivation for Multilevel Models

As we have seen, multilevel models allow analysts to assess the importance of cross-level effects. Specifically, the multilevel approach allows and/or forces researchers to hypothesize relationships at each level of analysis. Many different "units of analysis" within the same problem are possible, thus permitting modeling of complex systems. The ability to estimate cross-level effects is one advantage of multilevel modeling when compared to an alternate research strategy calling for the analysis of each level in isolation from the others.

As described in Chapter 1, multilevel models allow analysts to address problems of heterogeneity with samples of repeated measurements. Within the educational research literature, not accounting for heterogeneity from individuals is known as *aggregation bias*; see, for example, Raudenbush and Bryk (2002EP). Even if the interest is in understanding level-2 relationships, we will get a better picture by incorporating a level-1 model of individual effects. Moreover, multilevel modeling allows us to predict quantities at both level 1 and level 2; Section 5.3 describes this prediction problem.

Second and higher levels of multilevel models also provide us with an opportunity to estimate the variance structure using a parsimonious, parametric structure. Improved estimation of the variance structure provides a better understanding of the entire model and will often result in improved precision of our usual regression coefficient estimators. Moreover, as already discussed, often these relationships at the second and higher levels are of theoretical interest and may represent the main focus of the study. However, technical difficulties arise when testing certain hypotheses about variance components. These difficulties, and solutions, are presented in Section 5.4.

5.1.2 Multiple-Level Models

Extensions to more than two levels follow the same pattern as two-level models. To be explicit, we give a three-level model based on an example from Raudenbush and Bryk (2002EP). Consider modeling a student's achievement as the response y. The level-1 model is

$$y_{i,j,k} = \mathbf{z}'_{1,i,j,k}\boldsymbol{\beta}_{i,j} + \mathbf{x}'_{1,i,j,k}\boldsymbol{\beta}_1 + \varepsilon_{1,i,j,k}, \tag{5.8}$$

where there are $i = 1, \ldots, n$ schools, $j = 1, \ldots, J_i$ classrooms in the ith school, and $k = 1, \ldots, K_{i,j}$ students in the jth classroom (within the ith school). The explanatory variables $\mathbf{z}_{1,i,j,k}$ and $\mathbf{x}_{1,i,j,k}$ may depend on the student (gender, family income, and so on), classroom (teacher characteristics, classroom facilities, and so on) or school (organization, structure, location, and so on). The parameters that depend on either school i or classroom j appear as part of the $\boldsymbol{\beta}_{i,j}$ vector, whereas parameters that are constant appear in the $\boldsymbol{\beta}_1$ vector. This dependence is made explicit in the higher level model formulation. Conditional on the classroom and school, the disturbance term $\varepsilon_{1,i,j,k}$ is mean zero and has a variance that is constant over all students, classrooms, and schools.

The level-2 model describes the variability at the classroom level. The level-2 model is of the form

$$\boldsymbol{\beta}_{i,j} = \mathbf{Z}_{2,i,j}\boldsymbol{\gamma}_i + \mathbf{X}_{2,i,j}\boldsymbol{\beta}_2 + \boldsymbol{\epsilon}_{2,i,j}. \tag{5.9}$$

Analogous to level 1, the explanatory variables $\mathbf{Z}_{2,i,j}$ and $\mathbf{X}_{2,i,j}$ may depend on the classroom or school but not the student. The parameters associated with the $\mathbf{Z}_{2,i,j}$ explanatory variables, $\boldsymbol{\gamma}_i$, may depend on school i, whereas the parameters associated with the $\mathbf{X}_{2,i,j}$ explanatory variables are constant. Conditional on the school, the disturbance term $\boldsymbol{\epsilon}_{2,i,j}$ is mean zero and has a variance that is constant over classrooms and schools. The level-1 parameters $\boldsymbol{\beta}_{i,j}$ may be (i) varying but nonstochastic or (ii) stochastic. With this notation, we use a zero variance to model parameters that are varying but nonstochastic.

The level-3 model describes the variability at the school level. Again, the level-2 parameters $\boldsymbol{\gamma}_i$ may be varying but nonstochastic or stochastic. The level-3 model is of the form

$$\boldsymbol{\gamma}_i = \mathbf{X}_{3,i}\boldsymbol{\beta}_3 + \boldsymbol{\epsilon}_{3,i}. \tag{5.10}$$

Again, the explanatory variables $\mathbf{X}_{3,i}$ may depend on the school. Conditional on the school, the disturbance term $\boldsymbol{\epsilon}_{3,i}$ is mean zero and has a variance that is constant over schools.

Putting Equations (5.8)–(5.10) together, we have

$$\begin{aligned} y_{i,j,k} &= \mathbf{z}'_{1,i,j,k}(\mathbf{Z}_{2,i,j}(\mathbf{X}_{3,i}\boldsymbol{\beta}_3 + \boldsymbol{\epsilon}_{3,i}) + \mathbf{X}_{2,i,j}\boldsymbol{\beta}_2 + \boldsymbol{\epsilon}_{2,i,j}) + \mathbf{x}'_{1,i,j,k}\boldsymbol{\beta}_1 + \varepsilon_{1,i,j,k} \\ &= \mathbf{x}'_{1,i,j,k}\boldsymbol{\beta}_1 + \mathbf{z}'_{1,i,j,k}\mathbf{X}_{2,i,j}\boldsymbol{\beta}_2 + \mathbf{z}'_{1,i,j,k}\mathbf{Z}_{2,i,j}\mathbf{X}_{3,i}\boldsymbol{\beta}_3 + \mathbf{z}'_{1,i,j,k}\mathbf{Z}_{2,i,j}\boldsymbol{\epsilon}_{3,i} \\ &\quad + \mathbf{z}'_{1,i,j,k}\boldsymbol{\epsilon}_{2,i,j} + \varepsilon_{1,i,j,k} = \mathbf{x}'_{i,j,k}\boldsymbol{\beta} + \mathbf{z}'_{i,j,k}\boldsymbol{\alpha}_{i,j} + \varepsilon_{1,i,j,k}, \end{aligned} \tag{5.11}$$

where $\mathbf{x}'_{i,j,k} = (\mathbf{x}'_{1,i,j,k} \quad \mathbf{z}'_{1,i,j,k}\mathbf{X}_{2,i,j} \quad \mathbf{z}'_{1,i,j,k}\mathbf{Z}_{2,i,j}\mathbf{X}_{3,i})$, $\boldsymbol{\beta} = (\boldsymbol{\beta}'_1 \quad \boldsymbol{\beta}'_2 \quad \boldsymbol{\beta}'_3)'$, $\mathbf{z}'_{i,j,k} = (\mathbf{z}'_{1,i,j,k} \quad \mathbf{z}'_{1,i,j,k}\mathbf{Z}_{2,i,j})$, and $\boldsymbol{\alpha}_{i,j} = (\boldsymbol{\epsilon}'_{2,i,j} \quad \boldsymbol{\epsilon}'_{3,i})'$. We have already specified the usual assumption of homoscedasticity for each random quantity $\varepsilon_{1,i,j,k}$, $\boldsymbol{\epsilon}_{2,i,j}$, and $\boldsymbol{\epsilon}_{3,i}$. Moreover, it is customary to assume that these quantities are uncorrelated with one another. Our main point is that, as with the two-level model, Equation (5.11) expresses the three-level model as a linear mixed-effects model. (Converting the model in Equation (5.11) into the linear mixed-effects model in Equation (3.5) is a matter of defining vector expressions carefully. Section 5.3 provides further details.) Thus, parameter estimation is a direct consequence of our Chapter 3 results. Many variations of the basic assumptions that we have described are possible. In Section 5.2 on longitudinal multilevel models, we will give a more detailed description of an example of a three-level model. Appendix 5A extends the discussion to higher order multilevel models.

For applications, several statistical software packages exist (such as *HLM*, *MLwiN*, and *MIXREG*) that allow analysts to fit multilevel models without combining the several equations into a single expression such as Equation (5.11). However, these specialized packages may not have all of the features that the analyst wishes to display in his or her analysis. As pointed out by Singer (1998EP), an alternative, or supplementary, approach is to use a general-purpose mixed linear effects package (such as SAS *PROC MIXED*) and rely directly on the fundamental mixed linear model theory.

5.1.3 Multilevel Modeling in Other Fields

The field of educational research has been an area of active development of cross-sectional multilevel modeling although it by no means has a corner on the market. This subsection describes examples where these models have been used in other fields of study.

One type of study that is popular in economics uses data based on a *matched-pairs sample*. For example, we might select a set of families for a level-2 sample and, for each family, observe the behavior of siblings (or twins). The idea underlying this design is that by observing more than one family member we will be able to control for unobserved family characteristics. See Wooldridge (2002E) and Exercise 3.10 for further discussion of this design.

In insurance and actuarial science, it is possible to model claims distributions using a hierarchical framework. Typically, the level-2 unit of analysis is based on an insurance customer, and explanatory variables may include characteristics of the customer. The level-1 model uses claims amount as the response (typically over time) and time-varying explanatory variables such as exposures and time trends. For example, Klugman (1992O) gives a Bayesian perspective of this problem. For a frequentist perspective, see Frees et al. (1999O).

5.2 Longitudinal Multilevel Models

This section shows how to use the conditional modeling framework to represent longitudinal (time-ordered) data. The key change in the modeling setup is that we now will typically consider the individual as the level-2 unit of analysis and observations at different time points as the level-1 units. The goal is now also substantially different; typically, in longitudinal studies the assessment of change is the key research interest. As with Section 5.1, we begin with the two-level model and then discuss general multilevel extensions.

5.2.1 Two-Level Models

Following the notation established in Section 5.1, we consider level-1 models of the form

$$y_{it} = \mathbf{z}'_{1,it}\boldsymbol{\beta}_i + \mathbf{x}'_{1,it}\boldsymbol{\beta}_1 + \varepsilon_{it}. \tag{5.12}$$

This is a model of $t = 1, \ldots, T_i$ responses over time for the ith individual. The unit of analysis for the level-1 model is an observation at a point in time, not the individual as in Section 5.1. Thus, we use the subscript "t" as an index for time. Most other aspects of the model are as in Section 5.1.1; $\mathbf{z}_{1,it}$ and $\mathbf{x}_{1,it}$ represent sets of level-1 explanatory variables. The associated parameters that may depend on the ith individual appear as part of the $\boldsymbol{\beta}_i$ vector, whereas parameters that are constant appear in the $\boldsymbol{\beta}_1$ vector. Conditional on the subject, the disturbance term ε_{it} is a mean-zero random variable that is uncorrelated with $\boldsymbol{\beta}_i$.

An important feature of the longitudinal multilevel model that distinguishes it from its cross-sectional counterpart is that time generally enters the level-1 specification. There are a number of ways that this can happen. One way is to let one or more of the explanatory variables be a function of time. This is the approach historically taken in growth-curve modeling, described below. Another approach is to let one of the explanatory variables be a lagged-response variable. This approach is particularly prevalent in economics and will be further explored in Chapter 6. Yet another approach is to model the serial correlation through the variance–covariance matrix of the vector of disturbance $\boldsymbol{\epsilon}_i = (\varepsilon_{i1} \ldots \varepsilon_{iT_i})'$. Specifically, in Sections 2.5.1 and 3.3.1 we developed the notation $\mathrm{Var}\,\boldsymbol{\epsilon}_i = \mathbf{R}_i$ to represent the serial covariance structure. This approach is widely adopted in biostatistics and educational research and will be further developed here.

Like the cross-sectional model, the level-2 model can be represented as $\boldsymbol{\beta}_i = \mathbf{X}_{2,i}\boldsymbol{\beta}_2 + \boldsymbol{\alpha}_i$; see Equation (5.6). Now, however, we interpret the unobserved $\boldsymbol{\beta}_i$ to be the random coefficients associated with the ith individual. Thus, although the mathematical representation is similar to the cross-sectional setting, our interpretations of individual components of the model are quite different. Yet, as with Equation (5.7), we may still combine level-1 and level-2 models to get

$$y_{it} = \mathbf{z}'_{1,it}(\mathbf{X}_{2,i}\boldsymbol{\beta}_2 + \boldsymbol{\alpha}_i) + \mathbf{x}'_{1,it}\boldsymbol{\beta}_1 + \varepsilon_{it} = \mathbf{z}'_{it}\boldsymbol{\alpha}_i + \mathbf{x}'_{it}\boldsymbol{\beta} + \varepsilon_{it}, \quad (5.13)$$

using the notation $\mathbf{x}'_{it} = (\mathbf{x}'_{1,it}\ \ \mathbf{z}'_{1,it}\mathbf{X}_{2,i})$, $\mathbf{z}_{it} = \mathbf{z}_{1,it}$, and $\boldsymbol{\beta} = (\boldsymbol{\beta}'_1\ \ \boldsymbol{\beta}'_2)'$. This is the linear mixed-effects model introduced in Section 3.3.1.

Growth-Curve Models

To develop intuition, we now consider *growth-curve models*, models that have a long history of applications. The idea behind growth-curve models is that we seek to monitor the natural development or aging of an individual. This development is typically monitored without intervention and the goal is to assess differences among groups. In growth-curve modeling, one uses a polynomial function of age or time to track growth. Because growth-curve data may reflect observations from a development process, it is intuitively appealing to think of the expected response as a function of time. Parameters of the function vary by individual, so that one can summarize an individual's growth through the parameters. To illustrate, we now consider a classic example.

Example 5.2: Dental Data This example is originally due to Potthoff and Roy (1964B); see also Rao (1987B). Here, y is the distance, measured in millimeters, from the center of the pituitary to the pteryomaxillary fissure. Measurements were taken on eleven girls and sixteen boys at ages 8, 10, 12, and 14. Of interest is the relation between the distance and age, specifically, in how the distance grows with age and whether there is a difference between males and females.

Table 5.1. *Dental measurements in millimeters of eleven girls and sixteen boys*

	Age of girls				Age of boys			
Number	8	10	12	14	8	10	12	14
1	21	20	21.5	23	26	25	29	31
2	21	21.5	24	25.5	21.5	22.5	23	26.5
3	20.5	24	24.5	26	23	22.5	24	27.5
4	23.5	24.5	25	26.5	25.5	27.5	26.5	27
5	21.5	23	22.5	23.5	20	23.5	22.5	26
6	20	21	21	22.5	24.5	25.5	27	28.5
7	21.5	22.5	23	25	22	22	24.5	26.5
8	23	23	23.5	24	24	21.5	24.5	25.5
9	20	21	22	21.5	23	20.5	31	26
10	16.5	19	19	19.5	27.5	28	31	31.5
11	24.5	25	28	28	23	23	23.5	25
12					21.5	23.5	24	28
13					17	24.5	26	29.5
14					22.5	25.5	25.5	26
15					23	24.5	26	30
16					22	21.5	23.5	25

Source: Potthoff and Roy (1964B) and Rao (1987B).

Table 5.1 shows the data and Figure 5.1 gives a graphical impression of the growth over time. From Figure 5.1, we can see that the measurement length grows as each child ages, although it is difficult to detect differences between boys and girls. In Figure 5.1, we use open circular plotting symbols for girls and filled circular plotting symbols for boys. Figure 5.1 does show that the ninth boy has an unusual growth pattern; this pattern can also be seen in Table 5.1.

A level-1 model is

$$y_{it} = \beta_{0i} + \beta_{1i} z_{1,it} + \varepsilon_{it},$$

where $z_{1,it}$ is the age of the child i on occasion t. This model relates the dental measurement to the age of the child, with parameters that are specific to the child. Thus, we may interpret the quantity β_{1i} to be the growth rate for the ith child. A level-2 model is

$$\beta_{0i} = \beta_{00} + \beta_{01} \text{ GENDER}_i + \alpha_{0i}$$

and

$$\beta_{1i} = \beta_{10} + \beta_{11} \text{ GENDER}_i + \alpha_{1i},$$

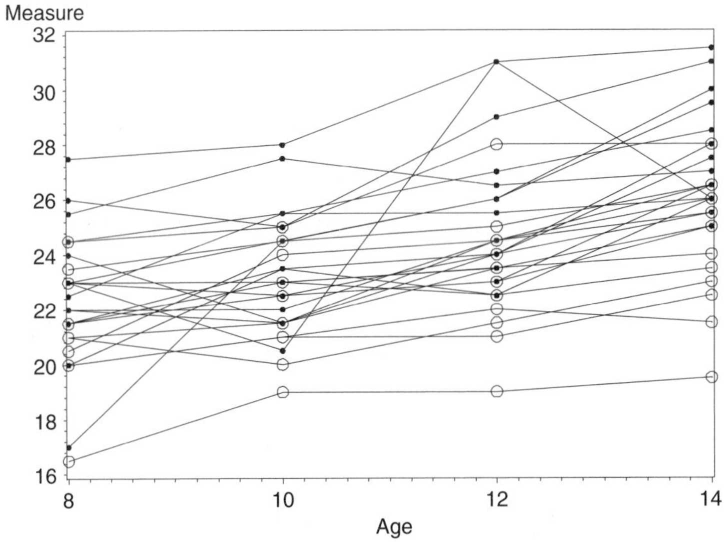

Figure 5.1. Multiple time-series plot of dental measurements. Open circles represent girls; filled circles represent boys.

where β_{00}, β_{01}, β_{10}, and β_{11} are fixed parameters to be estimated. Suppose that we use a binary variable for gender, say, coding the GENDER variable 1 for females and 0 for males. Then, we may interpret β_{10} to be the expected male growth rate and β_{11} to be the difference in growth rates between females and males.

Table 5.2 shows the parameter estimates for this model. Here, we see that the coefficient associated with linear growth is statistically significant, over all models. Moreover, the rate of increase for girls is lower than for boys. The estimated covariance between α_{0i} and α_{1i} (which is also the estimated covariance between β_{0i} and β_{1i}) turns out to be negative. One interpretation of the negative covariance between initial status and growth rate is that subjects who start at a low level tend to grow more quickly than those who start at higher levels, and vice versa.

For comparison purposes, Table 5.2 shows the parameter estimates with the ninth boy deleted. The effects of this subject deletion on the parameter estimates are small. Table 5.2 also shows parameter estimates of the error-components model. This model employs the same level-1 model but with level-2 models

$$\beta_{0i} = \beta_{00} + \beta_{01}\ \text{GENDER}_i + \alpha_{0i}$$

Table 5.2. *Dental data growth-curve-model parameter estimates*

	Error-components model		Growth-curve model		Growth-curve model deleting the ninth boy	
Variable	Parameter estimates	t-statistic	Parameter estimates	t-statistic	Parameter estimates	t-statistic
β_{00}	16.341	16.65	16.341	16.04	16.470	15.42
Age (β_{10})	0.784	10.12	0.784	9.12	0.772	8.57
GENDER (β_{01})	1.032	0.67	1.032	0.65	0.903	0.55
AGE*GENDER (β_{11})	−0.305	−2.51	−0.305	−2.26	−0.292	−2.11
Var ε_{it}	1.922		1.716		0.971	
Var α_{0i}	3.299		5.786		−11.005	
Var α_{1i}			0.033		0.073	
Cov(α_{0i}, α_{1i})			−0.290		−0.734	
−2 log likelihood	433.8		432.6		388.5	
AIC	445.8		448.6		404.5	

and

$$\beta_{1i} = \beta_{10} + \beta_{11} \text{ GENDER}_i.$$

With parameter estimates calculated using the full data set, there again is little change in the parameter estimates. Because the results appear to be robust to both unusual subjects and model selection, we have greater confidence in our interpretations.

5.2.2 Multiple-Level Models

Longitudinal versions of multiple-level models follow the same notation as the cross-sectional models in Section 5.1.2 except that the level-1 replications are over time. To illustrate, we consider a three-level model in the context of a social work application by Guo and Hussey (1999EP).

Guo and Hussey examined subjective assessments of children's behavior made by multiple raters at two or more time points. That is, the level-1 repeated measurements are over time t, where the assessment was made by rater j on child i. Raters assessed $n = 144$ seriously emotionally disturbed children receiving services through a large mental health treatment agency for children located in Cleveland, Ohio. For this study, the assessment is the response of interest, y; this response is the Deveroux Scale of Mental Disorders, a score made up of 111 items. Ratings were taken over a two-year period by parents and teachers; at each time point, assessments may be made by the parent, the teacher, or both. The time of the assessment was recorded as $\text{TIME}_{i,j,t}$, measured in days since the inception of the study. The variable $\text{PROGRAM}_{i,j,t}$ was recorded as a 1 if the child was in program residence at the time of the assessment and 0 if the child was in day treatment or day treatment combined with treatment foster care. The variable $\text{RATER}_{i,j}$ was recorded as a 1 if the rater was a teacher and 0 if the rater was a caretaker.

Analogous to Equation (5.8), the level-1 model is

$$y_{i,j,t} = \mathbf{z}'_{1,i,j,t}\boldsymbol{\beta}_{i,j} + \mathbf{x}'_{1,i,j,t}\boldsymbol{\beta}_1 + \varepsilon_{1,i,j,t}, \tag{5.14}$$

where there are $i = 1, \ldots, n$ children, $j = 1, \ldots, J_i$ raters, and $t = 1, \ldots, T_{i,j}$ evaluations. Specifically, Guo and Hussey (1999EP) used $\mathbf{x}_{1,i,j,t} = \text{PROGRAM}_{i,j,t}$ and $\mathbf{z}_{1,i,j,t} = (1 \ \text{TIME}_{i,j,t})'$. Thus, their level-1 model can be written as

$$y_{i,j,t} = \beta_{0,i,j} + \beta_{1,i,j} \text{ TIME}_{i,j,t} + \beta_1 \text{ PROGRAM}_{i,j,t} + \varepsilon_{1,i,j,t}.$$

The variables associated with the intercept and the coefficient for time may vary over child and rater whereas the program coefficient is constant over all observations.

The level-2 model is the same as Equation (5.9):

$$\boldsymbol{\beta}_{i,j} = \mathbf{Z}_{2,i,j}\boldsymbol{\gamma}_i + \mathbf{X}_{2,i,j}\boldsymbol{\beta}_2 + \boldsymbol{\epsilon}_{2,i,j},$$

where there are $i = 1, \ldots, n$ children and $j = 1, \ldots, J_i$ raters. The level-2 model of Guo and Hussey can be written as

$$\beta_{0,i,j} = \beta_{0,i,0} + \beta_{0,0,1}\ \mathrm{RATER}_{i,j} + \varepsilon_{2,i,j}$$

and

$$\beta_{1,i,j} = \beta_{2,0} + \beta_{2,1}\ \mathrm{RATER}_{i,j}.$$

Again, we leave it as an exercise for the reader to show how this formulation is a special case of Equation (5.9).

The level-3 model is the same as Equation (5.10):

$$\boldsymbol{\gamma}_i = \mathbf{X}_{3,i}\boldsymbol{\beta}_3 + \boldsymbol{\epsilon}_{3,i}.$$

To illustrate, we can write the level-3 model of Guo and Hussey as

$$\beta_{0,i,0} = \beta_{0,0,0} + \beta_{0,1,0}\ \mathrm{GENDER}_i + \varepsilon_{3,i},$$

where GENDER_i is a binary variable indicating the gender of the child.

As with the cross-sectional models in Section 5.1.2, one combines the three levels to form a single equation representation, as in Equation (5.14). The hierarchical framework allows analysts to develop hypotheses that are interesting to test. The combined model allows for simultaneous, over all levels, estimation of parameters, which is more efficient than estimating each level in isolation of the others.

5.3 Prediction

In Chapter 4, we distinguished between the concepts of *estimating* model parameters and *predicting* random variables. In multilevel models, the dependent variables at second and higher levels are unobserved random coefficients. Because it is often desirable to understand their behavior, we wish to predict these random coefficients. For example, if the unit of analysis at the second level is a school, we may wish to use predictions of second-level coefficients to rank schools. It may also be of interest to use predictions of second- (or higher) level coefficients for prediction in a first-level model. To illustrate, if we are studying a child's development over time, we may wish to make predictions about the future status of a child's development.

This subsection shows how to use the BLUPs developed in Chapter 4 for these prediction problems. Best linear unbiased predictors, by definition, have the smallest variance among all unbiased predictors. In Chapter 4, we showed that these predictors can also be interpreted as empirical Bayes estimators. Moreover, they often have desirable interpretations as *shrinkage* "estimators." Because we have expressed multilevel models in terms of linear mixed-effects models, we will not need to develop new theory but will be able to rely directly on the Chapter 4 results.

Two-Level Models

We begin our prediction discussion with the two-level model, introduced in Equations (5.5)–(5.7). To make the multilevel model notation consistent with Chapters 3 and 4, use $\mathbf{D} = \mathrm{Var}\,\boldsymbol{\alpha}_i = \mathrm{Var}\,\boldsymbol{\beta}_i$ and $\mathbf{R}_i = \mathrm{Var}\,\boldsymbol{\epsilon}_i$, where $\boldsymbol{\epsilon}_i = (\varepsilon_{i1} \ldots \varepsilon_{iT_i})'$. Suppose that we wish to predict $\boldsymbol{\beta}_i$. Using the results in Section 4.4.2, it is easy to check that the BLUP of $\boldsymbol{\beta}_i$ is

$$\mathbf{b}_{i,\mathrm{BLUP}} = \mathbf{a}_{i,\mathrm{BLUP}} + \mathbf{X}_{2,i}\mathbf{b}_{2,\mathrm{GLS}},$$

where $\mathbf{b}_{2,\mathrm{GLS}}$ is the GLS estimator of $\boldsymbol{\beta}_2$ and, from Equation (4.11),

$$\mathbf{a}_{i,\mathrm{BLUP}} = \mathbf{D}\mathbf{Z}_i'\mathbf{V}_i^{-1}(\mathbf{y}_i - \mathbf{X}_i\mathbf{b}_{\mathrm{GLS}}).$$

Recall that $\mathbf{z}_{it} = \mathbf{z}_{1,it}$, so that $\mathbf{Z}_i = (\mathbf{z}_{1,i1}, \mathbf{z}_{1,i2}, \ldots, \mathbf{z}_{1,iJ_i})'$. Further, $\mathbf{b}_{\mathrm{GLS}} = (\mathbf{b}_{1,\mathrm{GLS}}' \; \mathbf{b}_{2,\mathrm{GLS}}')'$, $\mathbf{V}_i = \mathbf{R}_i + \mathbf{Z}_i\mathbf{D}\mathbf{Z}_i'$, and $\mathbf{X}_i = (\mathbf{x}_{i1}, \mathbf{x}_{i2}, \ldots, \mathbf{x}_{iT_i})'$, where $\mathbf{x}_{it}' = (\mathbf{x}_{1,it}' \; \mathbf{z}_{1,it}'\mathbf{X}_{2,i})$. Thus, it is easy to compute these predictors.

Chapter 4 discussed interpretation in some special cases: the error-components and the random-coefficients models. Suppose that we have the error-components model, so that $\mathbf{z}_{ij} = \mathbf{z}_{1,ij} = 1$ and $\mathbf{R}_i$ is a scalar times the identity matrix. Further suppose that there are no level-1 explanatory variables. Then, one can check that the BLUP of the conditional mean of the level-1 response, $\mathrm{E}(y_{it} \mid \alpha_i) = \alpha_i + \mathbf{X}_{2,i}\boldsymbol{\beta}_2$, is

$$\begin{aligned} a_{i,\mathrm{BLUP}} + \mathbf{X}_{2,i}\mathbf{b}_{2,\mathrm{GLS}} &= \zeta_i(\bar{y}_i - \mathbf{X}_{2,i}\mathbf{b}_{2,\mathrm{GLS}}) + \mathbf{X}_{2,i}\mathbf{b}_{2,\mathrm{GLS}} \\ &= \zeta_i\bar{y}_i + (1 - \zeta_i)\mathbf{X}_{2,i}\mathbf{b}_{2,\mathrm{GLS}}, \end{aligned}$$

where

$$\zeta_i = \frac{T_i}{T_i + (\mathrm{Var}\,\varepsilon)/(\mathrm{Var}\,\alpha)}.$$

Thus, the predictor is a weighted average of the level-2 ith unit's average, $\bar{y}_i$, and the regression estimator, which is an estimator derived from all level-2 units. As noted in Section 5.1.1, Raudenbush and Bryk (2002EP) refer to this as the "means-as-outcomes" model.

As described in Section 4.4.4, one can also use the BLUP technology to predict the future development of a level-1 response. From Equation (4.14), we have that the forecast L lead times in the future of y_{iT_i} is

$$\hat{y}_{i,T_i+L} = \mathbf{z}'_{1,i,T_i+L}\mathbf{b}_{i,\mathrm{BLUP}} + \mathbf{x}'_{1,i,T_i+L}\mathbf{b}_{1,\mathrm{GLS}} + \mathrm{Cov}(\varepsilon_{i,T_i+L}, \boldsymbol{\epsilon}_i)\mathbf{R}_i^{-1}\mathbf{e}_{i,\mathrm{BLUP}},$$

where $\mathbf{e}_{i,\mathrm{BLUP}}$ is the vector of BLUP residuals, given in Equation (4.13a). As we saw in Section 4.4.4, in the case where the disturbances follow an autoregressive model of order 1 ($AR(1)$) with parameter ρ, we have

$$\hat{y}_{i,T_i+L} = \mathbf{z}'_{1,i,T_i+L}\mathbf{b}_{i,\mathrm{BLUP}} + \mathbf{x}'_{1,i,T_i+L}\mathbf{b}_{1,\mathrm{GLS}} + \rho^L e_{iT_i,\mathrm{BLUP}}.$$

To illustrate, consider the Example 5.2. Here, we have no serial correlation (so that $\mathbf{R}$ is a scalar times the identity matrix), no level-1 fixed parameters, and $T_i = 4$ observations for all children. Thus, the L-step forecast for the ith child is

$$\hat{y}_{i,4+L} = b_{0,i,\mathrm{BLUP}} + b_{1,i,\mathrm{BLUP}} z_{1,i,4+L},$$

where $z_{1,i,4+L}$ is the age of the child at time $4 + L$.

Multiple-Level Models

For models with three or more levels, the approach is the same as with two-level models although it becomes more difficult to interpret the results. Nonetheless, for applied work, the idea is straightforward.

Procedure for Forecasting Future Level-1 Responses

1. Hypothesize a model at each level.
2. Combine all level models into a single model.
3. Estimate the parameters of the single model, using GLS and variance components estimators, as described in Sections 3.4 and 3.5, respectively.
4. Determine BLUPs of each unobserved random coefficient for levels 2 and higher, as described in Section 4.4.
5. Use the parameter estimators and random-coefficients predictors to form forecasts of future level-1 responses.

To illustrate, let's see how this procedure works for the three-level longitudinal data model.

Step 1. We will use the level-1 model described in Equation (5.14), together with the level-2 and level-3 models in Equations (5.9) and (5.10), respectively. For the level-1 model, let $\mathbf{R}_{ij} = \mathrm{Var}\, \boldsymbol{\epsilon}_{1,i,j}$, where $\boldsymbol{\epsilon}_{1,i,j} = (\varepsilon_{1,i,j,1} \ldots \varepsilon_{1,i,j,T_{ij}})'$.

Step 2. The combined model is Equation (5.11), except that now we use a "t" subscript for time in lieu of the "k" subscript. Assuming the level-1, level-2, and level-3 random quantities are uncorrelated with one another, we define

$$\mathrm{Var}\, \boldsymbol{\alpha}_{i,j} = \mathrm{Var}\begin{pmatrix} \boldsymbol{\epsilon}_{2,i,j} \\ \boldsymbol{\epsilon}_{3,i} \end{pmatrix} = \begin{pmatrix} \mathrm{Var}\, \boldsymbol{\epsilon}_{2,i,j} & \mathbf{0} \\ \mathbf{0} & \mathrm{Var}\, \boldsymbol{\epsilon}_{3,i} \end{pmatrix} = \begin{pmatrix} \mathbf{D}_2 & \mathbf{0} \\ \mathbf{0} & \mathbf{D}_3 \end{pmatrix} = \mathbf{D}_V$$

and

$$\begin{aligned} \mathrm{Cov}(\boldsymbol{\alpha}_{i,j}, \boldsymbol{\alpha}_{i,k}) &= \begin{pmatrix} \mathrm{Cov}(\boldsymbol{\epsilon}_{2,i,j}, \boldsymbol{\epsilon}_{2,i,k}) & \mathrm{Cov}(\boldsymbol{\epsilon}_{2,i,j}, \boldsymbol{\epsilon}_{3,i}) \\ \mathrm{Cov}(\boldsymbol{\epsilon}_{3,i}, \boldsymbol{\epsilon}_{2,i,k}) & \mathrm{Var}\, \boldsymbol{\epsilon}_{3,i} \end{pmatrix} \\ &= \begin{pmatrix} \mathbf{0} & \mathbf{0} \\ \mathbf{0} & \mathbf{D}_3 \end{pmatrix} = \mathbf{D}_C. \end{aligned}$$

Stacking vectors, we write $\mathbf{y}_{i,j} = (y_{i,j,1} \ldots y_{i,j,T_{ij}})'$, $\mathbf{y}_i = (\mathbf{y}'_{i,1} \ldots \mathbf{y}'_{i,J_i})'$, $\boldsymbol{\epsilon}_i = (\boldsymbol{\epsilon}'_{1,i,1} \ldots \boldsymbol{\epsilon}'_{1,i,J_i})'$, and $\boldsymbol{\alpha}_i = (\boldsymbol{\alpha}'_{i,1} \ldots \boldsymbol{\alpha}'_{i,J_i})'$. Stacking matrices, we have $\mathbf{X}_{i,j} = (\mathbf{x}_{i,j,1} \ldots \mathbf{x}_{i,j,T_{i,j}})'$, $\mathbf{Z}_{i,j} = (\mathbf{z}_{i,j,1} \ldots \mathbf{z}_{i,j,T_{i,j}})'$, $\mathbf{X}_i = (\mathbf{X}'_{i,1} \ldots \mathbf{X}'_{i,J_i})'$, and

$$\mathbf{Z}_i = \begin{pmatrix} \mathbf{Z}_{i,1} & \mathbf{0} & \cdots & \mathbf{0} \\ \mathbf{0} & \mathbf{Z}_{i,2} & \cdots & \mathbf{0} \\ \vdots & \vdots & \ddots & \vdots \\ \mathbf{0} & \mathbf{0} & \cdots & \mathbf{Z}_{i,J_i} \end{pmatrix}.$$

With this notation, we may write Equation (5.11) in a linear mixed-effects model form as $\mathbf{y}_i = \mathbf{Z}_i \boldsymbol{\alpha}_i + \mathbf{X}_i \boldsymbol{\beta} + \boldsymbol{\epsilon}_i$. Note the form of $\mathbf{R}_i = \mathrm{Var}\, \boldsymbol{\epsilon}_i = \text{block diag}(\mathbf{R}_{i,1}, \ldots \mathbf{R}_{i,J_i})$ and

$$\mathbf{D} = \mathrm{Var}\, \boldsymbol{\alpha}_i = \begin{pmatrix} \mathbf{D}_V & \mathbf{D}_C & \cdots & \mathbf{D}_C \\ \mathbf{D}_C & \mathbf{D}_V & \cdots & \mathbf{D}_C \\ \vdots & \vdots & \ddots & \vdots \\ \mathbf{D}_C & \mathbf{D}_C & \cdots & \mathbf{D}_V \end{pmatrix}.$$

Step 3. Having coded the explanatory variables and the form of the variance matrices $\mathbf{D}$ and $\mathbf{R}_i$, parameter estimates follow directly from the results of Sections 3.4 and 3.5.

Step 4. The BLUPs are formed beginning with predictors for $\boldsymbol{\alpha}_i$ of the form $\mathbf{a}_{i,\mathrm{BLUP}} = \mathbf{D}\mathbf{Z}_i'\mathbf{V}_i^{-1}(\mathbf{y}_i - \mathbf{X}_i\mathbf{b}_{\mathrm{GLS}})$. This yields the BLUPs for $\boldsymbol{\alpha}_{i,j} = (\boldsymbol{\epsilon}_{2,i,j}' \quad \boldsymbol{\epsilon}_{3,i}')'$, say $\mathbf{a}_{i,j,\mathrm{BLUP}} = (\mathbf{e}_{2,i,j,\mathrm{BLUP}}' \quad \mathbf{e}_{3,i,\mathrm{BLUP}}')'$. These BLUPs allow us to predict the second- and higher level random coefficients through the relations

$$\mathbf{g}_{i,\mathrm{BLUP}} = \mathbf{X}_{3,i}\mathbf{b}_{3,\mathrm{GLS}} + \mathbf{e}_{3,i,\mathrm{BLUP}}$$

and

$$\mathbf{b}_{i,j,\mathrm{BLUP}} = \mathbf{Z}_{2,i,j}\mathbf{g}_{i,\mathrm{BLUP}} + \mathbf{X}_{2,i,j}\mathbf{b}_{2,\mathrm{GLS}} + \mathbf{e}_{2,i,j,\mathrm{BLUP}},$$

corresponding to Equations (5.9) and (5.10), respectively.

Step 5. If desired, we may forecast future level-1 responses. From Equation (4.14), for an L-step forecast, we have

$$\begin{aligned}\hat{y}_{i,j,T_{ij}+L} &= \mathbf{z}_{1,i,j,T_{ij}+L}'\mathbf{b}_{i,j,\mathrm{BLUP}} + \mathbf{x}_{1,i,j,T_{ij}+L}'\mathbf{b}_{1,\mathrm{GLS}} \\ &\quad + \mathrm{Cov}(\varepsilon_{i,j,T_{ij}+L}, \boldsymbol{\epsilon}_{1,i,j})\mathbf{R}_{ij}^{-1}\mathbf{e}_{1,i,j,\mathrm{BLUP}}.\end{aligned}$$

For $AR(1)$ level-1 disturbances, this simplifies to

$$\hat{y}_{i,j,T_{ij}+L} = \mathbf{z}_{1,i,j,T_{ij}+L}'\mathbf{b}_{i,j,\mathrm{BLUP}} + \mathbf{x}_{1,i,j,T_{ij}+L}'\mathbf{b}_{1,\mathrm{GLS}} + \rho^L e_{i,j,T_{ij},\mathrm{BLUP}}.$$

Please see Frees and Kim (2004O) for further details.

5.4 Testing Variance Components

Multilevel models implicitly provide a representation for the variance as a function of explanatory variables. To illustrate, consider the cross-sectional two-level model summarized in Equations (5.5)–(5.7). With Equation (5.7), we have

$$\mathrm{Var}\ y_{ij} = \mathbf{z}_{ij}'(\mathrm{Var}\ \boldsymbol{\alpha}_i)\mathbf{z}_{ij} + \mathrm{Var}\ \varepsilon_{ij}$$

and

$$\mathrm{Cov}(y_{ij}, y_{ik}) = \mathbf{z}_{ij}'(\mathrm{Var}\ \boldsymbol{\alpha}_i)\mathbf{z}_{ik}.$$

Thus, even if the random quantities $\boldsymbol{\alpha}_i$ and ε_{ij} are homoscedastic, the variance is a function of the explanatory variables $\mathbf{z}_{ij}$. Particularly in education and psychology, researchers wish to test theories by examining hypotheses concerning these variance functions.

Unfortunately, the usual likelihood ratio testing procedure is invalid for testing many variance components of interest. In particular, there is concern for testing parameters where the null hypothesis lies on the boundary of possible values. As a general rule, the standard hypothesis testing procedures favors the simpler null hypothesis more often than it should.

To illustrate the difficulties with boundary problems, let's consider the classic example of i.i.d. random variables $y_1, \ldots, y_n$, where each random variable is distributed normally with known mean zero and variance σ^2. Suppose that we wish to test the null hypothesis $\mathrm{H}_0 : \sigma^2 = \sigma_0^2$, where σ_0^2 is a known positive constant. It is easy to check that the MLE of σ^2 is $n^{-1} \sum_{i=1}^n y_i^2$. As we have seen, a standard method of testing hypotheses is the likelihood ratio test procedure (described in more detail in Appendix C.7). Here, one computes the likelihood ratio test statistic, which is twice the difference between the unconstrained maximum log likelihood and the maximum log likelihood under the null hypothesis, and compares this statistic to a chi-square distribution with one degree of freedom. Unfortunately, this procedure is not available when $\sigma_0^2 = 0$ because the log likelihoods are not well defined. Because $\sigma_0^2 = 0$ is on the boundary of the parameter space $[0, \infty)$, the regularity conditions of our usual test procedures are not valid.

However, $\mathrm{H}_0 : \sigma^2 = 0$ is still a testable hypothesis; a simple test is to reject H_0 if the MLE exceeds zero. This procedure will always reject the null hypothesis when $\sigma^2 > 0$ and accept it when $\sigma^2 = 0$. Thus, this test procedure has power 1 versus all alternatives and a significance level of zero, making it a very good test!

For an example closer to longitudinal data models, consider the Section 3.1 error-components model with variance parameters σ^2 and σ_α^2. In Exercise 5.4, we outline the proof to establish that the likelihood ratio test statistic for assessing $\mathrm{H}_0 : \sigma_\alpha^2 = 0$ is $(1/2)\chi_{(1)}^2$, where $\chi_{(1)}^2$ is a chi-square random variable with one degree of freedom. In the usual likelihood ratio procedure for testing one variable, the likelihood ratio test statistic has a $\chi_{(1)}^2$ distribution under the null hypothesis. This means that using nominal values, we will accept the null hypothesis more often than we should; thus, we will sometimes use a simpler model than suggested by the data.

The critical point of this exercise is that we define MLEs to be nonnegative, arguing that a negative estimator of variance components is not valid. Thus, the difficulty stems from the usual regularity conditions (see, for example, Serfling, 1980G) requiring that the hypotheses that we test lie on the interior of a parameter space. For most variance components, the parameter space is

$[0, \infty)$. By testing that the variance equals zero, we are on the boundary and the usual asymptotic results are not valid. This does not mean that tests for all variance components are invalid. For example, for testing most correlations and autocorrelations, the parameter space is $[-1, 1]$. Thus, for testing correlations (and covariances) equal to zero, we are in the interior of the parameter space and so the usual test procedures are valid.

In contrast, in Exercise 5.3, we allow negative variance estimators. In this case, by following the outline of the proof, you will see that the usual likelihood ratio test statistic for assessing $\mathrm{H}_0 : \sigma_\alpha^2 = 0$ is $\chi^2_{(1)}$, the customary distribution. Thus, it is important to know the constraints underlying the software package you are using.

A complete theory for testing variance components has yet to be developed. When only one variance parameter needs to be assessed for equality to zero, results similar to the error-components model discussed here have been worked out. For example, Balagi and Li (1990E) developed a test for a second (independent) error component representing time; this model will be described in Chapter 8. More generally, checking for the presence of an additional random effect in the model implicitly means checking that not only the variance but also the covariances are equal to zero. For example, for the linear mixed-effects model with a $q \times 1$ vector of variance components $\boldsymbol{\alpha}_i$, we might wish to assess the null hypothesis

$$\mathrm{H}_0 : \mathbf{D} = \mathrm{Var}\, \boldsymbol{\alpha}_i = \begin{pmatrix} \mathrm{Var}(\alpha_{i,1}, \ldots, \alpha_{i,q-1}) & \mathbf{0} \\ \mathbf{0}' & 0 \end{pmatrix}.$$

In this case, based on the work of Self and Liang (1987S), Stram and Lee (1994S) showed that the usual likelihood ratio test statistic has asymptotic distribution $(1/2)\chi^2_{(q-1)} + (1/2)\chi^2_{(q)}$, where $\chi^2_{(q-1)}$ and $\chi^2_{(q)}$ are independent chi-square random variables with $q - 1$ and q degrees of freedom, respectively. The usual procedure for testing means comparing the likelihood ratio test statistic to $\chi^2_{(q)}$ because we are testing a variance parameter and $q - 1$ covariance parameters. Thus, if one rejects using the usual procedure, one will reject using the mixture distribution corresponding to $(1/2)\chi^2_{(q-1)} + (1/2)\chi^2_{(q)}$. Put another way, the actual p-value (computed using the mixture distribution) is less than the nominal p-value (computed using the standard distribution). Based on this, we see that the standard hypothesis testing procedure favors the simpler null hypothesis more often than it should.

No general rules for checking for the presence of several additional random effects are available although simulation methods are always possible. The important point is that analysts should not quickly quote p-values associated with testing variance components without carefully considering the model and estimator.

Further Reading

There are many introductions to multilevel modeling available in the literature. Two of the more technical, and widely cited, references are Raudenbush and Bryk (2002EP) and Goldstein (2002EP). If you would like an introduction that employs the minimal amount of mathematics, consider Toon (2000EP). A review of multilevel software is by de Leeuw and Kreft (2001EP). Reinsel (1984S, 1985S) considers growth curve modeling in a longitudinal data context. Andrews (2001E) provides recent results on testing when a parameter is on the boundary of the null hypothesis.

Appendix 5A High-Order Multilevel Models

Despite their widespread application, standard treatments that introduce the multilevel model use at most only three levels, anticipating that users will be able to intuit patterns (and hopefully equation structures) to higher levels. In contrast, this appendix describes a high-order multilevel model using "k" levels.

To motivate the extensions, begin with the three-level model in Section 5.1.2. By extending Equation (5.8), the level-1 model is expressed as

$$y_{i_1,i_2,\ldots,i_k} = \mathbf{Z}^{(1)}_{i_1,\ldots,i_k} \boldsymbol{\beta}^{(1)}_{i_1,\ldots,i_{k-1}} + \mathbf{X}^{(1)}_{i_1,\ldots,i_k} \boldsymbol{\beta}_1 + \varepsilon^{(1)}_{i_1,\ldots,i_k}.$$

Here, we might use i_k as a time index, i_{k-1} as a student index, i_{k-2} as a classroom index, and so on. We denote the observation set by

$$\mathbf{i}(k) = \{(i_1, i_2, \ldots, i_k) : y_{i_1,i_2,\ldots,i_k} \text{ is observed}\}.$$

More generally, define

$$\mathbf{i}(k-s) = \{(i_1, \ldots, i_{k-s}) : y_{i_1,\ldots,i_{k-s},j_{k-s+1},\ldots,j_k} \text{ is observed for some } j_{k-s+1}, \ldots, j_k\},$$

for $s = 0, 1, \ldots, k-1$. We will let $i(k) = \{i_1, i_2, \ldots, i_k\}$ be a typical element of $\mathbf{i}(k)$ and use $i(k-s) = \{i_1, \ldots, i_{k-s}\}$ for the corresponding element of $\mathbf{i}(k-s)$.

With this additional notation, we are now in a position to provide a recursive specification of high-order multilevel models.

Recursive Specification of High-Order Multilevel Models

1. The level-1 model is

$$y_{i(k)} = \mathbf{Z}_{i(k)}^{(1)} \boldsymbol{\beta}_{i(k-1)}^{(1)} + \mathbf{X}_{i(k)}^{(1)} \boldsymbol{\beta}_1 + \varepsilon_{i(k)}^{(1)}, \quad i(k) \in \mathbf{i}(k). \tag{5A.1}$$

The level-1 fixed parameter vector $\boldsymbol{\beta}_1$ has dimension $K_1 \times 1$ and the level-1 vector of parameters that may vary over higher levels, $\boldsymbol{\beta}_{i(k-1)}^{(1)}$, has dimension $q_1 \times 1$.

2. For $g = 2, \ldots, k-1$, the level-g model is

$$\boldsymbol{\beta}_{i(k+1-g)}^{(g-1)} = \mathbf{Z}_{i(k+1-g)}^{(g)} \boldsymbol{\beta}_{i(k-g)}^{(g)} + \mathbf{X}_{i(k+1-g)}^{(g)} \boldsymbol{\beta}_g + \boldsymbol{\epsilon}_{i(k+1-g)}^{(g)}, \quad \text{for} \quad g = 2, \ldots, k-1. \tag{5A.2}$$

Here, the level-g fixed parameter vector $\boldsymbol{\beta}_g$ has dimension $K_g \times 1$ and the level-g varying parameter vector $\boldsymbol{\beta}_{i(k-g)}^{(g)}$ has dimension $q_g \times 1$. Thus, for the covariates, $\mathbf{Z}_{i(k+1-g)}^{(g)}$ has dimension $q_{g-1} \times q_g$ and $\mathbf{X}_{i(k+1-g)}^{(g)}$ has dimension $q_{g-1} \times K_g$.

3. The level-k model is

$$\boldsymbol{\beta}_{i(1)}^{(k-1)} = \mathbf{X}_{i(1)}^{(k)} \boldsymbol{\beta}_k + \boldsymbol{\epsilon}_{i(1)}^{(k)}. \tag{5A.3}$$

We assume that all disturbance terms $\boldsymbol{\epsilon}$ are mean zero and are uncorrelated with one another. Further, define

$$\mathbf{D}_g = \mathrm{Var}\left(\boldsymbol{\epsilon}_{i(k+1-g)}^{(g)}\right) = \sigma_g^2 \mathbf{I}_{q_{g-1}}, \quad \text{for} \quad g \geq 2.$$

We now show how to write the multilevel model as a linear mixed-effects model. We do this by recursively inserting the higher level models from Equation (5A.2) into the level-1 Equation (5A.1). This yields

$$\begin{aligned}
y_{i(k)} &= \varepsilon_{i(k)}^{(1)} + \mathbf{X}_{i(k)}^{(1)} \boldsymbol{\beta}_1 + \mathbf{Z}_{i(k)}^{(1)} \left(\mathbf{Z}_{i(k-1)}^{(2)} \boldsymbol{\beta}_{i(k-2)}^{(2)} + \mathbf{X}_{i(k-1)}^{(2)} \boldsymbol{\beta}_2 + \varepsilon_{i(k-1)}^{(2)} \right) \\
&= \varepsilon_{i(k)}^{(1)} + \mathbf{X}_{i(k)}^{(1)} \boldsymbol{\beta}_1 + \mathbf{Z}_{i(k)}^{(1)} \left(\boldsymbol{\epsilon}_{i(k-1)}^{(2)} + \mathbf{X}_{i(k-1)}^{(2)} \boldsymbol{\beta}_2 \right) \\
&\quad + \mathbf{Z}_{i(k)}^{(1)} \mathbf{Z}_{i(k-1)}^{(2)} \left(\mathbf{Z}_{i(k-2)}^{(3)} \boldsymbol{\beta}_{i(k-3)}^{(3)} + \mathbf{X}_{i(k-2)}^{(3)} \boldsymbol{\beta}_3 + \boldsymbol{\epsilon}_{i(k-2)}^{(3)} \right) \\
&= \cdots = \varepsilon_{i(k)}^{(1)} + \mathbf{X}_{i(k)}^{(1)} \boldsymbol{\beta}_1 + \sum_{s=1}^{k-1} \left(\prod_{j=1}^{s} \mathbf{Z}_{i(k+1-j)}^{(j)} \right) \left(\boldsymbol{\epsilon}_{i(k-s)}^{(s+1)} + \mathbf{X}_{i(k-s)}^{(s+1)} \boldsymbol{\beta}_{s+1} \right).
\end{aligned}$$

To simplify notation, define the $1 \times q_s$ vector

$$\mathbf{Z}_{s,i(k)} = \prod_{j=1}^{s} \mathbf{Z}_{i(k+1-j)}^{(j)}. \tag{5A.4}$$

Further, define $K = K_1 + \cdots + K_k$, the $K \times 1$ vector $\boldsymbol{\beta} = (\boldsymbol{\beta}_1', \boldsymbol{\beta}_2', \ldots, \boldsymbol{\beta}_k')'$, and the $1 \times K$ vector

$$\mathbf{X}_{i(k)} = \begin{pmatrix} \mathbf{X}_{i(k)}^{(1)} & \mathbf{Z}_{1,i(k)}\mathbf{X}_{i(k-1)}^{(2)} & \cdots & \mathbf{Z}_{k-1,i(k)}\mathbf{X}_{i(1)}^{(k)} \end{pmatrix}.$$

This yields

$$\mathbf{X}_{i(k)}\boldsymbol{\beta} = \mathbf{X}_{i(k)}^{(1)}\boldsymbol{\beta}_1 + \sum_{s=1}^{k-1} \mathbf{Z}_{s,i(k)}\mathbf{X}_{i(k-s)}^{(s+1)}\boldsymbol{\beta}_{s+1}.$$

Thus, we may express the multilevel model as

$$y_{i(k)} = \mathbf{X}_{i(k)}\boldsymbol{\beta} + \varepsilon_{i(k)}^{(1)} + \sum_{s=1}^{k-1} \mathbf{Z}_{s,i(k)}\boldsymbol{\epsilon}_{i(k-s)}^{(s+1)}. \tag{5A.5}$$

To write Equation (5A.5) as a mixed linear model, we require some additional notation. For a fixed set $\{i_1, \ldots, i_{k-1}\} = i(k-1)$, let $n(i(k-1))$ denote the number of observed responses of the form $y_{i_1,\ldots,i_{k-1},j}$, for some j. Denote the set of observed responses as

$$\mathbf{y}_{i(k-1)} = \begin{pmatrix} y_{i_1,\ldots,i_{k-1},1} \\ \vdots \\ y_{i_1,\ldots,i_{k-1},n(i(k-1))} \end{pmatrix} = \begin{pmatrix} y_{i(k-1),1} \\ \vdots \\ y_{i(k-1),n(i(k-1))} \end{pmatrix}.$$

For each $s = 1, \ldots, k-1$, consider a set $\{i_1, \ldots, i_{k-s}\} = i(k-s)$ and let $n(i(k-s))$ denote the number of observed responses of the form $\mathbf{y}_{i(k-s),j}$, for some j. Thus, we define

$$\mathbf{y}_{i(k-s)} = \begin{pmatrix} \mathbf{y}_{i(k-s),1} \\ \vdots \\ \mathbf{y}_{i(k-s),n(i(k-s))} \end{pmatrix}.$$

Finally, let $\mathbf{y} = (\mathbf{y}_1', \ldots, \mathbf{y}_{n(i(1))}')'$. Use a similar stacking scheme for $\mathbf{X}$ and $\boldsymbol{\epsilon}^{(s)}$, for $s = 1, \ldots, k$. We may also use this notation when stacking over the first level of $\mathbf{Z}$. Thus, define

$$\mathbf{Z}_{s,i(k-1)} = \begin{pmatrix} \mathbf{Z}_{s,i(k-1),1} \\ \vdots \\ \mathbf{Z}_{s,i(k-1),n(i(k-1))} \end{pmatrix}, \quad \text{for} \quad s = 1, \ldots, k-1.$$

With this notation, when stacking over the first level, we may express Equation (5A.5) as

$$\mathbf{y}_{i(k-1)} = \mathbf{X}_{i(k-1)}\boldsymbol{\beta} + \boldsymbol{\epsilon}^{(1)}_{i(k-1)} + \sum_{s=1}^{k-1} \mathbf{Z}_{s,i(k-1)}\boldsymbol{\epsilon}^{(s+1)}_{i(k-s)}.$$

For the next level, define

$$\mathbf{Z}_{s,i(k-2)} = \begin{pmatrix} \mathbf{Z}_{s,i(k-2),1} \\ \vdots \\ \mathbf{Z}_{s,i(k-2),n(i(k-2))} \end{pmatrix}, \quad \text{for} \quad s = 2, \ldots, k-1,$$

and

$$\mathbf{Z}_{1,i(k-2)} = \text{block diag}(\mathbf{Z}_{1,i(k-2),1} \quad \cdots \quad \mathbf{Z}_{1,i(k-2),n(i(k-2))}).$$

With this notation, we have

$$\begin{aligned} &\mathbf{Z}_{1,i(k-2)}\boldsymbol{\epsilon}^{(2)}_{i(k-2)} \\ &= \begin{pmatrix} \mathbf{Z}_{1,i(k-2),1} & \mathbf{0} & \cdots & \mathbf{0} \\ \mathbf{0} & \mathbf{Z}_{1,i(k-2),2} & \cdots & \mathbf{0} \\ \vdots & \vdots & \ddots & \vdots \\ \mathbf{0} & \mathbf{0} & \cdots & \mathbf{Z}_{1,i(k-2),n(i(k-2))} \end{pmatrix} \begin{pmatrix} \boldsymbol{\epsilon}^{(2)}_{i(k-2),1} \\ \boldsymbol{\epsilon}^{(2)}_{i(k-2),2} \\ \vdots \\ \boldsymbol{\epsilon}^{(2)}_{i(k-2),n(i(k-2))} \end{pmatrix} \\ &= \begin{pmatrix} \mathbf{Z}_{1,i(k-2),1}\boldsymbol{\epsilon}^{(2)}_{i(k-2),1} \\ \mathbf{Z}_{1,i(k-2),2}\boldsymbol{\epsilon}^{(2)}_{i(k-2),2} \\ \vdots \\ \mathbf{Z}_{1,i(k-2),n(i(k-2))}\boldsymbol{\epsilon}^{(2)}_{i(k-2),n(i(k-2))} \end{pmatrix}. \end{aligned}$$

Thus, we have

$$\mathbf{y}_{i(k-2)} = \mathbf{X}_{i(k-2)}\boldsymbol{\beta} + \boldsymbol{\epsilon}^{(1)}_{i(k-2)} + \mathbf{Z}_{1,i(k-2)}\boldsymbol{\epsilon}^{(2)}_{i(k-2)} + \sum_{s=2}^{k-1} \mathbf{Z}_{s,i(k-2)}\boldsymbol{\epsilon}^{(s+1)}_{i(k-s)}.$$

Continuing, at the gth stage, we have

$$\mathbf{Z}_{s,i(k-g)} = \begin{pmatrix} \begin{pmatrix} \mathbf{Z}_{s,i(k-g),1} \\ \vdots \\ \mathbf{Z}_{s,i(k-g),n(i(k-g))} \end{pmatrix} & \text{for} \quad s \geq g \\ \text{block diag}(\mathbf{Z}_{s,i(k-g),1} \quad \cdots \quad \mathbf{Z}_{s,i(k-g),n(i(k-g))}) & \text{for} \quad s < g \end{pmatrix}.$$

This yields

$$\mathbf{y}_{i(k-g)} = \mathbf{X}_{i(k-g)}\boldsymbol{\beta} + \boldsymbol{\epsilon}^{(1)}_{i(k-g)} + \sum_{s=1}^{g} \mathbf{Z}_{s,i(k-g)}\boldsymbol{\epsilon}^{(s+1)}_{i(k-g)} + \sum_{s=g+1}^{k-1} \mathbf{Z}_{s,i(k-g)}\boldsymbol{\epsilon}^{(s+1)}_{i(k-s)}.$$

Taking $g = k - 1$, we have

$$\mathbf{y}_{i(1)} = \mathbf{X}_{i(1)}\boldsymbol{\beta} + \boldsymbol{\epsilon}^{(1)}_{i(1)} + \sum_{s=1}^{k-1} \mathbf{Z}_{s,i(1)}\boldsymbol{\epsilon}^{(s+1)}_{i(1)}, \tag{5A.6}$$

an expression for the usual linear mixed-effects model.

The system of notation takes us directly from the multilevel model in Equations (5A.1)–(5A.3) to the linear mixed-effects model in Equation (5A.6). Properties of parameter estimates for linear mixed-effects models are well established. Thus, parameter estimators of the multilevel model also enjoy these properties. Moreover, by showing how to write multilevel models as linear mixed-effects models, no special statistical software is required. One may simply use software written for linear mixed-effect smodels for multilevel modeling. See Frees and Kim (2004EP) for further details.

Exercises and Extensions

Section 5.3

5.1 Two-level model Consider the two-level model described in Section 5.1.1 and suppose that we have the error-components model, so that $\mathbf{z}_{ij} = \mathbf{z}_{1,ij} = 1$ and $\mathbf{R}_i$ is a scalar times the identity matrix. Further suppose that there are no level-1 explanatory variables. Show that the BLUP of the conditional mean of the level-1 response, $\mathrm{E}(y_{it} \mid \alpha_i) = \alpha_i + \mathbf{X}_{2,i}\boldsymbol{\beta}_2$, is $\zeta_i \bar{y}_i + (1 - \zeta_i)\mathbf{X}_{2,i}\mathbf{b}_{2,\mathrm{GLS}}$, where

$$\zeta_i = \frac{T_i}{T_i + (\mathrm{Var}\ \varepsilon)/(\mathrm{Var}\ \alpha)}.$$

5.2 Random-intercepts three-level model Assume that we observe $i = 1, \ldots, n$ school districts. Within each school district, we observe $j = 1, \ldots, J$ students. For each student, we have $t = 1, \ldots, T_{ij}$ observations. Assume that the model is given by

$$y_{i,j,t} = \mathbf{x}'_{ijt}\boldsymbol{\beta} + \alpha_i + \nu_{i,j} + \varepsilon_{i,j,t}.$$

Here, assume that each of $\{\alpha_i\}$, $\{\nu_{i,1}\}, \ldots, \{\nu_{i,J}\}$, and $\{\varepsilon_{i,j,t}\}$ are i.i.d. as well as independent of one another. Also assume that $\{\alpha_i\}$, $\{\nu_{i,1}\}, \ldots, \{\nu_{i,J}\}$, and $\{\varepsilon_{i,j,t}\}$ are mean zero with variances $\sigma^2_\alpha, \sigma^2_{\upsilon 1}, \ldots, \sigma^2_{\upsilon J}$, and σ^2_ε, respectively.

Define $\mathbf{z}_{ij}$ to be a $T_{ij} \times (J+1)$ matrix with ones in the first and $j+1$st columns and zeros elsewhere. For example, we have

$$\mathbf{z}_{i2} = \begin{pmatrix} 1 & 0 & 1 & \cdots & 0 \\ 1 & 0 & 1 & \cdots & 0 \\ \vdots & \vdots & \vdots & \vdots & \vdots \\ 1 & 0 & 1 & \cdots & 0 \end{pmatrix}.$$

Further, define $\mathbf{Z}_i = (\mathbf{z}_{i1}' \; \mathbf{z}_{i2}' \; \cdots \; \mathbf{z}_{iJ}')'$, $\boldsymbol{\alpha}_i = (\alpha_i \, \upsilon_{i1} \; \cdots \; \upsilon_{iJ})'$, and

$$\mathbf{D} = \mathrm{Var}\, \boldsymbol{\alpha}_i = \begin{pmatrix} \sigma_\alpha^2 & \mathbf{0} \\ \mathbf{0} & \mathbf{D}_\upsilon \end{pmatrix},$$

where $\mathbf{D}_\upsilon = \mathrm{diag}(\sigma_{\upsilon 1}^2, \sigma_{\upsilon 2}^2, \ldots, \sigma_{\upsilon J}^2)$.

a. Define $\mathbf{y}_i$, $\mathbf{X}_i$, and $\boldsymbol{\epsilon}_i$ in terms of $\{y_{i,j,t}\}$, $\{\mathbf{x}_{ijt}\}$, and $\{\varepsilon_{i,j,t}\}$, so that we may write $\mathbf{y}_i = \mathbf{Z}_i \boldsymbol{\alpha}_i + \mathbf{X}_i \boldsymbol{\beta} + \boldsymbol{\epsilon}_i$, using the usual notation.
b. For the appropriate choice of $\mathbf{R}_i$, show that

$$\mathbf{Z}_i' \mathbf{R}_i^{-1} (\mathbf{y}_i - \mathbf{X}_i \mathbf{b}_{\mathrm{GLS}}) = \frac{1}{\sigma_\varepsilon^2} (T_i \bar{e}_i \quad T_{i1} \bar{e}_{i1} \quad \cdots \quad T_{iJ} \bar{e}_{iJ})',$$

where $e_{i,j,t} = y_{i,j,t} - \mathbf{x}_{ijt}' \mathbf{b}_{\mathrm{GLS}}$, $\bar{e}_{ij} = T_{ij}^{-1} \sum_{t=1}^{T_{ij}} e_{ijt}$, and $T_i \bar{e}_i = \sum_{j=1}^{J} \sum_{t=1}^{T_{ij}} e_{ijt}$.
c. Show that

$$\left(\mathbf{D}^{-1} + \mathbf{Z}_i' \mathbf{R}_i^{-1} \mathbf{Z}_i \right)^{-1} = \begin{pmatrix} \mathbf{C}_{11}^{-1} & -\mathbf{C}_{11}^{-1} \boldsymbol{\zeta}_\upsilon' \\ -\mathbf{C}_{11}^{-1} \boldsymbol{\zeta}_\upsilon & \mathbf{C}_{22}^{-1} \end{pmatrix},$$

where

$$\zeta_i = \frac{\sigma_\alpha^2 T_i}{\sigma_\varepsilon^2 + \sigma_\alpha^2 T_i}, \qquad \zeta_{\upsilon j} = \frac{\sigma_{\upsilon j}^2 T_{ij}}{\sigma_\varepsilon^2 + \sigma_{\upsilon j}^2 T_{ij}},$$
$$\boldsymbol{\zeta}_\upsilon = (\zeta_{\upsilon 1}, \zeta_{\upsilon 2}, \ldots, \zeta_{\upsilon J})',$$

$\mathbf{T}_{i,2} = \mathrm{diag}(T_{i1}, T_{i2}, \ldots, T_{iJ})$,

$$\mathbf{C}_{11}^{-1} = \frac{\sigma_\varepsilon^2 \zeta_i}{\sum_{j=1}^{J} T_{ij} (1 - \zeta_i \zeta_{\upsilon j})},$$

and

$$\mathbf{C}_{22}^{-1} = \left(\mathbf{D}_\upsilon^{-1} + \sigma_\varepsilon^{-2} \mathbf{T}_{i,2} \right)^{-1} + \mathbf{C}_{11}^{-1} \boldsymbol{\zeta}_\upsilon \boldsymbol{\zeta}_\upsilon'.$$

d. With the notation $\mathbf{a}_{i,\text{BLUP}} = (a_{i,\text{BLUP}} \quad \upsilon_{i1,\text{BLUP}} \quad \cdots \quad \upsilon_{iJ,\text{BLUP}})'$, show that

$$a_{i,\text{BLUP}} = \zeta_i \frac{\sum_{j=1}^{J} T_{ij}(1-\zeta_{\upsilon j})\bar{e}_{ij}}{\sum_{j=1}^{J} T_{ij}(1-\zeta_i \zeta_{\upsilon j})}$$

and

$$\upsilon_{ij,\text{ BLUP}} = \zeta_{\upsilon j}(\bar{e}_{ij} - a_{i,\text{BLUP}}).$$

Section 5.4

5.3 Maximum likelihood variance estimators without boundary conditions Consider the basic random-effects model and suppose that $T_i = T$, $K = 1$, and $x_{it} = 1$. Parts (a) and (b) are the same as Exercise 3.12(a) and (c). As there, we now ignore boundary conditions so that the estimator may become negative with positive probability.

a. Show that the MLE of σ^2 may be expressed as

$$\hat{\sigma}^2_{\varepsilon,\text{ML}} = \frac{1}{n(T-1)} \sum_{i=1}^{n} \sum_{t=1}^{T} (y_{it} - \bar{y}_i)^2.$$

b. Show that the MLE of σ_α^2 may be expressed as

$$\hat{\sigma}^2_{\alpha,\text{ML}} = \frac{1}{n} \sum_{i=1}^{n} (\bar{y}_i - \bar{y})^2 - \frac{1}{T}\hat{\sigma}^2_{\varepsilon,\text{ML}}.$$

c. Show that the maximum likelihood may be expressed as

$$L\left(\hat{\sigma}^2_{\alpha,\text{ML}}, \hat{\sigma}^2_{\varepsilon,\text{ML}}\right)$$
$$= -\frac{n}{2}\left\{T \ln(2\pi) + T + (T-1)\ln \hat{\sigma}^2_{\varepsilon,\text{ML}} + \ln\left(T\hat{\sigma}^2_{\alpha,\text{ML}} + \hat{\sigma}^2_{\varepsilon,\text{ML}}\right)\right\}.$$

d. Consider the null hypothesis $\text{H}_0 : \sigma_\alpha^2 = 0$. Under this null hypothesis, show that the MLE of σ^2 may be expressed as

$$\hat{\sigma}^2_{\varepsilon,\text{Reduced}} = \frac{1}{nT} \sum_{i=1}^{n} \sum_{t=1}^{T} (y_{it} - \bar{y})^2.$$

e. Under the null hypothesis $\text{H}_0 : \sigma_\alpha^2 = 0$, show that the maximum likelihood may be expressed as

$$L\left(0, \hat{\sigma}^2_{\varepsilon,\text{Reduced}}\right) = -\frac{n}{2}\left\{T \ln(2\pi) + T + T \ln\left(\frac{1}{nT} \sum_{i=1}^{n} \sum_{t=1}^{T} (y_{it} - \bar{y})^2\right)\right\}.$$

f. Use a second-order approximation of the logarithm function to show that twice the difference of log-likelihoods may be expressed as

$$2\big(L(\hat{\sigma}^2_{\alpha,\text{ML}}, \hat{\sigma}^2_{\varepsilon,\text{ML}}) - L(0, \hat{\sigma}^2_{\varepsilon,\text{Reduced}})\big) = \frac{1}{2nT(T-1)\sigma^4_\varepsilon}\left\{SSW - (T-1)\,SSB\right\}^2,$$

where $SSW = \sum_{i=1}^n \sum_{t=1}^T (y_{it} - \bar{y}_i)^2$ and $SSB = T\sum_{i=1}^n (\bar{y}_i - \bar{y})^2$.

g. Assuming normality of the responses and the null hypothesis $\text{H}_0 : \sigma^2_\alpha = 0$, show that

$$2\left(L\left(\hat{\sigma}^2_{\alpha,\text{ML}}, \hat{\sigma}^2_{\varepsilon,\text{ML}}\right) - L\left(0, \hat{\sigma}^2_{\varepsilon,\text{Reduced}}\right)\right) \to_D \chi^2_{(1)},$$

as $n \to \infty$.

5.4 Maximum likelihood variance estimators with boundary conditions Consider the basic random-effects model and suppose that $T_i = T$, $K = 1$, and $x_{it} = 1$. Unlike Exercise 5.3, we now impose boundary conditions so that variance estimators must be nonnegative.

a. Using the notation of Exercise 5.3, show that the MLEs of σ^2_α and σ^2 may be expressed as

$$\hat{\sigma}^2_{\alpha,\text{CML}} = \begin{cases} \hat{\sigma}^2_{\alpha,\text{ML}} & \text{if } \hat{\sigma}^2_{\alpha,\text{ML}} > 0, \\ 0 & \text{if } \hat{\sigma}^2_{\alpha,\text{ML}} \leq 0 \end{cases}$$

and

$$\hat{\sigma}^2_{\varepsilon,\text{CML}} = \begin{cases} \hat{\sigma}^2_{\varepsilon,\text{ML}} & \text{if } \hat{\sigma}^2_{\alpha,\text{ML}} > 0, \\ \hat{\sigma}^2_{\varepsilon,\text{Reduced}} & \text{if } \hat{\sigma}^2_{\alpha,\text{ML}} \leq 0. \end{cases}$$

An early reference for this result is Herbach (1959S). Here, CML stands for constrained maximum likelihood.

b. Show that the maximum likelihood may be expressed as

$$L\left(\hat{\sigma}^2_{\alpha,\text{CML}}, \hat{\sigma}^2_{\varepsilon,\text{CML}}\right) = \begin{cases} L\left(\hat{\sigma}^2_{\alpha,\text{ML}}, \hat{\sigma}^2_{\varepsilon,\text{ML}}\right) & \text{if } \hat{\sigma}^2_{\alpha,\text{ML}} > 0, \\ L\left(0, \hat{\sigma}^2_{\varepsilon,\text{Reduced}}\right) & \text{if } \hat{\sigma}^2_{\alpha,\text{ML}} \leq 0. \end{cases}$$

c. Define the cutoff $c_n = (T-1)SSB/SSW - 1$. Check that $c_n > 0$ if and only if $\hat{\sigma}^2_{\alpha,\text{ML}} > 0$. Confirm that we may express the likelihood ratio statistic as

$$
\begin{aligned}
&2\left(L\left(\hat{\sigma}^2_{\alpha,\text{CML}}, \sigma^2_{\varepsilon,\text{CML}}\right) - L\left(0, \hat{\sigma}^2_{\varepsilon,\text{Reduced}}\right)\right) \\
&\quad = \begin{cases} n\left[T\ln\left(1+\frac{c_n}{T}\right) - \ln\left(1+c_n\right)\right] & \text{if } c_n > 0, \\ 0 & \text{if } c_n \le 0. \end{cases}
\end{aligned}
$$

d. Assuming normality of the responses and the null hypothesis $\text{H}_0 : \sigma^2_\alpha = 0$, show that the cutoff $c_n \to_p 0$ as $n \to \infty$.

e. Assuming normality of the responses and the null hypothesis $H_0 : \sigma^2_\alpha = 0$, show that

$$\sqrt{n}c_n \to_D N\left(0, \frac{2T}{T-1}\right) \quad \text{as } n \to \infty.$$

f. Assume normality of the responses and the null hypothesis $\text{H}_0 : \sigma^2_\alpha = 0$. Show, for $a > 0$, that

$$
\begin{aligned}
&\text{Prob}\left[2\left(L\left(\hat{\sigma}^2_{\alpha,\text{CML}}, \sigma^2_{\varepsilon,\text{CML}}\right) - L\left(0, \hat{\sigma}^2_{\varepsilon,\text{Reduced}}\right)\right) > a\right] \\
&\quad \to 1 - \Phi\left(\sqrt{a}\right) \quad \text{as } n \to \infty,
\end{aligned}
$$

where Φ is the standard normal distribution function.

g. Assume normality of the responses and the null hypothesis $\text{H}_0 : \sigma^2_\alpha = 0$. Summarize the previous results to establish that the likelihood ratio test statistic asymptotically has a distribution that is 50% equal to 0 and 50% a chi-square distribution with one degree of freedom.

Empirical Exercise

5.5 Student achievement These data were gathered to assess the relationship between student achievement and education initiatives. Moreover, they can also be used to address related interesting questions, such as how one can rank the performance of schools or how one can forecast a child's future performance on achievement tests based on early test scores.

Webb et al. (2002EP) investigated relationships between student achievement and Texas school district participation in the National Science Foundation Statewide Systemic Initiatives program between 1994 and 2000. They focused on the effects of systemic reform on performance on a state mathematics test. We consider here a subset of these data to model trajectories of students' mathematics achievement over time. This subset consists of a random sample of 20 elementary schools in Dallas, with 20 students randomly selected from

Table 5E.1. *Variables for student achievement excercise*

Variable	Description
Level-1 variables (replications over time)	
GRADE	Grade when assessment was made (3–6)
YEAR	Year of assessment (1994–2000)
TIME	Observed repeated occasions for each student
RETAINED	Retained in grade for a particular year (1 = yes, 0 = no)
SWITCH_SCHOOLS	Switched schools in a particular year (1 = yes, 0 = no)
DISADVANTAGED	Economically disadvantaged (1 = free/reduced lunch, 0 = no)
TLI_MATH	Texas Learning Index on mathematics (assessment measure)
Level-2 variables (replications over child)	
CHILDID	Student identification number
MALE	Gender of students (1 = male, 0 = female)
ETHNICITY	White, black, Hispanic, other ("other" includes Asian as well as mixed races)
FIRST_COHORT	First observed cohort membership
LAST_COHORT	Last observed cohort membership
Level-3 variables (replications over school)	
SCHOOLID	School identification number
USI	Urban System Initiative cohort (1 = 1993, 2 = 1994, 3 = 1995)
MATH_SESSIONS	Number of teachers who attended mathematics sessions
N_TEACHERS	Total number of teachers in the school

Source: Webb et al. (2002EP).

each school. All available records for these 400 students during elementary school are included. In Dallas, grades 3 through 6 correspond to elementary school.

Although there exists a natural hierarchy at each time point (students are nested within schools), this hierarchy was not maintained completely over time. Several students switched schools (see variable SWITCH_SCHOOLS) and some students were not promoted (see variable RETAINED). To maintain the hierarchy of students within schools, a student was associated with a school at the time of selection. To maintain a hierarchy over time, a cohort variable was defined as 1, 2, 3, or 4 for those in grades 6, 5, 4, or 3, respectively, in 1994. For those in grade 3 in 1995, the cohort variable takes on a value of 5, and so on up to a 10 for those in grade 3 in 2000. The variable FIRST_COHORT attaches a student to a cohort during the first year of observation whereas the variable LAST_COHORT attaches a student to a cohort during the last year of observation. The variables used are listed in Table 5E.1.

a. *Basic summary statistics*
   i. Summarize the school-level variables. Produce a table to summarize the frequency of the USI variable. For MATH_SESSIONS and N_TEACHERS, provide the mean, median, standard deviation, minimum, and maximum.
   ii. Summarize the child-level variables. Produce tables to summarize the frequency of gender, ethnicity, and the cohort variables.
   iii. Provide basic relationships among level-2 and level-3 variables. Summarize the number of teachers by gender. Examine ethnicity by gender.
   iv. Summarize the level-1 variables. Produce means for the binary variables RETAINED, SWITCH_SCHOOLS, and DISAVANTAGED. For TLI_MATH, provide the mean, median, standard deviation, minimum, and maximum.
   v. Summarize numerically some basic relationships between TLI_MATH and the explanatory variables. Produce tables of means of TLI_MATH by GRADE, YEAR, RETAINED, SWITCH_SCHOOLS, and DISAVANTAGED.
   vi. Summarize graphically some basic relationships between TLI_MATH and the explanatory variables. Produce boxplots of TLI_MATH by GRADE, YEAR, RETAINED, SWITCH_SCHOOLS, and DISAVANTAGED. Comment on the trend over time and grade.
   vii. Produce a multiple time-series plot of TLI_MATH. Comment on the dramatic declines of some students in year-to-year test scores.
b. *Two-level error-components model*
   i. Ignoring the school-level information, run an error-components model using child as the second-level unit of analysis. Use the level-1 categorical variables GRADE and YEAR and binary variables RETAINED and SWITCH_SCHOOLS. Use the level-2 categorical variables ETHNICITY and the binary variable MALE.
   ii. Repeat your analysis in part (b)(i) but include the variable DISAVANTAGED. Describe the advantages and disadvantages of including this variable in the model specification.
   iii. Repeat your analysis in part (b)(i) but include an $AR(1)$ specification of the error. Does this improve the model specification?
   iv. Repeat your analysis in part (b)(iii) but include a (fixed) school-level categorical variable. Does this improve the model specification?

c. *Three-level model*
   i. Now incorporate school-level information into your model in (b)(i). At the first level, the random intercept varies by child and school. We also include GRADE, YEAR, RETAINED, and SWITCH_SCHOOLS as level-1 explanatory variables. For the second-level model, the random intercept varies by school and includes ETHNICITY and MALE as level-2 explanatory variables. At the third level, we include USI, MATH_SESSIONS, and N_TEACHERS as level-3 explanatory variables. Comment on the appropriateness of this fit.
   ii. Is the USI categorical variable statistically significant? Rerun the part (c)(i) model without USI and use a likelihood ratio test statistic to respond to this question.
   iii. Repeat your analysis in part (c)(i) but include an $AR(1)$ specification of the error. Does this improve the model specification?

Appendix 5A

5.6 BLUPs for a general multilevel model Consider the general multilevel model developed in Appendix 5A and the mixed linear model representation in Equation (5A.6). Let $\mathbf{V}_{i(1)} = \text{Var}\, \mathbf{y}_{i(1)}$.

a. Using best linear unbiased prediction introduced in Section 4.3, show that we can express the BLUPs of the residuals as

$$\mathbf{e}^{(g)}_{i(k+1-g),\text{BLUP}} = \text{Cov}\left(\boldsymbol{\epsilon}^{(g)}_{i(1)}, \boldsymbol{\epsilon}^{(g)}_{i(k+1-g)}\right)\mathbf{Z}'_{g-1,i(1)}\mathbf{V}^{-1}_{i(1)}\left(\mathbf{y}_{i(1)} - \mathbf{X}_{i(1)}\mathbf{b}_{\text{GLS}}\right),$$

   for $g = 2, \ldots, k$ and, for $g = 1$, as

$$\mathbf{e}^{(1)}_{i(k),\text{BLUP}} = \text{Cov}\left(\boldsymbol{\epsilon}^{(1)}_{i(1)}, \varepsilon^{(1)}_{i(k)}\right)\mathbf{V}^{-1}_{i(1)}\left(\mathbf{y}_{i(1)} - \mathbf{X}_{i(1)}\mathbf{b}_{\text{GLS}}\right).$$

b. Show that the BLUP of $\boldsymbol{\beta}^{(g-1)}_{i(k+1-g)}$ is

$$\mathbf{b}^{(g-1)}_{i(k+1-g),\text{BLUP}} = \mathbf{X}^{(g)}_{i(k+1-g)}\mathbf{b}_{g,\text{GLS}} + \mathbf{Z}^{(g)}_{i(k+1-g)}\mathbf{b}^{(g)}_{i(k-g),\text{BLUP}} + \mathbf{e}^{(g)}_{i(k+1-g),\text{BLUP}}.$$

c. Show that the BLUP forecast of $y_{i_1,i_2,\ldots,i_k+L}$ is

$$\begin{aligned}\hat{y}_{i_1,i_2,\ldots,i_k+L} &= \mathbf{Z}^{(1)}_{i_1,i_2,\ldots,i_k+L}\mathbf{b}^{(1)}_{i(k-1),\text{BLUP}} + \mathbf{X}^{(1)}_{i_1,i_2,\ldots,i_k+L}\mathbf{b}_{1,\text{GLS}} \\ &\quad + \text{Cov}\left(\varepsilon^{(1)}_{i_1,i_2,\ldots,i_k+L}, \boldsymbol{\epsilon}^{(1)}_{i(1)}\right)\mathbf{V}^{-1}_{i(1)}\left(\mathbf{y}_{i(1)} - \mathbf{X}_{i(1)}\mathbf{b}_{\text{GLS}}\right).\end{aligned}$$

6
Stochastic Regressors

Abstract. In many applications of interest, explanatory variables, or *regressors*, cannot be thought of as fixed quantities but, rather, they are modeled stochastically. In some applications, it can be difficult to determine which variables are being predicted and which are doing the prediction! This chapter summarizes several models that incorporate stochastic regressors. The first consideration is to identify under what circumstances we can safely condition on stochastic regressors and to use the results from prior chapters. We then discuss *exogeneity*, formalizing the idea that a regressor influences the response variable and not the other way around. Finally, this chapter introduces situations where more than one response is of interest, thus permitting us to investigate complex relationships among responses.

6.1 Stochastic Regressors in Nonlongitudinal Settings

Up to this point, we have assumed that the explanatory variables, $\mathbf{X}_i$ and $\mathbf{Z}_i$, are nonstochastic. This convention follows a long-standing tradition in the statistics literature. Pedagogically, this tradition allows for simpler verification of properties of estimators than the stochastic convention. Moreover, in classical experimental or laboratory settings, treating explanatory variables as nonstochastic allows for intuitive interpretations, such as when $\mathbf{X}$ is under the control of the analyst.

However, for other applications, such as the analysis of survey data drawn as a probability sample from a population, the assumption of nonstochastic variables is more difficult to interpret. For example, when drawing a sample of individuals to understand each individual's health care decisions, we may wish to explain his or her health care services utilization in terms of age, gender, race, and so on. These are plausible explanatory variables and it seems sensible to model them as stochastic in that the sample values are determined by a random draw from a population.

In some ways, the study of stochastic regressors subsumes that of nonstochastic regressors. First, with stochastic regressors, we can always adopt the convention that a stochastic quantity with zero variance is simply a deterministic, or nonstochastic, quantity. Second, we may make inferences about population relationships conditional on values of stochastic regressors, essentially treating them as fixed. However, the choice of variables on which we condition depends on the scientific interest of the problem, making the difference between fixed and stochastic regressors dramatic in some cases.

Understanding the best ways to use stochastic regressors in longitudinal settings remains a developing research area. Thus, before presenting techniques useful for longitudinal data, this section reviews known and proven methods that are useful in nonlongitudinal settings, either for the cross-section or the time dimension. Subsequent sections in this chapter focus on longitudinal settings that incorporate both the cross-section and time dimensions.

6.1.1 Endogenous Stochastic Regressors

An *exogenous variable* is one that can be taken as "given" for the purposes at hand. As we will see, exogeneity requirements vary depending on the context. An *endogenous variable* is one that fails the exogeneity requirement. In contrast, it is customary in economics to use the term "endogenous" to mean a variable that is determined within an economic system, whereas an exogenous variable is determined outside the system. Thus, the accepted econometric and statistical usage differs from the general economic meaning.

To develop exogeneity and endogeneity concepts, we begin by thinking of $\{(\mathbf{x}_i, y_i)\}$ as a set of observations from the same distribution. For example, this assumption is appropriate when the data arise from a survey where information is collected using a simple random-sampling mechanism. We suppress the t subscript because we are considering only one dimension in this section. Thus, i may represent either the cross-sectional identifier or time period. For nonlongitudinal data, we do not consider the $\mathbf{z}_i$ variables.

For independent observations, we can write the assumptions of the linear model as in Chapter 2, adding only that we are conditioning on the stochastic explanatory variables when writing down the moments of the response y. To begin, for the regression function, we assume that $\mathrm{E}(y_i \mid \mathbf{x}_i) = \mathbf{x}_i'\boldsymbol{\beta}$ and the conditional variance is $\mathrm{Var}\,(y_i \mid \mathbf{x}_i) = \sigma^2$.

To handle additional sampling mechanisms, we now introduce a more general setting. Specifically, we condition on *all* of the explanatory variables in the sample, not just the ones associated with the ith draw. Define $\mathbf{X} = (\mathbf{x}_1, \ldots, \mathbf{x}_n)$ and work with the following assumptions.

Assumptions of the Linear Regression Model with Strictly Exogenous Regressors

SE1. $\mathrm{E}(y_i \mid \mathbf{X}) = \mathbf{x}_i' \boldsymbol{\beta}$.
SE2. $\{\mathbf{x}_1, \ldots, \mathbf{x}_n\}$ are stochastic variables.
SE3. $\mathrm{Var}\,(y_i \mid \mathbf{X}) = \sigma^2$.
SE4. $\{y_i \mid \mathbf{X}\}$ are independent random variables.
SE5. $\{y_i\}$ are normally distributed, conditional on $\{\mathbf{X}\}$

Assuming for the moment that $\{(\mathbf{x}_i, y_i)\}$ are mutually independent, then $\mathrm{E}(y_i \mid \mathbf{X}) = \mathrm{E}(y_i \mid \mathbf{x}_i)$, $\mathrm{Var}\,(y_i \mid \mathbf{X}) = \mathrm{Var}\,(y_i \mid \mathbf{x}_i)$, and $(y_i \mid \mathbf{x}_i)$ are independent. Thus, Assumptions SE1–SE4 are certainly useful in the random-sampling context.

Moreover, Assumptions SE1–SE4 are the appropriate stochastic regressor generalization of the fixed regressors model assumptions that ensure that we retain most of the desirable properties of the OLS estimators of $\boldsymbol{\beta}$. For example, the unbiasedness and the Gauss–Markov properties of OLS estimators of $\boldsymbol{\beta}$ hold under SE1–SE4. Moreover, assuming conditional normality of the responses, then the usual t- and F-statistics have their customary distributions, regardless as to whether or not $\mathbf{X}$ is stochastic; see, for example Greene (2002E) or Goldberger (1991E).

The usual OLS estimators also have desirable asymptotic properties under Assumptions SE1–SE4. For many social science applications, data sets are large and researchers are primarily interested in asymptotic properties of estimators. If achieving desirable asymptotic properties is the goal, then the stochastic regressor model assumptions SE1–SE4 can be relaxed, thus permitting a wider scope of applications.

For discussion purposes, we now focus on the first assumption. Using the linear model framework, we define the disturbance term to be $\varepsilon_i = y_i - \mathbf{x}_i' \boldsymbol{\beta}$ and write SE1 as $\mathrm{E}(\varepsilon_i \mid \mathbf{X}) = 0$. This assumption on the regressors is known as *strict exogeneity* in the econometrics literature (see, for example, Hayashi, 2000E). If the index i represents independent cross-sectional draws, such as with simple random sampling, then strict exogeneity is an appropriate assumption. However, if the index i represents time, then the strict exogeneity is not useful for many applications; it assumes that the time i disturbance term is orthogonal to all regressors in the past, present, and the future.

An alternative assumption is as follows:

SE1p. $\mathrm{E}(\varepsilon_i \mathbf{x}_i) = \mathrm{E}((y_i - \mathbf{x}_i' \boldsymbol{\beta}) \mathbf{x}_i) = \mathbf{0}$.

If SE1p holds, then the regressors are said to be *predetermined*. Because SE1p implies zero covariance between the regressors and the disturbances, we say that predetermined regressors are uncorrelated with contemporaneous disturbances. Another way of expressing Assumption SE1p is through the *linear projection*

$$LP(\varepsilon_i \mid \mathbf{x}_i) = 0.$$

See Appendix 6A for definitions and properties of linear projections. This alternative method will be useful as we explore longitudinal extensions of the notion of endogeneity in Section 6.3.

Assumption SE1p is weaker than SE1. Only the weaker assumption (and conditions analogous to those in SE2–SE4) is required for the asymptotic property of consistency of the OLS estimators of $\boldsymbol{\beta}$. We will be more specific in our discussion of longitudinal data beginning in Section 6.2. For specifics regarding nonlongitudinal data settings, see, for example, Hayashi (2000E).

To reiterate, the strict exogeneity assumption SE1 is sufficient for the OLS estimators of $\boldsymbol{\beta}$ to retain finite sample properties such as unbiasedness whereas only the weaker predetermined assumption SE1p is required for consistency. For asymptotic normality, we require an assumption that is somewhat stronger than SE1p. A sufficient condition is the following:

SE1m. $\mathrm{E}(\varepsilon_i \mid \varepsilon_{i-1}, \ldots, \varepsilon_1, \mathbf{x}_i, \ldots, \mathbf{x}_1) = 0$ for all i.

When SE1m holds, then $\{\varepsilon_i\}$ satisfies the requirements for a *martingale difference* sequence. We note, using the law of iterated expectations, that SE1m implies SE1p. For time-series data where the index i represents time, we see that both Assumptions SE1p and SE1m do not rule out the possibility that the current error term ε_i will be related to future regressors, as does the strict exogeneity assumption SE1.

6.1.2 Weak and Strong Exogeneity

We began Section 6.1.1 by expressing two types of exogeneity, strict exogeneity and predeterminedness, in terms of conditional means. This is appropriate for linear models because it gives precisely the conditions needed for inference and is directly testable. Here we begin by generalizing these concepts to assumptions regarding the entire distribution, not just the mean function. Although stronger than the conditional mean versions, these assumptions are directly applicable to nonlinear models. Moreover, we use this distribution framework to introduce two new types of exogeneity, weak and strong exogeneity.

A stronger version of strict exogeneity in SE1 is

SE1′. ε_i is independent of $\mathbf{X}$.

Here, we are using the convention that the zero-mean disturbances are defined as $\varepsilon_i = y_i - \mathbf{x}_i' \boldsymbol{\beta}$. Note that SE1′ implies SE1; SE1′ is a requirement on the joint distribution of ε_i and $\mathbf{X}$, not just the conditional mean. Similarly, a stronger version of SE1p is

SE1p′. ε_i is independent of $\mathbf{x}_i$.

A drawback of SE1 and SE1p is that the reference to parameter estimability is only implicit. An alternative set of definitions introduced by Engle, Hendry, and Richard (1983E) explicitly defines exogeneity in terms of parametric likelihood functions. Intuitively, a set of variables is said to be *weakly exogenous* if, when we condition on them, there is no loss of information for the parameters of interest. If, in addition, the variables are "not caused" by the endogenous variables, then they are said to be *strongly exogenous*. Weak exogeneity is sufficient for efficient estimation. Strong exogeneity is required for conditional predictions in forecasting of endogenous variables.

Specifically, suppose that we have random variables $(\mathbf{x}_1, y_1), \ldots, (\mathbf{x}_T, y_T)$ with joint probability density (or mass) function for $f(y_1, \ldots, y_T, \mathbf{x}_1, \ldots, \mathbf{x}_T)$. Using t for the (time) index, we can always write this conditionally as

$$
\begin{aligned}
f(y_1, \ldots, y_T, \mathbf{x}_1, \ldots, \mathbf{x}_T) \\
&= \prod_{t=1}^{T} f(y_t, \mathbf{x}_t \mid y_1, \ldots, y_{t-1}, \mathbf{x}_1, \ldots, \mathbf{x}_{t-1}) \\
&= \prod_{t=1}^{T} \{ f(y_t \mid y_1, \ldots, y_{t-1}, \mathbf{x}_1, \ldots, \mathbf{x}_t) f(\mathbf{x}_t \mid y_1, \ldots, y_{t-1}, \mathbf{x}_1, \ldots, \mathbf{x}_{t-1}) \}.
\end{aligned}
$$

Here, when $t = 1$ the conditional distributions are the marginal distributions of y_1 and $\mathbf{x}_1$, as appropriate. Now, suppose that this joint distribution is characterized by vectors of parameters $\boldsymbol{\theta}$ and $\boldsymbol{\psi}$ such that

$$
\begin{aligned}
\text{SE1w. } & f(y_1, \ldots, y_T, \mathbf{x}_1, \ldots, \mathbf{x}_T) \\
&= \left(\prod_{t=1}^{T} f(y_t \mid y_1, \ldots, y_{t-1}, \mathbf{x}_1, \ldots, \mathbf{x}_t, \boldsymbol{\theta}) \right) \\
&\quad \times \left(\prod_{t=1}^{T} f(\mathbf{x}_t \mid y_1, \ldots, y_{t-1}, \mathbf{x}_1, \ldots, \mathbf{x}_{t-1}, \boldsymbol{\psi}) \right).
\end{aligned}
$$

In this case, we can ignore the second term for inference about $\boldsymbol{\theta}$, treating the $\mathbf{x}$ variables as essentially fixed. If the relationship SE1w holds, then we say that the explanatory variables are *weakly exogenous*.

Suppose, in addition, that

$$f(\mathbf{x}_t \mid y_1, \ldots, y_{t-1}, \mathbf{x}_1, \ldots, \mathbf{x}_{t-1}, \psi) = f(\mathbf{x}_t \mid \mathbf{x}_1, \ldots, \mathbf{x}_{t-1}, \psi); \quad (6.1\text{a})$$

that is, conditional on $\mathbf{x}_1, \ldots, \mathbf{x}_{t-1}$, the distribution of $\mathbf{x}_t$ does not depend on past values of y, $y_1, \ldots, y_{t-1}$. Then, we say that $\{y_1, \ldots, y_{t-1}\}$ does not *Granger-cause* $\mathbf{x}_t$. This condition, together with SE1w, suffices for *strong exogeneity*.We note that Engle et al. (1983E) also introduce a *super exogeneity* assumption for policy analysis purposes; we will not consider this type of exogeneity.

6.1.3 Causal Effects

Issues of when variables are endogenous or exogenous are important to researchers who use statistical models as part of their arguments for assessing whether causal relationships hold. Researchers are interested in causal effects, often more so than measures of association among variables.

Traditionally, statistics has contributed to making causal statements primarily through randomization. This tradition goes back to the work of Fisher and Neyman in the context of agricultural experiments. In Fisher's work, treatments were randomly allocated to experimental units (plots of land). Because of this random assignment, differences in responses (crop yields from the land) could be reasonably ascribed to treatments without fear of underlying systematic influences from unknown factors. Data that arise from this random-assignment mechanism are known as *experimental*.

In contrast, most data from the social sciences are *observational*, where it is not possible to use random mechanisms to randomly allocate observations according to variables of interest. However, it is possible to use random mechanisms to gather data through probability samples and thus to estimate stochastic relationships among variables of interest. The primary example of this is the use of a simple random-sample mechanism to collect data and estimate a conditional mean through regression methods. The important point is that this regression function measures relationships developed through the data-gathering mechanism, not necessarily the relationships of interest to researchers.

In the economics literature, Goldberger (1972E) defines a *structural model* as a stochastic model representing a causal relationship, not a relationship that simply captures statistical associations. In contrast, a sampling-based model is derived from our knowledge of the mechanisms used to gather the data. The sampling-based model directly generates statistics that can be used to estimate quantities of interest and thus is also known as an *estimable* model. To illustrate, let us suppose that $\{(\mathbf{x}_i, y_i)\}$ represents a random sample from a population. Then, we can always estimate $\mathrm{E}(y \mid \mathbf{x})$ nonparametrically. Moreover, we might assume that $\mathrm{E}(y \mid \mathbf{x}) = \mathbf{x}' \boldsymbol{\beta}$, for some vector $\boldsymbol{\beta}$. This requires no appeal to the theory from an underlying functional field. We use only the assumption of the data-generating mechanism and thus refer to this as a sampling-based model.

As an example of a structural model, Duncan (1969EP) considers the following model equations that relate one's self-esteem (y_{it}, $t = 1, 2$) to delinquency (x_{it}, $t = 1, 2$):

$$y_{i2} = \beta_0 + \beta_1 \, y_{i1} + \beta_2 \, x_{i1} + \varepsilon_{i1},$$

$$x_{i2} = \gamma_0 + \gamma_1 \, y_{i1} + \gamma_2 \, x_{i1} + \varepsilon_{i2}.$$

In this model, current period ($t = 2$) self-esteem and delinquency are affected by the prior period's self-esteem and delinquency. This model specification relies on theory from the functional field. This is an example of a *structural equations model*, which will be discussed in more detail in Sections 6.4 and 6.5.

Particularly for observational data, causal statements are based primarily on substantive hypotheses, which the researcher carefully develops. Causal inference is theoretically driven. Causal processes generally cannot be demonstrated directly from the data; the data can only present relevant empirical evidence serving as a link in a chain of reasoning about causal mechanisms.

Longitudinal data are much more useful in establishing causal relationships than (cross-sectional) regression data. This is because, for most disciplines, the "causal" variable must precede the "effect" variable in time. For example, Lazarsfeld and Fiske (1938O) considered the effect of radio advertising on product sales. Traditionally, hearing radio advertisements was thought to increase the likelihood of purchasing a product. Lazarsfeld and Fiske considered whether those who bought the product would be more likely to hear the advertisement, thus positing a reverse in the direction of causality. They proposed repeatedly interviewing a set of people (the "panel") to clarify the issue.

Notions of randomization have been extended by Rubin (1974EP, 1976G, 1978G, 1990G) to observational data through the concept of *potential outcomes*. This is an area that is rapidly developing; we refer to Angrist, Imbens, and Rubin (1996G) for further discussions.

6.1.4 Instrumental Variable Estimation

According to Wooldridge (2002E, p. 83), instrumental variable estimation "is probably second only to ordinary least squares in terms of methods used in empirical economics research." Instrumental variable estimation is a general technique that is widely used in economics and related fields to handle problems associated with the disconnect between the structural model and a sampling-based model.

To introduce instrumental variable estimation, we assume that the index i represents cross-sectional draws and that these draws are independent. The instrumental variable technique can be used in instances where the structural model is specified by a linear equation of the form

$$y_i = \mathbf{x}_i'\boldsymbol{\beta} + \varepsilon_i, \tag{6.1}$$

yet not all of the regressors are predetermined; that is, $\mathrm{E}(\varepsilon_i \mathbf{x}_i) \neq \mathbf{0}$. The instrumental variable technique employs a set of predetermined variables, $\mathbf{w}_i$, that are correlated with the regressors specified in the structural model. Specifically, we assume

IV1. $\mathrm{E}(\varepsilon_i \mathbf{w}_i) = \mathrm{E}((y_i - \mathbf{x}_i'\boldsymbol{\beta})\mathbf{w}_i) = \mathbf{0}$

and

IV2. $\mathrm{E}(\mathbf{w}_i \mathbf{w}_i')$ is invertible.

With these additional variables, an instrumental variable estimator of $\boldsymbol{\beta}$ is

$$\mathbf{b}_{\mathrm{IV}} = (\mathbf{X}'\mathbf{P}_{\mathbf{W}}\mathbf{X})^{-1}\mathbf{X}'\mathbf{P}_{\mathbf{W}}\mathbf{y},$$

where $\mathbf{P}_{\mathbf{W}} = \mathbf{W}(\mathbf{W}'\mathbf{W})^{-1}\mathbf{W}'$ is a projection matrix and $\mathbf{W} = (\mathbf{w}_1, \ldots, \mathbf{w}_n)'$ is the matrix of instrumental variables. Instrumental variable estimators can be expressed as special cases of the generalized method of moment estimators; see Appendix C.6 for further details.

To illustrate, we now describe three commonly encountered situations where the instrumental variable technique has proven to be useful.

The first concerns situations where important variables have been omitted from the sampling model. In this situation, we write the structural regression function as $\mathrm{E}(y_i \mid \mathbf{x}_i, \mathbf{u}_i) = \mathbf{x}_i'\boldsymbol{\beta} + \boldsymbol{\gamma}'\mathbf{u}_i$, where $\mathbf{u}_i$ represents important unobserved variables. However, the sampling-based model uses only $\mathrm{E}(y_i \mid \mathbf{x}_i) = \mathbf{x}_i'\boldsymbol{\beta}$, thus omitting the unobserved variables. For example, in his discussion of omitted-variable bias, Wooldridge (2002E) presents an application by Card (1995E) concerning a cross section of men where studying (logarithmic) wages in relation to years of education is of interest. Additional control variables include years of experience (and its square), regional indicators, and racial indicators. The concern is that the structural model omits an important variable, the man's "ability" ($\mathbf{u}$), which is correlated with years of education. Card introduces a variable to indicate whether a man grew up in the vicinity of a four-year college as an instrument for years of education. The motivation behind this choice is that this variable should be correlated with education and yet uncorrelated with ability. In our notation, we would define $\mathbf{w}_i$ to be the same set of explanatory variables used in the structural equation model but with the vicinity variable replacing the years of education variable. Assuming positive correlation between the vicinity and years of education variables, we expect Assumption IV2 to hold. Moreover, assuming vicinity to be uncorrelated with ability, we expect Assumption IV1 to hold.

The second situation where the instrumental variable technique has proven useful concerns important explanatory variables that have been measured with error. Here, the structural model is given as in Equation (6.1) but estimation is based on the model

$$y_i = \mathbf{x}_i^{*\prime}\beta + \varepsilon_i, \tag{6.2}$$

where $\mathbf{x}_i^* = \mathbf{x}_i + \boldsymbol{\eta}_i$ and $\boldsymbol{\eta}_i$ is an error term. That is, the observed explanatory variables $\mathbf{x}_i^*$ are measured with error, yet the underlying theory is based on the "true" explanatory variables $\mathbf{x}_i$. Measurement error causes difficulties because even if the structural model explanatory variables are predetermined, such that $\mathrm{E}((y_i - \mathbf{x}_i'\beta)\mathbf{x}_i) = 0$, this does not guarantee that the observed variables will be because $\mathrm{E}((y_i - \mathbf{x}_i^{*\prime}\beta)\mathbf{x}_i^*) \neq \mathbf{0}$. For example, in Card's (1995E) returns to schooling example, it is often maintained that records for years of education are fraught with errors owing to lack of recall or other reasons. One strategy is to replace years of education by a more reliable instrument such as completion of high school or not. As with omitted variables, the goal is to select instruments that are highly related to the suspect endogenous variables yet are unrelated to model deviations.

A third important application of instrumental variable techniques regards the endogeneity induced by systems of equations. We will discuss this topic further in Section 6.4.

In many situations, instrumental variable estimators can be easily computed using two-stage least squares. In the first stage, one regresses each endogenous regressor on the set of exogenous explanatory variables and calculates fitted values of the form $\hat{\mathbf{X}} = \mathbf{P}_\mathbf{W}\mathbf{X}$. In the second stage, one regresses the dependent variable on the fitted values using ordinary least squares to get the instrumental variable estimator, that is, $(\hat{\mathbf{X}}'\hat{\mathbf{X}})^{-1}\hat{\mathbf{X}}'\mathbf{y} = \mathbf{b}_{\mathrm{IV}}$. However, Wooldridge (2002E, p. 98) recommends for empirical work that researchers use statistical packages that explicitly incorporate a two-stage least-squares routine; this is because some of the sums of squares produced in the second stage that would ordinarily be used for hypothesis testing are not appropriate in the two-stage setting.

The choice of instruments is the most difficult decision faced by empirical researchers using instrumental variable estimation. Theoretical results are available concerning the optimal choice of instruments (White, 1984E). For practical implementation of these results, empirical researchers should essentially try to choose instruments that are highly correlated with the endogenous explanatory variables. Higher correlation means that the bias as well as standard error of $\mathbf{b}_{\mathrm{IV}}$ will be lower (Bound, Jaeger, and Baker, 1995E). For additional background reading, we refer the reader to virtually any graduate econometrics text (see, for example, Greene, 2002E, Hayashi, 2000E, or Wooldridge, 2002E).

6.2 Stochastic Regressors in Longitudinal Settings

This section describes estimation in longitudinal settings that can be readily handled using techniques already described in the text. Section 6.3 follows by considering more complex models that require specialized techniques.

6.2.1 Longitudinal Data Models without Heterogeneity Terms

As we will see, the introduction of heterogeneity terms α_i complicates the endogeneity questions in longitudinal and panel data models considerably. Conversely, without heterogeneity terms, longitudinal and panel data models carry few features that would not allow us to apply the Section 6.1 techniques directly. To begin, we may write a model with strictly exogenous regressors as follows:

Assumptions of the Longitudinal Data Model with Strictly Exogenous Regressors

SE1. $\mathrm{E}(y_{it} \mid \mathbf{X}) = \mathbf{x}_{it}' \beta$.
SE2. $\{\mathbf{x}_{it}\}$ are stochastic variables.
SE3. $\mathrm{Var}(\mathbf{y}_i \mid \mathbf{X}) = \mathbf{R}_i$.
SE4. $\{\mathbf{y}_i \mid \mathbf{X}\}$ are independent random vectors.
SE5. $\{\mathbf{y}_i\}$ are normally distributed, conditional on $\{\mathbf{X}\}$.

Recall that $\mathbf{X} = \{\mathbf{X}_1, \ldots, \mathbf{X}_n\}$ is the complete set of regressors over all subjects and time periods. Because this set of assumptions includes those in the Section 6.1.1 nonlongitudinal setting, we still refer to the set as Assumptions SE1–SE5.

With longitudinal data, we have repeatedly noted the important fact that observations from the same subject tend to be related. Often, we have used the heterogeneity term α_i to account for this relationship. However, one can also use the covariance structure of the disturbances ($\mathbf{R}_i$) to account for these dependencies; see Section 7.1. Thus, SE3 allows analysts to choose a correlation structure such as arises from an autoregressive or compound symmetry structure to account for these intrasubject correlations. This formulation, employing strictly exogenous variables, means that the usual least-squares estimators have desirable finite, as well as asymptotic, properties.

As we saw in Section 6.1.1, the strict exogeneity assumption does not permit lagged dependent variables, another widely used approach for incorporating intrasubject relationships among observations. Nonetheless, without heterogeneity terms, we can weaken the assumptions on the regressors to the assumption of predetermined regressors, as in Section 6.1.1, and still achieve consistent regression estimators. With the longitudinal data notation, this assumption can be written as

SE1p. $\mathrm{E}(\varepsilon_{it}\, \mathbf{x}_{it}) = \mathrm{E}((y_{it} - \mathbf{x}_{it}'\boldsymbol{\beta})\mathbf{x}_{it}) = \mathbf{0}$.

Using linear projection notation (Appendix 6A), we can also express this assumption as $LP\,(\varepsilon_{it} \mid \mathbf{x}_{it}) = 0$, assuming $\mathrm{E}\mathbf{x}_{it}\mathbf{x}_{it}'$ is invertible. Writing the corresponding martingale difference sequence assumption that allows for asymptotic normality is slightly more cumbersome because of the two indices for the observations in the conditioning set. We leave this as an exercise for the reader.

The important point is to emphasize that longitudinal and panel data models have the same endogeneity concerns as the cross-sectional models. Moreover, often the analyst may use well-known techniques for handling endogeneity developed in cross-sectional analysis for longitudinal data. However, when these techniques are employed, the longitudinal data models should not possess heterogeneity terms. Instead, devices such as invoking a correlation structure for the conditional response or lagging the dependent variable can be used to account for heterogeneity in longitudinal data, thus allowing the analyst to focus on endogeneity concerns.

6.2.2 Longitudinal Data Models with Heterogeneity Terms and Strictly Exogenous Regressors

As we saw in Section 6.1 for nonlongitudinal data, the remedies that account for endogenous stochastic regressors require knowledge of a functional field. The formulation of an underlying structural model is by definition field-specific and this formulation affects the determination of the best model estimators. For longitudinal data, in many disciplines it is customary to incorporate a subject-specific heterogeneity term to account for intrasubject correlations, either from knowledge of the underlying data-generating process being studied or by tradition within the field of study. Thus, it is often important to understand the effects of regressors when a heterogeneity term α_i is present in the model.

To define "endogeneity" in the panel and longitudinal data context, we again begin with the simpler concept of strict exogeneity. Recall the linear mixed-effects model

$$y_{it} = \mathbf{z}'_{it}\boldsymbol{\alpha}_i + \mathbf{x}'_{it}\boldsymbol{\beta} + \varepsilon_{it}$$

and its vector version

$$\mathbf{y}_i = \mathbf{Z}_i\boldsymbol{\alpha}_i + \mathbf{X}_i\boldsymbol{\beta} + \boldsymbol{\epsilon}_i.$$

To simplify the notation, let $\mathbf{X}^* = \{\mathbf{X}_1, \mathbf{Z}_1, \ldots, \mathbf{X}_n, \mathbf{Z}_n\}$ be the collection of all observed explanatory variables and $\boldsymbol{\alpha} = (\boldsymbol{\alpha}'_1, \ldots, \boldsymbol{\alpha}'_n)'$ be the collection of all subject-specific terms. We now consider the following assumptions:

Assumptions of the Linear Mixed-Effects Model with Strictly Exogenous Regressors Conditional on the Unobserved Effect

SEC1. $\mathrm{E}(\mathbf{y}_i \mid \boldsymbol{\alpha}, \mathbf{X}^*) = \mathbf{Z}_i\boldsymbol{\alpha}_i + \mathbf{X}_i\boldsymbol{\beta}$.
SEC2. $\{\mathbf{X}^*\}$ are stochastic variables.
SEC3. $\mathrm{Var}\,(\mathbf{y}_i \mid \boldsymbol{\alpha}, \mathbf{X}^*) = \mathbf{R}_i$.
SEC4. $\{\mathbf{y}_i\}$ are independent random vectors, conditional on $\{\boldsymbol{\alpha}\}$ and $\{\mathbf{X}^*\}$.
SEC5. $\{\mathbf{y}_i\}$ are normally distributed, conditional on $\{\boldsymbol{\alpha}\}$ and $\{\mathbf{X}^*\}$.
SEC6. $\mathrm{E}(\boldsymbol{\alpha}_i \mid X^*) = \mathbf{0}$ and $\mathrm{Var}\,(\boldsymbol{\alpha}_i \mid \mathbf{X}^*) = \mathbf{D}$. Further, $\{\boldsymbol{\alpha}_1, \ldots, \boldsymbol{\alpha}_n\}$ are mutually independent, conditional on $\{\mathbf{X}^*\}$.
SEC7. $\{\boldsymbol{\alpha}_i\}$ are normally distributed, conditional on $\{\mathbf{X}^*\}$.

These assumptions are readily supported by a random-sampling scheme. For example, suppose that $(\mathbf{x}_1, \mathbf{z}_1, \mathbf{y}_1), \ldots, (\mathbf{x}_n, \mathbf{z}_n, \mathbf{y}_n)$ represents a random sample from a population. Each draw $(\mathbf{x}_i, \mathbf{z}_i, \mathbf{y}_i)$ has associated with it an unobserved, latent vector $\boldsymbol{\alpha}_i$ that is part of the conditional regression function. Then, because $\{(\boldsymbol{\alpha}_i, \mathbf{x}_i, \mathbf{z}_i, \mathbf{y}_i)\}$ are i.i.d., we immediately have SEC2 and SEC4, as well as the conditional independence of $\{\boldsymbol{\alpha}_i\}$ in SEC6. Assumptions SEC1 and SEC2, and the first part of SEC6, are moment conditions and thus depend on the conditional distributions of the draws. Further, assumptions SEC5 and SEC7 are also assumptions about the conditional distribution of a draw.

Assumption SEC1 is a stronger assumption than strict exogeneity (SE1). Using the disturbance-term notation, we may rewrite this as $\mathrm{E}(\boldsymbol{\epsilon}_i \mid \boldsymbol{\alpha}, \mathbf{X}^*) = \mathbf{0}$. By the law of iterated expectations, this implies that both $\mathrm{E}(\boldsymbol{\epsilon}_i \mid \mathbf{X}^*) = \mathbf{0}$ and

$\mathrm{E}(\epsilon_i \boldsymbol{\alpha}') = \mathbf{0}$ hold. That is, this condition requires both that the regressors are strictly exogenous and that the unobserved effects are uncorrelated with the disturbance terms. In the context of an error-components model with random sampling (the case where $q = 1$, $z_{it} = 1$, and random variables from different subjects are independent), SEC1 may be expressed as

$$\mathrm{E}(y_{it} \mid \alpha_i, \mathbf{x}_{i1}, \ldots, \mathbf{x}_{iT_i}) = \alpha_i + \mathbf{x}_{it}'\boldsymbol{\beta}, \quad \text{for each } t.$$

Chamberlain (1982E, 1984E) introduced conditional strict exogeneity in this context.

The first part of Assumption SEC6, $\mathrm{E}(\boldsymbol{\alpha}_i \mid \mathbf{X}^*) = 0$, also implies that the unobserved, time-constant effects and the regressors are uncorrelated. Many econometric panel data applications use an error-components model such as in Section 3.1. In this case, it is customary to interpret α_i to be an unobserved, time-constant effect that influences the expected response. This is motivated by the relation $\mathrm{E}(y_{it} \mid \alpha_i, \mathbf{X}^*) = \alpha_i + \mathbf{x}_{it}'\boldsymbol{\beta}$. In this case, we interpret this part of Assumption SEC6 to mean that this unobserved effect is not correlated with the regressors. Sections 7.2 and 7.3 will discuss ways of testing and relaxing this assumption.

Example: Tax Liability (continued) Section 3.2 introduced an example where we use a random sample of taxpayers and examine their tax liability (y) in terms of demographic and economic characteristics, summarized in Table 3.1. Because the data were gathered using a random-sampling mechanism, we can interpret the regressors as stochastic and assume that observable variables from different taxpayers are mutually independent. In this context, the assumption of strict exogeneity implies that we are assuming that tax liability will not affect any of the explanatory variables. For example, the demographic characteristics such as number of dependents and marital status may affect the tax liability, but the reverse implication is not true. In particular, note that the total personal income is based on positive income items from the tax return; exogeneity concerns dictated using this variable in contrast to an alternative such as net income, a variable that may be affected by the prior year's tax liability.

One potentially troubling variable is the use of the tax preparer; it may be reasonable to assume that the tax preparer variable is predetermined, although not strictly exogenous. That is, we may be willing to assume that this year's tax liability does not affect our decision to use a tax preparer because we do not know the tax liability prior to this choice, making the variable predetermined. However, it seems plausible that the prior year's tax liability will affect our decision to retain a tax preparer, thus failing the strict exogeneity test. In a model without heterogeneity terms, consistency may be achieved by assuming only that the regressors are predetermined.

For a model with heterogeneity terms, consider the error-components model in Section 3.2. Here, we interpret the heterogeneity terms to be unobserved subject-specific (taxpayer) characteristics, such as "aggressiveness," that would influence the expected tax liability. For strict exogeneity conditional on the unobserved effects, one needs to argue that the regressors are strictly exogenous and that the disturbances, representing "unexpected" tax liabilities, are uncorrelated with the unobserved effects. Moreover, Assumption SEC6 employs the condition that the unobserved effects are uncorrelated with the observed regressor variables. One may be concerned that individuals with high earnings potential who have historically high levels of tax liability (relative to their control variables) may be more likely to use a tax preparer, thus violating this assumption.

As in Chapter 3, the assumptions based on distributions conditional on unobserved effects lead to the following conditions that are the basis of statistical inference.

Observables Representation of the Linear Mixed-Effects Model with Strictly Exogenous Regressors Conditional on the Unobserved Effect

SE1. $\mathrm{E}(\mathbf{y}_i \mid \mathbf{X}^*) = \mathbf{X}_i \boldsymbol{\beta}$.

SE2. $\{\mathbf{X}^*\}$ are stochastic variables.

SE3a. $\mathrm{Var}\,(\mathbf{y}_i \mid \mathbf{X}^*) = \mathbf{Z}_i \mathbf{D} \mathbf{Z}_i' + \mathbf{R}_i$.

SE4. $\{\mathbf{y}_i\}$ are independent random vectors, conditional on $\{\mathbf{X}^*\}$.

SE5. $\{\mathbf{y}_i\}$ are normally distributed, conditional on $\{\mathbf{X}^*\}$.

These conditions are virtually identical to the assumptions of the longitudinal data mixed model with strictly exogenous regressors, which does not contain heterogeneity terms. The difference is in the conditional variance component, SE3. In particular, the inference procedures described in Chapters 3 and 4 can be readily used in this situation.

Fixed-Effects Estimation

As we saw in the previous example that discussed exogeneity in terms of the income tax liability, there are times when the analyst is concerned with Assumption SEC6. Among other things, this assumption implies that the unobserved effects are uncorrelated with the observed regressors. Although readily accepted as the norm in the biostatistics literature, this assumption is often questioned in the economics literature. Fortunately, Assumptions SEC1–SEC4 (and SEC5, as needed) are sufficient to allow for consistent (as well as asymptotic normality) estimation, using the fixed-effects estimators described in Chapter 2. Intuitively, this is because the fixed-effects estimation procedures "sweep out" the heterogeneity terms and thus do not rely on the assumption that they are uncorrelated with observed regressors. See Mundlak (1978aE) for an early contribution; Section 7.2 provides further details.

These observations suggest a strategy that is commonly used by analysts. If there is no concern that unobserved effects may be correlated with observed regressors, use the more efficient inference procedures in Chapter 3 based on mixed models and random effects. If there is a concern, use the more robust fixed-effects estimators. Some analysts prefer to test the assumption of correlation between unobserved and observed effects by examining the difference between these two estimators. This is the subject of Sections 7.2 and 7.3, where we will examine inference for the unobserved, or "omitted," variables.

In some applications, researchers have partial information about the first part of Assumption SEC6. Specifically, we may rearrange the observables into two pieces: $\mathbf{o}_i = (\mathbf{o}_i^{(1)} \mathbf{o}_i^{(2)})$, where $\mathrm{Cov}(\boldsymbol{\alpha}_i, \mathbf{o}_i^{(1)}) \neq \mathbf{0}$ and $\mathrm{Cov}(\boldsymbol{\alpha}_i, \mathbf{o}_i^{(2)}) = \mathbf{0}$. That is, the first piece of $\mathbf{o}_i$ is correlated with the unobservables whereas the second piece is not. In this case, estimators that are neither fixed nor random-effects have been developed in the literature. This idea, due to Hausman and Taylor (1981E), is further pursued in Section 7.3.

6.3 Longitudinal Data Models with Heterogeneity Terms and Sequentially Exogenous Regressors

For some economic applications such as production function modeling (Keane and Runkle, 1992E), the assumption of strict exogeneity even when conditioning on unobserved heterogeneity terms is limiting. This is because strict exogeneity rules out current values of the response (y_{it}) feeding back and influencing future values of the explanatory variables (such as $\mathbf{x}_{i,t+1}$). An alternative assumption introduced by Chamberlain (1992E) allows for this feedback. To introduce this assumption, we follow Chamberlain and assume random sampling so that random variables from different subjects are independent. Following this econometrics literature, we assume $q = 1$ and $z_{it} = 1$ so that the heterogeneity term is an intercept. We say that the regressors are *sequentially exogenous conditional on the unobserved effects* if

$$\mathrm{E}(\varepsilon_{it} \mid \alpha_i, \mathbf{x}_{i1}, \ldots, \mathbf{x}_{it}) = 0. \tag{6.3}$$

This implies

$$\mathrm{E}(y_{it} \mid \alpha_i, \mathbf{x}_{i1}, \ldots, \mathbf{x}_{it}) = \alpha_i + \mathbf{x}_{it}'\boldsymbol{\beta}.$$

After controlling for α_i and $\mathbf{x}_{it}$, past values of regressors do not affect the expected value of y_{it}.

Lagged–Dependent Variable Model

In addition to feedback models, this formulation allows us to consider lagged dependent variables as regressors. For example, the equation

$$y_{it} = \alpha_i + \gamma y_{i,t-1} + \mathbf{x}_{it}'\boldsymbol{\beta} + \varepsilon_{it} \tag{6.4}$$

satisfies Equation (6.3) by using the set of regressors $\mathbf{o}_{it} = (1, y_{i,t-1}, \mathbf{x}_{it}')'$ and $\mathrm{E}(\varepsilon_{it} \mid \alpha_i, y_{i,1}, \ldots, y_{i,t-1}, \mathbf{x}_{i,1}, \ldots, \mathbf{x}_{i,t}) = 0$. The explanatory variable $y_{i,t-1}$ is not strictly exogenous so that the Section 6.2.2 discussion does not apply.

As will be discussed in Section 8.1, this model differs from the autoregressive error structure, the common approach in the longitudinal biomedical literature. Judged by the number of applications, this is an important dynamic panel data model in econometrics. The model is appealing because it is easy to interpret the lagged dependent variable in the context of economic modeling. For example, if we think of y as the demand of a product, it is easy to think of situations where a strong demand in the prior period ($y_{i,t-1}$) has a positive influence on the current demand (y_{it}), suggesting that γ be a positive parameter.

Estimation Difficulties

Estimation of the model in Equation (6.4) is difficult because the parameter γ appears in both the mean and variance structure. It appears in the variance structure because

$$\begin{aligned}\mathrm{Cov}\,(y_{it}, y_{i,t-1}) &= \mathrm{Cov}\,(\alpha_i + \gamma y_{i,t-1} + \mathbf{x}_{it}'\boldsymbol{\beta} + \varepsilon_{it}, y_{i,t-1}) \\ &= \mathrm{Cov}\,(\alpha_i, y_{i,t-1}) + \gamma \,\mathrm{Var}\,(y_{i,t-1}).\end{aligned}$$

To see that it appears in the mean structure, consider Equation (6.4). By recursive substitution, we have

$$\begin{aligned}\mathrm{E}y_{it} &= \gamma \;\mathrm{E}y_{i,t-1} + \mathbf{x}_{it}'\boldsymbol{\beta} = \gamma(\gamma \;\mathrm{E}y_{i,t-2} + \mathbf{x}_{i,t-1}'\boldsymbol{\beta}) + \mathbf{x}_{it}'\boldsymbol{\beta} \\ &= \cdots = \left(\mathbf{x}_{it}' + \gamma \mathbf{x}_{i,t-1}' + \cdots + \gamma^{t-2}\mathbf{x}_{i,2}'\right)\boldsymbol{\beta} + \gamma^{t-1}\;\mathrm{E}y_{i,1}.\end{aligned}$$

Thus, $\mathrm{E}y_{it}$ clearly depends on γ.

Special estimation techniques are required for the model in Equation (6.4); it is not possible to treat lagged dependent variables as explanatory variables, either using a fixed- or random-effects formulation for the heterogeneity terms α_i. We first examine the fixed-effects form, beginning with an example from Hsiao (2002E).

Special Case: (Hsiao, 2002E) Suppose that α_i are treated as fixed parameters and, for simplicity, take $K = 1$ and $x_{it,1} = 1$ so that Equation (6.4) reduces to

$$y_{it} = \alpha_i^* + \gamma y_{i,t-1} + \varepsilon_{it}, \tag{6.5}$$

where $\alpha_i^* = \alpha_i + \beta$. The OLS estimator of γ is then

$$\begin{aligned}\hat{\gamma} &= \frac{\sum_{i=1}^n \sum_{t=2}^T (y_{it} - \bar{y}_i)(y_{i,t-1} - \bar{y}_{i,-1})}{\sum_{i=1}^n \sum_{t=2}^T (y_{i,t-1} - \bar{y}_{i,-1})^2} \\ &= \gamma + \frac{\sum_{i=1}^n \sum_{t=2}^T \varepsilon_{it}(y_{i,t-1} - \bar{y}_{i,-1})}{\sum_{i=1}^n \sum_{t=2}^T (y_{i,t-1} - \bar{y}_{i,-1})^2},\end{aligned}$$

where $\bar{y}_{i,-1} = (\sum_{t=1}^{T-1} y_t)/(T-1)$. Now, we can argue that $\mathrm{E}(\varepsilon_{it} y_{i,t-1}) = 0$ by conditioning on information available at time $t-1$. However, it is not true that $\mathrm{E}(\varepsilon_{it} \bar{y}_{i,-1}) = 0$, suggesting that $\hat{\gamma}$ is biased. In fact, Hsiao demonstrates that the asymptotic bias is

$$\lim_{n\to\infty} \mathrm{E}\hat{\gamma} - \gamma = \frac{-\frac{1+\gamma}{T-1}\left(1 - \frac{1-\gamma^T}{T(1-\gamma)}\right)}{1 - \frac{2\gamma}{1-\gamma(T-1)}\left(1 - \frac{1-\gamma^T}{T(1-\gamma)}\right)}.$$

This bias is small for large T and tends to zero as T tends to infinity. Notice that the bias is nonzero even when $\gamma = 0$.

To see the estimation difficulties in the context of the random-effects model, we now consider the model in Equation (6.4), where α_i are treated as random variables that are independent of the error terms ε_{it}. It is tempting to treat lagged response variables as explanatory variables and use the usual GLS estimators. However, this procedure also induces bias. To see this, we note that it is clear that y_{it} is a function of α_i and, thus, so is $y_{i,t-1}$. However, GLS estimation procedures implicitly assume independence of the random-effects and explanatory variables. Thus, this estimation procedure is not optimal.

Although the usual GLS estimators are not desirable, alternative estimators are available. For example, taking first differences of the model in Equation

(6.5) yields

$$y_{it} - y_{i,t-1} = \gamma(y_{i,t-1} - y_{i,t-2}) + \varepsilon_{it} - \varepsilon_{i,t-1},$$

eliminating the heterogeneity term. Note that $y_{i,t-1}$ and $\varepsilon_{i,t-1}$ are clearly dependent; thus, using ordinary least squares with regressors $\Delta y_{i,t-1} = y_{i,t-1} - y_{i,t-2}$ produces biased estimators of γ. We can, however, use $\Delta y_{i,t-2}$ as an instrument for $\Delta y_{i,t-1}$, because $\Delta y_{i,t-2}$ is independent of the (differenced) disturbance term $\varepsilon_{it} - \varepsilon_{i,t-1}$. This approach of differencing and using instrumental variables is due to Anderson and Hsiao (1982E). Of course, this estimator is not efficient, because the differenced error terms will usually be correlated.

Thus, first-differencing proves to be a useful device for handling the heterogeneity term. To illustrate how first-differencing by itself can fail, consider the following special case.

Special Case: Feedback Consider the error components $y_{it} = \alpha_i + \mathbf{x}'_{it}\boldsymbol{\beta} + \varepsilon_{it}$, where $\{\varepsilon_{it}\}$ are i.i.d. with variance σ^2. Suppose that the current regressors are influenced by the "feedback" from the prior period's disturbance through the relation $\mathbf{x}_{it} = \mathbf{x}_{i,t-1} + \boldsymbol{\upsilon}_i \varepsilon_{i,t-1}$, where $\{\boldsymbol{\upsilon}_i\}$ is an independently and identically distributed random vector that is independent of $\{\varepsilon_{it}\}$. Taking differences of the model, we have

$$\Delta y_{it} = y_{it} - y_{i,t-1} = \Delta \mathbf{x}'_{it}\boldsymbol{\beta} + \Delta\varepsilon_{it},$$

where $\Delta\varepsilon_{it} = \varepsilon_{it} - \varepsilon_{i,t-1}$ and $\Delta\mathbf{x}_{it} = \mathbf{x}_{it} - \mathbf{x}_{i,t-1} = \boldsymbol{\upsilon}_i\varepsilon_{i,t-1}$. Using first differences, we get the OLS estimator of $\boldsymbol{\beta}$,

$$\begin{aligned}\mathbf{b}_{\text{FD}} &= \left(\sum_{i=1}^{n} \Delta\mathbf{X}_i \Delta\mathbf{X}'_i\right)^{-1} \sum_{i=1}^{n} \Delta\mathbf{X}_i \Delta\mathbf{y}_i \\ &= \boldsymbol{\beta} + \left(\sum_{i=1}^{n} \Delta\mathbf{X}_i \Delta\mathbf{X}'_i\right)^{-1} \sum_{i=1}^{n} \Delta\mathbf{X}_i \Delta\boldsymbol{\epsilon}_i,\end{aligned}$$

where $\Delta\mathbf{y}_i = (\Delta y_{i2}, \ldots, \Delta y_{i,T})'$, $\Delta\boldsymbol{\epsilon}_i = (\Delta\varepsilon_{i2}, \ldots, \Delta\varepsilon_{i,T})'$, and

$$\Delta\mathbf{X}_i = (\Delta\mathbf{x}_{i2}, \ldots, \Delta\mathbf{x}_{i,T}) = (\boldsymbol{\upsilon}_i\varepsilon_{i,2}, \ldots, \boldsymbol{\upsilon}_i\varepsilon_{i,T}).$$

Straightforward calculations show that

$$\lim_{n\to\infty} \frac{1}{n}\sum_{i=1}^{n} \Delta\mathbf{X}_i \Delta\mathbf{X}'_i = \lim_{n\to\infty} \frac{1}{n}\sum_{i=1}^{n} \boldsymbol{\upsilon}_i\boldsymbol{\upsilon}'_i \sum_{t=2}^{T} \varepsilon_{it}^2 = (T-1)\sigma^2 \,\mathrm{E}\,\boldsymbol{\upsilon}_i\boldsymbol{\upsilon}'_i$$

and

$$\lim_{n\to\infty} \frac{1}{n} \sum_{i=1}^{n} \Delta \mathbf{X}_i \Delta \epsilon_i = \lim_{n\to\infty} \frac{1}{n} \sum_{i=1}^{n} \left(\boldsymbol{v}_i \varepsilon_{i1} \cdots \boldsymbol{v}_i \varepsilon_{i,T-1} \right) \begin{pmatrix} \varepsilon_{i2} & - & \varepsilon_{i1} \\ & \vdots & \\ \varepsilon_{i,T} & - & \varepsilon_{i,T-1} \end{pmatrix}$$

$$= -(T-1)\sigma^2 \,\mathrm{E}\, \boldsymbol{v}_i ,$$

both with probability one. With probability one, this yields the asymptotic bias

$$\lim_{n\to\infty} \mathbf{b}_{\mathrm{FD}} - \boldsymbol{\beta} = -(\mathrm{E}\, \boldsymbol{v}_i \boldsymbol{v}_i')^{-1} \,\mathrm{E}\, \boldsymbol{v}_i .$$

One strategy for handling sequentially exogenous regressors with heterogeneity terms is to use a transform, such as first-differencing or fixed effects, to sweep out the heterogeneity and then use instrumental variable estimation. One such treatment has been developed by Arellano and Bond (1991E). For this treatment, we assume that the responses follow the model equation

$$y_{it} = \alpha_i + \mathbf{x}_{it}' \boldsymbol{\beta} + \varepsilon_{it},$$

yet the regressors are potentially endogenous. We also assume that there exist two sets of instrumental variables. The variables of first set, of the form $\mathbf{w}_{1,it}$, are strictly exogenous so that

$$LP(\varepsilon_{it} \mid \mathbf{w}_{1,i1}, \ldots, \mathbf{w}_{1,iT_i}) = 0, \quad t = 1, \ldots, T_i. \tag{6.6}$$

The second set, of the form $\mathbf{w}_{2,it}$ satisfies the following sequential exogeneity conditions:

$$LP(\varepsilon_{it} \mid \mathbf{w}_{2,i1}, \ldots, \mathbf{w}_{2,it}) = 0, \quad t = 1, \ldots, T_i. \tag{6.7}$$

The dimensions of $\{\mathbf{w}_{1,it}\}$ and $\{\mathbf{w}_{2,it}\}$ are $p_1 \times 1$ and $p_2 \times 1$, respectively. Because we will remove the heterogeneity term via transformation, we need not specify this in our linear projections. Note that Equation (6.7) implies that current disturbances are uncorrelated with current as well as past instruments.

Time-constant heterogeneity parameters are handled via sweeping out their effects, so let $\mathbf{K}_i$ be a $(T_i - 1) \times T_i$ upper triangular matrix such that $\mathbf{K}_i \mathbf{1}_i = \mathbf{0}_i$. For example, Arellano and Bover (1995E) recommend the matrix (suppressing the i subscript)

$$\mathbf{K}_{\mathrm{FOD}} = \mathrm{diag}\left(\sqrt{\frac{T}{T-1}} \cdots \sqrt{\frac{1}{2}}\right)$$
$$\times \begin{pmatrix} 1 & -\frac{1}{T-1} & -\frac{1}{T-1} & \cdots & -\frac{1}{T-1} & -\frac{1}{T-1} & -\frac{1}{T-1} \\ 0 & 1 & -\frac{1}{T-2} & \cdots & -\frac{1}{T-2} & -\frac{1}{T-2} & -\frac{1}{T-2} \\ \vdots & \vdots & \vdots & \ddots & \vdots & \vdots & \vdots \\ 0 & 0 & 0 & \cdots & 1 & -\frac{1}{2} & -\frac{1}{2} \\ 0 & 0 & 0 & \cdots & 0 & 1 & -1 \end{pmatrix}.$$

If we define $\boldsymbol{\epsilon}_{i,\mathrm{FOD}} = \mathbf{K}_{\mathrm{FOD}}\boldsymbol{\epsilon}_i$, the tth row becomes

$$\varepsilon_{it,\mathrm{FOD}} = \sqrt{\frac{T-t}{T-t+1}}\left(\varepsilon_{it} - \frac{1}{T-t}(\varepsilon_{i,t+1} + \cdots + \varepsilon_{i,T})\right),$$

for $t = 1, \ldots, T-1$. These are known as *forward orthogonal deviations*. If the original disturbances are serially uncorrelated and of constant variance, then so are orthogonal deviations. Preserving this structure is the advantage of forward orthogonal deviations when compared to simple differences.

To define the instrumental variable estimator, let $\mathbf{W}_i^*$ be a block diagonal matrix with the tth block given by $(\mathbf{w}'_{1,i}\mathbf{w}'_{2,i1} \ldots \mathbf{w}'_{2,it})$ where $\mathbf{w}_{1,i} = (\mathbf{w}'_{1,i1}, \ldots, \mathbf{w}'_{1,iT_i})'$. That is, define

$$\mathbf{W}_i^* = \begin{pmatrix} \mathbf{w}'_{1,i}\,\mathbf{w}'_{2,i1} & \mathbf{0} & \cdots & \mathbf{0} \\ \mathbf{0} & (\mathbf{w}'_{1,i}\,\mathbf{w}'_{2,i1}\quad \mathbf{w}'_{2,i2}) & \cdots & \mathbf{0} \\ \vdots & \vdots & \ddots & \vdots \\ \mathbf{0} & \mathbf{0} & \cdots & (\mathbf{w}'_{1,i}\,\mathbf{w}'_{2,i1}\quad \mathbf{w}'_{2,i2}\quad \cdots \quad \mathbf{w}'_{2,i,T_i-1}) \end{pmatrix}.$$

With this notation, it can be shown that the sequentially exogeneity assumption in Equations (6.6) and (6.7) implies $\mathrm{E}\mathbf{W}_i^{*\prime}\,\mathbf{K}_i\boldsymbol{\epsilon}_i = \mathbf{0}_i$. Let $\mathbf{W}_i = (\mathbf{W}_i^* : \mathbf{0}_i)$, where $\mathbf{W}_i$ has dimensions $(T_i - 1) \times (p_1 T + p_2 T(T+1)/2)$, and $T = \max(T_1, \ldots, T_n)$. This zero matrix augmentation is needed when we have unbalanced data; it is not needed if $T_i = T$.

Special Case: Feedback (continued) A natural set of instruments is to choose $\mathbf{w}_{2,it} = \mathbf{x}_{it}$. For simplicity, we also use first differences in our choice of $\mathbf{K}$. Thus,

$$\mathbf{K}_{\mathrm{FD}}\boldsymbol{\epsilon}_i = \begin{pmatrix} -1 & 1 & 0 & \cdots & 0 & 0 & 0 \\ 0 & -1 & 1 & \cdots & 0 & 0 & 0 \\ \vdots & \vdots & \vdots & \ddots & \vdots & \vdots & \vdots \\ 0 & 0 & 0 & \cdots & -1 & 1 & 0 \\ 0 & 0 & 0 & \cdots & 0 & -1 & 1 \end{pmatrix} \begin{pmatrix} \varepsilon_{i1} \\ \vdots \\ \varepsilon_{i,T} \end{pmatrix} = \begin{pmatrix} \varepsilon_{i2} - \varepsilon_{i1} \\ \vdots \\ \varepsilon_{i,T} - \varepsilon_{i,T-1} \end{pmatrix}.$$

With these choices, the tth block of $\mathrm{E}\mathbf{W}_i' \mathbf{K}_{\mathrm{FD}} \boldsymbol{\epsilon}_i$ is

$$\mathrm{E}\begin{pmatrix} \mathbf{x}_{i1} \\ \vdots \\ \mathbf{x}_{it} \end{pmatrix} \left(\varepsilon_{i,t+1} - \varepsilon_{i,t} \right) = \begin{pmatrix} \mathbf{0} \\ \vdots \\ \mathbf{0} \end{pmatrix},$$

so the sequential exogeneity assumption in Equation (6.7) is satisfied.

Special Case: Lagged dependent variable Consider the model in Equation (6.5). In this case, one can choose $w_{2,it} = y_{i,t-1}$ and, for simplicity, use first differences in our choice of $\mathbf{K}$. With these choices, the tth block of $\mathrm{E}\mathbf{W}_i' K_{\mathrm{FD}}\, \boldsymbol{\epsilon}_i$ is

$$\mathrm{E}\begin{pmatrix} y_{i1} \\ \vdots \\ y_{i,t-1} \end{pmatrix} \left(\varepsilon_{i,t+1} - \varepsilon_{i,t} \right) = \begin{pmatrix} 0 \\ \vdots \\ 0 \end{pmatrix},$$

so the sequentially exogeneity assumption in Equation (6.7) is satisfied.

Now, define the matrices $\mathbf{M}_{WX} = \sum_{i=1}^n \mathbf{W}_i'\mathbf{K}_i\mathbf{X}_i$ and $\mathbf{M}_{Wy} = \sum_{i=1}^n \mathbf{W}_i'\mathbf{K}_i\mathbf{y}_i$. With this notation, we define the instrumental variable estimator as

$$\mathbf{b}_{\mathrm{IV}} = \left(\mathbf{M}_{WX}' \boldsymbol{\Sigma}_{\mathrm{IV}}^{-1} \mathbf{M}_{WX} \right)^{-1} \mathbf{M}_{WX}' \boldsymbol{\Sigma}_{\mathrm{IV}}^{-1} \mathbf{M}_{Wy}, \qquad (6.8)$$

where $\boldsymbol{\Sigma}_{\mathrm{IV}} = \mathrm{E}(\mathbf{W}_i' \mathbf{K}_i \boldsymbol{\epsilon}_i \boldsymbol{\epsilon}_i' \mathbf{K}_i' \mathbf{W}_i)$. This estimator is consistent and asymptotically normal, with asymptotic covariance matrix

$$\mathrm{Var}\, \mathbf{b}_{\mathrm{IV}} = \left(\mathbf{M}_{WX}' \boldsymbol{\Sigma}_{\mathrm{IV}}^{-1} \mathbf{M}_{WX} \right)^{-1}.$$

Both the estimator and the asymptotic variance rely on the unknown matrix $\boldsymbol{\Sigma}_{\mathrm{IV}}$. To compute a first-stage estimator, we may assume that the disturbances are serially uncorrelated and homoscedastic and thus use

$$\mathbf{b}_{\mathrm{IV}} = \left(\mathbf{M}_{WX}' \hat{\boldsymbol{\Sigma}}_{\mathrm{IV},1}^{-1} \mathbf{M}_{WX} \right)^{-1} \mathbf{M}_{WX}' \hat{\boldsymbol{\Sigma}}_{\mathrm{IV},1}^{-1} \mathbf{M}_{Wy},$$

where $\hat{\boldsymbol{\Sigma}}_{\mathrm{IV},1} = \sum_{i=1}^n \mathbf{W}_i' \mathbf{K}_i \mathbf{K}_i' \mathbf{W}_i$. As is the usual case with GLS estimators, this estimator is invariant to the estimator of the scale $\mathrm{Var}\, \varepsilon$. To estimate this scale parameter, use the residuals $e_{it} = y_{it} - \mathbf{x}_{it}' \mathbf{b}_{\mathrm{IV}}$. An estimator of $\mathrm{Var}\, \mathbf{b}_{\mathrm{IV}}$ that is robust to the assumption of no serial correlation and homoscedasticity of the disturbances is

$$\hat{\boldsymbol{\Sigma}}_{\mathrm{IV},2} = \frac{1}{n} \sum_{i=1}^n \mathbf{W}_i' \mathbf{K}_i\, \mathbf{e}_i \mathbf{e}_i' \mathbf{K}_i' \mathbf{W}_i,$$

where $\mathbf{e}_i = (e_{i1}, \ldots, e_{iT_i})'$.

Table 6.1. *Comparison among instrumental variable estimators*

	Strictly exogenous demographic and economic variables		Strictly exogenous demographic variables and sequentially exogenous economic variables		
Variable	Parameter estimates	Model-based t-statistic	Parameter estimates	Model-based t-statistic	Robust t-statistic
Lag LNTAX	0.205	4.26	0.108	3.13	2.48
Demographic					
MS	−0.351	−0.94	−0.149	−0.42	−0.49
HH	−1.236	−2.70	−1.357	−3.11	−1.71
AGE	−0.160	−0.34	0.010	0.02	0.02
DEPEND	0.026	0.21	0.084	0.73	0.68
Economic					
LNTPI	0.547	4.53	0.340	1.91	1.07
MR	0.116	8.40	0.143	7.34	5.21
EMP	0.431	1.22	0.285	0.48	0.36
PREP	−0.272	−1.20	−0.287	−0.78	−0.68
INTERCEPT	0.178	4.21	0.215	4.90	3.41

Example: Tax Liability (continued) In Section 6.2.2, we suggested that a heterogeneity term may be due to an individual's earning potential and that this may be correlated with the variable that indicates use of a professional tax preparer. Moreover, there was concern that tax liabilities from one year may influence the taxpayer's choice in subsequent tax years of whether or not to use a professional tax preparer. Instrumental variable estimators provide protection against these endogeneity concerns.

Table 6.1 summarizes the fit of two dynamic models. Both models use heterogeneity terms and lagged dependent variables. One model assumes that all demographic and economic variables are strictly exogenous (so are used as $\mathbf{w}_{1,it}$); the other model assumes that demographic variables are strictly exogenous (so are used as $\mathbf{w}_{1,it}$) but that the economic variables are sequentially exogenous (so are used as $\mathbf{w}_{2,it}$). For the second model, we also present robust t-statistics (based on the variance–covariance matrix $\hat{\Sigma}_{\mathrm{IV},2}$) in addition to the usual "model-based" t-statistics (based on the variance–covariance matrix $\hat{\Sigma}_{\mathrm{IV},1}$). These models were fit using the statistical econometrics package STATA.

Table 6.1 shows that the lagged dependent variable was statistically significant for both models and methods of calculating t-statistics. Of the demographic variables, only the head-of-household variable (HH) was statistically significant, and this was not even true under the model treating economic variables as sequentially exogenous and using robust t-statistics. Of the economic variables, neither EMP nor PREP was statistically significant, whereas MR was statistically significant. The other measure of income, LNTPI, was statistically significant when treated as strictly exogenous but not when treated as sequentially exogenous. Because the main purpose of this study was to study the effect of PREP on LNTAX, the effects of LNTPI were not investigated further.

As with other instrumental variable procedures, a test of the exogeneity assumption $\mathrm{E}\mathbf{W}_i'\mathbf{K}_i\boldsymbol{\epsilon}_i = \mathbf{0}_i$ is available. The robust version of the test statistic is

$$TS_{\mathrm{IV}} = \left(\sum_{i=1}^{n} \mathbf{e}_i'\mathbf{K}_i'\mathbf{W}_i\right) \hat{\Sigma}_{\mathrm{IV},2}^{-1} \left(\sum_{i=1}^{n} \mathbf{W}_i'\,\mathbf{K}_i\,\mathbf{e}_i\right).$$

Under the null hypothesis of $\mathrm{E}\mathbf{W}_i'\mathbf{K}_i\boldsymbol{\epsilon}_i = \mathbf{0}_i$, this test statistic has an asymptotic chi-square distribution with $(p_1T + p_2T(T+1)/2) - K$ degrees of freedom (see Arellano and Honoré, 2001E). Moreover, one can use incremental versions of this test statistic to assess the exogeneity of selected variables in the same manner as with the partial F-test. This is important because the number of moment conditions increases substantially as one considers modeling a variable as strictly exogenous (which uses T moment conditions) compared to the less restrictive sequential exogeneity assumption (which uses $T(T+1)/2$ moment conditions). For additional discussion on testing exogeneity using instrumental variable estimators, we refer to Arellano (2003E), Baltagi (2001E), and Wooldridge (2002E).

6.4 Multivariate Responses

As with Section 6.1, we begin by reviewing ideas from a nonlongitudinal setting, specifically, cross-sectional data. Section 6.4.1 introduces multivariate responses in the context of multivariate regression. Section 6.4.2 describes the relation between multivariate regression and sets of regression equations. Section 6.4.3 introduces simultaneous equations methods. Section 6.4.4 then applies these ideas to systems of equations with error components that have been proposed in the econometrics literature.

6.4.1 Multivariate Regression

A classic method of modeling multivariate responses is through a *multivariate regression model*. We start with a general expression

$$\mathbf{Y} = \mathbf{X}\mathbf{\Gamma}' + \boldsymbol{\epsilon}, \tag{6.9}$$

where $\mathbf{Y}$ is an $n \times G$ matrix of responses, $\mathbf{X}$ is an $n \times K$ matrix of explanatory variables, $\mathbf{\Gamma}$ is a $G \times K$ matrix of parameters, and $\boldsymbol{\epsilon}$ is an $n \times G$ matrix of disturbances. To provide intuition, consider the (transpose of the) ith row of Equation (6.9),

$$\mathbf{y}_i = \mathbf{\Gamma}\mathbf{x}_i + \boldsymbol{\epsilon}_i, \quad i = 1, \ldots, n. \tag{6.10}$$

Here, G responses are measured for each subject, $\mathbf{y}_i = (y_{1i}, \ldots, y_{Gi})'$, whereas the vector of explanatory variables, $\mathbf{x}_i$, is $K \times 1$. We initially assume that the disturbance terms are i.i.d. with variance–covariance matrix $\mathrm{Var}\,\boldsymbol{\epsilon}_i = \mathbf{\Sigma}$.

Example 6.1: Supply and Demand To illustrate, consider a sample of n countries; for each country, we measure $G = 2$ responses. Specifically, we wish to relate y_1, the price of a good provided by suppliers, and y_2, the quantity of this good demanded, to several exogenous measures ($\mathbf{x}$s) such as income and the price of substitute goods. We assume that price and quantity may be related through

$$\mathrm{Var}\,\mathbf{y}_i = \mathrm{Var}\begin{pmatrix} y_{1i} \\ y_{2i} \end{pmatrix} = \begin{pmatrix} \sigma_1^2 & \sigma_{12} \\ \sigma_{12} & \sigma_2^2 \end{pmatrix} = \mathbf{\Sigma}.$$

Specifically, σ_{12} measures the covariance between price and quantity. Using $\mathbf{\Gamma} = (\boldsymbol{\beta}_1\ \boldsymbol{\beta}_2)'$, we can express Equation (6.10) as

$$\begin{aligned} y_{1i} &= \boldsymbol{\beta}_1'\mathbf{x}_i + \varepsilon_{1i} \quad \text{(price)}, \\ y_{2i} &= \boldsymbol{\beta}_2'\mathbf{x}_i + \varepsilon_{2i} \quad \text{(quantity)}. \end{aligned}$$

The OLS regression coefficient estimator of $\mathbf{\Gamma}$ is

$$\mathbf{G}_{\mathrm{OLS}} = \left(\sum_{i=1}^{n} \mathbf{y}_i\mathbf{x}_i'\right)\left(\sum_{i=1}^{n} \mathbf{x}_i\mathbf{x}_i'\right)^{-1}. \tag{6.11}$$

Somewhat surprisingly, this estimator is also the GLS estimator and, hence, the MLE.

Now, let $\boldsymbol{\beta}_g'$ be the gth row of $\boldsymbol{\Gamma}$ so that $\boldsymbol{\Gamma} = (\boldsymbol{\beta}_1\, \boldsymbol{\beta}_2 \ldots \boldsymbol{\beta}_G)'$. With this notation, the gth row of Equation (6.10) is

$$y_{gi} = \boldsymbol{\beta}_g' \mathbf{x}_i + \varepsilon_{gi}, \quad i = 1, \ldots, n.$$

We can calculate the OLS estimator of $\boldsymbol{\beta}_g$ as

$$\mathbf{b}_{g,\mathrm{OLS}} = \left(\sum_{i=1}^{n} \mathbf{x}_i \mathbf{x}_i' \right)^{-1} \sum_{i=1}^{n} \mathbf{x}_i y_{ig}, \quad g = 1, \ldots, G.$$

Thus, the estimator $\mathbf{G}_{\mathrm{OLS}}$ can be calculated on a row-by-row basis, that is, using standard (univariate response) multiple linear regression software. Nonetheless, the multivariate model structure has important features. To illustrate, by considering sets of responses simultaneously in Equation (6.10), we can account for relationships among responses in the covariance of the regression estimators. For example, with Equation (6.10), it is straightforward to show that

$$\mathrm{Cov}(\mathbf{b}_{g,\mathrm{OLS}}, \mathbf{b}_{k,\mathrm{OLS}}) = \sigma_{gk} \left(\sum_{i=1}^{n} \mathbf{x}_i \mathbf{x}_i' \right)^{-1}.$$

Multivariate regression analysis can be applied directly to longitudinal data by treating each observation over time as one of the G responses. This allows for a variety of serial correlation matrices through the variance matrix $\boldsymbol{\Sigma}$. However, this perspective requires balanced data ($G = T$).

6.4.2 Seemingly Unrelated Regressions

Even for cross-sectional data, a drawback of the classic multivariate regression is that the same set of explanatory variables is required for each type of response. To underscore this point, we return to our supply-and-demand example.

Example 6.1: Supply and Demand (continued) Suppose now that the expected price of a good (y_1) depends linearly on x_1, the purchasers' income; similarly, quantity (y_2) depends on x_2, the suppliers' wage rate. That is, we wish to consider

$$\begin{aligned} y_{1i} &= \beta_{10} + \beta_{11} x_{1i} + \varepsilon_{1i} \quad \text{(price)}, \\ y_{2i} &= \beta_{20} + \beta_{21} x_{2i} + \varepsilon_{2i}, \quad \text{(quantity)} \end{aligned}$$

so that different explanatory variables are used in different equations.

Let us reorganize our observed variables so that $\mathbf{y}_g = (y_{1g}, \ldots, y_{ng})'$ is the vector of the gth country and $\mathbf{X}_g$ is the $n \times K_g$ matrix of explanatory variables.

This yields two sets of regression equations,

$$\mathbf{y}_1 = \mathbf{X}_1\boldsymbol{\beta}_1 + \boldsymbol{\epsilon}_1 \quad \text{(price)}$$

and

$$\mathbf{y}_2 = \mathbf{X}_2\boldsymbol{\beta}_2 + \boldsymbol{\epsilon}_2 \quad \text{(quantity)},$$

representing $2n$ equations. Here, $\mathrm{Var}\,\mathbf{y}_1 = \sigma_1^2\mathbf{I}_n$, $\mathrm{Var}\,\mathbf{y}_2 = \sigma_2^2\mathbf{I}_n$, and $\mathrm{Cov}(\mathbf{y}_1, \mathbf{y}_2) = \sigma_{12}\mathbf{I}_n$. Thus, there is zero covariance between different countries yet a common covariance within a country between price and quantity of the good.

If we run separate regressions, we get regression coefficient estimators

$$\mathbf{b}_{g,\mathrm{OLS}} = (\mathbf{X}_g'\mathbf{X}_g)^{-1}\mathbf{X}_g'\mathbf{y}_g, \qquad g = 1, 2.$$

These are OLS estimators; they do not account for the information in σ_{12}.

The seemingly unrelated regression technique attributed to Zellner (1962E) combines different sets of regression equations and uses GLS estimation. Specifically, suppose that we start with G sets of regression equations of the form $\mathbf{y}_g = \mathbf{X}_g\boldsymbol{\beta}_g + \boldsymbol{\epsilon}_g$. To see how to combine these, we work with $G = 2$ and define

$$\mathbf{y} = \begin{pmatrix} \mathbf{y}_1 \\ \mathbf{y}_2 \end{pmatrix}, \quad \mathbf{X} = \begin{pmatrix} \mathbf{X}_1 & \mathbf{0} \\ \mathbf{0} & \mathbf{X}_2 \end{pmatrix}, \quad \boldsymbol{\beta} = \begin{pmatrix} \boldsymbol{\beta}_1 \\ \boldsymbol{\beta}_2 \end{pmatrix}, \quad \boldsymbol{\epsilon} = \begin{pmatrix} \boldsymbol{\epsilon}_1 \\ \boldsymbol{\epsilon}_2 \end{pmatrix}. \tag{6.12}$$

Thus, we have

$$\mathrm{Var}\,\mathbf{y} = \mathrm{Var}\,\boldsymbol{\epsilon} = \begin{pmatrix} \sigma_1^2\mathbf{I}_n & \sigma_{12}\mathbf{I}_n \\ \sigma_{12}\mathbf{I}_n & \sigma_2^2\mathbf{I}_n \end{pmatrix}$$

and, with this, the GLS estimator is

$$\mathbf{b}_{\mathrm{GLS}} = (\mathbf{X}'(\mathrm{Var}\,\mathbf{y})^{-1}\mathbf{X})^{-1}\mathbf{X}'(\mathrm{Var}\,\mathbf{y})^{-1}\mathbf{y}.$$

These are known as the *seemingly unrelated regression* (SUR) estimators. It is easy to check that

$$\mathbf{b}_{\mathrm{GLS}} = \begin{pmatrix} \mathbf{b}_{1,\mathrm{OLS}} \\ \mathbf{b}_{2,\mathrm{OLS}} \end{pmatrix}$$

if either $\sigma_{12} = 0$ or $\mathbf{X}_1 = \mathbf{X}_2$ holds. In either case, we have that the GLS estimator is equivalent to the OLS estimator.

On one hand, the SUR setup can be viewed as a special case of multiple linear, and hence multivariate, regression, with Equations (6.12). On the other hand, SURs can be viewed as a way of extending multivariate regressions to allow for explanatory variables that depend on the type of response. As we will see, another way of allowing type-specific explanatory variables is to restrict the parameter matrix.

6.4.3 Simultaneous-Equations Models

In our supply-and-demand example, we assumed that price and quantity were potentially related through covariance terms. However, researchers often wish to estimate alternative models that allow for relationships among responses directly through the regression equations.

Example 6.1: Supply and Demand (continued) Consider a supply-and-demand model that yields the following two sets of equations:

$$\begin{aligned} y_{1i} &= \beta_1 y_{2i} + \gamma_{10} + \gamma_{11} x_{1i} + \varepsilon_{1i} \quad \text{(price)}, \\ y_{2i} &= \beta_2 y_{1i} + \gamma_{20} + \gamma_{21} x_{2i} + \varepsilon_{2i} \quad \text{(quantity)}. \end{aligned} \tag{6.13}$$

Here, we assume that quantity linearly affects price, and vice versa. As before, both xs are assumed to be exogenous for these equations.

In Section 6.1.3, we saw that using only ordinary least squares in a single equation produced biased estimators owing to the endogenous regressor variables. That is, when examining the price equation, we see that the quantity variable (y_{2i}) is clearly endogenous because it is influenced by price, as is seen in the quantity equation. One can use similar reasoning to argue that SUR estimators also yield biased and inconsistent estimators; seemingly unrelated regression techniques improve upon the efficiency of ordinary least squares but do not change the nature of the bias in estimators.

To introduce estimators for Equations (6.13), we collect the dependent variables with the matrix

$$\mathbf{B} = \begin{pmatrix} 0 & \beta_1 \\ \beta_2 & 0 \end{pmatrix}.$$

Thus, we may express Equation (6.13) as $\mathbf{y}_i = \mathbf{B}\mathbf{y}_i + \mathbf{\Gamma}\mathbf{x}_i + \boldsymbol{\epsilon}_i$, where $\boldsymbol{\epsilon}_i = (\varepsilon_{1i}\ \varepsilon_{2i})'$, $\mathbf{x}_i = (1\ x_{1i}\ x_{2i})'$, and

$$\mathbf{\Gamma} = \begin{pmatrix} \gamma_{10} & \gamma_{11} & 0 \\ \gamma_{20} & 0 & \gamma_{21} \end{pmatrix}.$$

This expression for ($\mathbf{y}_i$) looks very much like the multivariate regression model in Equation (6.10), the difference being that we now have included a set of endogenous regressors, $\mathbf{B}\mathbf{y}_i$. As noted in Section 6.4.2, we have incorporated different regressors in different equations by defining a "combined" set of explanatory variables and imposing the appropriate restrictions on the matrix of coefficients, $\boldsymbol{\beta}$.

In this subsection we consider systems of regression equations where responses from one equation may serve as endogenous regressors in another equation. Specifically, we consider model equations of the form

$$\mathbf{y}_i = \mathbf{B}\mathbf{y}_i + \boldsymbol{\Gamma}\mathbf{x}_i + \boldsymbol{\epsilon}_i. \tag{6.14}$$

Here, we assume that $\mathbf{I} - \mathbf{B}$ is a $G \times G$ nonsingular matrix, $\boldsymbol{\Gamma}$ has dimension $G \times K$, and the vector of explanatory variables, $\mathbf{x}_i$, is of dimension $K \times I$ and $\mathrm{Var}\, \boldsymbol{\epsilon}_i = \boldsymbol{\Sigma}$. With these assumptions, we may write the so-called reduced form

$$\mathbf{y}_i = \boldsymbol{\Pi}\mathbf{x}_i + \boldsymbol{\eta}_i,$$

where $\boldsymbol{\Pi} = (\mathbf{I} - \mathbf{B})^{-1}\boldsymbol{\Gamma}$, $\boldsymbol{\eta}_i = (\mathbf{I} - \mathbf{B})^{-1}\boldsymbol{\epsilon}_i$, and $\mathrm{Var}\, \boldsymbol{\eta}_i = \mathrm{Var}\, [(\mathbf{I} - \mathbf{B})^{-1}\boldsymbol{\epsilon}_i] = (\mathbf{I} - \mathbf{B})^{-1}\boldsymbol{\Sigma}((\mathbf{I} - \mathbf{B})^{-1})' = \boldsymbol{\Omega}$. For example, in our supply-and-demand example, we have

$$(\mathbf{I} - \mathbf{B})^{-1} = \frac{1}{1 - \beta_1\beta_2}\begin{pmatrix} 1 & \beta_1 \\ \beta_2 & 1 \end{pmatrix}$$

and thus

$$\boldsymbol{\Pi} = \frac{1}{1 - \beta_1\beta_2}\begin{pmatrix} \gamma_{10} + \beta_1\gamma_{20} & \gamma_{11} & \beta_1\gamma_{21} \\ \beta_2\gamma_{10} + \gamma_{20} & \beta_2\gamma_{11} & \gamma_{21} \end{pmatrix}.$$

The reduced form is simply a multivariate regression model as in Equation (6.10). We will assume sufficient conditions on the observables to consistently estimate the reduced-form coefficients $\boldsymbol{\Pi}$ and the corresponding variance–covariance matrix $\boldsymbol{\Omega}$. Thus, we will have information on the GK elements in $\boldsymbol{\Pi}$ and the $G(G+1)/2$ elements of $\boldsymbol{\Omega}$. However, this information in and of itself will not allow us to properly identify all the elements of $\boldsymbol{\Gamma}$, $\mathbf{B}$, and $\boldsymbol{\Sigma}$. There are GK, G^2, and $G(G+1)/2$ elements in these matrices, respectively. Additional restrictions, generally from economic theory, are required. To illustrate, note that, in our example, there are six structural parameters of interest in $\boldsymbol{\Gamma}$ and six elements of $\boldsymbol{\Pi}$. Thus, we need to check to see whether this provides sufficient information to recover the relevant structural parameters. This process is known as *identification*. Detailed treatments of this topic are available in many sources; see, for example, Greene (2002E), Hayashi (2000E), and Wooldridge (2002E).

Estimates of $\mathbf{\Pi}$ allow us to recover the structural parameters in $\mathbf{\Gamma}$ and $\mathbf{B}$. This method of estimating the structural parameters is known as *indirect least squares*. Alternatively, it is possible to estimate Equation (6.14) directly using maximum likelihood theory. However, this becomes complex because the parameters in $\mathbf{B}$ appear in both the mean and variance.

Not surprisingly, many alternative estimation procedures are available. A commonly used method is *two-stage least squares*, introduced in Section 6.1.3. For the first stage of this procedure, one uses all the exogenous variables to fit the responses. That is, using Equation (6.11), calculate

$$\hat{\mathbf{y}}_i = \mathbf{G}_{\mathrm{OLS}}\mathbf{x}_i = \left(\sum_{i=1}^{n} \mathbf{y}_i \mathbf{x}_i'\right)\left(\sum_{i=1}^{n} \mathbf{x}_i \mathbf{x}_i'\right)^{-1} \mathbf{x}_i. \tag{6.15}$$

For the second stage, assume that we can write the gth row of Equation (6.14) as

$$y_{gi} = \mathbf{B}_g' \mathbf{y}_{i(g)} + \boldsymbol{\beta}_g' \mathbf{x}_i + \varepsilon_{gi}, \tag{6.16}$$

where $\mathbf{y}_{i(g)}$ is $\mathbf{y}_i$ with the gth row omitted and $\mathbf{B}_g$ is the transpose of the gth row of $\mathbf{B}$, omitting the diagonal element. Then, we may calculate OLS estimators corresponding to the equation

$$y_{gi} = \mathbf{B}_g' \hat{\mathbf{y}}_{i(g)} + \boldsymbol{\beta}_g' \mathbf{x}_i + \text{residual}. \tag{6.17}$$

The fitted values, $\hat{\mathbf{y}}_{i(g)}$, are determined from $\hat{\mathbf{y}}_i$ in Equation (6.15), after removing the gth row. Using ordinary least squares in Equation (6.16) is inappropriate because of the endogenous regressors in $\mathbf{y}_{i(g)}$. Because the fitted values, $\hat{\mathbf{y}}_{i(g)}$, are linear combinations of the exogenous variables, there is no such endogeneity problem.

Using Equation (6.17), we can express the two-stage least-squares estimators of the structural parameters as

$$\begin{pmatrix} \hat{\mathbf{B}}_g \\ \hat{\boldsymbol{\beta}}_g \end{pmatrix} = \left(\sum_{i=1}^{n} \begin{pmatrix} \hat{\mathbf{y}}_{i(g)} \\ \mathbf{x}_i \end{pmatrix}\begin{pmatrix} \hat{\mathbf{y}}_{i(g)} \\ \mathbf{x}_i \end{pmatrix}'\right)^{-1} \sum_{i=1}^{n} \begin{pmatrix} \hat{\mathbf{y}}_{i(g)} \\ \mathbf{x}_i \end{pmatrix} y_{gi}, \quad g = 1, \ldots, G. \tag{6.18}$$

Note that, for this estimation methodology to work, the number of exogenous variables excluded from the gth row must be at least as large as the number of endogenous variables that appear in $\mathbf{B}_g' \mathbf{y}_{i(g)}$.

Example 6.1: Supply and Demand (continued) To illustrate, we return to our supply-and-demand example. Then, for $g = 1$, we have $\mathbf{B}_g = \gamma_1, \mathbf{y}_{gi} = y_{2i}, \boldsymbol{\beta}_g' = (\gamma_{10} \;\; \gamma_{11} \, 0)$, and $\mathbf{x}_i = (1 \, x_{1i} \;\; x_{2i})'$. We calculate fitted values for y_{1i} as

$$\hat{y}_{1i} = \left(\sum_{i=1}^{n} y_{1i} \mathbf{x}_i' \right) \left(\sum_{i=1}^{n} \mathbf{x}_i \mathbf{x}_i' \right)^{-1} \mathbf{x}_i .$$

Similar calculations hold for $g = 2$. Then, straightforward substitution into Equation (6.18) yields the two-stage least-squares estimators.

6.4.4 Systems of Equations with Error Components

In this section we describe panel data extensions to systems of equations involving error components to model the heterogeneity. We first examine seemingly unrelated regression models and then simultaneous-equations models. Only one-way error components are dealt with. Interested readers will find additional details and summaries of extensions to two-way error components in Baltgi (2001E) and Krishnakumar (Chapter 9 of Mátyás and Sevestre, 1996E).

Seemingly Unrelated Regression Models with Error Components

Like SURs in the cross-sectional setting, we now wish to study several error-components models simultaneously. Following the notation of Section 3.1, our interest is in studying situations that can be represented by

$$y_{git} = \alpha_{gi} + \mathbf{x}_{git}' \boldsymbol{\beta}_g + \varepsilon_{git}, \quad g = 1, \ldots, G.$$

In our supply-and-demand example, g represented the price and quantity equations and i represented the country. We now assume that we follow countries over time, so that $t = 1, \ldots, T_i$. Assuming that the xs are the only exogenous variables and that type and country-specific random effects, α_{gi}, are independent of the disturbance terms, one can always use OLS estimators of $\boldsymbol{\beta}_g$; they are unbiased and consistent.

To compute the more efficient GLS estimators, we begin by stacking over the G responses,

$$\begin{pmatrix} y_{1it} \\ \vdots \\ y_{Git} \end{pmatrix} = \begin{pmatrix} \alpha_{1i} \\ \vdots \\ \alpha_{Gi} \end{pmatrix} + \begin{pmatrix} \mathbf{x}_{1it}' & \mathbf{0} & \mathbf{0} \\ \mathbf{0} & \ddots & \mathbf{0} \\ \mathbf{0} & \mathbf{0} & \mathbf{x}_{Git}' \end{pmatrix} \begin{pmatrix} \boldsymbol{\beta}_1 \\ \vdots \\ \boldsymbol{\beta}_G \end{pmatrix} + \begin{pmatrix} \varepsilon_{1it} \\ \vdots \\ \varepsilon_{Git} \end{pmatrix},$$

which we write as

$$\mathbf{y}_{it} = \boldsymbol{\alpha}_i + \mathbf{X}_{it}\boldsymbol{\beta} + \boldsymbol{\epsilon}_{it}. \tag{6.19}$$

Here, we assume that $\boldsymbol{\beta}$ has dimension $K \times 1$ so that $\mathbf{X}_{it}$ has dimension $G \times K$. Following conventional SURs, we may allow for covariances among responses through the notation $\mathrm{Var}\, \boldsymbol{\epsilon}_{it} = \boldsymbol{\Sigma}$. We may also allow for covariances through $\mathrm{Var}\, \boldsymbol{\alpha}_i = \mathbf{D}$. Stacking over t, we have

$$\begin{pmatrix} \mathbf{y}_{i1} \\ \vdots \\ \mathbf{y}_{iT_i} \end{pmatrix} = \begin{pmatrix} \boldsymbol{\alpha}_i \\ \vdots \\ \boldsymbol{\alpha}_i \end{pmatrix} + \begin{pmatrix} \mathbf{X}_{i1} \\ \vdots \\ \mathbf{X}_{iT_i} \end{pmatrix} \boldsymbol{\beta} + \begin{pmatrix} \boldsymbol{\epsilon}_{i1} \\ \vdots \\ \boldsymbol{\epsilon}_{iT_i} \end{pmatrix},$$

which we write as

$$\mathbf{y}_i = \boldsymbol{\alpha}_i \otimes \mathbf{1}_{T_i} + \mathbf{X}_i\boldsymbol{\beta} + \boldsymbol{\epsilon}_i.$$

With this notation, note that $\mathrm{Var}\, \boldsymbol{\epsilon}_i = \mathit{block\ diag}(\mathrm{Var}\, \boldsymbol{\epsilon}_{i1}, \ldots, \mathrm{Var}\, \boldsymbol{\epsilon}_{iT_i}) = \boldsymbol{\Sigma} \otimes \mathbf{I}_{T_i}$ and that $\mathrm{Var}\,(\alpha_i \otimes \mathbf{1}_{T_i}) = \mathrm{E}(\boldsymbol{\alpha}_i \otimes \mathbf{1}_{T_i})(\boldsymbol{\alpha}_i \otimes \mathbf{1}_{T_i})' = \mathbf{D} \otimes \mathbf{J}_{T_i}$. Thus, the GLS estimator for $\boldsymbol{\beta}$ is

$$\mathbf{b}_{\mathrm{GLS}} = \left(\sum_{i=1}^{n} \mathbf{X}_i'(\mathbf{D} \otimes \mathbf{J}_{T_i} + \Sigma \otimes \mathbf{I}_{T_i})^{-1}\mathbf{X}_i \right)^{-1} \times \left(\sum_{i=1}^{n} \mathbf{X}_i'(\mathbf{D} \otimes \mathbf{J}_{T_i} + \Sigma \otimes \mathbf{I}_{T_i})^{-1}\mathbf{y}_i \right). \tag{6.20}$$

Avery (1977E) and Baltagi (1980E) considered this model.

Simultaneous-Equations Models with Error Components

Extending the notation of Equation (6.14), we now consider

$$\mathbf{y}_{it} = \boldsymbol{\alpha}_i + \mathbf{B}\mathbf{y}_{it} + \boldsymbol{\Gamma}\mathbf{x}_{it} + \boldsymbol{\epsilon}_{it}. \tag{6.21}$$

The subject-specific term is $\boldsymbol{\alpha}_i = (\alpha_{1i}, \alpha_{2i}, \ldots, \alpha_{Gi})'$, which has mean zero and variance–covariance matrix $\mathrm{Var}\, \boldsymbol{\alpha}_i = \mathbf{D}$. We may rewrite Equation (6.21) in reduced form as

$$\mathbf{y}_{it} = \boldsymbol{\Pi}\mathbf{x}_{it} + \boldsymbol{\eta}_{it},$$

where $\boldsymbol{\Pi} = (\mathbf{I} - \mathbf{B})^{-1}\boldsymbol{\Gamma}$ and $\boldsymbol{\eta}_{it} = (\mathbf{I} - \mathbf{B})^{-1}(\boldsymbol{\alpha}_i + \boldsymbol{\epsilon}_{it})$. With this formulation, we see that the panel data mean effects are the same as those for the model in Equation (6.14) without subject-specific effects. Specifically, as pointed out by Hausman and Taylor (1981E), without additional restrictions on the variance or covariance parameters, the identification issues are the same with and without subject-specific effects. In addition, estimation of the reduced form is similar; details are provided in the review by Krishnakumar (Chapter 9 of Mátyás and Sevestre, 1996E).

We now consider direct estimation of the structural parameters using two-stage least squares. To begin, note that the two-stage least-squares estimators described in Equation (6.18) still provide unbiased, consistent estimators. However, they do not account for the error-components variance structure, and thus, they may be inefficient. Nonetheless, these estimators can be used to calculate estimators of the variance components that will be used in the following estimation procedure.

For the first stage, we need to calculate fitted values of the responses, using only the exogenous variables as regressors. Note that

$$(\mathbf{I}-\mathbf{B})^{-1}\,\mathbf{\Gamma}\mathbf{x}_{it} = \begin{pmatrix} \boldsymbol{\beta}_1^{\prime *}\mathbf{x}_{it} \\ \vdots \\ \boldsymbol{\beta}_G^{\prime *}\mathbf{x}_{it} \end{pmatrix} = \begin{pmatrix} \mathbf{x}_{it}^{\prime} & \mathbf{0} & \mathbf{0} \\ \mathbf{0} & \ddots & \mathbf{0} \\ \mathbf{0} & \mathbf{0} & \mathbf{x}_{it}^{\prime} \end{pmatrix} \begin{pmatrix} \boldsymbol{\beta}_1^{*} \\ \vdots \\ \boldsymbol{\beta}_G^{*} \end{pmatrix} = \mathbf{X}_{it}^{*}\boldsymbol{\beta}^{*}.$$

Thus, with Equation (6.21), we may express the reduced form as

$$\mathbf{y}_{it} = (\mathbf{I}-\mathbf{B})^{-1}\boldsymbol{\alpha}_i + (\mathbf{I}-\mathbf{B})^{-1}\mathbf{\Gamma}\mathbf{x}_{it} + (\mathbf{I}-\mathbf{B})^{-1}\boldsymbol{\epsilon}_{it} = \boldsymbol{\alpha}_i^{*} + \mathbf{X}_{it}^{*}\boldsymbol{\beta}^{*} + \boldsymbol{\epsilon}_{it}^{*}.$$

This has the same form as the seemingly unrelated regression with the error-components model in Equation (6.19). Thus, one can use Equation (6.20) to get fitted regression coefficients and thus fitted values. Alternatively, we have seen that ordinary least squares provides unbiased and consistent estimates for this model. Thus, this technique would also serve for computing the first-stage fitted values.

For the second stage, write the model in Equation (6.21) in the same fashion as Equation (6.16) to get

$$y_{git} = \mathbf{y}_{it,(g)}^{\prime}\mathbf{B}_g + \mathbf{x}_{it}^{\prime}\boldsymbol{\beta}_g + \alpha_{gi} + \varepsilon_{git}.$$

Recall that $\mathbf{B}_g$ is a $(G-1)\times 1$ vector of parameters and $\mathbf{y}_{it(g)}$ is $\mathbf{y}_{it}$ with the gth row omitted. Stacking over $t = 1, \ldots, T$ yields

$$\mathbf{y}_{gi} = \mathbf{Y}_{i(g)}\mathbf{B}_g + \mathbf{X}_i\boldsymbol{\beta}_g + \alpha_{gi}\mathbf{1}_T + \boldsymbol{\epsilon}_{gi} = (\,\mathbf{Y}_{i(g)}\ \mathbf{X}_i\,)\begin{pmatrix} \mathbf{B}_g \\ \boldsymbol{\beta}_g \end{pmatrix} + \alpha_{gi}\mathbf{1}_T + \boldsymbol{\epsilon}_{gi}.$$

Let $\mathrm{Var}\,(\alpha_{gi}\mathbf{1}_T + \epsilon_{gi}) = \sigma_\alpha^2 \mathbf{J}_T + \sigma^2 \mathbf{I}_T$. Replacing $\mathbf{Y}_{i(g)}$ by $\hat{\mathbf{Y}}_{i(g)}$ yields the two-stage least-squares estimators

$$\begin{pmatrix} \hat{\mathbf{B}}_g \\ \hat{\beta}_g \end{pmatrix} = \left(\sum_{i=1}^{n} \begin{pmatrix} \hat{\mathbf{Y}}_{i(g)} \\ \mathbf{X}_i \end{pmatrix}' \left(\sigma_\alpha^2 \mathbf{J}_T + \sigma^2 \mathbf{I}_T\right)^{-1} \begin{pmatrix} \hat{\mathbf{Y}}_{i(g)} \\ \mathbf{X}_i \end{pmatrix} \right)^{-1}$$
$$\times \sum_{i=1}^{n} \begin{pmatrix} \hat{\mathbf{Y}}_{i(g)} \\ \mathbf{X}_i \end{pmatrix}' \left(\sigma_\alpha^2 \mathbf{J}_T + \sigma^2 \mathbf{I}_T\right)^{-1} \mathbf{y}_{gi}, \quad g = 1, \ldots, G.$$

6.5 Simultaneous-Equations Models with Latent Variables

Simultaneous-equations models with latent variables comprise a broad framework for handling complex systems of equations, as well as longitudinal data models. This framework, which originated in the work of Jöreskog, Keesling, and Wiley (see Bollen, 1989EP, p. 6), is widely applied in sociology, psychology, and educational sciences. Like the introductions to simultaneous equations in Sections 6.4.3 and 6.4.4, the systems of equations may be used for different types of responses (multivariate), different times of responses (longitudinal), or both. The estimation techniques preferred in psychology and education are known as *covariance structure analysis*.

To keep our treatment self-contained, we first outline the framework in Section 6.5.1 in the cross-sectional context. Section 6.5.2 then describes longitudinal data applications.

6.5.1 Cross-Sectional Models

We begin by assuming that $(\mathbf{x}_i, \mathbf{y}_i)$ are i.i.d. observable draws from a population. We treat $\mathbf{x}$ as the exogenous vector and $\mathbf{y}$ as the endogenous vector.

The $\mathbf{x}$-*measurement equation* is

$$\mathbf{x}_i = \boldsymbol{\tau}_x + \boldsymbol{\Lambda}_x \boldsymbol{\xi}_i + \boldsymbol{\delta}_i, \tag{6.22}$$

where $\boldsymbol{\delta}$ is the disturbance term, $\boldsymbol{\xi}$ is a vector of latent (unobserved) variables, and $\boldsymbol{\Lambda}_x$ is a matrix of regression coefficients that relates $\boldsymbol{\xi}$ to $\mathbf{x}$. In a measurement error context, we might use $\boldsymbol{\Lambda}_x = \mathbf{I}$ and interpret $\boldsymbol{\tau}_x + \boldsymbol{\xi}$ to be the "true" values of the observables that are corrupted by the "error" $\boldsymbol{\delta}$. In other applications, ideas from classic factor analysis drive the model setup. That is, there may be many observable exogenous measurements that are driven by relatively few underlying latent variables. Thus, the dimension of $\mathbf{x}$ may be large relative to the dimension of $\boldsymbol{\xi}$. In this case, we can reduce the dimension of the problem by focusing on the latent variables. Moreover, the latent variables more closely correspond to social science theory than do the observables.

To complete the specification of this measurement model, we assume that $E\boldsymbol{\xi}_i = \boldsymbol{\mu}_\xi$ and $\mathrm{E}\boldsymbol{\delta}_i = \mathbf{0}$ so that $\mathrm{E}\mathbf{x}_i = \boldsymbol{\tau}_x + \boldsymbol{\Lambda}_x\boldsymbol{\mu}_\xi$. Further, we use $\mathrm{Var}\,\boldsymbol{\delta}_i = \boldsymbol{\Theta}_\delta$ and $\mathrm{Var}\,\boldsymbol{\xi}_i = \boldsymbol{\Phi}$, so that $\mathrm{Var}\,\mathbf{x}_i = \boldsymbol{\Lambda}_x\boldsymbol{\Phi}\boldsymbol{\Lambda}_x' + \boldsymbol{\Theta}_\delta$. That is, $\boldsymbol{\xi}$ and $\boldsymbol{\delta}$ are mutually uncorrelated.

Specification of the **y**-*measurement equation* is similar. Define

$$\mathbf{y}_i = \boldsymbol{\tau}_y + \boldsymbol{\Lambda}_y\boldsymbol{\eta}_i + \boldsymbol{\epsilon}_i, \tag{6.23}$$

where $\boldsymbol{\epsilon}$ is the disturbance term, $\boldsymbol{\eta}$ is a vector of latent variables, and $\boldsymbol{\Lambda}_y$ is a matrix of regression coefficients that relates $\boldsymbol{\eta}$ to $\mathbf{y}$. We use $\boldsymbol{\Theta}_\varepsilon = \mathrm{Var}\,\boldsymbol{\epsilon}_i$ and assume that $\boldsymbol{\eta}$ and $\boldsymbol{\epsilon}$ are mutually uncorrelated. Note that Equation (6.23) is not a usual multiple linear regression equation because regressor $\boldsymbol{\eta}$ is unobserved.

To link the exogenous and endogenous latent variables, we have the *structural equation*

$$\boldsymbol{\eta}_i = \boldsymbol{\tau}_\eta + \mathbf{B}\boldsymbol{\eta}_i + \boldsymbol{\Gamma}\boldsymbol{\xi}_i + \boldsymbol{\varsigma}_i, \tag{6.24}$$

where $\boldsymbol{\tau}_\eta$ is a fixed intercept, $\mathbf{B}$ and $\boldsymbol{\Gamma}$ are matrices of regression parameters, and $\boldsymbol{\varsigma}$ is a mean-zero disturbance term with second moment $\mathrm{Var}\,\boldsymbol{\varsigma} = \boldsymbol{\Psi}$.

A drawback of a structural equation model with latent variables, summarized in Equations (6.22)–(6.24), is that it is overparameterized and too unwieldy to use, without further restrictions. The corresponding advantage is that this model formulation captures a number of different models under a single structure, thus making it a desirable framework for analysis. Before describing important applications of the model, we first summarize the mean and variance parameters that are useful in this formulation.

Mean Parameters

With $\mathrm{E}\boldsymbol{\xi}_i = \boldsymbol{\mu}_\xi$, from Equation (6.24), we have

$$\mathrm{E}\boldsymbol{\eta}_i = \boldsymbol{\tau}_\eta + \mathbf{B}\mathrm{E}\boldsymbol{\eta}_i + \mathrm{E}(\boldsymbol{\Gamma}\boldsymbol{\xi}_i + \boldsymbol{\varsigma}_i) = \boldsymbol{\tau}_\eta + \mathbf{B}\mathrm{E}\boldsymbol{\eta}_i + \boldsymbol{\Gamma}\boldsymbol{\mu}_\xi.$$

Thus, $\mathrm{E}\boldsymbol{\eta}_i = (\mathbf{I}-\mathbf{B})^{-1}(\boldsymbol{\tau}_\eta + \boldsymbol{\Gamma}\boldsymbol{\mu}_\xi)$, assuming that $\mathbf{I}-\mathbf{B}$ is invertible. Summarizing, we have

$$\begin{pmatrix} \mathrm{E}\,\mathbf{x} \\ \mathrm{E}\,\mathbf{y} \end{pmatrix} = \begin{pmatrix} \boldsymbol{\tau}_x + \boldsymbol{\Lambda}_x\boldsymbol{\mu}_\xi \\ \boldsymbol{\tau}_y + \boldsymbol{\Lambda}_y(\mathbf{I}-\mathbf{B})^{-1}\left(\boldsymbol{\tau}_\eta + \boldsymbol{\Gamma}\boldsymbol{\mu}_\xi\right) \end{pmatrix}. \tag{6.25}$$

Covariance Parameters

From Equation (6.24), we have $\eta_i = (\mathbf{I}-\mathbf{B})^{-1}(\tau_\eta + \Gamma\xi_i + \varsigma_i)$ and

$$\begin{aligned}\mathrm{Var}\ \eta &= \mathrm{Var}\left((\mathbf{I}-\mathbf{B})^{-1}(\Gamma\xi+\zeta)\right) = (\mathbf{I}-\mathbf{B})^{-1}\,\mathrm{Var}\left((\Gamma\xi+\zeta)\right)(\mathbf{I}-\mathbf{B})^{-1\prime} \\ &= (\mathbf{I}-\mathbf{B})^{-1}\left(\Gamma\Phi\Gamma'+\Psi\right)(\mathbf{I}-\mathbf{B})^{-1\prime}.\end{aligned}$$

Thus, with Equation (6.23), we have

$$\begin{aligned}\mathrm{Var}\ \mathbf{y} &= \mathbf{\Lambda}_y\,(\mathrm{Var}\ \eta)\,\mathbf{\Lambda}_y' + \mathbf{\Theta}_\epsilon \\ &= \mathbf{\Lambda}_y\,(\mathbf{I}-\mathbf{B})^{-1}\,(\Gamma\Phi\Gamma'+\Psi)\,(\mathbf{I}-\mathbf{B})^{-1\prime}\,\mathbf{\Lambda}_y' + \mathbf{\Theta}_\epsilon.\end{aligned}$$

With Equation (6.22), we have

$$\begin{aligned}\mathrm{Cov}(\mathbf{y},\mathbf{x}) &= \mathrm{Cov}(\mathbf{\Lambda}_y\eta+\epsilon,\ \mathbf{\Lambda}_x\xi+\delta) = \mathbf{\Lambda}_y\mathrm{Cov}(\eta,\xi)\mathbf{\Lambda}_x' \\ &= \mathbf{\Lambda}_y\mathrm{Cov}\left((\mathbf{I}-\mathbf{B})^{-1}\,\Gamma\xi,\xi\right)\mathbf{\Lambda}_y' = \mathbf{\Lambda}_y\,(\mathbf{I}-\mathbf{B})^{-1}\,\Gamma\Phi\mathbf{\Lambda}_x'.\end{aligned}$$

Summarizing, we have

$$\begin{pmatrix}\mathrm{Var}\ \mathbf{y} & \mathrm{Cov}(\mathbf{y},\mathbf{x}) \\ \mathrm{Cov}(\mathbf{y},\mathbf{x})' & \mathrm{Var}\,\mathbf{x}\end{pmatrix}$$

$$= \begin{pmatrix}\mathbf{\Lambda}_y\,(\mathbf{I}-\mathbf{B})^{-1}\left(\Gamma\Phi\Gamma'+\Psi\right)(\mathbf{I}-\mathbf{B})^{-1\prime}\,\mathbf{\Lambda}_y'+\mathbf{\Theta}_\epsilon & \mathbf{\Lambda}_y\,(\mathbf{I}-\mathbf{B})^{-1}\,\Gamma\Phi\Lambda_x' \\ \mathbf{\Lambda}_x\Phi\Gamma'\,(\mathbf{I}-\mathbf{B})^{-1\prime}\,\mathbf{\Lambda}_y' & \mathbf{\Lambda}_x\Phi\Lambda_x'+\mathbf{\Theta}_\delta\end{pmatrix}. \tag{6.26}$$

Identification Issues

With the random-sampling assumption, one can consistently estimate the means and covariances of the observables, specifically, the left-hand sides of Equations (6.25) and (6.26). The model parameters are given in terms of the right-hand sides of these equations. There are generally more parameters that can be uniquely identified by the data. *Identification* is demonstrated by showing that the unknown parameters are functions only of the means and covariances and that these functions lead to unique solutions. In this case, we say that the unknown parameters are *identified*. Otherwise, they are said to be *underidentified*.

There are many approaches available for this process. We will illustrate a few in conjunction with some special cases. Detailed broad treatments of this topic are available in many sources; see, for example, Bollen (1989EP).

Special Cases

As noted, the model summarized in Equations (6.22)–(6.24) is overparameterized and too unwieldy to use, although it does encompass many special cases that are directly relevant for applications. To provide focus and intuition, we now summarize a few of these special cases.

- Consider only the **x**-measurement equation. This is the classic factor analysis model (see, for example, Johnson and Wichern, 1999G).
- Assume that both **x** and **y** are used directly in the structural equation model without any additional latent variables. (That is, assume $\mathbf{x}_i = \boldsymbol{\xi}_i$ and $\mathbf{y}_i = \boldsymbol{\eta}_i$.) Then, Equation (6.24) represents a structural equation model based on observables, introduced in Section 6.4.3.
- Moreover, assuming the $\mathbf{B} = \mathbf{0}$, the structural equation model with observables reduces to the multivariate regression model.
- Assume that **y** is used directly in the structural equation model but that **x** is measured with error so that $\mathbf{x}_i = \boldsymbol{\tau}_x + \boldsymbol{\xi}_i + \boldsymbol{\delta}_i$. Assuming no feedback effects (for **y**, so that $\mathbf{B} = \mathbf{0}$), then Equation (6.24) represents the classic *errors-in-variables* model.

Many other special cases appear in the literature. Our focus is on the longitudinal special cases in Section 6.5.2.

Path Diagrams

The popularity of structural equation models with latent variables in education and psychology is due in part to *path diagrams*. Path diagrams, due to Wright (1918B), are pictorial representations of systems of equations. These diagrams show the relations among all variables, including disturbances and errors. These graphical relations allow many users to readily understand the consequences of modeling relationships. Moreover, statistical software routines have been developed to allow analysts to specify the model graphically, without resorting to algebraic representations. Table 6.2 summarizes the primary symbols used to make path diagrams.

Estimation Techniques

Estimation is typically done using maximum likelihood assuming normality, sometimes using instrumental variable estimation for initial values. Descriptions of alternative techniques, including generalized least squares and "unweighted" least squares, can be found in Bollen (1989EP).

Interestingly, the likelihood estimation is customarily done by maximizing the likelihood over *all* the observables. Specifically, assuming that $(\mathbf{x}_i, \mathbf{y}_i)$ are

Table 6.2. *Primary symbols used in path diagrams*

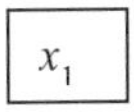	Rectangular or square box signifies an observed variable.
	Circle or ellipse signifies a latent variable.
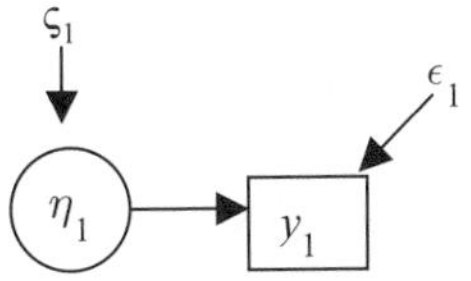	Unenclosed variable signifies a disturbance term (error in either the structural equation or measurement equation). Straight arrow signifies an assumption that the variable at the base of the arrow "causes" the variable at the head of the arrow.
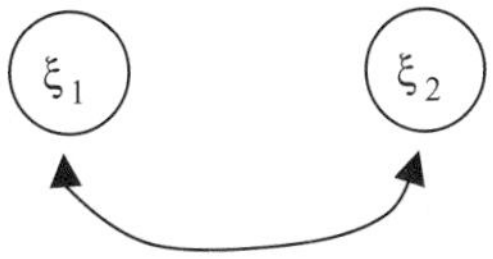	Curved two-headed arrow signifies an association between two variables.
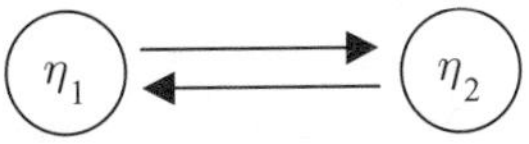	Two straight single-headed arrows connecting two variables signifies a feedback relation or reciprocal causation.

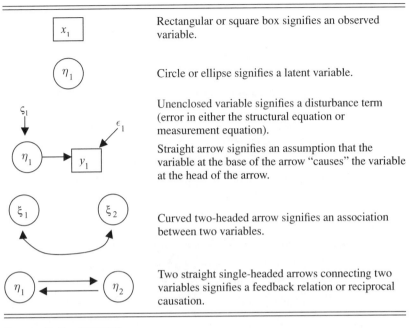

Source: Bollen (1989EP).

jointly multivariate normal with moments given in Equations (6.25) and (6.26), one maximizes the likelihood over the entire sample.

In contrast, most of the maximum likelihood estimation presented in this text has been for the likelihood of the response (or endogenous) variables, *conditional* on the exogenous observables. Specifically, suppose that observables consist of exogenous variables $\mathbf{x}$ and endogenous variables $\mathbf{y}$. Let $\boldsymbol{\theta}_1$ be a vector of parameters that indexes the conditional distribution of the endogenous variables given the exogenous variables, say $p_1(\mathbf{y} \mid \mathbf{x}, \boldsymbol{\theta}_1)$. Assume that there is another set of parameters, $\boldsymbol{\theta}_2$, that are unrelated to $\boldsymbol{\theta}_1$ and that indexes the distribution of the exogenous variables, say $p_2(\mathbf{x}, \boldsymbol{\theta}_2)$. With this setup, the complete likelihood is given by $p_1(\mathbf{y} \mid \mathbf{x}, \boldsymbol{\theta}_1) p_2(\mathbf{x}, \boldsymbol{\theta}_2)$. If our interest is only in the parameters that influence the relationship between $\mathbf{x}$ and $\mathbf{y}$, we can be content with maximizing the likelihood with respect to $\boldsymbol{\theta}_1$. Thus, the distribution of the exogenous variables, $p_2(\mathbf{x}, \boldsymbol{\theta}_2)$, is not relevant to the interest at hand and may be ignored. Because of this philosophy, in our prior examples in this text, we did not concern ourselves with the sampling distribution of the xs. (See Engle et al. 1983E.) Section 7.4.2 will discuss this further.

Although the requirement that the two sets of parameters be unrelated is a restrictive assumption (which is generally not tested), it provides the analyst with some important freedoms. With this assumption, the sampling distribution of the exogenous variables does not provide information about the conditional relationship under investigation. Thus, we need not make restrictive assumptions about the shape of this sampling distribution of the exogenous variables. As a consequence, we need not model the exogenous variables as multivariate normal or even require that they be continuous. For example, a major distinction between the multiple linear regression model and the general linear model is that the latter formulation easily handles categorical regressors. The general linear model is about the conditional relationship between the response and the regressors, imposing few restrictions on the behavior of the regressors.

For the structural equation model with latent variables, the parameters associated with the distribution of exogenous variables are $\boldsymbol{\theta}_2 = \{\boldsymbol{\tau}_x, \boldsymbol{\mu}_\xi, \boldsymbol{\Lambda}_x, \boldsymbol{\Theta}_\delta, \boldsymbol{\Phi}\}$. Assuming, for example, multivariate normality, one can use Equations (6.25) and (6.26) to compute the conditional likelihood of $\mathbf{y} \mid \mathbf{x}$ (Appendix B). However, it is difficult to write down a set of parameters $\boldsymbol{\theta}_1$ that are a subset of the full model parameters that are not related to $\boldsymbol{\theta}_2$. Thus, maximum likelihood for the structural equations model with latent variables requires maximizing the full likelihood over all observables. This is in contrast to our first look at structural equations models in Section 6.5.3 (without the measurement equations of Sections 6.5.1 and 6.5.2), where we could isolate the conditional model parameters from the sampling distribution of the exogenous variables.

In summary, maximum likelihood estimation for structural equation models with latent variables customarily employs maximum likelihood estimation, where the likelihood is with respect to all the observables. A full model of both the exogenous and endogenous variables is specified; maximum likelihood estimators are well known to have many optimality properties. A consequence of this specification of the exogenous variables is that it is difficult to handle categorical variables; multivariate distributions for discrete variables are much less well understood than the continuous-variable case.

6.5.2 Longitudinal Data Applications

The structural equation model with latent variables provides a broad framework for modeling longitudinal data. This section provides a series of examples to demonstrate the breadth of this framework.

Special Case: Autoregressive Model Suppose that y_{it} represents the reading ability of the ith child; assume that we observe the set of n children over

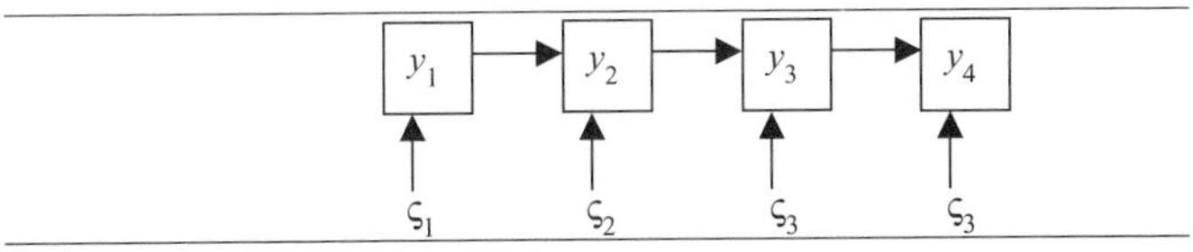

Figure 6.1. Path diagram for the model in Equation (6.27a).

$t = 1, \ldots, 4$ time periods. We might represent a child's reading ability as

$$y_{it} = \beta_0 + \beta_1 y_{i,t-1} + \varsigma_{it}, \quad \text{for} \quad t = 2, 3, 4, \tag{6.27a}$$

using $t = 1$ as the baseline measurement. We summarizes these relations using

$$\boldsymbol{\eta}_i = \begin{pmatrix} y_{i1} \\ y_{i2} \\ y_{i3} \\ y_{i4} \end{pmatrix} = \begin{pmatrix} \beta_0 \\ \beta_0 \\ \beta_0 \\ \beta_0 \end{pmatrix} + \begin{pmatrix} 0 & 0 & 0 & 0 \\ \beta_1 & 0 & 0 & 0 \\ 0 & \beta_1 & 0 & 0 \\ 0 & 0 & \beta_1 & 0 \end{pmatrix} \begin{pmatrix} y_{i1} \\ y_{i2} \\ y_{i3} \\ y_{i4} \end{pmatrix} + \begin{pmatrix} \varsigma_{i1} \\ \varsigma_{i2} \\ \varsigma_{i3} \\ \varsigma_{i4} \end{pmatrix}$$
$$= \boldsymbol{\tau}_\eta + \mathbf{B}\boldsymbol{\eta}_i + \boldsymbol{\varsigma}_i. \tag{6.27b}$$

Thus, this model is a special case of Equations (6.22)–(6.24) by choosing $\mathbf{y}_i = \boldsymbol{\eta}_i$ and $\boldsymbol{\Gamma} = \mathbf{0}$. Graphically, we can express this model as shown in Figure 6.1.

This basic model could be extended in several ways. For example, one may wish to consider evaluation at unequally spaced time points, such as fourth, sixth, eighth, and twelfth grades. This suggests using slope coefficients that depend on time. In addition, one could also use more than one lag predictor variable. For some other extensions, see the following continuation of this basic example.

Special Case: Autoregressive Model with Latent Variables and Multiple Indicators Suppose now that "reading ability" is considered a latent variable denoted by η_{it} and that we have two variables, y_{1it} and y_{2it}, that measure this ability, known as *indicators*. The indicators follow a measurement-error model,

$$y_{jit} = \lambda_0 + \lambda_1 \eta_{i,t} + \varepsilon_{jit}, \tag{6.28a}$$

and the latent reading variable follows an autoregressive model,

$$\eta_{it} = \beta_0 + \beta_1 \eta_{i,t-1} + \varsigma_{it}. \tag{6.28b}$$

With the notation $\mathbf{y}_i = (y_{1i1}, y_{2i1}, y_{1i2}, y_{2i2}, y_{1i3}, y_{2i3}, y_{1i4}, y_{2i4})'$, $\boldsymbol{\epsilon}_i$ defined similarly, and $\boldsymbol{\eta}_i = (\eta_{i1}, \eta_{i1}, \eta_{i2}, \eta_{i2}, \eta_{i3}, \eta_{i3}, \eta_{i4}, \eta_{i4})'$, we can express Equation (6.28a) as in Equation (6.23). Here, $\boldsymbol{\tau}_y$ is λ_0 times a vector of ones and

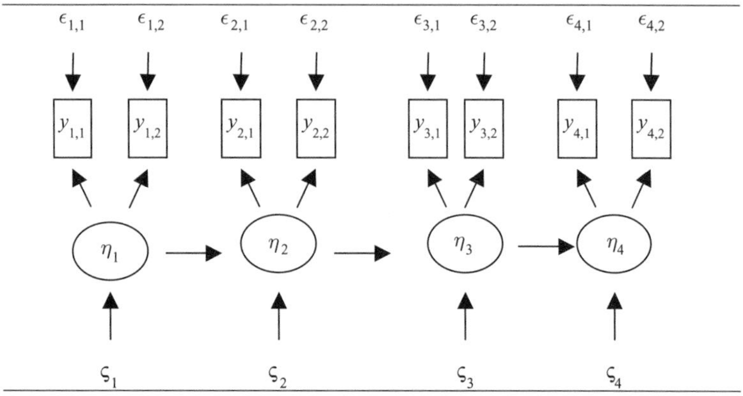

Figure 6.2. Path diagram for the model in Equations (6.28a) and (6.28b).

$\mathbf{\Lambda}_y$ is λ_1 times an identity matrix. Equation (6.28b) can be expressed as the structural equation (6.24) using notation similar to Equation (6.27b).

Graphically, we can express Equations (6.28a) and (6.28b) as shown in Figure 6.2.

Special Case: Autoregressive Model with Latent Variables, Multiple Indicators, and Predictor Variables Continuing with this example, we might also wish to incorporate exogenous predictor variables, such as $x_{1,it}$, the number of hours reading, and $x_{2,it}$, a rating of emotional support given in a child's home. One way of doing this is to simply add this as a predictor in the structural equation, replacing Equation (6.28b) by

$$\eta_{it} = \beta_0 + \beta_1 \eta_{i,t-1} + \gamma_1 x_{1,i,t} + \gamma_2 x_{2,i,t} + \varsigma_{it}.$$

Jöreskog and Goldberger (1975S) introduced a model with multiple indicators of a latent variable that could be explained by multiple causes; this is now known as a MIMIC model.

Growth-Curve Models

Integrating structural equation modeling with latent variables and longitudinal data modeling has been the subject of extensive research in recent years; see, for example, Duncan et al. (1999EP). One widely adopted approach concerns growth-curve modeling.

Example 5.1: Dental Data (continued) In the Section 5.2 dental data example, we used y_{it} to represent the dental measurement of the ith child, measured at four times corresponding to ages 8, 10, 12, and 14 years of age. Using structural equation modeling notation, we could represent the **y**-measurement equation as

$$\mathbf{y}_i = \begin{pmatrix} y_{i1} \\ y_{i2} \\ y_{i3} \\ y_{i4} \end{pmatrix} = \begin{pmatrix} 1 & 8 \\ 1 & 10 \\ 1 & 12 \\ 1 & 14 \end{pmatrix} \begin{pmatrix} \beta_{0i} \\ \beta_{1i} \end{pmatrix} + \begin{pmatrix} \varepsilon_{i1} \\ \varepsilon_{i2} \\ \varepsilon_{i3} \\ \varepsilon_{i4} \end{pmatrix} = \boldsymbol{\Lambda}_y \boldsymbol{\eta}_i + \boldsymbol{\epsilon}_i .$$

Note that this representation assumes that all children are evaluated at the same ages and that these ages correspond to a known "parameter" matrix $\boldsymbol{\Lambda}_y$. Alternatively, one could form groups so that all children within a group are measured at the same set of ages and let $\boldsymbol{\Lambda}_y$ vary by group.

Taking $\boldsymbol{\tau}_\eta = 0$ and $\mathbf{B} = \mathbf{0}$ and using the **x**-measurement directly (without error), we can write the structural equation as

$$\boldsymbol{\eta}_i = \begin{pmatrix} \beta_{0i} \\ \beta_{1i} \end{pmatrix} = \begin{pmatrix} \beta_{00} & \beta_{01} \\ \beta_{10} & \beta_{11} \end{pmatrix} \begin{pmatrix} 1 \\ \text{GENDER}_i \end{pmatrix} + \begin{pmatrix} \varsigma_{0i} \\ \varsigma_{1i} \end{pmatrix} = \boldsymbol{\Gamma} \mathbf{x}_i + \boldsymbol{\varsigma}_i .$$

Thus, this model serves to express intercepts and growth rates associated with each child, β_{0i} and β_{1i}, as a function of gender.

Willet and Sayer (1994EP) introduced growth-curve modeling in the context of structural equations with latent variables. There are several advantages and disadvantages when using structural equations to model growth curves compared to our Chapter 5 multilevel models. The main advantage of structural equation models is the ease of incorporating multivariate responses. For example, in our dental example, there may be more than one dental measurement of interest, or it may be of interest to model dental and visual acuity measurements simultaneously.

The main disadvantage of structural equation models also relates to its multivariate response nature: It is difficult to handle unbalanced structure with this approach. If children came into the clinic for measurements at different ages, this would complicate the design considerably. Moreover, if not all observations were not available, issues of missing data are more difficult to deal with in this context. Finally, we have seen that structural equations with latent variables implicitly assume continuous data that can be approximated by multivariate normality; if the predictor variables are categorical (such as gender), this poses additional problems.

Further Reading

Other introductions to the concept of exogeneity can be found in most graduate econometrics texts; see, for example, Greene (2002E) and Hayashi (2000E). The text by Wooldridge (2002E) gives an introduction with a special emphasis on panel data. Arellano and Honoré (2001E) provide a more sophisticated overview of panel data exogeneity. The collection of chapters in Mátyás and Sevestre (1996E) provides another perspective, as well as an introduction to structural equations with error components. For other methods for handling endogenous regressors with heterogeneity terms, we refer to Arellano (2003E), Baltagi (2001E), and Wooldridge (2002E).

There are many sources that introduce structural equations with latent variables. Bollen (1989EP) is a widely cited source that has been available for many years.

Appendix 6A Linear Projections

Suppose that $\{\mathbf{x}'\ y\}$ is a random vector with finite second moments. Then, the *linear projection* of y onto $\mathbf{x}$ is $\mathbf{x}'(\mathrm{E}(\mathbf{x}\mathbf{x}'))^{-1}\mathrm{E}(\mathbf{x}y)$, provided that $\mathrm{E}(\mathbf{x}\mathbf{x}')$ is invertible. To ease notation, we define $\boldsymbol{\beta} = (\mathrm{E}(\mathbf{x}\mathbf{x}'))^{-1}\mathrm{E}(\mathbf{x}y)$ and denote this projection as

$$LP(y \mid \mathbf{x}) = \mathbf{x}'\boldsymbol{\beta}.$$

Others, such as Goldberger (1991E), use $\mathrm{E}^*(y \mid \mathbf{x})$ to denote this projection. By the linearity of expectations, it is easy to check that $LP(. \mid \mathbf{x})$ is a linear operator in y. As a consequence, if we define ε through the equation

$$y = \mathbf{x}'\boldsymbol{\beta} + \varepsilon,$$

then $\mathrm{E}^*(\varepsilon \mid \mathbf{x}) = 0$. Note that this result is a consequence of the definition, not a model assumption that requires checking.

Suppose that we now *assume* that the model is of the form $y = \mathbf{x}'\boldsymbol{\beta} + \varepsilon$ and that $\mathrm{E}(\mathbf{x}\mathbf{x}')$ is invertible. Then, the condition $\mathrm{E}(\varepsilon\mathbf{x}) = \mathbf{0}$ is equivalent to checking the condition that $LP(\varepsilon \mid \mathbf{x}) = 0$. That is, checking for a correlation between the disturbance term and the predictors is equivalent to checking that the linear projection of the disturbance term on the predictors is zero.

Linear projections can be justified as minimum mean-square predictors. That is, the choice of $\boldsymbol{\beta}$ that minimizes $\mathrm{E}(y - \mathbf{x}'\boldsymbol{\beta})^2$ is $\boldsymbol{\beta} = (\mathrm{E}(\mathbf{x}\mathbf{x}'))^{-1}\mathrm{E}(\mathbf{x}y)$. As an example, suppose that we have a data set of the form $\{\mathbf{x}_i'\ y_i\}, i = 1, \ldots, n$, and we define the expectation operator to be that probability distribution that assigns $1/n$ probability to each outcome (the empirical distribution). Then, we are attempting to minimize

$$E(y - \mathbf{x}'\beta)^2 = \frac{1}{n}\sum_{i=1}^{n}(y_i - \mathbf{x}_i'\boldsymbol{\beta})^2.$$

The solution is

$$(\mathrm{E}(\mathbf{x}\mathbf{x}'))^{-1}\mathrm{E}(\mathbf{x}y) = \left(\frac{1}{n}\sum_{i=1}^{n}\mathbf{x}_i\mathbf{x}_i'\right)^{-1}\left(\frac{1}{n}\sum_{i=1}^{n}\mathbf{x}_i y_i\right),$$

the familiar OLS estimator.

7
Modeling Issues

Abstract. As introduced in Chapter 1, longitudinal and panel data are often heterogeneous and may suffer from problems of attrition. This chapter describes models for handling these tendencies, as well as models designed to handle omitted-variable bias.

Heterogeneity may be induced by (1) fixed effects, (2) random effects, or (3) within-subject covariances. In practice, distinguishing among these mechanisms can be difficult, although, as the chapter points out, it is not always necessary. The chapter also describes the well-known *Hausman test* for distinguishing between estimators based on fixed versus random effects. As pointed out by Mundlak (1978aE), the Hausman test provides a test of the significance of time-constant omitted variables, certain types of which are handled well by longitudinal and panel data.

This ability to deal with omitted variables is one of the important benefits of using longitudinal and panel data; in contrast, *attrition* is one of the main drawbacks. The chapter reviews methods for detecting biases arising from attrition and introduces models that provide corrections for attrition difficulties.

7.1 Heterogeneity

Heterogeneity is a common feature of many longitudinal and panel data sets. When we think of longitudinal data, we think of repeated measurements on subjects. This text emphasizes repeated observations over time, although other types of clustering are of interest. For example, one could model the family unit as a "subject" and have individual measurements of family members as the repeated observations. Similarly, one could have a geographic area (such as a state) as the subject and have individual measurements of towns as the repeated observations. Regardless of the nature of the repetition, the common theme is that different observations from the same subject, or observational unit, tend to be related to one another. In contrast, the word "heterogeneity" refers to things that are unlike, or dissimilar. Thus, when discussing heterogeneity in the context of longitudinal data analysis, we mean that observations from different subjects tend to be dissimilar whereas observations from the same subject tend to be similar. We refer to models without heterogeneity components as *homogenous*.

Two Approaches to Modeling Heterogeneity

In multivariate analysis, there are many methods for quantifying relationships among random variables. The goal of each method is to understand the joint distribution function of random variables; distribution functions provide all the details on the possible outcomes of random variables, both in isolation of one another and as they occur jointly. There are several methods for constructing multivariate distributions; see Hougaard (1987G) and Hutchinson and Lai (1990G) for detailed reviews. For applied longitudinal data analysis, we focus on two different methods of generating jointly dependent distribution functions, through (1) common variables and (2) covariances.

With the "variables-in-common" technique for generating multivariate distributions, a common element serves to induce dependencies among several random variables. We have already used this modeling technique extensively, beginning in Chapters 2 and 3. Here, we used the vector of parameters $\boldsymbol{\alpha}_i$ to denote time-constant characteristics of a subject. In Chapter 2, the fixed parameters induced similarities among different observations from the same subject through the mean function. In Chapter 3, the random vectors $\boldsymbol{\alpha}_i$ induced similarities through the covariance function. In each case, $\boldsymbol{\alpha}_i$ is common to all observations within a subject and thus induces a dependency among these observations.

Although subject-specific common variables are widely used for modeling heterogeneity, they do not cover all the important longitudinal data applications. We have already discussed time-specific variables in Chapter 2 (denoted as λ_t) and will extend this discussion to cross-classified data in Chapter 8, that is, by incorporating both subject-specific and time-specific common variables. Another important area of application involves clustered data, described in Chapter 5.

We can also account for heterogeneity by directly modeling the covariance among observations within a subject. To illustrate, in Section 3.1 on error components, we saw that a common random α_i induced a positive covariance among observations within a subject. We also saw that we could model this feature using the compound symmetry correlation matrix. The advantage of the compound symmetry covariance approach is that it also allows for models of negative dependence. Thus, modeling the joint relation among observations directly using covariances can be simpler than following an approach using common variables and may also cover additional distributions. Moreover, for serial correlations, modeling covariances directly is much simpler than alternative approaches. We know that for normally distributed data, modeling the covariance function, together with the mean, is sufficient to specify the entire joint distribution function. Although this is not true in general, correctly specifying the first two moments suffices for much applied work. We take up this issue further in Chapters 9 and 10 when discussing the generalized estimating equations approach.

Practical Identification of Heterogeneity May Be Difficult

For many longitudinal data sets, an analyst could consider myriad alternative strategies for modeling the heterogeneity. One could use subject-specific intercepts that may be fixed or random. One could use subject-specific slopes that, again, may be fixed or random. Alternatively, one could use covariance specifications to model the tendency for observations from the same subject to be related. As the following illustration from Jones (1993S) shows, it may be difficult to distinguish among these alternative models when only using the data to aid model specification.

Figure 7.1 shows panels of times-series plots for $n = 3$ subjects. The data are generated with no serial correlation over time but with three different subject-specific parameters, $\alpha_1 = 0$, $\alpha_2 = 2$, and $\alpha_3 = -2$. With perfect knowledge of the subject-specific parameters, one would correctly use a scalar times the identity matrix for the covariance structure. However, if these subject-specific variables are ignored, a correlation analysis shows a strong positive serial correlation. That is, from the first panel in Figure 7.1, we see that observations tend to oscillate about the overall mean of zero in a random fashion. However, the second panel shows that all observations are above zero and the third panel indicates that almost all observations are below zero. Thus, an analysis without subject-specific terms would indicate strong positive autocorrelation. Although not the "correct" formulation, a time-series model such as the $AR(1)$ model would serve to capture the heterogeneity in the data.

Theoretical Identification with Heterogeneity May Be Impossible

Thus, identifying the correct model formulation to represent heterogeneity can be difficult. Moreover, in the presence of heterogeneity, identifying all the model components may be impossible. For example, our training in linear model theory has led us to believe that with N observations and p linear parameters, we require only that $N > p$ to identify variance components. Although this is

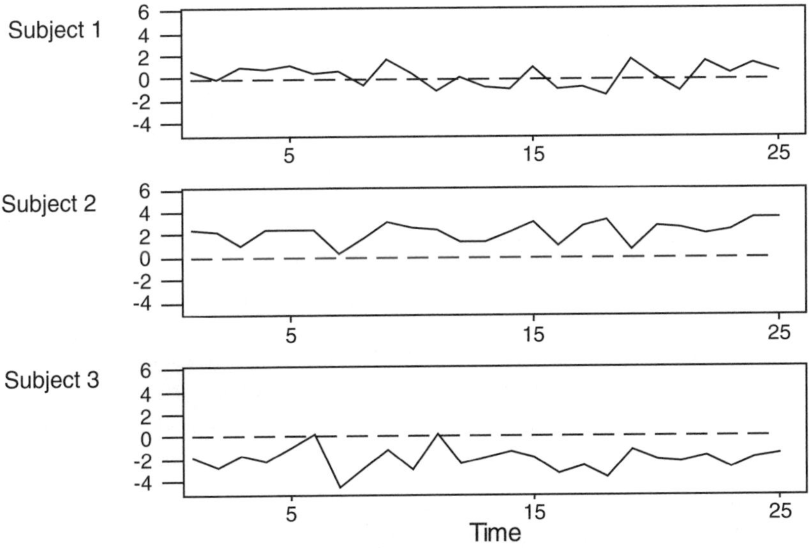

Figure 7.1. Time-series plots showing how different subject-specific parameters can induce positive serial correlation.

true in cross-sectional regression, the more complex longitudinal data setting requires additional assumptions. The following example, due to Neyman and Scott (1948E), illustrates some of these complexities.

Example 7.1: Neyman–Scott on Identification of Variance Components Consider the fixed-effects model

$$y_{it} = \alpha_i + \varepsilon_{it}, \quad i = 1, \ldots, n, \; t = 1, 2,$$

where Var $\varepsilon_{it} = \sigma^2$ and $\mathrm{Cov}(\varepsilon_{i1}, \varepsilon_{i2}) = \sigma^2 \rho$. The OLS estimator of α_i is $\bar{y}_i = (y_{i1} + y_{i2})/2$. Thus, the residuals are $e_{i1} = y_{i1} - \bar{y}_i = (y_{i1} - y_{i2})/2$ and $e_{i2} = y_{i2} - \bar{y}_i = (y_{i2} - y_{i1})/2 = -e_{i1}$. Because of these relations, ρ cannot be estimated, despite having $2n - n = n$ degrees of freedom available for estimating the variance components.

Estimation of Regression Coefficients without Complete Identification Is Possible

Fortunately, complete model identification is not required for all inference purposes. To illustrate, if our main goal is to estimate or test hypotheses about the regression coefficients, then we do not require knowledge of all aspects of the model. For example, consider the one-way fixed-effects model

$$\mathbf{y}_i = \alpha_i \mathbf{1}_i + \mathbf{X}_i \boldsymbol{\beta} + \boldsymbol{\epsilon}_i .$$

For the balanced case, $T_i = T$, Kiefer (1980E) showed how to consistently estimate all components of Var $\boldsymbol{\epsilon}_i = \mathbf{R}$ that are needed for inference about $\boldsymbol{\beta}$. That is, by applying the common transformation matrix $\mathbf{Q} = \mathbf{I} - T^{-1}\mathbf{J}$ to each equation we get

$$\mathbf{y}_i^* = \mathbf{Q}\mathbf{y}_i = \mathbf{Q}\mathbf{X}_i \boldsymbol{\beta} + \mathbf{Q}\boldsymbol{\epsilon}_i = \mathbf{X}_i^* \boldsymbol{\beta} + \boldsymbol{\epsilon}_i^* ,$$

because $\mathbf{Q1} = \mathbf{0}$. Note that Var $\boldsymbol{\epsilon}_i^* = \mathbf{QRQ} = \mathbf{R}^*$. With this transformed equation, the population parameters $\boldsymbol{\beta}$ can be consistently (and root-n) estimated. Further, elements of $\mathbf{R}^*$ can be consistently estimated and used to get feasible GLS estimators.

Example 7.1: Neyman–Scott (continued) Here, we have $T = 2$,

$$\mathbf{Q} = \begin{pmatrix} 1 & 0 \\ 0 & 1 \end{pmatrix} - \frac{1}{2}\begin{pmatrix} 1 & 1 \\ 1 & 1 \end{pmatrix} = \frac{1}{2}\begin{pmatrix} 1 & -1 \\ -1 & 1 \end{pmatrix},$$

and

$$\mathbf{R} = \begin{pmatrix} \sigma^2 & \sigma^2\rho \\ \sigma^2\rho & \sigma^2 \end{pmatrix} = \sigma^2 \begin{pmatrix} 1 & \rho \\ \rho & 1 \end{pmatrix}.$$

Thus,

$$\begin{aligned} \mathbf{R}^* = \mathbf{QRQ} &= \frac{\sigma^2}{4}\begin{pmatrix} 1 & -1 \\ -1 & 1 \end{pmatrix}\begin{pmatrix} 1 & \rho \\ \rho & 1 \end{pmatrix}\begin{pmatrix} 1 & -1 \\ -1 & 1 \end{pmatrix} \\ &= \frac{\sigma^2(1-\rho)}{2}\begin{pmatrix} 1 & -1 \\ -1 & 1 \end{pmatrix}. \end{aligned}$$

Using moment-based estimators, we can estimate the quantity $\sigma^2(1-\rho)$ but cannot separate the terms σ^2 and ρ.

This example shows that, in the balanced basic fixed-effects model, feasible GLS estimation is possible, even without complete identification of all variance components. More generally, consider the case where we have unbalanced data and variable slopes, represented with the model

$$\mathbf{y}_i = \mathbf{Z}_i \boldsymbol{\alpha}_i + \mathbf{X}_i \boldsymbol{\beta} + \boldsymbol{\epsilon}_i ,$$

where Var $\boldsymbol{\epsilon}_i = \mathbf{R}_i$. For this model, in Section 2.5.3 we introduced the transformation $\mathbf{Q}_i = \mathbf{I}_i - \mathbf{Z}_i(\mathbf{Z}_i'\mathbf{Z}_i)^{-1}\mathbf{Z}_i'$. Applying this transform to the model yields

$$\mathbf{y}_i^* = \mathbf{Q}_i\mathbf{y}_i = \mathbf{Q}_i\mathbf{X}_i\boldsymbol{\beta} + \mathbf{Q}_i\boldsymbol{\epsilon}_i = \mathbf{X}_i^*\boldsymbol{\beta} + \boldsymbol{\epsilon}_i^*,$$

because $\mathbf{Q}_i\mathbf{Z}_i = \mathbf{0}_i$. Note that Var $\boldsymbol{\epsilon}_i^* = \mathbf{Q}_i\mathbf{R}_i\mathbf{Q}_i = \mathbf{R}_i^*$. For this model, we see that an OLS estimator of $\boldsymbol{\beta}$ is unbiased and (root-n) consistent. Without knowledge of the variance components in $\mathbf{R}_i^*$, one can still use robust standard errors to get asymptotically correct confidence intervals and tests of hypotheses.

The case for feasible generalized least squares is more complex. Now, if $\mathbf{R}_i = \sigma^2\mathbf{I}_i$, then $\mathbf{R}_i^*$ is known up to a constant; in this case, the usual GLS estimator, given in Equation (2.16), is applicable. For other situations, Kung (1996O) provided sufficient conditions for the identification and estimation of a feasible GLS estimator.

7.2 Comparing Fixed- and Random-Effects Estimators

In Section 3.1, we introduced the sampling and inferential bases for choosing a random-effects model. However, there are many instances when these bases do not provide sufficient guidance to dictate which type of estimator, fixed or random-effects, the analyst should employ. In Section 3.1, we saw that the random-effects estimator is derived using generalized least squares and thus has minimum variance among all unbiased linear estimators. However, in Section 6.2 we saw that fixed-effects estimators do not rely on Assumption SEC6, zero correlation between the time-constant heterogeneity variables and the regressor variables. Often, analysts look to features of the data to provide additional guidance. This section introduces the well-known *Hausman test* for deciding whether to use a fixed- or random-effects estimator. The Hausman (1978E) test is based on an interpretation due to Mundlak (1978aE) that the fixed-effects estimator is robust to certain omitted-variable model specifications. Throughout this section, we maintain Assumption SEC1, the strict exogeneity of the regressor variables conditional on the unobserved effects.

To introduce the Hausman test, we first return to a version of our Section 3.1 error-components model

$$y_{it} = \alpha_i + \mathbf{x}_{it}'\boldsymbol{\beta} + \varepsilon_{it} + u_i. \qquad (3.1)^*$$

Here, as in Section 3.1, α_i is a random variable that is uncorrelated with the disturbance term ε_{it} (and uncorrelated with the explanatory variables $\mathbf{x}_{it}$; see Chapter 6). However, we have also added u_i, a term for unobserved omitted variables. Unlike α_i, the concern is that the u_i quantity may represent a fixed effect, or a random effect that is correlated with either the disturbance terms or explanatory variables. If u_i is present, then the heterogeneity term $\alpha_i^* = \alpha_i + u_i$ does not satisfy the usual assumptions required for unbiased and consistent regression coefficient estimators.

We do, however, restrict the omitted variables to be constant in time; this assumption allows us to derive unbiased and consistent estimators, even in the presence of omitted variables. Taking averages over time in Equation (3.1)*, we have

$$\bar{y}_i = \alpha_i + \bar{\mathbf{x}}_i'\beta + \bar{\varepsilon}_i + u_i.$$

Subtracting this from Equation (3.1)* yields

$$y_{it} - \bar{y}_i = (x_{it} - \bar{\mathbf{x}}_i)'\beta + \varepsilon_{it} - \bar{\varepsilon}_i.$$

Based on these deviations, we have removed the effects of the unobserved variable u_i. Thus, the Equation (2.6) fixed-effects estimator

$$\mathbf{b}_{\mathrm{FE}} = \left(\sum_{i=1}^{n}\sum_{t=1}^{T_i}(\mathbf{x}_{it} - \bar{\mathbf{x}}_i)(\mathbf{x}_{it} - \bar{\mathbf{x}}_i)'\right)^{-1}\left(\sum_{i=1}^{n}\sum_{t=1}^{T_i}(\mathbf{x}_{it} - \bar{\mathbf{x}}_i)(y_{it} - \bar{y}_i)\right)$$

is not corrupted by u_i; it turns out to be unbiased and consistent even in the presence of omitted variables. For notational purposes, we have added the subscript FE to suggest the motivation of this estimator even though there may not be any fixed effects (α_i) in Equation (3.1)*.

The Hausman test statistic compares the robust estimator $\mathbf{b}_{\mathrm{FE}}$ to the GLS estimator $\mathbf{b}_{\mathrm{EC}}$. Under the null hypothesis of no omitted variables, $\mathbf{b}_{\mathrm{EC}}$ is more efficient (because it is a GLS estimator). Under the alternative hypothesis of (time-constant) omitted variables, $\mathbf{b}_{\mathrm{FE}}$ is still unbiased and consistent. Hausman (1978E) showed that the statistic

$$\chi^2_{\mathrm{FE}} = (\mathbf{b}_{\mathrm{FE}} - \mathbf{b}_{\mathrm{EC}})'\left(\mathrm{Var}\left(\mathbf{b}_{\mathrm{FE}}\right) - \mathrm{Var}\left(\mathbf{b}_{\mathrm{EC}}\right)\right)^{-1}(\mathbf{b}_{\mathrm{FE}} - \mathbf{b}_{\mathrm{EC}})$$

has an asymptotic chi-square distribution with K degrees of freedom under the null hypothesis. It is a widely used statistic for detecting omitted variables.

Example 7.2: Income Tax Payments To illustrate the performance of the fixed-effects estimators and omitted-variable tests, we examine data on determinants of income tax payments introduced in Section 3.2. Specifically, we begin with the error-components model with $K = 8$ coefficients estimated using generalized least squares. The parameter estimates and the corresponding t-statistics appear in Table 7.1.

Table 7.1. *Comparison of random-effects estimators to robust alternatives, for a model with variable intercepts but no variable slopes (error components)*

	Robust fixed-effects estimation		Random-effects estimation	
Variable	Parameter estimates	t-statistic	Parameter estimates	t-statistic
LNTPI	0.717	9.30	0.760	10.91
MR	0.122	13.55	0.115	15.83
MS	0.072	0.28	0.037	0.21
HH	−0.707	−2.17	−0.689	−2.98
AGE	0.002	0.01	0.021	0.10
EMP	−0.244	−0.99	−0.505	−3.02
PREP	−0.030	−0.18	−0.022	−0.19
DEPEND	−0.069	−0.83	−0.113	−1.91
$\chi^2_{FE} = 6.021$				

Also in Table 7.1 are the corresponding fixed-effects estimators. The fixed-effects estimators are robust to time-constant omitted variables. The standard errors and corresponding t-statistics are computed in a straightforward fashion using the procedures described in Chapter 2. Comparing the fixed- and random-effects estimators in Table 7.1, we see that the coefficients are qualitatively similar. For each variable, the estimators have the same sign and similar orders of magnitude. They also indicate the same order of statistical significance. For example, the two measures of income, LNTPI and MR, are both strongly statistically significant under both estimation procedures. The one exception is the EMP variable, which is strongly statistically significantly negative using the random-effects estimation but is not statistically significant using the fixed-effects estimation.

To assess the overall differences among the coefficient estimates, we may use the omitted-variable test statistic due to Hausman (1978E). As indicated in Table 7.1, this test statistic is $\chi^2_{FE} = 6.021$. Comparing this test statistic to a chi-square distribution with $K = 8$ degrees of freedom, we find that the p-value associated with this test statistic is $\text{Prob}(\chi^2 > 6.021) = .6448$. This does not provide enough evidence to indicate a serious problem with omitted variables. Thus, the random-effects estimator is preferred.

Because of the complexities and the widespread usage of this test in the econometrics literature, we split the remainder of the discussion into two parts. The first part (Section 7.2.1) introduces some important additional ideas in the context of a special case. Here, we show a relationship between fixed- and random-effects estimators, introduce Mundlak's alternative hypothesis, derive the fixed-effects estimator under this hypothesis, and discuss Hausman's test of omitted variables. The second part (Section 7.2.2) extends the discussion to incorporate (1) unbalanced data, (2) many variables, (3) variable slopes, and (4) potential serial correlation and heteroscedasticity. Section 7.3 will discuss alternative sampling bases.

7.2.1 A Special Case

Consider the Section 3.1 error-components model with $K = 2$ so that

$$y_{it} = \alpha_i + \beta_0 + \beta_1 x_{it,1} + \varepsilon_{it}, \tag{7.1}$$

where both $\{\alpha_i\}$ and $\{\varepsilon_{it}\}$ are i.i.d. as well as independent of one another. For simplicity, we also assume balanced data so that $T_i = T$. If we ignore the variability in $\{\alpha_i\}$, the usual OLS estimator of β_1 is

$$b_{1,\mathrm{HOM}} = \frac{\sum_{i=1}^{n}\sum_{t=1}^{T}(x_{it}-\bar{x})(y_{it}-\bar{y})}{\sum_{i=1}^{n}\sum_{t=1}^{T}(x_{it}-\bar{x})^2}.$$

Because this estimator excludes the heterogeneity component $\{\alpha_i\}$, we label it using the subscript HOM for homogeneous. In contrast, from Exercise 7.1, an expression for the GLS estimator is

$$b_{1,\mathrm{EC}} = \frac{\sum_{i=1}^{n}\sum_{t=1}^{T}(x_{it}^*-\bar{x}^*)(y_{it}^*-\bar{y}^*)}{\sum_{i=1}^{n}\sum_{t=1}^{T}(x_{it}^*-\bar{x}^*)^2},$$

where

$$x_{it}^* = x_{it} - \bar{x}_i\left(1-\left(\frac{\sigma^2}{T\sigma_\alpha^2+\sigma^2}\right)^{1/2}\right)$$

and

$$y_{it}^* = y_{it} - \bar{y}_i\left(1-\left(\frac{\sigma^2}{T\sigma_\alpha^2+\sigma^2}\right)^{1/2}\right).$$

As described in Section 3.1, both estimators are unbiased, consistent, and asymptotically normal. Because $b_{1,\mathrm{EC}}$ is a GLS estimator, it has a smaller variance than $b_{1,\mathrm{HOM}}$. That is, $\mathrm{Var}\, b_{1,\mathrm{EC}} \le \mathrm{Var}\, b_{1,\mathrm{HOM}}$, where

$$\mathrm{Var}\, b_{1,\mathrm{EC}} = \sigma^2\left(\sum_{i=1}^{n}\sum_{t=1}^{T}(x_{it}^*-\bar{x}^*)^2\right)^{-1}. \tag{7.2}$$

Also note that $b_{1,\mathrm{EC}}$ and $b_{1,\mathrm{HOM}}$ are approximately equivalent when the heterogeneity variance is small. Formally, because $x_{it}^* \rightarrow x_{it}$ and $y_{it}^* \rightarrow y_{it}$ as $\sigma_\alpha^2 \rightarrow 0$, we have that $b_{1,\mathrm{EC}} \rightarrow b_{1,\mathrm{HOM}}$ as $\sigma_\alpha^2 \rightarrow 0$.

This section also considers an alternative estimator

$$b_{1,\mathrm{FE}} = \frac{\sum_{i=1}^n \sum_{t=1}^T (x_{it} - \bar{x}_i)(y_{it} - \bar{y}_i)}{\sum_{i=1}^n \sum_{t=1}^T (x_{it} - \bar{x}_i)^2}. \tag{7.3}$$

This estimator could be derived from the model in Equation (7.1) by assuming that the terms $\{\alpha_i\}$ are fixed, not random, components. In the notation of Chapter 2, we may assume that $\alpha_i^* = \alpha_i + \beta_0$ are the fixed components that are not centered about zero. An important point of this section is that the estimator defined in Equation (7.3) is unbiased, consistent, and asymptotically normal under the model that *includes random effects* in Equation (7.1). Further, straightforward calculations show that

$$\mathrm{Var}\, b_{1,\mathrm{FE}} = \sigma^2 \left(\sum_{i=1}^n \sum_{t=1}^T (x_{it} - \bar{x}_i)^2 \right)^{-1}. \tag{7.4}$$

We note that $b_{1,\mathrm{EC}}$ and $b_{1,\mathrm{FE}}$ are approximately equivalent when the heterogeneity variance is large. Formally, because $x_{it}^* \rightarrow x_{it} - \bar{x}_i$ and $y_{it}^* \rightarrow y_{it} - \bar{y}_i$ as $\sigma_\alpha^2 \rightarrow \infty$, we have that $b_{1,\mathrm{EC}} \rightarrow b_{1,\mathrm{FE}}$ as $\sigma_\alpha^2 \rightarrow \infty$.

To relate the random- and fixed-effects estimators, we define the so-called between groups estimator,

$$b_{1,\mathrm{B}} = \frac{\sum_{i=1}^n (\bar{x}_i - \bar{x})(\bar{y}_i - \bar{y})}{\sum_{i=1}^n (\bar{x}_i - \bar{x})^2}. \tag{7.5}$$

This estimator can be motivated by averaging all observations from a subject and then computing an OLS estimator using the data $\{\bar{x}_i, \bar{y}_i\}_{i=1}^n$. As with the other estimators, this estimator is unbiased, consistent, and asymptotically normal under the Equation (7.1) model. Further, straightforward calculations show that

$$\mathrm{Var}\, b_{1,\mathrm{B}} = \left(T\sigma_\alpha^2 + \sigma^2\right) \left(T \sum_{i=1}^n (\bar{x}_i - \bar{x})^2 \right)^{-1}. \tag{7.6}$$

To interpret the relations among $b_{1,\mathrm{EC}}$, $b_{1,\mathrm{FE}}$, and $b_{1,\mathrm{B}}$, we cite the following decomposition due to Maddala (1971E);

$$b_{1,\mathrm{EC}} = (1 - \Delta) b_{1,\mathrm{FE}} + \Delta b_{1,\mathrm{B}}. \tag{7.7}$$

Here, the term $\Delta = \mathrm{Var}\, b_{1,\mathrm{EC}} / \mathrm{Var}\, b_{1,\mathrm{B}}$ measures the relative precision of the two estimators of β. Because $b_{1,\mathrm{EC}}$ is the GLS estimator, we have that $\mathrm{Var}\, b_{1,\mathrm{EC}} \le \mathrm{Var}\, b_{1,\mathrm{B}}$ so that $0 \le \Delta \le 1$.

Omitted Variables: Model of Correlated Effects

Thus, assuming the random-effects model in Equation (7.1) is an adequate representation, we expect each of the four estimators of β_1 to be close to one another. However, in many data sets these estimators can differ dramatically. To explain these differences, Mundlak (1978aE) proposed what we will call a model of "correlated effects." Here, we interpret α_i to represent time-constant, or "permanent," characteristics of $\mathbf{y}_i$ that are unobserved and hence "omitted." Mundlak introduced the possibility that $\{\alpha_i\}$ are correlated with the observed variables $\mathbf{x}_i$. That is, the latent variables α_i fail the exogeneity assumption SE6 described in Section 6.2.2. To express the relationship between α_i and $\mathbf{x}_i$, we consider the function $\mathrm{E}[\alpha_i \mid \mathbf{x}_i]$. Specifically, for our special case, Mundlak assumed that $\alpha_i = \eta_i + \gamma \bar{x}_{i,1}$, where $\{\eta_i\}$ is an i.i.d. sequence that is independent of $\{\varepsilon_{it}\}$. Thus, the *model of correlated effects* is

$$y_{it} = \eta_i + \beta_0 + \beta_1 x_{it,1} + \gamma \bar{x}_{i,1} + \varepsilon_{it}. \tag{7.8}$$

Under this model, one can show that the GLS estimator of β_1 is $b_{1,\mathrm{FE}}$. Further, the estimator $b_{1,\mathrm{FE}}$ is unbiased, consistent, and asymptotically normal. In contrast, the estimators $b_{1,\mathrm{HOM}}$, $b_{1,\mathrm{B}}$, and $b_{1,\mathrm{EC}}$ are biased and inconsistent.

To compare the model of correlated effects in Equation (7.8) with the baseline model in Equation (7.1), we need only examine the null hypothesis $\mathrm{H}_0 : \gamma = 0$. This is customarily done using the Hausman (1978E) test statistic

$$\chi^2_{\mathrm{FE}} = \frac{(b_{1,\mathrm{EC}} - b_{1,\mathrm{FE}})^2}{\mathrm{Var}\, b_{1,\mathrm{FE}} - \mathrm{Var}\, b_{1,\mathrm{EC}}}. \tag{7.9}$$

Under the null hypothesis of the model in Equation (7.1), this test statistic has an asymptotic (as $n \to \infty$) chi-square distribution with one degree of freedom. This provides the basis for comparing the two models. Moreover, we see that the test statistic will be large when there is a large difference between the fixed- and random-effects estimators. In addition, it is straightforward to construct the test statistic based on a fit of the random-effects model in Equation (7.1) (to get $b_{1,\mathrm{EC}}$ and $\mathrm{Var}\, b_{1,\mathrm{EC}}$) and a fit of the corresponding fixed-effects model (to get $b_{1,\mathrm{FE}}$ and $\mathrm{Var}\, b_{1,\mathrm{FE}}$). Thus, one need not construct the "augmented" variable $\bar{x}_{i,1}$ in Equation (7.8).

7.2.2 General Case

Extension to the general case follows directly. To incorporate intercepts, we use a modification of the linear mixed-effects model in Equation (3.5). Thus, we assume that $\mathrm{E}\boldsymbol{\alpha}_i = \boldsymbol{\alpha}$ and rewrite the model as

$$\mathbf{y}_i = \mathbf{Z}_i \boldsymbol{\alpha} + \mathbf{X}_i \boldsymbol{\beta} + \boldsymbol{\epsilon}_i^*, \tag{7.10}$$

where $\boldsymbol{\epsilon}_i^* = \boldsymbol{\epsilon}_i + \mathbf{Z}_i(\boldsymbol{\alpha}_i - \boldsymbol{\alpha})$ and $\mathrm{Var}\, \boldsymbol{\epsilon}_i^* = \mathbf{Z}_i \mathbf{D} \mathbf{Z}_i' + \mathbf{R}_i = \mathbf{V}_i$. Straightforward calculations (Exercise 7.3) show that the GLS estimator of $\boldsymbol{\beta}$ is

$$\mathbf{b}_{\mathrm{GLS}} = \mathbf{C}_{\mathrm{GLS}}^{-1} \sum_{i=1}^{n} \left\{ \mathbf{X}_i' \mathbf{V}_i^{-1} - \left(\sum_{i=1}^{n} \mathbf{X}_i' \mathbf{V}_i^{-1} \mathbf{Z}_i \right) \left(\sum_{i=1}^{n} \mathbf{Z}_i' \mathbf{V}_i^{-1} \mathbf{Z}_i \right)^{-1} \mathbf{Z}_i' \mathbf{V}_i^{-1} \right\} \mathbf{y}_i$$

with

$$\mathbf{C}_{\mathrm{GLS}} = \left(\sum_{i=1}^{n} \mathbf{X}_i' \mathbf{V}_i^{-1} \mathbf{X}_i \right) - \left(\sum_{i=1}^{n} \mathbf{X}_i' \mathbf{V}_i^{-1} \mathbf{Z}_i \right) \left(\sum_{i=1}^{n} \mathbf{Z}_i' \mathbf{V}_i^{-1} \mathbf{Z}_i \right)^{-1} \left(\sum_{i=1}^{n} \mathbf{Z}_i' \mathbf{V}_i^{-1} \mathbf{X}_i \right).$$

From Equation (2.16), we have that the corresponding fixed-effects estimator is

$$\mathbf{b}_{\mathrm{FE}} = \mathbf{C}_{\mathrm{FE}}^{-1} \sum_{i=1}^{n} \mathbf{X}_i' \mathbf{R}_i^{-1/2} \mathbf{Q}_{Z,i} \mathbf{R}_i^{-1/2} \mathbf{y}_i,$$

where $\mathbf{C}_{\mathrm{FE}} = \sum_{i=1}^{n} \mathbf{X}_i' \mathbf{R}_i^{-1/2} \mathbf{Q}_{Z,i} \mathbf{R}_i^{-1/2} \mathbf{X}_i$ and $\mathbf{Q}_{Z,i} = \mathbf{I}_i - \mathbf{R}_i^{-1/2} \mathbf{Z}_i (\mathbf{Z}_i' \mathbf{R}_i^{-1} \mathbf{Z}_i)^{-1} \mathbf{Z}_i' \mathbf{R}_i^{-1/2}$. From Exercise 7.5, the between-groups estimator is

$$\mathbf{b}_{\mathrm{B}} = \mathbf{C}_{\mathrm{B}}^{-1} \sum_{i=1}^{n} \left\{ \mathbf{X}_i' \mathbf{V}_i^{-1} \mathbf{Z}_i \left(\mathbf{Z}_i' \mathbf{V}_i^{-1} \mathbf{Z}_i \right)^{-1} \mathbf{Z}_i' \mathbf{V}_i^{-1} - \left(\sum_{i=1}^{n} \mathbf{X}_i' \mathbf{V}_i^{-1} \mathbf{Z}_i \right) \left(\sum_{i=1}^{n} \mathbf{Z}' \mathbf{V}_i^{-1} \mathbf{Z}_i \right)^{-1} \mathbf{Z}_i' \mathbf{V}_i^{-1} \right\} \mathbf{y}_i,$$

where

$$\mathbf{C}_B = \sum_{i=1}^{n} \mathbf{X}_i' \mathbf{V}_i^{-1} \mathbf{Z}_i \left(\mathbf{Z}_i' \mathbf{V}_i^{-1} \mathbf{Z}_i \right)^{-1} \mathbf{Z}_i' \mathbf{V}_i^{-1} \mathbf{X}_i - \left(\sum_{i=1}^{n} \mathbf{X}_i' \mathbf{V}_i^{-1} \mathbf{Z}_i \right) \left(\sum_{i=1}^{n} \mathbf{Z}_i' \mathbf{V}_i^{-1} \mathbf{Z}_i \right)^{-1} \left(\sum_{i=1}^{n} \mathbf{Z}_i' \mathbf{V}_i^{-1} \mathbf{X}_i \right).$$

To relate these three estimators, we use the extension of Maddala's (1971E) result,

$$\mathbf{b}_{\mathrm{GLS}} = (\mathbf{I} - \boldsymbol{\Delta})\, \mathbf{b}_{\mathrm{FE}} + \boldsymbol{\Delta} \mathbf{b}_{\mathrm{B}},$$

where $\mathbf{\Delta} = (\mathrm{Var}\,\mathbf{b}_{\mathrm{GLS}})(\mathrm{Var}\,\mathbf{b}_{\mathrm{B}})^{-1}$, $\mathrm{Var}\,\mathbf{b}_{\mathrm{B}} = \mathbf{C}_{\mathrm{B}}^{-1}$, and $\mathrm{Var}\,\mathbf{b}_{\mathrm{GLS}} = \mathbf{C}_{\mathrm{GLS}}^{-1}$ (see Exercise 7.5). Again, the matrix $\mathbf{\Delta}$ is a weight matrix that quantifies the relative precision of the two estimators, $\mathbf{b}_{\mathrm{GLS}}$ and $\mathbf{b}_{\mathrm{B}}$.

Correlated-Effects Model

For a model of correlated effects that describes the correlation between $\{\alpha_i\}$ and $\{\mathbf{X}_i\}$, let $\mathbf{x}_i = \mathrm{vec}(\mathbf{X}_i')$, a $KT_i \times 1$ vector built by stacking vectors $\{\mathbf{x}_{i1}', \ldots, \mathbf{x}_{iT_i}'\}$. For notation, we denote the observed independent variables by $\mathbf{o}_{it} = (\mathbf{z}_{it}', \mathbf{x}_{it}')'$, a $(q+K) \times 1$ vector of observed effects, and let $\mathbf{o}_i$ be the associated column vector, that is, $\mathbf{o}_i = (\mathbf{o}_{i1}', \ldots, \mathbf{o}_{iT_i}')'$. We assume that the relationship between α_i and $\mathbf{X}_i$ can be described through the conditional moments

$$\mathrm{E}\,[\alpha_i \mid \mathbf{o}_i] = \mathbf{\Sigma}_{\alpha x}\mathbf{\Sigma}_x^{-1}(\mathbf{x}_i - \mathrm{E}\mathbf{x}_i), \qquad \mathrm{Var}\,[\alpha_i \mid \mathbf{o}_i] = \mathbf{D}^*, \tag{7.11}$$

where $\mathbf{\Sigma}_{\alpha x} = \mathrm{Cov}(\alpha_i, \mathbf{x}_i) = \mathrm{E}(\alpha_i(\mathbf{x}_{i1}', \ldots, \mathbf{x}_{iT_i}'))$ and $\mathbf{\Sigma}_x = \mathrm{Var}\,\mathbf{x}_i$. Equation (7.11) can be motivated by joint normality of α_i and $\mathbf{o}_i$ but is also useful when some components of $\mathbf{o}_i$ are categorical. With Equation (7.11), we have

$$\mathrm{E}\,[\mathbf{y}_i \mid \mathbf{o}_i] = \mathrm{E}\mathrm{E}\,[\mathbf{y}_i \mid \alpha_i, \mathbf{o}_i] \mid \mathbf{o}_i = \mathbf{Z}_i\mathbf{\Sigma}_{\alpha x}\mathbf{\Sigma}_x^{-1}(\mathbf{x}_i - \mathrm{E}\mathbf{x}_i) + \mathbf{X}_i\boldsymbol{\beta} \tag{7.12}$$

and

$$\begin{aligned}\mathrm{Var}\,[\mathbf{y}_i \mid \mathbf{o}_i] &= \mathrm{E}\,(\mathrm{Var}\,[\mathbf{y}_i \mid \alpha_i, \mathbf{o}_i] \mid \mathbf{o}_i) + \mathrm{Var}\,(\mathrm{E}\,[\mathbf{y}_i \mid \alpha_i, \mathbf{o}_i] \mid \mathbf{o}_i)\\ &= \mathbf{R}_i + \mathbf{Z}_i\mathbf{D}^*\mathbf{Z}_i'.\end{aligned} \tag{7.13}$$

The correlated effects alter the form of the regression function in Equation (7.12) but not the conditional variance in Equation (7.13).

Now, under the model of correlated effects summarized in Equations (7.12) and (7.13), it is easy to see that the random-effects estimator is generally biased. In contrast, the fixed-effects estimator is unbiased and has variance

$$\mathrm{Var}\,\mathbf{b}_{\mathrm{FE}} = \left(\sum_{i=1}^{n} \mathbf{X}_i'\,\mathbf{R}_i^{-1/2}\mathbf{Q}_i\mathbf{R}_i^{-1/2}\mathbf{X}_i\right)^{-1},$$

which is the same as under the fixed-effects model formulation; see Section 2.5.3.

Again, an extension of the Hausman (1978E) test allows us to compare the baseline model and the model of correlated effects. The test statistic is

$$\chi_{\mathrm{FE}}^2 = (\mathbf{b}_{\mathrm{FE}} - \mathbf{b}_{\mathrm{GLS}})'\,(\mathrm{Var}\,(\mathbf{b}_{\mathrm{FE}}) - \mathrm{Var}\,(\mathbf{b}_{\mathrm{GLS}}))^{-1}\,(\mathbf{b}_{\mathrm{FE}} - \mathbf{b}_{\mathrm{GLS}}). \tag{7.14}$$

Under the null hypothesis of the model in Equation (7.10), this test statistic has an asymptotic (as $n \to \infty$) chi-square distribution with K degrees of freedom.

As in Section 7.2.1, this test statistic is intuitively pleasing in that large differences between the fixed- and random-effects estimators allow us to reject the null hypothesis of no correlated effects.

In summary, the fixed-effects estimator is easy to compute and is robust to omitted-variable bias. The estimator has desirable properties under a variation of the random-effects model that we call a model of correlated effects. Under this model of correlated-effects formulation, the many subject-specific fixed parameters generally associated with fixed-effects models need not be computed. In addition to the estimator itself, standard errors associated with this estimator are easy to compute. Further, the Equation (7.14) test statistic provides a simple method for assessing the adequacy of the random-effects model; this could lead to further follow-up investigations, which may in turn lead to an improved model specification.

Example 7.2: Income Tax Payments (continued) The preceding analysis is based on the error-components model. Many additional features could be fit to the data. After additional model exploration, we retained the variable intercepts and also used subject-specific, variable slopes for the income variables, LNTPI and MR. In addition, to accommodate serial patterns in tax liabilities, we specified an $AR(1)$ component for the errors. Part of the rationale comes from the nature of tax liabilities. That is, we hypothesize that the tax liability increases with income. Moreover, because individuals have different attitudes toward tax-sheltering programs, live in different states that have their own tax programs (which affect the amount of the federal tax), and so on, we expect coefficients associated with income to differ among individuals. A similar argument can be made for MR because this is simply a nonlinear function of income.

The random-effects (estimated GLS) and robust fixed-effects estimators are given in Table 7.2. We now see some important differences in the two estimation methodologies. For example, examining the coefficients associated with the MS and EMP variables, we see that the random-effects estimator indicates that each variable is strongly statistically negatively significant whereas the robust estimators do not signal statistical significance. This is also true, but to a lesser extent, of the coefficient associated with AGE.

To compare the vectors of random-effects estimators to those of fixed-effects estimators, we use the omitted-variable test statistic, $\chi^2_{\text{FE}} = 13.628$. Using $K = 6$ degrees of freedom, the p-value associated with this test statistic is $\text{Prob}(\chi^2 > 13.628) = .0341$. In contrast to the error-components model, this provides evidence to indicate a serious problem with omitted variables. The result of the hypothesis test suggests using the robust estimators. Interestingly, both random- and fixed-effects estimation indicate that use of a tax preparer

Table 7.2. *Comparison of random-effects estimators to robust alternatives, for a model with variable intercepts and two variable slopes*

	Robust fixed-effects estimation		Random-effects estimation	
Variable	Parameter estimates	t-statistic	Parameter estimates	t-statistic
LNTPI				
MR				
MS	−0.197	−0.46	−0.603	−3.86
HH	−1.870	−4.41	−0.729	−3.75
AGE	−0.464	−1.28	−0.359	−2.15
EMP	−0.198	−0.68	−0.661	−5.05
PREP	−0.474	−2.51	−0.300	−3.21
DEPEND	−0.304	−2.56	−0.138	−2.84
$AR(1)$	0.454	3.76	0.153	3.38
$\chi^2_{\mathrm{FE}} = 13.628$				

(PREP) significantly lowers the tax liability of a taxpayer, when controlling for income and demographic characteristics. This was not a finding in the error-components model.

By introducing two variable slopes, the number of estimator comparisons dropped from eight to six. Examining Equation (7.10), we see that the variables included in the random-effects formulation are no longer included in the "$\mathbf{X}_i \boldsymbol{\beta}$" portion. Thus, in Equation (7.10), the number of rows of $\boldsymbol{\beta}$, K, refers to the number of variables *not associated with the random-effects portion*; we can think of these variables as associated with only fixed effects. This implies, among other things, that the Section 7.2.3 omitted-variable test is not available for the random-coefficients model where there are no variables associated with only fixed effects. Thus, Section 7.3 introduces a test that will allow us to consider the random-coefficients model.

7.3 Omitted Variables

Particularly in the social sciences where observational, in lieu of experimental, data are predominant, problems of omitted variables abound. The possibility of unobserved, omitted variables that affect both the response and explanatory variables encourage analysts to distinguish between "cause" and "effect." We have already seen one approach for handling cause-and-effect analysis through multiple systems of equations in Sections 6.4 and 6.5. Fortunately, the structure of longitudinal data allows us to construct estimators that are less susceptible to bias arising from omitted variables than common alternatives. For example, in Section 7.2 we saw that fixed-effects estimators are robust to certain types of time-constant omitted variables. This section introduces estimators that are robust to other types of omitted variables. These omitted-variable robust estimators do not provide protection from all types of omitted variables; they are sensitive to the nature of the variables being omitted. Thus, as a matter of practice, analysts should always attempt to collect as much information as possible regarding the nature of the omitted variables.

Specifically, Section 7.2 showed how a fixed-effects estimator is robust to Assumption SEC6, zero correlation between the time-constant heterogeneity variables and the regressor variables. Unfortunately, the fixed-effects transform sweeps out time-constant variables and these variables may be the focus of a study. To remedy this, in this section we show how to use partial information about the relationship between the unobserved heterogeneity variables and the regressor variables. This idea of partial information is due to Hausman and Taylor (1981E), who developed an instrumental variable estimation procedure. We consider a broader class of omitted variables models that, under certain circumstances, also allow for time-varying omitted variables. Here, we will see that the fixed-effects estimator is a special case of a class called "augmented regression" estimators. This class not only provides extensions but also gives a basis for providing heteroscedasticity-consistent standard errors of the estimators.

To set the stage for additional analyses, we first return to a version of the Section 3.1 error-components model

$$y_{it} = \alpha_i + \mathbf{x}_{it}^{(1)\prime} \boldsymbol{\beta}_1 + \mathbf{x}_i^{(2)\prime} \boldsymbol{\beta}_2 + \varepsilon_{it} + u_i. \qquad (3.1)^{**}$$

We have now split up the $\mathbf{x}_{it}'\boldsymbol{\beta}$ into two portions, one for time-varying explanatory variables ($\mathbf{x}_{it}^{(1)}$ and $\boldsymbol{\beta}_1$) and one for time-constant explanatory variables ($\mathbf{x}_i^{(2)}$ and $\boldsymbol{\beta}_2$). As before, the u_i term presents unobserved, omitted variables, which may be a fixed effect or a random effect that is correlated with either the disturbance terms or explanatory variables. As pointed out in Section 2.3, if an explanatory variable is constant in time, then the fixed-effects estimator is no longer estimable. Thus, the techniques introduced in Section 7.2 no longer immediately apply. However, by examining deviations from the mean,

$$y_{it} - \bar{y}_i = \left(\mathbf{x}_{it}^{(1)} - \bar{\mathbf{x}}_i^{(1)}\right)' \boldsymbol{\beta}_1 + \varepsilon_{it} - \bar{\varepsilon}_i,$$

we see that we can still derive unbiased and consistent estimators of β_1, even in the presence of omitted variables. For example, one such estimator is

$$\mathbf{b}_{1,\mathrm{FE}} = \left(\sum_{i=1}^{n} \sum_{t=1}^{T_i} \left(\mathbf{x}_{it}^{(1)} - \bar{\mathbf{x}}_i^{(1)}\right)\left(\mathbf{x}_{it}^{(1)} - \bar{\mathbf{x}}_i^{(1)}\right)' \right)^{-1} \times \left(\sum_{i=1}^{n} \sum_{t=1}^{T_i} \left(\mathbf{x}_{it}^{(1)} - \bar{\mathbf{x}}_i^{(1)}\right)(y_{it} - \bar{y}_i) \right).$$

Moreover, with the additional assumption that u_i is not correlated with $\mathbf{x}_i^{(2)}$, we will be able to provide consistent estimators of β_2. This is a strong assumption; still, the interesting aspect is that with longitudinal and panel data, we can derive estimators with desirable properties even in the presence of omitted variables.

When hypothesizing relationships among variables, the breakdown between time-varying and time-constant variables can be artificial. Thus, in our discussion to follow, we refer to explanatory variables that either are or are not related to omitted variables.

7.3.1 Models of Omitted Variables

To describe how variations in sampling may induce omitted-variable bias, it is helpful to review the sampling basis for the model. The sampling can be described in two stages; see, for example, Section 3.1 and Ware (1985S). Specifically, we have the following:

Stage 1. Draw a random sample of n subjects from a population. The vector of subject-specific parameters $\boldsymbol{\alpha}_i$ is associated with the ith subject.

Stage 2. Conditional on $\boldsymbol{\alpha}_i$, draw realizations of $\{y_{it}, \mathbf{z}_{it}, \mathbf{x}_{it}\}$, for $t = 1, \ldots, T_i$ for the ith subject. Summarize these draws as $\{\mathbf{y}_i, \mathbf{Z}_i, \mathbf{X}_i\}$.

We follow notation introduced in Section 7.2 and denote the observed independent variables by $\mathbf{o}_{it} = (\mathbf{z}_{it}', \mathbf{x}_{it}')'$, a $(q + K) \times 1$ vector of observed effects, and let $\mathbf{o}_i$ be the associated column vector, that is, $\mathbf{o}_i = (\mathbf{o}_{i1}', \ldots, \mathbf{o}_{iT_i}')'$.

We now use an "unobserved-variable model" also considered by Palta and Yao (1991B), Palta, Yao, and Velu (1994B), and others; Frees (2001S) provides a summary. (In the biological sciences, omitted variables are known as *unmeasured confounders*.) Here, we assume that the latent vector $\boldsymbol{\alpha}_i$ is independent of the observed variables $\mathbf{o}_i$, yet we have omitted important, possibly time-varying, variables in the second sampling stage. Specifically, assume at the second stage we have the following:

Stage 2*. Conditional on α_i, draw realizations $\{\mathbf{y}_i, \mathbf{Z}_i, \mathbf{X}_i, \mathbf{U}_i\}$, for the ith subject. Here, $\mathbf{o}_i = \{\mathbf{Z}_i, \mathbf{X}_i\}$ represents observed, independent variables whereas $\mathbf{U}_i$ represents the unobserved, independent variables. Moments of the dependent variables are specified as

$$\mathrm{E}[\mathbf{y}_i \mid \alpha_i, \mathbf{o}_i, \mathbf{U}_i] = \mathbf{Z}_i\alpha_i + \mathbf{X}_i\beta + \mathbf{U}_i\gamma \tag{7.15}$$

and

$$\mathrm{Var}\,[\mathbf{y}_i \mid \alpha_i, \mathbf{o}_i, \mathbf{U}_i] = \mathbf{R}_i. \tag{7.16}$$

Thus, γ is a $g \times 1$ vector of parameters that signals the presence of omitted variables $\mathbf{U}_i$. Using the same notation convention as $\mathbf{o}_i$, let $\mathbf{u}_i$ be the column vector associated with $\mathbf{U}_i$.

To estimate the model parameters, simplifying assumptions are required. One route is to make full distributional assumptions, such as multivariate normality to permit estimation using maximum likelihood methods. In the following, we instead use assumptions on the sampling design because this permits estimation procedures that link back to Section 7.2 without the necessity of assuming that a distribution follows a particular parametric model.

To specify the model used for inference, we use $\Sigma_{uo} = \mathrm{Cov}(\mathbf{u}_i, \mathbf{o}_i)$ to capture the correlation between unobserved and observed effects. For simplicity, we drop the i subscript on the $T_i g \times T_i(q + K)$ matrix Σ_{uo}. Now, we only need be concerned with those observed variables that are related to the unobservables. Specifically, we may rearrange the observables into two pieces: $\mathbf{o}_i = (\mathbf{o}_i^{(1)}\mathbf{o}_i^{(2)})$, where $\Sigma_{uo} = (\mathrm{Cov}(\mathbf{u}_i, \mathbf{o}_i^{(1)})\ \mathrm{Cov}(\mathbf{u}_i, \mathbf{o}_i^{(2)})) = (\Sigma_{uo}^{(1)}\ \mathbf{0})$. That is, the first piece of $\mathbf{o}_i$ may be correlated to the unobservables whereas the second piece is not. To prevent an indirect relation between $\mathbf{u}_i$ and $\mathbf{o}_i^{(2)}$, we also assume that $\mathbf{o}_i^{(1)}$ and $\mathbf{o}_i^{(2)}$ have zero covariance. With these conventions, we assume that

$$\mathrm{E}\,[\mathbf{u}_i \mid \alpha_i, \mathbf{o}_i] = \Sigma_{uo}^{(1)}\left(\mathrm{Var}\,\mathbf{o}_i^{(1)}\right)^{-1}\left(\mathbf{o}_i^{(1)} - \mathrm{E}\mathbf{o}_i^{(1)}\right). \tag{7.17}$$

A sufficient condition for (7.17) is joint normality of $\mathbf{o}_i^{(1)}$ and $\mathbf{u}_i$, conditional on α_i. An advantage of assuming Equation (7.17) directly is that it also allows us to handle categorical variables within $\mathbf{o}_i$. An implication of (7.17) is that $\mathrm{Cov}(\mathbf{u}_i, \mathbf{o}_i^{(2)}) = \mathbf{0}$; however, no implicit distributional assumptions for $\mathbf{o}_i^{(2)}$ are required.

For the sampling design of the observables, we assume that the explanatory variables are generated from an error-components model. Specificially, we assume that

$$\mathrm{Cov}\left(\mathbf{o}_{is}^{(1)}, \mathbf{o}_{it}^{(1)}\right) = \begin{cases} \boldsymbol{\Sigma}_{o,1} + \boldsymbol{\Sigma}_{o,2} & \text{for} \quad s = t, \\ \boldsymbol{\Sigma}_{o,2} & \text{for} \quad s \neq t, \end{cases}$$

so that

$$\mathrm{Var}\ \mathbf{o}_i^{(1)} = (\mathbf{I}_i \otimes \boldsymbol{\Sigma}_{o,1}) + (\mathbf{J}_i \otimes \boldsymbol{\Sigma}_{o,2}).$$

We further assume that the covariance between $\mathbf{u}_{is}$ and $\mathbf{o}_{it}$ is constant in time. Thus, $\mathrm{Cov}(\mathbf{u}_{is}, \mathbf{o}_{it}^{(1)}) = \boldsymbol{\Sigma}_{uo,2}$ for all s, t. This yields $\mathrm{Cov}(\mathbf{u}_i, \mathbf{o}_i^{(1)}) = \mathbf{J}_i \otimes \boldsymbol{\Sigma}_{uo,2}$. Then, from Frees (2001S), we have

$$\begin{aligned} \mathrm{E}\left[y_{it} \mid \boldsymbol{\alpha}_i, \mathbf{o}_i\right] &= \mathbf{z}_{it}'\boldsymbol{\alpha}_i + \mathbf{x}_{it}'\boldsymbol{\beta} + \boldsymbol{\gamma}'\Sigma_{uo,2}\left(\frac{1}{T_i}\Sigma_{o,1} + \Sigma_{o,2}\right)^{-1}\left(\bar{\mathbf{o}}_i^{(1)} - \mathrm{E}\bar{\mathbf{o}}_i^{(1)}\right) \\ &= \beta_0^* + \mathbf{z}_{it}'\boldsymbol{\alpha}_i + \mathbf{x}_{it}'\boldsymbol{\beta} + \bar{\mathbf{z}}_i'^{(1)}\boldsymbol{\gamma}_1^* + \bar{\mathbf{x}}_i'^{(1)}\boldsymbol{\gamma}_2^*, \end{aligned} \tag{7.18}$$

where

$$\beta_0^* = -\boldsymbol{\gamma}'\boldsymbol{\Sigma}_{uo,2}\left(\frac{1}{T_i}\boldsymbol{\Sigma}_{o,1} + \boldsymbol{\Sigma}_{o,2}\right)^{-1}\mathrm{E}\bar{\mathbf{o}}_i.$$

In Equation (7.18), we have listed only those explanatory variables in $\bar{\mathbf{z}}_i^{(1)}$ and $\bar{\mathbf{x}}_i^{(1)}$ that we hypothesize will be correlated with the unobserved variables. This is a testable hypothesis. Further, Equation (7.18) suggests that, by incorporating the terms $\bar{\mathbf{z}}_i^{(1)}$ and $\bar{\mathbf{x}}_i^{(1)}$ as regression terms in the analysis, we may avoid omitted-variable bias. This is the topic of Section 7.3.2.

Special Case: Error-Components Model with Time-Constant Explanatory Variables With the error-components model, we have $q = 1$ and $z_{it} = 1$, so that Equation (7.18) reduces to

$$\mathrm{E}[y_{it} \mid \boldsymbol{\alpha}_i, \mathbf{x}_i] = \beta_0^* + \alpha_i + \mathbf{x}_{it}'\boldsymbol{\beta} + \bar{\mathbf{x}}_i'^{(1)}\boldsymbol{\gamma}_2^*.$$

To estimate the coefficients of this conditional regression equation, we require that the explanatory variables not be linear combinations of one another. In particular, if there are any time-constant variables in $\mathbf{x}_{it}$, they must not be included in $\mathbf{x}_i^{(1)}$. In other words, we require that time-constant variables be uncorrelated with omitted variables.

7.3.2 Augmented Regression Estimation

We now consider an "augmented" regression model of the form

$$\mathrm{E}[\mathbf{y}_i \mid \mathbf{o}_i] = \mathbf{X}_i\boldsymbol{\beta} + \mathbf{G}_i\boldsymbol{\gamma} \tag{7.19}$$

and

$$\mathrm{Var}\,[\mathbf{y}_i \mid \mathbf{o}_i] = \mathbf{R}_i + \mathbf{Z}_i\mathbf{D}\mathbf{Z}_i' = \mathbf{V}_i, \tag{7.20}$$

where $\boldsymbol{\gamma}$ is a $g \times 1$ vector of coefficients, $\mathbf{G}_i$ is a known function of independent variables $\{\mathbf{o}_i\}$, such as in Equation (7.18), and $\mathbf{D}$ is a positive definite variance–covariance matrix to be estimated. One may simply use weighted least squares to determine estimates of $\boldsymbol{\beta}$ and $\boldsymbol{\gamma}$. Explicitly, we use minimizers of the weighted sum of squares

$$\mathrm{WSS}\,(\boldsymbol{\beta}, \boldsymbol{\gamma}) = \sum_{i=1}^{n} (\mathbf{y}_i - (\mathbf{X}_i\boldsymbol{\beta} + \mathbf{G}_i\boldsymbol{\gamma}))' \, \mathbf{W}_i^{-1} \, (\mathbf{y}_i - (\mathbf{X}_i\boldsymbol{\beta} + \mathbf{G}_i\boldsymbol{\gamma})) \tag{7.21}$$

to define estimators of $\boldsymbol{\beta}$ and $\boldsymbol{\gamma}$. We denote these estimators as $\mathbf{b}_{\mathrm{AR}}$ and $\hat{\boldsymbol{\gamma}}_{\mathrm{AR}}$, respectively, where the AR subscript denotes "artificial regression" as in Davidson and MacKinnon (1990E) or "augmented regression" as in Arellano (1993E). The point is that no specialized software is required for the omitted-variable estimator ($\hat{\boldsymbol{\gamma}}_{\mathrm{AR}}$) or omitted variable bias–corrected regression coefficient estimator ($\mathbf{b}_{\mathrm{AR}}$).

Different choices of the $\mathbf{G}_i$ permit us to accommodate different data features. For example, using $\mathbf{W}_i = \mathbf{V}_i$ and omitting $\mathbf{G}_i$, we see that $\mathbf{b}_{\mathrm{AR}}$ reduces to $\mathbf{b}_{\mathrm{GLS}}$. Further, Frees (2001S) shows that $\mathbf{b}_{\mathrm{AR}}$ reduces to $\mathbf{b}_{\mathrm{FE}}$ when $\mathbf{W}_i = \mathbf{V}_i$ and $\mathbf{G}_i = \mathbf{Z}_i(\mathbf{Z}_i'\mathbf{R}_i^{-1}\mathbf{Z}_i)^{-1}\mathbf{Z}_i'\mathbf{R}_i^{-1}\mathbf{X}_i$. As another example, consider the random-coefficients design. Here, we assume that $q = K$ and $\mathbf{x}_{it} = \mathbf{z}_{it}$. Thus, from Equation (7.18), we have

$$\mathrm{E}[y_{it} \mid \boldsymbol{\alpha}_i, \mathbf{o}_i] = \beta_0^* + \mathbf{x}_{it}'(\boldsymbol{\alpha}_i + \boldsymbol{\beta}) + \bar{x}_i^{\prime(1)}\gamma_2^*.$$

Hypothesizing that all variables are potentially related to omitted variables, we require $g = K$ and $\mathbf{G}_i = \mathbf{1}_i\bar{\mathbf{x}}_i'$.

An advantage of the augmented regression formulation, compared to the Section 7.2.2 omitted-variable test, is that it permits a direct assessment of the hypothesis of omitted variables $\mathrm{H}_0 : \boldsymbol{\gamma} = \mathbf{0}$. This can be done directly using a Wald test statistic of the form $\chi^2_{\mathrm{AR}} = \hat{\boldsymbol{\gamma}}_{\mathrm{AR}}'(\mathrm{Var}\,\hat{\boldsymbol{\gamma}}_{\mathrm{AR}})^{-1}\hat{\boldsymbol{\gamma}}_{\mathrm{AR}}$. Here, $\mathrm{Var}\,\hat{\boldsymbol{\gamma}}_{\mathrm{AR}}$ can be estimated based on the variance specification in Equation (7.16) or using a robust alternative as introduced in Section 3.4.

Table 7.3. *Comparison of random-effects estimators to robust alternatives*

Variable	Model with variable intercepts and two variable slopes				Model with variable intercepts and eight variable slopes (random coefficients)			
	Random-effects estimation		Argumented regression estimation		Random-effects estimation		Argumented regression estimation	
	Parameter estimates	t-statistic	Parameter estimates	t-statistic	Parameter estimates	t-statistic	Parameter estimates	t-statistic
LNTPI	2.270	13.40	2.351	11.94	2.230	12.05	2.407	13.06
MR	0.005	0.46	0.031	2.77	0.009	0.91	0.029	2.85
MS	−0.603	−3.86	−0.563	−3.11	−0.567	−3.53	−0.670	−2.96
HH	−0.729	−3.75	−1.056	−2.67	−0.719	−2.97	−1.132	−3.42
AGE	−0.359	−2.15	−0.386	−1.08	−0.229	−1.54	−0.485	−1.76
EMP	−0.661	−5.05	−0.352	−1.01	−0.557	−2.58	−0.334	−1.22
PREP	−0.300	−3.21	−0.296	−2.15	−0.294	−3.24	−0.311	−2.32
DEPEND	−0.138	−2.84	−0.177	−2.19	−0.171	−3.56	−0.170	−2.44
LNTPIAVG			0.964	6.84			0.682	4.29
MRAVG			−0.109	−7.08			−0.091	−6.66
MSAVG			−0.265	−1.07			−0.057	−0.21
HHAVG			0.398	0.95			0.565	1.40
AGEAVG			−0.053	−0.13			0.256	0.79
EMPAVG			−0.489	−1.22			−0.351	−1.21
DEPAVG			0.039	0.22			0.089	0.53
PREPAVG			−0.007	−0.07			−0.038	−0.43
$AR(1)$	0.153	3.38	0.118	2.70	0.006	0.13	−0.043	−0.96
χ^2_{AR}	57.511				56.629			

Example 7.2: Income Tax Payments (continued) We continue with the example from Section 7.2.3 by first considering the model with variable intercepts, two variable slopes, and an $AR(1)$ serial correlation coefficient for the errors. The usual random-effects (GLS) estimators are presented in Table 7.3. These coefficients also appear in Tables 7.1 and 7.2, where we learned that this model suffered from omitted-variable bias. Table 7.3 presents the fits from the augmented regression model, where we have augmented the regression using the averages from all the explanatory variables $\bar{\mathbf{x}}_i$. Table 7.3 shows that the averages of both of the income variables, LNTPI and MR, are statistically significantly different from zero. Further, from the overall test of $\mathrm{H}_0 : \boldsymbol{\gamma} = \mathbf{0}$, the test statistic is $\chi^2_{\mathrm{AR}} = 57.511$

7.4 Sampling, Selectivity Bias, and Attrition

7.4.1 Incomplete and Rotating Panels

Historically, the primary approaches to panel and longitudinal data analysis in the econometrics and biostatistics literatures assumed *balanced data* of the following form:

$$\boxed{\begin{array}{ccc} y_{11} & \cdots & y_{n1} \\ \vdots & \cdots & \vdots \\ y_{1T} & \cdots & y_{nT} \end{array}}$$

This form suggested using techniques from multivariate analysis as described by, for example, Rao (1987B) and Chamberlain (1982E).

However, there are many ways in which a complete, balanced set of observations may not be available because of delayed entry, early exit, and intermittent nonresponse. This section begins by considering unbalanced situations where the lack of balance is *planned* or designed by the analyst. Section 7.4.2 then discusses unbalanced situations that are not planned. When unplanned, we call the nonresponses *missing data*. We will be particularly concerned with situations in which the mechanisms for missingness are related to the response, discussed in Section 7.4.3.

To accommodate missing data, for subject i we use the notation T_i for the number of observations and t_{ij} to denote the time of the jth observation. Thus, the times for the available set of observations are $\{t_{i,1}, \ldots, t_{i,T_i}\}$. For this chapter, we assume that these times are discrete and a subset of $\{1, 2, \ldots, T\}$. Chapter 6 described the continuous-time case. Define $\mathbf{M}_i$ to be the $T_i \times T$ design matrix

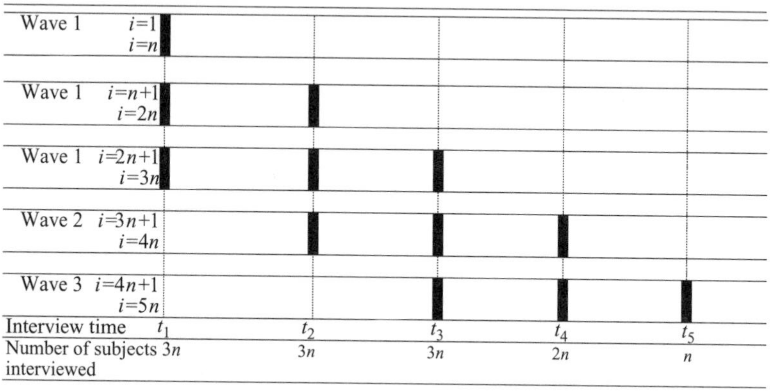

Figure 7.2. Three waves of a rotating panel. The solid lines indicate the individuals that are interviewed at each of five interview times.

that has a "one" in the $t_{i,j}$th column and jth row and is zero otherwise, $j = 1, \ldots, T_i$. Specifically, with this design matrix, we have

$$\begin{pmatrix} t_{i,1} \\ t_{i,2} \\ \vdots \\ t_{i,T_i} \end{pmatrix} = \mathbf{M}_i \begin{pmatrix} 1 \\ 2 \\ \vdots \\ T \end{pmatrix}. \tag{7.22}$$

Using this design matrix, we have that most of the formulas in Chapters 1–6 carry through for planned missing data. The case of time-varying parameters turns out to be more complex; Appendix 8A.1 will illustrate this point in the context of the two-way model. For simplicity, in this text we use the notation $\{1, 2, \ldots, T_i\}$ to denote an unbalanced observation set.

Panel surveys of people provide a good illustration of planned missing data; these studies are often designed to have incomplete observations. This is because we know that people become tired of responding to surveys on a regular basis. Thus, panel surveys are typically designed to "replenish" a certain proportion of the sample at each interview time. Specifically, for a rolling or rotating panel, a fixed proportion of individuals enter and exit the panel during each interview time.

Figure 7.2 illustrates three waves of a rotating panel. At time t_1, the first $3n$ individuals that comprise Wave 1 are interviewed. At time t_2, n individuals from Wave 1 are dropped and n new individuals that comprise Wave 2 are added. At time t_3, n individuals from Wave 1 are dropped and n new individuals that comprise Wave 3 are added. This pattern is continued, with n individuals dropped and added at each interview time. Thus, at each time, $3n$ individuals are interviewed. Each person stays in the survey for at most three time periods.

7.4.2 Unplanned Nonresponse

Few difficulties arise when the data are not balanced as a result of a planned design. However, when the data are unbalanced owing to unforeseen events, this lack of balance represents a potential source of bias. To provide some specific examples, we again return to panel surveys of people. Verbeek and Nijman (in Chapter 18 of Mátyás and P. Sevestre, 1996E) provide the following list of types of unplanned nonresponse:

- *Initial nonresponse.* A subject contacted cannot, or will not, participate. Because of limited information, this potential problem is often ignored in the analysis.
- *Unit nonresponse.* A subject contacted cannot, or will not, participate even after repeated attempts (in subsequent waves) to include the subject.
- *Wave nonresponse.* A subject does not respond for one or more time periods but does respond in the preceding and subsequent times. (For example, the subject may be on vacation.)
- *Attrition.* A subject leaves the panel after participating in at least one survey.

It is also of concern to deal with survey-item nonresponse where the items are treated as covariates in the analysis. Here, information on one or more variables is missing. For example, individuals may not wish to report income or age. However, we restrict consideration to missing responses.

To understand the mechanisms that lead to unplanned nonresponse, we model it stochastically. Let r_{ij} be an indicator variable for the ijth observation, with a one indicating that this response is observed and a zero indicating that the response is missing. Let $\mathbf{r} = (r_{11}, \ldots, r_{1T}, \ldots, r_{n1}, \ldots, r_{nT})'$ summarize the data availability for all subjects. The interest is in whether or not the responses influence the missing data mechanism. For notation, we use $\mathbf{Y} = (y_{11}, \ldots, y_{1T}, \ldots, y_{n1}, \ldots, y_{nT})'$ to be the collection of all potentially observed responses.

Missing-Data Models

In the case where $\mathbf{Y}$ does not affect the distribution of $\mathbf{r}$, we follow Rubin (1976G) and call this case *missing completely at random* (MCAR). Specifically, the missing data are MCAR if $f(\mathbf{r} \mid \mathbf{Y}) = f(\mathbf{r})$, where $f(.)$ is a generic probability mass function. An extension of this idea is in Little (1995G), where the adjective "covariate dependent" is added when $\mathbf{Y}$ does not affect the distribution of $\mathbf{r}$, conditional on the covariates. If the covariates are summarized as $\{\mathbf{X}, \mathbf{Z}\}$, then the condition corresponds to the relation $f(\mathbf{r} \mid \mathbf{Y}, \mathbf{X}, \mathbf{Z}) = f(\mathbf{r} \mid \mathbf{X}, \mathbf{Z})$. To illustrate this point, consider an example of Little and Rubin (1987G) where $\mathbf{X}$ corresponds to age and $\mathbf{Y}$ corresponds to income of all potential observations.

If the probability of being missing does not depend on income, then the missing data are MCAR. If the probability of being missing varies by age but does not by income over observations within an age group, then the missing data are covariate dependent MCAR. Under the latter specification, it is possible for the missing data to vary by income. For example, younger people may be less likely to respond to a survey. This shows that the "missing at random" feature depends on the purpose of the analysis. Specifically, it is possible that an analysis of the joint effects of age and income may encounter serious patterns of missing data whereas an analysis of income controlled for age suffers no serious bias patterns.

When the data are MCAR, Little and Rubin (1987G, Chapter 3) describe several approaches for handling partially missing data. One option is to treat the available data as if nonresponses were planned and use unbalanced estimation techniques. Another option is to utilize only subjects with a complete set of observations by discarding observations from subjects with missing responses. A third option is to impute values for missing responses. Little and Rubin note that each option is generally easy to carry out and may be satisfactory with small amounts of missing data. However, the second and third options may not be efficient. Further, each option implicitly relies heavily on the MCAR assumption.

Little and Rubin (1987G) advocate modeling the missing-data mechanisms; they call this the *model-based* approach. To illustrate, consider a likelihood approach using a selection model for the missing-data mechanism. Now, partition $\mathbf{Y}$ into observed and missing components using the notation $\mathbf{Y} = (\mathbf{Y}_{\text{obs}}, \mathbf{Y}_{\text{miss}})$. With the likelihood approach, we base inference on the observed random variables. Thus, we use a likelihood proportional to the joint function $f(\mathbf{r}, \mathbf{Y}_{\text{obs}})$. We also specify a *selection model* by specifying the conditional mass function $f(\mathbf{r} \mid \mathbf{Y})$.

Suppose that the observed responses and the selection model distributions are characterized by a vectors of parameters $\boldsymbol{\theta}$ and $\boldsymbol{\psi}$, respectively. Then, with the relation $f(\mathbf{r}, \mathbf{Y}_{\text{obs}}, \boldsymbol{\theta}, \boldsymbol{\psi}) = f(\mathbf{Y}_{\text{obs}}, \boldsymbol{\theta}) \times f(\mathbf{r} \mid \mathbf{Y}_{\text{obs}}, \boldsymbol{\psi})$, we may express the log likelihood of the observed random variables as

$$L(\boldsymbol{\theta}, \boldsymbol{\psi}) = \log f(\mathbf{r}, \mathbf{Y}_{\text{obs}}, \boldsymbol{\theta}, \boldsymbol{\psi}) = \log f(\mathbf{Y}_{\text{obs}}, \boldsymbol{\theta}) + \log f(\mathbf{r} \mid \mathbf{Y}_{\text{obs}}, \boldsymbol{\psi}).$$

In the case where the data are MCAR, then $f(\mathbf{r} \mid \mathbf{Y}_{\text{obs}}, \boldsymbol{\psi}) = f(\mathbf{r} \mid \boldsymbol{\psi})$ does not depend on $\mathbf{Y}_{\text{obs}}$. Little and Rubin (1987G) also consider the case where the selection mechanism model distribution does not depend on $\mathbf{Y}_{\text{miss}}$ but may depend on $\mathbf{Y}_{\text{obs}}$. In this case, they call the data *missing at random* (MAR).

In both the MAR and MCAR cases, we see that the likelihood may be maximized over the parameters, separately for each case. In particular, if one is only interested in the MLE of $\boldsymbol{\theta}$, then the selection model mechanism may be "ignored." Hence, both situations are often referred to as the *ignorable case*.

Example 7.2: Income Tax Payments (continued) Let y represent tax liability and x represent income. Consider the following five selection mechanisms.

- The taxpayer is not selected (missing) with probability ψ without regard to the level of tax liability. In this case, the selection mechanism is MCAR.
- The taxpayer is not selected if the tax liability is less than \$100. In this case, the selection mechanism depends on the observed and missing response. The selection mechanism cannot be ignored.
- The taxpayer is not selected if the income is less than \$20,000. In this case, the selection mechanism is MCAR, covariate dependent. That is, assuming that the purpose of the analysis is to understand tax liabilities conditional on knowledge of income, stratifying based on income does not seriously bias the analysis.
- The probability of a taxpayer being selected decreases with tax liability. For example, suppose the probability of being selected is logit $(-\psi y_i)$. In this case, the selection mechanism depends on the observed and missing response. The selection mechanism cannot be ignored.
- The taxpayer is followed over $T = 2$ periods. In the second period, a taxpayer is not selected if the first-period tax is less than \$100. In this case, the selection mechanism is MAR. That is, the selection mechanism is based on an observed response.

The second and fourth selection mechanisms represent situations where the selection mechanism must be explicitly modeled; these are known as *nonignorable cases*. In these situations without explicit adjustments, procedures that ignore the selection effect may produce seriously biased results. To illustrate a correction for selection bias in a simple case, we outline an example due to Little and Rubin (1987G). Section 7.4.3 describes additional mechanisms.

Example 7.3: Historical Heights Little and Rubin (1987G) discuss data due to Wachter and Trusell (1982G) on y, the height of men recruited to serve in the military. The sample is subject to censoring in that minimum height standards were imposed for admission to the military. Thus, the selection mechanism is

$$r_i = \begin{cases} 1 & \text{if } y_i > c_i, \\ 0 & \text{otherwise,} \end{cases}$$

where c_i is the known minimum height standard imposed at the time of recruitment. The selection mechanism is nonignorable because it depends on the individual's height.

For this example, additional information is available to provide reliable model inference. Specifically, based on other studies of male heights, we may assume that the population of heights is normally distributed. Thus, the likelihood of the observables can be written down and inference may proceed directly. To illustrate, suppose that $c_i = c$ is constant. Let μ and σ denote the mean and standard deviation of y, respectively. Further suppose that we have a random sample of $n + m$ men in which m men fall below the minimum standard height c and we observe $\mathbf{Y}_{\text{obs}} = (y_1, \ldots, y_n)'$. The joint distribution for observables is

$$\begin{aligned} f(\mathbf{r}, \mathbf{Y}_{\text{obs}}, \mu, \sigma) &= f(\mathbf{Y}_{\text{obs}}, \mu, \sigma) \times f(\mathbf{r} \mid \mathbf{Y}_{\text{obs}}) \\ &= \prod_{i=1}^{n} \{ f(y_i \mid y_i > c) \times \text{Prob}(y_i > c) \} \times (\text{Prob}(y \le c))^m . \end{aligned}$$

Now, let ϕ and Φ represent the density and distribution function for the standard normal distribution, respectively. Thus, the log-likelihood is

$$\begin{aligned} L(\mu, \sigma) &= \log f(\mathbf{r}, \mathbf{Y}_{\text{obs}}, \mu, \sigma) \\ &= \sum_{i=1}^{n} \log \left(\frac{1}{\sigma} \phi \left(\frac{y_i - \mu}{\sigma} \right) \right) + m \log \left(\Phi \left(\frac{c - \mu}{\sigma} \right) \right). \end{aligned}$$

This is easy to maximize in μ and σ. If one ignored the censoring mechanisms, then one would derive estimates of the observed data from the "log likelihood,"

$$\sum_{i=1}^{n} \log \left(\frac{1}{\sigma} \phi \left(\frac{y_i - \mu}{\sigma} \right) \right),$$

yielding different, and biased, results.

7.4.3 Nonignorable Missing Data

For nonignorable missing data, Little (1995G) recommends the following:

- Avoid missing responses whenever possible by using appropriate follow-up procedures.
- Collect covariates that are useful for predicting missing values.
- Collect as much information as possible regarding the nature of the missing-data mechanism.

For the third point, if little is known about the missing-data mechanism, then it is difficult to employ a robust statistical procedure to correct for the selection bias.

There are many models of missing-data mechanisms. A general overview appears in Little and Rubin (1987G). Verbeek and Nijman (in Chapter 18 of Mátyás and P. Sevestre, 1996E) survey more recent econometric panel data literature. Little (1995G) surveys the problem of attrition. Rather than survey this developing literature, we give a few models of nonignorable missing data.

Heckman Two-Stage Procedure

Heckman (1976E) developed this procedure in the context of cross-sectional data. Because it relies on correlations of unobserved variables, it is also applicable to fixed-effects panel data models. Thus, assume that the response follows a one-way fixed-effects model. As introduced in Chapter 2, this model can be expressed as

$$y_{it} = \alpha_i + \mathbf{x}_{it}'\boldsymbol{\beta} + \varepsilon_{it}.$$

Further assume that the sampling response mechanism is governed by the latent (unobserved) variable r_{it}^*, where

$$r_{it}^* = \mathbf{w}_{it}'\boldsymbol{\gamma} + \eta_{it}.$$

The variables in $\mathbf{w}_{it}$ may or may not include the variables in $\mathbf{x}_{it}$. We observe y_{it} if $r_{it}^* \geq 0$, that is, if r_{it}^* crosses the threshold 0. Thus, we observe

$$r_{it} = \begin{cases} 1 & \text{if } r_{it}^* \geq 0, \\ 0 & \text{otherwise.} \end{cases}$$

To complete the specification, we assume that $\{(\varepsilon_{it}, \eta_{it})\}$ are i.i.d and that the joint distribution of $(\varepsilon_{it}, \eta_{it})'$ is bivariate normal with means zero, variances σ^2 and σ_η^2, and correlation ρ. Note that, if the correlation parameter ρ equals zero, then the response and selection models are independent. In this case, the data are MCAR and the usual estimation procedures are unbiased and asymptotically efficient.

Under these assumptions, basic multivariate normal calculations show that

$$\mathrm{E}(y_{it} \mid r_{it}^* \geq 0) = \alpha_i + \mathbf{x}_{it}'\boldsymbol{\beta} + \beta_\lambda \lambda(\mathbf{w}_{it}'\boldsymbol{\gamma}),$$

where $\beta_\lambda = \rho\sigma$ and $\lambda(a) = \phi(a)/\Phi(a)$, with λ as the inverse of the "Mills ratio." This calculation suggests the following two-step procedure for estimating the parameters of interest.

Heckman's Two-Stage Procedure

1. Use the data $\{(r_{it}, \mathbf{w}_{it})\}$ and a probit regression model to estimate γ. Call this estimator $\mathbf{g}_{\mathrm{H}}$.
2. Use the estimator from stage 1 to create a new explanatory variable, $x_{it,K+1} = \lambda(\mathbf{w}_{it}'\mathbf{g}_{\mathrm{H}})$. Run a one-way fixed-effects model using the K explanatory variables $\mathbf{x}_{it}$ as well as the additional explanatory variable $x_{it,K+1}$. Use $\mathbf{b}_{\mathrm{H}}$ and $b_{\lambda,\mathrm{H}}$ to denote the estimators of $\boldsymbol{\beta}$ and β_λ, respectively.

Section 9.1.1 will introduce probit regressions. We also note that the two-step method does not work in the absence of covariates to predict the response and, for practical purposes, requires variables in $\mathbf{w}$ that are not in $\mathbf{x}$ (see Little and Rubin, 1987).

To test for selection bias, we may test the null hypothesis $\mathrm{H}_0 : \beta_\lambda = 0$ in the second stage due to the relation $\beta_\lambda = \rho\sigma$. When conducting this test, one should use heteroscedasticity-corrected standard errors. This is because the conditional variance $\mathrm{Var}(y_{it} \mid r_{it}^* \geq 0)$ depends on the observation. Specifically, $\mathrm{Var}(y_{it} \mid r_{it}^* \geq 0) = \sigma^2(1 - \rho^2\delta_{it})$, where $\delta_{it} = \lambda_{it}(\lambda_{it} + \mathbf{w}_{it}'\gamma)$ and $\lambda_{it} = \phi(\mathbf{w}_{it}'\gamma)/\Phi(\mathbf{w}_{it}'\gamma)$.

This procedure assumes normality for the selection latent variables to form the augmented variables. Other distribution forms are available in the literature, including the logistic and uniform distributions. A deeper criticism, raised by Little (1995G), is that the procedure relies heavily on assumptions that cannot be tested using the data available. This criticism is analogous to the historical-heights example where we relied heavily on the normal curve to infer the distribution of heights below the censoring point. Despite these criticisms, Heckman's procedure is widely used in the social sciences.

Hausman and Wise Procedure

To see how to extend the Heckman procedure to error-components panel data models, we now describe a procedure originally due to Hausman and Wise 1979E); see also the development in Verbeek and Nijman (in Chapter 18 of Mátyás and Sevestre, 1996E). For simplicity, we work with the error-components model described in Section 3.1, $y_{it} = \alpha_i + \mathbf{x}_{it}'\boldsymbol{\beta} + \varepsilon_{it}$. We also assume that the sampling response mechanism is governed by the latent variable of the form

$$r_{it}^* = \xi_i + \mathbf{w}_{it}'\gamma + \eta_{it}.$$

This is also an error-components model. The variables in $\mathbf{w}_{it}$ may or may not include the variables in $\mathbf{x}_{it}$. As before, r_{it} indicates whether y_{it} is observed and

$$r_{it} = \begin{cases} 1 & \text{if } r_{it}^* \geq 0, \\ 0 & \text{otherwise.} \end{cases}$$

The random variables α_i, ε_{it}, ξ_i, and η_{it} are assumed to be jointly normal, each with mean zero and variance

$$\text{Var}\begin{pmatrix} \alpha_i \\ \varepsilon_{it} \\ \xi_i \\ \eta_{it} \end{pmatrix} = \begin{pmatrix} \sigma_\alpha^2 & 0 & \sigma_{\alpha\xi} & 0 \\ 0 & \sigma_\varepsilon^2 & 0 & \sigma_{\varepsilon\eta} \\ \sigma_{\alpha\xi} & 0 & \sigma_\xi^2 & 0 \\ 0 & \sigma_{\varepsilon\eta} & 0 & \sigma_\eta^2 \end{pmatrix}.$$

If $\sigma_{\alpha\xi} = \sigma_{\varepsilon\eta} = 0$, then the selection process is independent of the observation process.

It is easy to check that $\mathbf{b}_{\text{EC}}$ is unbiased and consistent if $\text{E}(y_{it} \mid \mathbf{r}_i) = \mathbf{x}_{it}'\boldsymbol{\beta}$. Under conditional normality, one can check that

$$\text{E}(y_{it} \mid \mathbf{r}_i) = \mathbf{x}_{it}'\boldsymbol{\beta} + \frac{\sigma_{\alpha\xi}}{T\sigma_\xi^2 + \sigma_\eta^2} g_{it} + \frac{\sigma_{\varepsilon\eta}}{\sigma_\eta^2}\left(g_{it} - \frac{\sigma_\xi^2}{T\sigma_\xi^2 + \sigma_\eta^2}\sum_{s=1}^{T} g_{is}\right),$$

where $g_{it} = \text{E}(\xi_i + \eta_{it} \mid \mathbf{r}_i)$. The calculation of g_{it} involves the multivariate normal distribution and requires numerical integration. This calculation is straightforward although computationally intensive. Following this calculation, testing for ignorability and producing bias-corrected estimators proceeds as in the Heckman case. Chapter 9 will discuss the fitting of binary dependent responses to error-components models. For other additional details, we refer the reader to Verbeek and Nijman (in Chapter 18 of Mátyás and Sevestre, 1996E).

EM Algorithm

Thus far we have focused on introducing specific models of nonignorable nonresponse. General robust models of nonresponse are not available. Rather, a more appropriate strategy is to focus on a specific situation, collect as much information as possible regarding the nature of the selection problem, and then develop a model for this specific selection problem.

The *EM algorithm* is a computational device for computing model parameters. Although specific to each model, it has found applications in a wide variety of models involving missing data. Computationally, the algorithm iterates between the "E," for conditional expectation, and the "M," for maximization, steps. The E step finds the conditional expectation of the missing data given the observed data and current values of the estimated parameters. This is analogous to the time-honored tradition of imputing missing data. A key innovation of the

EM algorithm is that one imputes sufficient statistics for missing values, not the individual data points. For the M step, one updates parameter estimates by maximizing an observed log-likelihood. Both the sufficient statistics and the log-likelihood depend on the model specification.

Many introductions of the EM algorithm are available in the literature. Little and Rubin (1987G) provide a detailed treatment.

Exercises and Extensions

Section 7.2

7.1 *Fuller–Battese (1973G) transform* Consider the Section 3.1 error-components model with $\mathbf{y}_i = \alpha_i \mathbf{1}_i + \mathbf{X}_i \boldsymbol{\beta} + \boldsymbol{\epsilon}_i$ and $\mathrm{Var}\, \mathbf{y}_i = \mathbf{V}_i = \sigma^2 \mathbf{I}_i + \sigma_\alpha^2 \mathbf{J}_i$. Recall from Section 2.5.3 that $\mathbf{Q}_i = \mathbf{I}_i - \mathbf{Z}_i(\mathbf{Z}_i'\mathbf{Z}_i)^{-1}\mathbf{Z}_i'$.

a. Define

$$\psi_i = \left(\frac{\sigma^2}{T_i \sigma_\alpha^2 + \sigma^2} \right)^{1/2}$$

and $\mathbf{P}_i = \mathbf{I}_i - \mathbf{Q}_i$. Show that $\mathbf{V}_i^{-1/2} = \sigma^{-1}(\mathbf{P}_i + \psi_i \mathbf{Q}_i)$.

b. Transform the error-components model using $\mathbf{y}_i^* = (\mathbf{P}_i + \psi_i \mathbf{Q}_i)\mathbf{y}_i$ and $\mathbf{X}_i^* = (\mathbf{P}_i + \psi_i \mathbf{Q}_i)\mathbf{X}_i$. Show that the GLS estimator of $\boldsymbol{\beta}$ is $\mathbf{b}_{\mathrm{EC}} = (\sum_{i=1}^n \mathbf{X}_i^{*\prime} \mathbf{X}_i^*)^{-1} \sum_{i=1}^n \mathbf{X}_i^{*\prime} \mathbf{y}_i^*$.

c. Now consider the special case of the error-components model in Equation (7.1). Show that the GLS estimator of β_1 is

$$b_{1,\mathrm{EC}} = \frac{\left(\sum_i \sum_{t=1}^{T_i} x_{it}^* y_{it}^*\right)\left(\sum_i \psi_i^2 T_i\right) - \left(\sum_i \psi_i T_i \bar{y}_i^*\right)\left(\sum_i \psi_i T_i \bar{x}_i^*\right)}{\left(\sum_i \psi_i^2 T_i\right)\left(\sum_i \sum_{t=1}^{T_i} x_{it}^{*2}\right) - \left(\sum_i \psi_i T_i \bar{x}_i^*\right)^2},$$

where $x_{it}^* = x_{it} - \bar{x}_i(1 - \psi_i)$ and $y_{it}^* = y_{it} - \bar{y}_i(1 - \psi_i)$.

d. Now consider the balanced data case so that $T_i = T$ for each i. Show that

$$b_{1,\mathrm{EC}} = \frac{\sum_{i=1}^n \sum_{t=1}^T (x_{it}^* - \bar{x}^*)(y_{it}^* - \bar{y}^*)}{\sum_{i=1}^n \sum_{t=1}^T (x_{it}^* - \bar{x}^*)^2}.$$

e. For the model considered in part (d), show that variance of $b_{1,\mathrm{EC}}$ is as given in Equation (7.2).

7.2 *Maddala's decomposition* Consider the special case of the error-components model in Equation (7.1).

a. Show Equation (7.6). That is, show that

$$\operatorname{Var} b_{1,\mathrm{B}} = \left(T\sigma_\alpha^2 + \sigma^2\right)\left(T\sum_{i=1}^{n}(\bar{x}_i - \bar{x})^2\right)^{-1}$$

where

$$b_{1,\mathrm{B}} = \frac{\sum_{i=1}^{n}(\bar{x}_i - \bar{x})(\bar{y}_i - \bar{y})}{\sum_{i=1}^{n}(\bar{x}_i - \bar{x})^2},$$

as in Equation (7.5).

b. Show Equation (7.4). That is, show that

$$\operatorname{Var} b_{1,\mathrm{FE}} = \sigma^2\left(\sum_{i=1}^{n}\sum_{t=1}^{T}(x_{it} - \bar{x}_i)^2\right)^{-1}$$

where

$$b_{1,\mathrm{FE}} = \frac{\sum_{i=1}^{n}\sum_{t=1}^{T}(x_{it} - \bar{x}_i)(y_{it} - \bar{y}_i)}{\sum_{i=1}^{n}\sum_{t=1}^{T}(x_{it} - \bar{x}_i)^2},$$

as in Equation (7.3).

c. Use parts (a) and (b) and the expressions for $b_{1,\mathrm{EC}}$ and $\operatorname{Var} b_{1,\mathrm{EC}}$ in Section 7.2.1 to show that

$$\frac{1}{\operatorname{Var} b_{1,\mathrm{EC}}} = \frac{1}{\operatorname{Var} b_{1,\mathrm{FE}}} + \frac{1}{\operatorname{Var} b_{1,\mathrm{B}}}.$$

d. Show Equation (7.7). That is, with parts (a)–(c) and the expressions for $b_{1,\mathrm{EC}}$ and $\operatorname{Var} b_{1,\mathrm{EC}}$ in Section 7.2.1, show that $b_{1,\mathrm{EC}} = (1 - \Delta)b_{1,\mathrm{FE}} + \Delta b_{1,\mathrm{B}}$, where $\Delta = \operatorname{Var} b_{1,\mathrm{EC}}/\operatorname{Var} b_{1,\mathrm{B}}$.

7.3 Mixed linear model estimation with intercepts Consider the linear mixed-effects model described in Section 3.3, where $\boldsymbol{\alpha}_i$ are treated as random, with mean $\mathrm{E}\boldsymbol{\alpha}_i = \boldsymbol{\alpha}$ and variance–covariance matrix $\operatorname{Var}\boldsymbol{\alpha}_i = \mathbf{D}$, independent of the error term. Then, we may rewrite the model as

$$\mathbf{y}_i = \mathbf{Z}_i\boldsymbol{\alpha} + \mathbf{X}_i\boldsymbol{\beta} + \boldsymbol{\epsilon}_i^*,$$

where $\boldsymbol{\epsilon}_i^* = \boldsymbol{\epsilon}_i + \mathbf{Z}_i(\boldsymbol{\alpha}_i - \boldsymbol{\alpha})$ and $\operatorname{Var}\boldsymbol{\epsilon}_i^* = \mathbf{Z}_i\mathbf{D}\mathbf{Z}_i' + \mathbf{R}_i = \mathbf{V}_i$, a positive definite $T_i \times T_i$ matrix.

a. Show that we can express the GLS estimator of $\boldsymbol{\beta}$ as

$$\mathbf{b}_{\mathrm{GLS}} = \mathbf{C}_{\mathrm{GLS}}^{-1}\sum_{i=1}^{n}\left\{\mathbf{X}_i'\mathbf{V}_i^{-1} - \left(\sum_{i=1}^{n}\mathbf{X}_i'\mathbf{V}_i^{-1}\mathbf{Z}_i\right)\left(\sum_{i=1}^{n}\mathbf{Z}_i'\mathbf{V}_i^{-1}\mathbf{Z}_i\right)^{-1}\mathbf{Z}_i'\mathbf{V}_i^{-1}\right\}\mathbf{y}_i$$

with

$$\mathbf{C}_{\text{GLS}} = \left(\sum_{i=1}^{n} \mathbf{X}_i' \mathbf{V}_i^{-1} \mathbf{X}_i \right) - \left(\sum_{i=1}^{n} \mathbf{X}_i' \mathbf{V}_i^{-1} \mathbf{Z}_i \right) \left(\sum_{i=1}^{n} \mathbf{Z}_i' \mathbf{V}_i^{-1} \mathbf{Z}_i \right)^{-1} \left(\sum_{i=1}^{n} \mathbf{Z}_i' \mathbf{V}_i^{-1} \mathbf{X}_i \right).$$

b. Show that Var $\mathbf{b}_{\text{GLS}} = \mathbf{C}_{\text{GLS}}^{-1}$.

c. Now consider the error-components model so that $q = 1, \mathbf{D} = \sigma_\alpha^2, z_{it} = 1$, and $\mathbf{Z}_i = \mathbf{1}_i$. Use part (a) to show that

$$\mathbf{b}_{\text{EC}} = \left(\sum_{i=1}^{n} (\mathbf{X}_i' \mathbf{Q}_i \mathbf{X}_i + (1 - \zeta_i) T_i (\bar{\mathbf{x}}_i - \bar{\mathbf{x}}_w)(\bar{\mathbf{x}}_i - \bar{\mathbf{x}}_w)') \right)^{-1} \times \sum_{i=1}^{n} \{ \mathbf{X}_i' \mathbf{Q}_i \mathbf{y}_i + (1 - \zeta_i) T_i (\bar{\mathbf{x}}_i - \bar{\mathbf{x}}_w)(\bar{y}_i - \bar{y}_w) \},$$

where $\mathbf{Q}_i = \mathbf{I}_i - (1/T_i)\mathbf{J}_i$, $\bar{\mathbf{x}}_w = (\sum_{i=1}^n \zeta_i \bar{\mathbf{x}}_i)/(\sum_{i=1}^n \zeta_i,)$ and $\bar{y}_w = (\sum_{i=1}^n \zeta_i \bar{y}_i)/(\sum_{i=1}^n \zeta_i)$.

d. Consider part (c) and assume in addition that $K = 1$. Show that

$$b_{\text{EC}} = \frac{\sum_{i=1}^n \left\{ \sum_{t=1}^{T_i} (x_{it} - \bar{x}_i)(y_{it} - \bar{y}_i) + (1 - \zeta_i) T_i (\bar{x}_i - \bar{x}_w)(\bar{y}_i - \bar{y}_w) \right\}}{\sum_{i=1}^n \left(\sum_{t=1}^{T_i} (x_{it} - \bar{x}_i)^2 + (1 - \zeta_i) T_i (\bar{x}_i - \bar{x}_w)^2 \right)}.$$

e. As in part (a), show that the GLS estimator of α is

$$\mathbf{a}_{\text{GLS}} = \left(\sum_{i=1}^{n} \mathbf{Z}' \mathbf{V}_i^{-1} \mathbf{Z}_i \right)^{-1} \left\{ \left(\sum_{i=1}^{n} \mathbf{Z}_i' \mathbf{V}_i^{-1} \mathbf{y}_i \right) - \left(\sum_{i=1}^{n} \mathbf{Z}_i' \mathbf{V}_i^{-1} \mathbf{X}_i \right) \mathbf{b}_{\text{GLS}} \right\}.$$

f. For the case considered in part (c) with $q = 1, \mathbf{D} = \sigma_\alpha^2, z_{it} = 1$, and $\mathbf{Z}_i = \mathbf{1}_i$, show that

$$a_{\text{EC}} = \bar{y}_w - \bar{\mathbf{x}}_w' \mathbf{b}_{\text{EC}},$$

where $\mathbf{b}_{\text{EC}}$ is given in part (c).

7.4 Robust estimation Consider the linear mixed-effects model described in Exercise 7.3. Let $\mathbf{C}_{\text{FE}} = \sum_{i=1}^n \mathbf{X}_i' \mathbf{R}_i^{-1/2} \mathbf{Q}_{Z,i} \mathbf{R}_i^{-1/2} \mathbf{X}_i$, where $\mathbf{Q}_{Z,i} = \mathbf{I}_i - \mathbf{R}_i^{-1/2} \mathbf{Z}_i (\mathbf{Z}_i' \mathbf{R}_i^{-1} \mathbf{Z}_i)^{-1} \mathbf{Z}_i' \mathbf{R}_i^{-1/2}$. Recall that

$$\mathbf{b}_{\text{FE}} = \mathbf{C}_{\text{FE}}^{-1} \sum_{i=1}^{n} \mathbf{X}_i' \mathbf{R}_i^{-1/2} \mathbf{Q}_{Z,i} \mathbf{R}_i^{-1/2} \mathbf{y}_i.$$

a. Show that $\mathrm{E}\,\mathbf{b}_{\mathrm{FE}} = \boldsymbol{\beta}$.
b. Show that $\mathrm{Var}\,\mathbf{b}_{\mathrm{FE}} = \mathbf{C}_{\mathrm{FE}}^{-1}$.

7.5 Decomposing the random-effects estimator Consider the linear mixed-effects model described in Exercise 7.3. An alternative estimator of $\boldsymbol{\beta}$ is the "between-groups" estimator, given as

$$\mathrm{b}_B = \mathbf{C}_{\mathrm{B}}^{-1} \sum_{i=1}^{n} \left\{ \mathbf{X}_i' \mathbf{V}_i^{-1} \mathbf{Z}_i \left(\mathbf{Z}_i' \mathbf{V}_i^{-1} \mathbf{Z}_i \right)^{-1} \mathbf{Z}_i' \mathbf{V}_i^{-1} - \left(\sum_{i=1}^{n} \mathbf{X}_i' \mathbf{V}_i^{-1} \mathbf{Z}_i \right) \left(\sum_{i=1}^{n} \mathbf{Z}' \mathbf{V}_i^{-1} \mathbf{Z}_i \right)^{-1} \mathbf{Z}_i' \mathbf{V}_i^{-1} \right\} \mathbf{y}_i,$$

where

$$\mathbf{C}_{\mathrm{B}} = \sum_{i=1}^{n} \mathbf{X}_i' \mathbf{V}_i^{-1} \mathbf{Z}_i \left(\mathbf{Z}_i' \mathbf{V}_i^{-1} \mathbf{Z}_i \right)^{-1} \mathbf{Z}_i' \mathbf{V}_i^{-1} \mathbf{X}_i - \left(\sum_{i=1}^{n} \mathbf{X}_i' \mathbf{V}_i^{-1} \mathbf{Z}_i \right) \left(\sum_{i=1}^{n} \mathbf{Z}_i' \mathbf{V}_i^{-1} \mathbf{Z}_i \right)^{-1} \left(\sum_{i=1}^{n} \mathbf{Z}_i' \mathbf{V}_i^{-1} \mathbf{X}_i \right).$$

a. Show that $\mathrm{Var}\,\mathbf{b}_{\mathrm{B}} = \mathbf{C}_{\mathrm{B}}^{-1}$.
b. Now consider the error-components model so that $q = 1$, $\mathbf{D} = \sigma_\alpha^2$, $z_{it} = 1$, and $\mathbf{Z}_i = \mathbf{1}_i$. Use part (a) to show that

$$\mathbf{b}_{\mathrm{B}} = \left(\sum_{i=1}^{n} T_i \,(1 - \zeta_i)\,(\bar{\mathbf{x}}_i - \bar{\mathbf{x}}_w)\,(\bar{\mathbf{x}}_i - \bar{\mathbf{x}}_w)' \right)^{-1} \times \sum_{i=1}^{n} T_i (1 - \zeta_i)(\bar{\mathbf{x}}_i - \bar{\mathbf{x}}_w)(\bar{y}_i - \bar{y}_w).$$

c. Show that an alternative form for $\mathbf{b}_{\mathrm{B}}$ is

$$\mathbf{b}_{\mathrm{B}} = \left(\sum_{i=1}^{n} \zeta_i \,(\bar{\mathbf{x}}_i - \bar{\mathbf{x}}_w)\,(\bar{\mathbf{x}}_i - \bar{\mathbf{x}}_w)' \right)^{-1} \sum_{i=1}^{n} \zeta_i (\bar{\mathbf{x}}_i - \bar{\mathbf{x}}_w)(\bar{y}_i - \bar{y}_w).$$

d. Use Equation (A.4) of Appendix A.5 to establish

$$\mathbf{Z}_i' \mathbf{V}_i^{-1} \mathbf{Z}_i = \left(\mathbf{D} + \left(\mathbf{Z}_i' \mathbf{R}_i^{-1} \mathbf{Z}_i \right)^{-1} \right)^{-1}.$$

e. Use part (d) to establish

$$\mathbf{V}_i^{-1} - \mathbf{V}_i^{-1} \mathbf{Z}_i \left(\mathbf{Z}_i' \mathbf{V}_i^{-1} \mathbf{Z}_i \right)^{-1} \mathbf{Z}_i' \mathbf{V}_i^{-1} = \mathbf{R}_i^{-1} - \mathbf{R}_i^{-1} \mathbf{Z}_i \left(\mathbf{Z}_i' \mathbf{R}_i^{-1} \mathbf{Z}_i \right)^{-1} \mathbf{Z}_i' \mathbf{R}_i^{-1}.$$

f. Use Exercises 7.3(a), 7.4, and parts (a)–(e) to show that $\mathbf{C}_{\mathrm{B}} + \mathbf{C}_{\mathrm{FE}} = \mathbf{C}_{\mathrm{GLS}}$; that is, show that

$$(\mathrm{Var}\,\mathbf{b}_{\mathrm{B}})^{-1} + (\mathrm{Var}\,\mathbf{b}_{\mathrm{FE}})^{-1} = (\mathrm{Var}\,\mathbf{b}_{\mathrm{GLS}})^{-1}.$$

g. Prove Maddala's decomposition:

$$\mathbf{b}_{\mathrm{GLS}} = (\mathbf{I} - \boldsymbol{\Delta})\,\mathbf{b}_{\mathrm{FE}} + \boldsymbol{\Delta} b_{\mathrm{B}},$$

where $\boldsymbol{\Delta} = (\mathrm{Var}\,\mathbf{b}_{\mathrm{GLS}})\,(\mathrm{Var}\,\mathbf{b}_{\mathrm{B}})^{-1}$.

7.6 Omitted-variable test Consider the linear mixed-effects model described in Exercises 7.3, 7.4, and 7.5.

a. Show that $\mathrm{Cov}\,(\mathbf{b}_{\mathrm{GLS}}, \mathbf{b}_{\mathrm{FE}}) = \mathrm{Var}\,(\mathbf{b}_{\mathrm{GLS}})$.
b. Use part (a) to show that

$$\mathrm{Var}\,(\mathbf{b}_{\mathrm{FE}} - \mathbf{b}_{\mathrm{GLS}}) = \mathrm{Var}\,(\mathbf{b}_{\mathrm{FE}}) - \mathrm{Var}\,(\mathbf{b}_{\mathrm{GLS}}).$$

c. Show that

$$\chi^2_{\mathrm{FE}} = (\mathbf{b}_{\mathrm{FE}} - \mathbf{b}_{\mathrm{GLS}})'\,(\mathrm{Var}\,(\mathbf{b}_{\mathrm{FE}}) - \mathrm{Var}\,(\mathbf{b}_{\mathrm{GLS}}))^{-1}\,(\mathbf{b}_{\mathrm{FE}} - \mathbf{b}_{\mathrm{GLS}})$$

has an asymptotic (as $n \to \infty$) chi-square distribution with K degrees of freedom.

8
Dynamic Models

Abstract. This chapter considers models of longitudinal data sets with longer time dimensions than were considered in earlier chapters. With many observations per subject, analysts have several options for introducing more complex dynamic model features that address questions of interest or that represent important tendencies of the data (or both). One option is based on the serial correlation structure; this chapter extends the basic structures that were introduced in Chapter 2. Another dynamic option is to allow parameters to vary over time. Moreover, for a data set with a long time dimension relative to the number of subjects, we have an opportunity to model the cross-sectional correlation, an important issue in many studies. The chapter also considers the Kalman filter approach, which allows the analyst to incorporate many of these features simultaneously. Throughout, the assumption of exogeneity of the explanatory variables is maintained. Chapter 6 considered lagged dependent variables as explanatory variables, another way of introducing dynamic features into the model.

8.1 Introduction

Because longitudinal data vary over time as well as in the cross section, we have opportunities to model the *dynamic*, or temporal, patterns in the data. For the data analyst, when is it important to consider dynamic aspects of a problem?

Part of the answer to this question rests on the purpose of the analysis. If the main inferential task is forecasting of future observations as introduced in Chapter 4, then the dynamic aspect is critical. In this instance, every opportunity for understanding dynamic aspects should be explored. In contrast, in other problems the focus is on understanding relations among variables. Here, the dynamic aspects may be less critical. This is because many models still provide the basis for constructing unbiased estimators and reliable testing procedures when dynamic aspects are ignored, at the price of efficiency. For example, for problems with large sample sizes (in the cross section), efficiency may not be an important issue. Nonetheless, understanding the dynamic correlation structure is important for achieving efficient parameter estimators; this aspect can be vital, especially for data sets with many observations over time. The importance of dynamics is influenced by the size of the data set, both through

- the choice of the statistical model and
- the type of approximations used to establish properties of parameter estimators.

For many longitudinal data sets, the number of subjects (n) is large relative to the number of observations per subject (T). This suggests the use of regression analysis techniques; these methods are designed to reveal relationships among variables, observed and unobserved, and to account for subject-level heterogeneity. In contrast, for other problems, T is large relative to n. This suggests borrowing from other statistical methodologies, such as multivariate time-series. Here, although relationships among variables are important, understanding temporal patterns is the focus of this methodology. We remark that the modeling techniques presented in Chapters 1–5 are based on the linear model. In contrast, Section 8.5 presents a modeling technique from the multivariate time-series literature, the Kalman filter.

The sample size also influences the properties of our estimators. For longitudinal data sets where n is large compared to T, this suggests the use of asymptotic approximations where T is bounded and n tends to infinity. However, for other data sets, we may achieve more reliable approximations by considering instances where n and T approach infinity together or where n is bounded and T tends to infinity. For many models, this distinction is not an important one for applications. However, for some models, such as the fixed-effects lagged–dependent variable model in Section 6.3, the difference is critical. There, the approach where T is bounded and n tends to infinity leads to biased parameter estimators.

This chapter deals with problems where the dynamic aspect is important, either because of the inferential purposes underlying the problem or the nature of the data set. We now outline several approaches that are available for incorporating dynamic aspects into a longitudinal data model.

Perhaps the easiest way for handling dynamics is to let one of the explanatory variables be a proxy for time. For example, we might use $x_{it,j} = t$ for a *linear trend in time model*. Another technique is to use "time dummy variables," that is, binary variables that indicate the presence or absence of a period effect. To illustrate, in Chapter 2, we introduced the two-way model

$$y_{it} = \alpha_i + \lambda_t + \mathbf{x}'_{it}\boldsymbol{\beta} + \varepsilon_{it}. \tag{8.1}$$

Here, the parameters $\{\lambda_t\}$ are time-specific quantities that do not depend on subjects. Chapter 2 considered the case where $\{\lambda_t\}$ were fixed parameters.

In Chapter 3, we allowed $\{\lambda_t\}$ to be random. Section 8.3 extends this idea by allowing several parameters in the longitudinal data model to vary with time. To illustrate, one example that we will consider is

$$y_{it} = \mathbf{x}'_{it}\boldsymbol{\beta}_t + \varepsilon_{it},$$

that is, where regression parameters $\boldsymbol{\beta}$ vary over time.

Unlike cross-sectional data, with longitudinal data we also have the ability to accommodate temporal trends by looking at changes in either the response or the explanatory variables. This technique is straightforward and natural in some areas of application. For example, when examining stock prices, because of financial economics theory, we examine proportional changes in prices, which are simply returns. As another example, we may wish to analyze the model

$$\Delta y_{it} = \alpha_i + \mathbf{x}'_{it}\boldsymbol{\beta} + \varepsilon_{it}, \tag{8.2}$$

where $\Delta y_{it} = y_{it} - y_{i,t-1}$ is the change, or difference, in y_{it}. In general, one must be wary of this approach because you lose n (initial) observations when differencing.

Rewriting Equation (8.2), we have

$$y_{it} = \alpha_i + y_{i,t-1} + \mathbf{x}'_{it}\boldsymbol{\beta} + \varepsilon_{it}.$$

A generalization of this is

$$y_{it} = \alpha_i + \gamma y_{i,t-1} + \mathbf{x}'_{it}\boldsymbol{\beta} + \varepsilon_{it}, \tag{8.3}$$

where γ is a parameter to be estimated. If $\gamma = 1$, then the model in Equation (8.3) reduces to the model in Equation (8.2). If $\gamma = 0$, then the model in Equation (8.3) reduces to our "usual" one-way model. Thus, the parameter γ is a measure of the relationship between y_{it} and $y_{i,t-1}$. Because it measures the regression of $y_{i,t-1}$ on $y_{i,t}$, it is called an *autoregressive* parameter. The model in Equation (8.3) is an example of a *lagged–dependent variable* model, which was introduced in Section 6.3.

Another way to formulate an autoregressive model is

$$y_{it} = \alpha_i + \mathbf{x}'_{it}\boldsymbol{\beta} + \varepsilon_{it}, \tag{8.4}$$

where $\varepsilon_{it} = \rho\varepsilon_{i,t-1} + \eta_{it}$. Here, the autoregression is on the disturbance term, not the response. The models in Equations (8.3) and (8.4) are similar, yet they differ in some important aspects. To see this, use Equation (8.4) twice to get

$$\begin{aligned} y_{it} - \rho y_{i,t-1} &= (\alpha_i + \mathbf{x}_{it}'\boldsymbol{\beta} + \varepsilon_{it}) - \rho(\alpha_i + \mathbf{x}_{i,t-1}'\boldsymbol{\beta} + \varepsilon_{i,t-1}) \\ &= \alpha_i^* + (\mathbf{x}_{it} - \rho \mathbf{x}_{i,t-1})'\boldsymbol{\beta} + \eta_{it}, \end{aligned}$$

where $\alpha_i^* = \alpha_i(1-\rho)$. Thus, Equation (8.4) is similar to Equation (8.3) with $\gamma = \rho$; the difference lies in the variable associated with $\boldsymbol{\beta}$. Section 8.2 explores further the modeling strategy of assuming serial correlation directly on the disturbance terms in lieu of the response. There, Section 8.2 notes that, because of the assumption of bounded T, one need not assume stationarity of errors. This strategy was used implicitly in Chapters 1–5 for handling the dynamics of longitudinal data.

Finally, Section 8.5 shows how to adapt the Kalman filter technique to longitudinal data analysis. The Kalman filter approach is a flexible technique that allows analysts to incorporate time-varying parameters and broad patterns of serial correlation structures into the model. Further, we will show how to use this technique to simultaneously model cross-sectional, heterogeneity, and temporal aspects as well as *spatial* patterns.

8.2 Serial Correlation Models

One approach for handling the dynamics is through the specification of the covariance structure of the disturbance term, ε. This section examines stationary and nonstationary specifications of the correlation structure for equally spaced data and then introduces options for data that may not be equally spaced.

8.2.1 Covariance Structures

Recall from Section 2.5.1 that $\mathbf{R} = \mathrm{Var}\,\boldsymbol{\epsilon}$ is a $T \times T$ temporal variance–covariance matrix. Here, the element in the rth row and sth column is denoted by $\mathbf{R}_{rs}$. For the ith subject, we define $\mathrm{Var}\,\boldsymbol{\epsilon}_i = \mathbf{R}_i(\boldsymbol{\tau})$, a $T_i \times T_i$ submatrix of $\mathbf{R}$ that can be determined by removing the rows and columns of $\mathbf{R}$ that correspond to responses not observed. We denote this dependence of $\mathbf{R}$ on parameters using $\mathbf{R}(\boldsymbol{\tau})$, where $\boldsymbol{\tau}$ is the vector of unknown parameters, called variance components. Section 2.5.1 introduced four specifications of $\mathbf{R}$: (i) no correlation, (ii) compound symmetry, (iii) autoregressive of order one, and (iv) unstructured.

The autoregressive model of order one is a standard representation used in time-series analysis. This field of study also suggests alternative correlation structures. For example, one could entertain autoregressive models of higher order. Further, moving-average models suggest the "Toeplitz" specification of $\mathbf{R}$:

- $\mathbf{R}_{rs} = \sigma_{|r-s|}$. This defines elements of a Toeplitz matrix.
- $\mathbf{R}_{rs} = \sigma_{|r-s|}$ for $|r-s| < band$ and $\mathbf{R}_{rs} = 0$ for $|r-s| \geq band$. This is the banded Toeplitz matrix.

When the band is "q" + 1, this Toeplitz specification corresponds to a moving-average model of order q, also known as an *MA*(q) structure. More complex autoregressive, moving-average models may be handled in a similar fashion; see, for example, Jennrich and Schlucter (1986B).

The Toeplitz specification suggests a general linear variance structure of the form

$$\mathbf{R} = \tau_1 \mathbf{R}_1 + \tau_2 \mathbf{R}_2 + \cdots + \tau_{\dim(\boldsymbol{\tau})} \mathbf{R}_{\dim(\boldsymbol{\tau})},$$

where $\dim(\boldsymbol{\tau})$ is the dimension of $\boldsymbol{\tau}$ and $\mathbf{R}_1, \mathbf{R}_2, \ldots, \mathbf{R}_{\dim(\boldsymbol{\tau})}$ are known matrices. As pointed out in Section 3.5.3 on MIVQUE estimation, this general structure accommodates many, although not all (such as autoregressive), covariance structures.

Another broad covariance structure suggested by the multivariate analysis literature is the factor-analytic structure of the form $\mathbf{R} = \boldsymbol{\Lambda}\boldsymbol{\Lambda}' + \boldsymbol{\Psi}$, where $\boldsymbol{\Lambda}$ is a matrix of unknown factor loadings and $\boldsymbol{\Psi}$ is an unknown diagonal matrix. An important advantage of the factor-analytic specification is that it easily allows the data analyst to ensure that the estimated variance matrix will be positive (or nonnegative) definite, which can be important in some applications.

The covariance structures were described in the context of specification of $\mathbf{R}$, although they also apply to specification of $\mathrm{Var}\, \boldsymbol{\alpha}_i = \mathbf{D}$.

8.2.2 Nonstationary Structures

With large n and a bounded T, we need not restrict ourselves to stationary models. For example, we have already considered the unstructured model for $\mathbf{R}$. Making this specification imposes no additional restrictions on $\mathbf{R}$, including stationarity.

The primary advantage of stationary models is that they provide parsimonious representations for the correlation structure. However, parsimonious nonstationary models are also possible. To illustrate, suppose that the subject-level dynamics are specified through a random-walk model $\varepsilon_{it} = \varepsilon_{i,t-1} + \eta_{it}$, where $\{\eta_{it}\}$ is an i.i.d. sequence with $\mathrm{Var}\, \eta_{it} = \sigma_\eta^2$, which is independent of $\{\varepsilon_{i0}\}$. With the notation $\mathrm{Var}\, \varepsilon_{i0} = \sigma_0^2$, we have $\mathrm{Var}\, \varepsilon_{it} = \sigma_0^2 + t\sigma_\eta^2$ and $\mathrm{Cov}\,(\varepsilon_{ir}, \varepsilon_{is}) =$

$\operatorname{Var} \varepsilon_{ir} = \sigma_0^2 + r\sigma_\eta^2$, for $r < s$. This yields $\mathbf{R} = \sigma_0^2 \mathbf{J} + \sigma_\eta^2 \mathbf{R}_{\mathrm{RW}}$, where

$$\mathbf{R}_{\mathrm{RW}} = \begin{pmatrix} 1 & 1 & 1 & \ldots & 1 \\ 1 & 2 & 2 & \ldots & 2 \\ 1 & 2 & 3 & \ldots & 3 \\ \vdots & \vdots & \vdots & \ddots & \vdots \\ 1 & 2 & 3 & \ldots & T \end{pmatrix}.$$

Note that $\mathbf{R}$ is a function of only two unknown parameters. Further, this representation allows us to specify a nonstationary model without differencing the data (and thus without losing the initial set of observations). As shown in Exercise 4.8, this matrix has a simple inverse that can speed computations when T is large.

More generally, consider $\varepsilon_{it} = \rho\varepsilon_{i,t-1} + \eta_{it}$, which is similar to the $AR(1)$ specification except that we no longer require stationarity so that we may have $|\rho| \geq 1$. To specify covariances, we first define the function

$$S_t(\rho) = 1 + \rho^2 + \cdots + \rho^{2(t-1)} = \begin{cases} t & \text{if } |\rho| = 1, \\ \frac{1-\rho^{2t}}{1-\rho^2} & \text{if } |\rho| \neq 1. \end{cases}$$

Straightforward calculations show that $\operatorname{Var} \varepsilon_{it} = \sigma_0^2 + \sigma_\eta^2 S_t(\rho)$ and $\operatorname{Cov}(\varepsilon_{ir}, \varepsilon_{is}) = \rho^{s-r} \operatorname{Var} \varepsilon_{ir}$, for $r < s$. This yields $\mathbf{R} = \sigma_0^2 \mathbf{R}_{AR}(\rho) + \sigma_\eta^2 \mathbf{R}_{\mathrm{RW}}(\rho)$, where

$$\mathbf{R}_{AR}(\rho) = \begin{pmatrix} 1 & \rho & \rho^2 & \ldots & \rho^{T-1} \\ \rho & 1 & \rho & \ldots & \rho^{T-2} \\ \rho^2 & \rho & 1 & \ldots & \rho^{T-3} \\ \vdots & \vdots & \vdots & \ddots & \vdots \\ \rho^{T-1} & \rho^{T-2} & \rho^{T-3} & \ldots & 1 \end{pmatrix}$$

and

$$\mathbf{R}_{\mathrm{RW}}(\rho) = \begin{pmatrix} S_1(\rho) & \rho S_1(\rho) & \rho^2 S_1(\rho) & \ldots & \rho^{T-1} S_1(\rho) \\ \rho S_1(\rho) & S_2(\rho) & \rho S_2(\rho) & \ldots & \rho^{T-2} S_2(\rho) \\ \rho^2 S_1(\rho) & \rho S_2(\rho) & S_3(\rho) & \ldots & \rho^{T-3} S_3(\rho) \\ \vdots & \vdots & \vdots & \ddots & \vdots \\ \rho^{T-1} S_1(\rho) & \rho^{T-2} S_2(\rho) & \rho^{T-3} S_3(\rho) & \ldots & S_T(\rho) \end{pmatrix}.$$

A simpler expression assumes that ε_{i0} is a constant, either known or a parameter to be estimated. (Section 8.5 will discuss an estimation method using the Kalman filter.) In this case, we have $\sigma_0^2 = 0$ and $\mathbf{R}^{-1} = (1/\sigma_\eta^2)\mathbf{R}_{\mathrm{RW}}^{-1}(\rho)$. It is easy to see that the Choleksy square root of $\mathbf{R}_{\mathrm{RW}}^{-1}(\rho)$ is

$$\mathbf{R}_{\mathrm{RW}}^{-1/2}(\rho) = \begin{pmatrix} 1 & 0 & 0 & \dots & 0 & 0 \\ -\rho & 1 & 0 & \dots & 0 & 0 \\ 0 & -\rho & 1 & \dots & 0 & 0 \\ \vdots & \vdots & \vdots & \ddots & \vdots & \vdots \\ 0 & 0 & 0 & \dots & 1 & 0 \\ 0 & 0 & 0 & \dots & -\rho & 1 \end{pmatrix}.$$

This suggests using the transformation

$$\begin{pmatrix} y_{i1}^* \\ y_{i2}^* \\ \vdots \\ y_{i,T_i}^* \end{pmatrix} = \mathbf{R}_{\mathrm{RW}}^{-1/2}(\rho) \begin{pmatrix} y_{i1} \\ y_{i2} \\ \vdots \\ y_{i,T_i} \end{pmatrix} = \begin{pmatrix} y_{i1} \\ y_{i2} - \rho y_{i1} \\ \vdots \\ y_{i,T_i} - \rho y_{i,T_i-1} \end{pmatrix},$$

which is the same as the Prais–Winston transform except for the first row. The Prais–Winston transform is the usual one for a stationary specification. The point of this example is that we do not require $|\rho| < 1$ and thus do not require stationarity.

There is a developing literature on unit-root tests to handle nonstationarity in panel data models. We refer the reader to Baltagi (2001E) and Hsiao (2002E) for introductions.

8.2.3 Continuous-Time Correlation Models

When data are not equally spaced in time, a natural formulation is to still consider subjects drawn from a population, yet with responses as realizations of a continuous-time stochastic process. The continuous-time stochastic-process setting is natural in the context of biomedical applications where, for example, we can envision patients arriving at a clinic for testing at irregularly spaced intervals. Specifically, for each subject i, we denote the set of responses as $\{y_i(t)$, for $t \in R\}$. Here, t denotes that time of the observation that we allow to extend over the entire real line. In this context, it is convenient to use the subscript "j" for the order of observations within a subject while still using "i" for the subject. Observations of the ith subject are taken at time t_{ij} so that $y_{ij} = y_i(t_{ij})$ denotes the jth response of the ith subject. Similarly, let $\boldsymbol{x}_{ij}$ and z_{ij} denote sets of explanatory variables at time t_{ij}. We may then model the disturbance terms, $\varepsilon_{ij} = y_{ij} - (\mathbf{z}_{ij}'\boldsymbol{\alpha}_i + \mathbf{x}_{ij}'\boldsymbol{\beta})$, as realizations from a continuous-time stochastic process $\varepsilon_i(t)$. In many applications, this is assumed to be a mean-zero Gaussian process, which permits a wide choice of correlation structures.

Particularly for unequally spaced data, a parametric formulation for the correlation structure is useful. In this setting, we have $\mathbf{R}_{rs} = \mathrm{Cov}(\varepsilon_{ir}, \varepsilon_{is}) = \sigma^2 \rho(|t_{ir} - t_{is}|)$, where ρ is the correlation function of $\{\varepsilon_i(t)\}$. Two widely used choices include the *exponential correlation model*

$$\rho(u) = \exp(-\phi u), \quad \text{for} \quad \phi > 0, \tag{8.5}$$

and the *Gaussian correlation model*

$$\rho(u) = \exp(-\phi u^2), \quad \text{for} \quad \phi > 0. \tag{8.6}$$

Diggle et al. (2002S) provide additional details regarding the continuous-time model.

Another advantage of continuous-time stochastic-process models is that they easily permit indexing by orderings other than time. By far, the most interesting ordering other than time is a *spatial* ordering. Spatial orderings are of interest when we wish to model phenomena where responses that are close to one another geographically tend to be related to one another.

For some applications, it is straightforward to incorporate spatial correlations into our models. This is done by allowing d_{ij} to be some measure of spatial or geographical location of the jth observation of the ith subject. Then, using a measure such as Euclidean distance, we interpret $|d_{ij} - d_{ik}|$ to be the distance between the jth and kth observations of the ith subject. One could use the correlation structure in either Equation (8.5) or (8.6).

Another straightforward approach that handles other applications is to reverse the role of i and j, allowing i to represent the time period (or replication) and j to represent the subject. To illustrate, suppose that we consider observing purchases of insurance in each of the fifty states in the United States over ten years. Suppose that most of the heterogeneity is due to the period effects; that is, changes in insurance purchases are influenced by changes in the country-wide economy. Because insurance is regulated at the state level, we expect each state to have different experiences owing to local regulation. Further, we may be concerned that states close to one another share similar economic environments and thus will be more related to one another than states that are geographically distant.

With this reversal of notation, the vector $\mathbf{y}_i$ represents all the subjects in the ith time period and the term $\boldsymbol{\alpha}_i$ represents temporal heterogeneity. However, in the basic linear mixed-effects model, this approach essentially ignores cross-sectional heterogeneity and treats the model as successive independent cross sections. More details on this approach are in Section 8.2.

To see how to allow for cross-sectional heterogeneity, temporal dynamics, and spatial correlations simultaneously, consider a basic two-way model

$$y_{it} = \alpha_i + \lambda_t + \mathbf{x}_{it}'\boldsymbol{\beta} + \varepsilon_{it},$$

where, for simplicity, we assume balanced data. Stacking over i, we have

$$\begin{pmatrix} y_{1t} \\ y_{2t} \\ \vdots \\ y_{nt} \end{pmatrix} = \begin{pmatrix} \alpha_1 \\ \alpha_2 \\ \vdots \\ \alpha_n \end{pmatrix} + \mathbf{1}_n \lambda_t + \begin{pmatrix} \mathbf{x}'_{1t} \\ \mathbf{x}'_{2t} \\ \vdots \\ \mathbf{x}'_{nt} \end{pmatrix} \boldsymbol{\beta} + \begin{pmatrix} \varepsilon_{1t} \\ \varepsilon_{2t} \\ \vdots \\ \varepsilon_{nt} \end{pmatrix},$$

where $\mathbf{1}_n$ is an $n \times 1$ vector of ones. We rewrite this as

$$\mathbf{y}_t = \boldsymbol{\alpha} + \mathbf{1}_n \lambda_t + \mathbf{X}_t \boldsymbol{\beta} + \boldsymbol{\epsilon}_t. \tag{8.7}$$

Define $\mathbf{H} = \operatorname{Var} \boldsymbol{\epsilon}_t$ to be the spatial variance matrix, which we assume does not vary over time. Specifically, the ijth element of $\mathbf{H}$ is $\mathbf{H}_{ij} = \operatorname{Cov}(\varepsilon_{it}, \varepsilon_{jt}) = \sigma^2 \rho(|d_i - d_j|)$, where d_i is a measure of geographic location. Assuming that $\{\lambda_t\}$ is i.i.d. with variance σ_λ^2, we have

$$\operatorname{Var} \mathbf{y}_t = \operatorname{Var} \boldsymbol{\alpha} + \sigma_\lambda^2 \mathbf{1}_n \mathbf{1}'_n + \operatorname{Var} \boldsymbol{\epsilon}_t = \sigma_\alpha^2 \mathbf{I}_n + \sigma_\lambda^2 \mathbf{J}_n + \mathbf{H} = \sigma_\alpha^2 \mathbf{I}_n + \mathbf{V_H}.$$

Stacking over t, we may express Equation (8.7) as a special case of the mixed linear model, with $\mathbf{y} = (\mathbf{y}'_1, \ldots, \mathbf{y}'_T)'$. Because $\operatorname{Cov}(\mathbf{y}_r, \mathbf{y}_s) = \sigma_\alpha^2 \mathbf{I}_n$ for $r \neq s$, we have $\mathbf{V} = \operatorname{Var} \mathbf{y} = \sigma_\alpha^2 \mathbf{I}_n \otimes \mathbf{J}_T + \mathbf{V_H} \otimes \mathbf{I}_T$. It is easy to verify that

$$\mathbf{V}^{-1} = \left(\left(\sigma_\alpha^2 \mathbf{I}_n + T^{-1} \mathbf{V_H} \right)^{-1} - T \mathbf{V_H}^{-1} \right) \otimes \mathbf{J}_T + \mathbf{V_H}^{-1} \otimes \mathbf{I}_T.$$

Thus, it is straightforward to compute the regression coefficient estimator and the likelihood, as in Equation (3.20). For example, with $\mathbf{X} = (\mathbf{X}'_1, \ldots, \mathbf{X}'_T)'$, the GLS estimator of $\boldsymbol{\beta}$ is

$$\begin{aligned} \mathbf{b}_{\text{GLS}} &= (\mathbf{X}'\mathbf{V}^{-1}\mathbf{X})^{-1} \mathbf{X}'\mathbf{V}^{-1}\mathbf{y} \\ &= \left(\sum_{r=1}^{T} \sum_{s=1}^{T} \mathbf{X}'_r \left(\left(\sigma_\alpha^2 \mathbf{I}_n + T^{-1} \mathbf{V_H} \right)^{-1} - T \mathbf{V_H}^{-1} \right) \mathbf{X}_s + \sum_{t=1}^{T} \mathbf{X}'_t \mathbf{V_H}^{-1} \mathbf{X}_t \right)^{-1} \\ &\quad \times \left(\sum_{r=1}^{T} \sum_{s=1}^{T} \mathbf{X}'_r \left(\left(\sigma_\alpha^2 \mathbf{I}_n + T^{-1} \mathbf{V_H} \right)^{-1} - T \mathbf{V_H}^{-1} \right) \mathbf{y}_s + \sum_{t=1}^{T} \mathbf{X}'_t \mathbf{V_H}^{-1} \mathbf{y}_t \right). \end{aligned}$$

Returning to the simpler case of no subject heterogeneity, suppose that $\sigma_\alpha^2 = 0$. In this case, we have

$$\mathbf{b}_{\text{GLS}} = \left(\sum_{t=1}^{T} \mathbf{X}'_t \mathbf{V_H}^{-1} \mathbf{X}_t \right)^{-1} \left(\sum_{t=1}^{T} \mathbf{X}'_t \mathbf{V_H}^{-1} \mathbf{y}_t \right).$$

8.3 Cross-Sectional Correlations and Time-Series Cross-Section Models

Cross-sectional correlations are particularly important in studies of governmental units, such as states or nations. In some fields, such as political science, when T is large relative to n the data are referred to as *time-series cross-section* data. This nomenclature distinguishes this setup from the panel context, where n is large relative to T. For example, according to Beck and Katz (1995O), time-series cross-section data would typically range from 10 to 100 subjects with each subject observed over a long period, perhaps 20 to 50 years; many cross-national studies have ratios of n to T that are close to 1. Such studies involve economic, social, or political comparisons of countries or states; because of the linkages among governmental units, the interest is in models that permit substantial contemporaneous correlation.

Following the political science literature, we consider a time-series cross-section (TSCS) model of the form

$$\mathbf{y}_i = \mathbf{X}_i \boldsymbol{\beta} + \boldsymbol{\epsilon}_i, \quad i = 1, \ldots, n, \tag{8.8}$$

which summarizes T_i responses over time. Unlike prior chapters, we allow for correlation across different subjects through the notation $\mathrm{Cov}(\boldsymbol{\epsilon}_i, \boldsymbol{\epsilon}_j) = \mathbf{V}_{ij}$. Because n is not large relative to T, fixed-effects heterogeneity terms could easily be incorporated into the regression coefficients $\boldsymbol{\beta}$ by using binary indicator (dummy) variables. Incorporation of random-effects heterogeneity terms would involve an extension of the current discussion; we follow the literature and ignore this aspect, for now. Stimson (1985O) surveys a range of models of interest in political science.

To complete the specification of the TSCS model, we need to make an assumption about the form of $\mathbf{V}_{ij}$. There are four basic specifications of cross-sectional covariances:

1. $$\mathbf{V}_{ij} = \begin{cases} \sigma^2 \mathbf{I}_i & i = j, \\ \mathbf{0} & i \neq j. \end{cases}$$

   This is the traditional model setup in which ordinary least squares is efficient.

2. $$\mathbf{V}_{ij} = \begin{cases} \sigma_i^2 \mathbf{I}_i & i = j, \\ \mathbf{0} & i \neq j. \end{cases}$$

   This specification permits heterogeneity across subjects.

3.
$$\mathrm{Cov}(\varepsilon_{it}, \varepsilon_{js}) = \begin{cases} \sigma_{ij} & t = s, \\ 0 & t \neq s. \end{cases}$$

This specification permits cross-sectional correlations across subjects. However, observations from different time points are uncorrelated.

4. $\mathrm{Cov}(\varepsilon_{it}, \varepsilon_{js}) = \sigma_{ij}$ for $t = s$ and $\varepsilon_{i,t} = \rho_i \varepsilon_{i,t-1} + \eta_{it}$. This specification permits contemporaneous cross-correlations as well as intrasubject serial correlation through an $AR(1)$ model. Moreover, with some mild additional assumptions, the model has an easy to interpret cross-lag correlation function of the form $\mathrm{Cov}(\varepsilon_{it}, \varepsilon_{js}) = \sigma_{ij}\rho_j^{t-s}$ for $s < t$.

The TSCS model is estimated using feasible GLS procedures. At the first stage, OLS residuals are used to estimate the variance parameters. One can think of the model as n separate regression equations and use seemingly unrelated regression techniques, described in Section 6.4.2, to compute estimators. It was in the context of seemingly unrelated regressions that Parks (1967S) proposed the contemporaneous cross-correlation with intrasubject serial $AR(1)$ correlation model.

Generalized least-squares estimation in a regression context has drawbacks that are well documented; see, for example, Carroll and Ruppert (1988G). That is, GLS estimators are more efficient than OLS estimators when the variance parameters are known. However, because variance parameters are rarely known, one must use instead feasible GLS estimators. Asymptotically, feasible GLS are just as efficient as GLS estimators. However, in finite samples, feasible GLS estimators may be more or less efficient than OLS estimators, depending on the regression design and distribution of disturbances. For the TSCS model, which allows for cross-sectional covariances, there are $n(n+1)/2$ variance parameters. Moreover, for the Parks model, there are additional an n serial correlation parameters. As documented by Beck and Katz (1995O) in the TSCS context, having this many variance parameters means that feasible GLS estimators are inefficient in regression designs that are typically of interest in political science applications.

Thus, Beck and Katz (1995O) recommend using OLS estimators of regression coefficients. To account for the cross-sectional correlations, they recommend using standard errors that are robust to the presence of cross-sectional correlations that they call *panel-corrected* standard errors. In our notation, this is equivalent to the robust standard errors introduced in Section 2.5.3 without the subject-specific fixed effects yet reversing the roles of i and t. That is, for the asymptotic theory, we now require independence over time but allow for (cross-sectional) correlation across subjects.

Specifically, for balanced data, one computes panel-corrected standard errors as follows:

Procedure for Computing Panel-Corrected Standard Errors

1. Calculate OLS estimators of $\boldsymbol{\beta}$, $\mathbf{b}_{\text{OLS}}$, and the corresponding residuals, $e_{it} = y_{it} - \mathbf{x}'_{it}\mathbf{b}_{\text{OLS}}$.
2. Define the estimator of the (ij)th cross-sectional covariance to be $\hat{\sigma}_{ij} = T^{-1}\sum_{t=1}^{T} e_{it}e_{jt}$.
3. Estimate the variance of $\mathbf{b}_{\text{OLS}}$ using $(\sum_{i=1}^{n}\mathbf{X}'_i\mathbf{X}_i)^{-1}(\sum_{i=1}^{n}\sum_{j=1}^{n}\hat{\sigma}_{ij}\mathbf{X}'_i\mathbf{X}_j)(\sum_{i=1}^{n}\mathbf{X}'_i\mathbf{X}_i)^{-1}$.

For unbalanced data, steps 2 and 3 need to be modified to align data from the same time periods.

Beck and Katz (1995O) provide simulation studies that establish that the robust t-statistics resulting from the use of panel-corrected standard errors are preferable to the ordinary t-statistics, using either OLS or feasible GLS. They also argue that this procedure can be used with serial $AR(1)$ correlation, by first applying a (Prais–Winston) transformation to the data to induce independence over time. Using simulation, they demonstrate that this procedure is superior to the feasible GLS estimator using the Parks model. For general applications, we caution the reader that, by reversing the roles of i and t, one now relies heavily on the independence over time (instead of subjects). The presence of even mild serial correlation means that the usual same asymptotic approximations are no longer valid. Thus, although panel-corrected standard errors are indeed robust to the presence of cross-sectional correlations, to use these procedures one must be especially careful about modeling the dynamic aspects of the data.

8.4 Time-Varying Coefficients

8.4.1 The Model

Beginning with the basic two-way model in Equation (8.1), more generally, we use subject-varying terms

$$z_{\alpha,i,t,1}\alpha_{i,1} + \cdots + z_{\alpha,i,t,q}\alpha_{i,q} = \mathbf{z}'_{\alpha,i,t}\boldsymbol{\alpha}_i$$

and time-varying terms

$$z_{\lambda,i,t,1}\lambda_{t,1} + \cdots + z_{\lambda,i,t,r}\lambda_{t,r} = \mathbf{z}'_{\lambda,i,t}\boldsymbol{\lambda}_t.$$

With these terms, we define the longitudinal data mixed model with time-varying coefficients as

$$y_{it} = \mathbf{z}'_{\alpha,i,t}\boldsymbol{\alpha}_i + \mathbf{z}'_{\lambda,i,t}\boldsymbol{\lambda}_t + \mathbf{x}'_{it}\boldsymbol{\beta} + \varepsilon_{it}, \quad t = 1, \ldots, T_i, \ i = 1, \ldots, n. \tag{8.9}$$

Here, $\boldsymbol{\alpha}_i = (\alpha_{i,1}, \ldots, \alpha_{i,q})'$ is a $q \times 1$ vector of subject-specific terms and $\mathbf{z}_{\alpha,i,t} = (z_{\alpha,i,t,1}, \ldots, z_{\alpha,i,t,q})'$ is the corresponding vector of covariates. Similarly, $\boldsymbol{\lambda}_t = (\lambda_{t,1}, \ldots, \lambda_{t,r})'$ is an $r \times 1$ vector of time-specific terms and $\mathbf{z}_{\lambda,i,t} = (z_{\lambda,i,t,1}, \ldots, z_{\lambda,i,t,r})'$ is the corresponding vector of covariates. We use the notation $t = 1, \ldots, T_i$ to indicate the unbalanced nature of the data.

A more compact form of Equation (8.9) can be given by stacking over t. This yields a matrix form of the longitudinal data mixed model

$$\mathbf{y}_i = \mathbf{Z}_{\alpha,i}\boldsymbol{\alpha}_i + \mathbf{Z}_{\lambda,i}\boldsymbol{\lambda} + \mathbf{X}_i\boldsymbol{\beta} + \boldsymbol{\epsilon}_i, \quad i = 1, \ldots, n. \tag{8.10}$$

This expression uses matrices of covariates $\mathbf{X}_i = (\mathbf{x}_{i1} \ \ \mathbf{x}_{i2} \ \ \cdots \ \ \mathbf{x}_{iT_i})'$, of dimension $T_i \times K$, $\mathbf{Z}_{\alpha,i} = (\mathbf{z}_{\alpha,i,1} \ \ \mathbf{z}_{\alpha,i,2} \ \ \cdots \ \ \mathbf{z}_{\alpha,i,T_i})'$, of dimension $T_i \times q$, and

$$\mathbf{Z}_{\lambda,i} = \begin{pmatrix} \mathbf{z}'_{\lambda,i,1} & \mathbf{0} & \ldots & \mathbf{0} \\ \mathbf{0} & \mathbf{z}'_{\lambda,i,2} & \ldots & \mathbf{0} \\ \vdots & \vdots & \ddots & \vdots \\ \mathbf{0} & \mathbf{0} & \ldots & \mathbf{z}'_{\lambda,i,T_i} \end{pmatrix} : \mathbf{0}_i,$$

of dimension $T_i \times rT$, where $\mathbf{0}_i$ is a $T_i \times r(T - T_i)$ zero matrix. Finally, $\boldsymbol{\lambda} = (\boldsymbol{\lambda}'_1, \ldots, \boldsymbol{\lambda}'_T)'$ is the $rT \times 1$ vector of time-specific coefficients.

We assume that sources of variability, $\{\boldsymbol{\epsilon}_i\}$, $\{\boldsymbol{\alpha}_i\}$, and $\{\boldsymbol{\lambda}_t\}$, are mutually independent and mean zero. The nonzero means are accounted for in the $\boldsymbol{\beta}$ parameters. The disturbances are independent between subjects, yet we allow for serial correlation and heteroscedasticity through the notation $\mathrm{Var}\,\boldsymbol{\epsilon}_i = \sigma^2\mathbf{R}_i$. Further, we assume that the subject-specific effects $\{\boldsymbol{\alpha}_i\}$ are random with variance–covariance matrix $\sigma^2\mathbf{D}$, a $q \times q$ positive definite matrix. Time-specific effects $\boldsymbol{\lambda}$ have variance–covariance matrix $\sigma^2\boldsymbol{\Sigma}_\lambda$, an $rT \times rT$ positive definite matrix. For each variance component, we separate out the scale parameter σ^2 to simplify the estimation calculations described in Appendix 8A.2. With this notation, we may express the variance of each subject as $\mathrm{Var}\,\mathbf{y}_i = \sigma^2(\mathbf{V}_{\alpha,i} + \mathbf{Z}_{\lambda,i}\boldsymbol{\Sigma}_\lambda\mathbf{Z}'_{\lambda,i})$, where $\mathbf{V}_{\alpha,i} = \mathbf{Z}_{\alpha,i}\mathbf{D}\mathbf{Z}'_{\alpha,i} + \mathbf{R}_i$.

To see how the model in Equation (8.10) is a special case of the mixed linear model, take $\mathbf{y} = (\mathbf{y}'_1, \mathbf{y}'_2, \ldots, \mathbf{y}'_n)'$, $\boldsymbol{\epsilon} = (\boldsymbol{\epsilon}'_1, \boldsymbol{\epsilon}'_2, \ldots, \boldsymbol{\epsilon}'_n)'$,

$\alpha = (\alpha'_1, \alpha'_2, \ldots, \alpha'_n)'$,

$$\mathbf{X} = \begin{pmatrix} \mathbf{X}_1 \\ \mathbf{X}_2 \\ \mathbf{X}_3 \\ \vdots \\ \mathbf{X}_n \end{pmatrix}, \qquad \mathbf{Z}_\lambda = \begin{pmatrix} \mathbf{Z}_{\lambda,1} \\ \mathbf{Z}_{\lambda,2} \\ \mathbf{Z}_{\lambda,3} \\ \vdots \\ \mathbf{Z}_{\lambda,n} \end{pmatrix},$$

and

$$\mathbf{Z}_\alpha = \begin{pmatrix} \mathbf{Z}_{\alpha,1} & \mathbf{0} & \mathbf{0} & \ldots & \mathbf{0} \\ \mathbf{0} & \mathbf{Z}_{\alpha,2} & \mathbf{0} & \ldots & \mathbf{0} \\ \mathbf{0} & \mathbf{0} & \mathbf{Z}_{\alpha,3} & \ldots & \mathbf{0} \\ \vdots & \vdots & \vdots & \ddots & \vdots \\ \mathbf{0} & \mathbf{0} & \mathbf{0} & \ldots & \mathbf{Z}_{\alpha,n} \end{pmatrix}.$$

With these choices, we can express the model in Equation (8.10) as a mixed linear model, given by

$$\mathbf{y} = \mathbf{Z}_\alpha \boldsymbol{\alpha} + \mathbf{Z}_\lambda \boldsymbol{\lambda} + \mathbf{X}\boldsymbol{\beta} + \boldsymbol{\epsilon}.$$

8.4.2 Estimation

By writing Equation (8.10) as a mixed linear model, we may appeal to the many estimation results for this latter class of models. For example, for known variance parameters, direct calculations show that the GLS estimator of $\boldsymbol{\beta}$ is

$$\mathbf{b}_{\text{GLS}} = (\mathbf{X}'\mathbf{V}^{-1}\mathbf{X})^{-1}\mathbf{X}'\mathbf{V}^{-1}\mathbf{y},$$

where $\text{Var}\, \mathbf{y} = \sigma^2 \mathbf{V}$. Further, there is a variety of ways of calculating estimators of variance components. These include maximum likelihood, restricted maximum likelihood, and several unbiased estimation techniques. For smaller data sets, one may use mixed linear model software directly. For larger data sets, direct appeal to such software may be computationally burdensome. This is because the time-varying random variables $\boldsymbol{\lambda}_t$ are common to all subjects, obliterating the independence among subjects. However, computational shortcuts are available and are described in detail in Appendix 8A.2.

Example 8.1: Forecasting Wisconsin Lottery Sales (continued) Table 8.1 reports the estimation results from fitting the two-way error-components model in Equation (8.1), with and without an $AR(1)$ term. For comparison purposes, the fitted coefficients for the one-way model with an $AR(1)$ term are also presented in

Table 8.1. *Lottery model coefficient estimates based on in-sample data of $n = 50$ ZIP codes and $T = 35$ weeks*[a]

Variable	One-way error-components model with $AR(1)$ term		Two-way error-components model		Two-way error-components model with $AR(1)$ term	
	Parameter estimate	t-statistic	Parameter estimate	t-statistic	Parameter estimate	t-statistic
Intercept	13.821	2.18	16.477	2.39	15.897	2.31
PERPERHH	−1.085	−1.36	−1.210	−1.43	−1.180	−1.40
MEDSCHYR	−0.821	−2.53	−0.981	−2.79	−0.948	−2.70
MEDHVL	0.014	0.81	0.001	0.71	0.001	0.75
PRCRENT	0.032	1.53	0.028	1.44	0.029	1.49
PRC55P	−0.070	−1.01	−0.071	−1.00	−0.072	−1.02
HHMEDAGE	0.118	1.09	0.118	1.06	0.120	1.08
MEDINC	0.043	1.58	0.004	1.59	0.004	1.59
POPULATN	0.057	2.73	0.001	5.45	0.001	4.26
NRETAIL	0.021	0.20	−0.009	−1.07	−0.003	−0.26
Var $\alpha(\sigma_\alpha^2)$	0.528	—	0.564	—	0.554	—
Var $\varepsilon(\sigma^2)$	0.279	—	0.022	—	0.024	—
Var $\lambda(\sigma_\lambda^2)$	—	—	0.241	—	0.241	—
$AR(1)$ corr(ρ)	0.555	25.88	—	—	0.518	25.54
AIC	2,270.97		−1,109.61		−1,574.02	

[a] The response is (natural) logarithmic sales.

this table. As in Table 4.3, we see that the goodness of fit statistic, *AIC*, indicates that the more complex two-way models provide an improved fit compared to the one-way models. As with the one-way models, the autocorrelation coefficient is statistically significant even with the time-varying parameter λ_t. In each of the three models in Table 8.1, only the population size (POP) and education levels (MEDSCHYR) have a significant effect on lottery sales.

8.4.3 Forecasting

For forecasting, we wish to predict

$$y_{i,T_i+L} = \mathbf{z}'_{\alpha,i,T_i+L}\boldsymbol{\alpha}_i + \mathbf{z}'_{\lambda,i,T_i+L}\boldsymbol{\lambda}_{T_i+L} + \mathbf{x}'_{i,T_i+L}\boldsymbol{\beta} + \epsilon_{i,T_i+L}, \qquad (8.11)$$

for L lead time units in the future. We use Chapter 4 results for best linear unbiased prediction. To calculate these predictors, we use the sum of squares $\mathbf{S}_{ZZ} = \sum_{i=1}^{n} \mathbf{Z}'_{\lambda,i}\mathbf{V}_{\alpha,i}^{-1}\mathbf{Z}_{\lambda,i}$. The details of the derivation of BLUPs are in Appendix 8A.3.

As an intermediate step, it is useful to provide the BLUPs for $\boldsymbol{\lambda}$, $\boldsymbol{\epsilon}_i$, and $\boldsymbol{\alpha}_i$. The BLUP of $\boldsymbol{\lambda}$ is

$$\boldsymbol{\lambda}_{\text{BLUP}} = \left(\mathbf{S}_{ZZ} + \boldsymbol{\Sigma}_{\lambda}^{-1}\right)^{-1} \sum_{i=1}^{n} \mathbf{Z}'_{\lambda,i}\mathbf{V}_{\alpha,i}^{-1}\mathbf{e}_{i,\text{GLS}}, \qquad (8.12)$$

where we use the vector of residuals $\mathbf{e}_{i,\text{GLS}} = \mathbf{y}_i - \mathbf{X}_i\mathbf{b}_{\text{GLS}}$. The BLUP of $\boldsymbol{\epsilon}_t$ is

$$\mathbf{e}_{i,\text{BLUP}} = \mathbf{R}_i\mathbf{V}_{\alpha,i}^{-1}(\mathbf{e}_{i,\text{GLS}} - \mathbf{Z}_{\lambda,i}\boldsymbol{\lambda}_{\text{BLUP}}), \qquad (8.13)$$

and the BLUP of $\boldsymbol{\alpha}_i$ is

$$\mathbf{a}_{i,\text{BLUP}} = \mathbf{D}\mathbf{Z}'_{\alpha,i}\mathbf{V}_{\alpha,i}^{-1}(\mathbf{e}_{i,\text{GLS}} - \mathbf{Z}_{\lambda,i}\boldsymbol{\lambda}_{\text{BLUP}}). \qquad (8.14)$$

We remark that the BLUP of $\boldsymbol{\epsilon}_t$ can also be expressed as

$$\mathbf{e}_{i,\text{BLUP}} = \mathbf{y}_i - (\mathbf{Z}_{\alpha,i}\mathbf{a}_{i,\text{BLUP}} + \mathbf{Z}_{\lambda,i}\boldsymbol{\lambda}_{\text{BLUP}} + \mathbf{X}_i\mathbf{b}_{\text{GLS}}).$$

With these quantities, the BLUP forecast of y_{i,T_i+L} is

$$\begin{aligned}\hat{y}_{i,T_i+L} = {} & \mathbf{x}'_{i,T_i+L}\mathbf{b}_{\text{GLS}} + \mathbf{z}'_{\alpha,i,T_i+L}\boldsymbol{\alpha}_{i,\text{BLUP}} \\ & + \mathbf{z}'_{\lambda,i,T_i+L}\text{Cov}(\boldsymbol{\lambda}_{T_i+L}, \boldsymbol{\lambda})\left(\sigma^2\boldsymbol{\Sigma}_{\lambda}\right)^{-1}\boldsymbol{\lambda}_{\text{BLUP}} \\ & + \text{Cov}(\varepsilon_{i,T_i+L}, \boldsymbol{\epsilon}_i)\left(\sigma^2\mathbf{R}_i\right)^{-1}\mathbf{e}_{i,\text{BLUP}}. \end{aligned} \qquad (8.15)$$

An expression for the variance of the forecast error, $\text{Var}(\hat{y}_{i,T_i+L} - y_{i,T_i+L})$, is given in Appendix 8A.3 as Equation (8A.20).

Equations (8.12)–(8.15) provide sufficient structure to calculate forecasts for a wide variety of models. Additional computational details appear in Appendix

8A.3. Still, it is instructive to interpret the BLUP forecast in a number of special cases. We first consider the case of independently distributed time-specific components $\{\boldsymbol{\lambda}_t\}$.

Example 8.2: Independent Time-Specific Components We consider the special case where $\{\boldsymbol{\lambda}_t\}$ are independent and assume that $T_i + L > T$, so that $\mathrm{Cov}(\boldsymbol{\lambda}_{T_i+L}, \boldsymbol{\lambda}) = \mathbf{0}$. Thus, from Equation (8.15), we have the BLUP forecast of y_{i,T_i+L}:

$$\hat{y}_{i,T_i+L} = \mathbf{x}'_{i,T_i+L}\mathbf{b}_{\mathrm{GLS}} + \mathbf{z}'_{\alpha,i,\,T_i+L}\mathbf{a}_{i,\mathrm{BLUP}} + \mathrm{Cov}(\varepsilon_{i,T_i+L}, \boldsymbol{\epsilon}_i)\left(\sigma^2\mathbf{R}_i\right)^{-1}\mathbf{e}_{i,\mathrm{BLUP}}. \tag{8.16}$$

This is similar in appearance to the forecast formula in Equation (4.14). However, note that even when $\{\boldsymbol{\lambda}_t\}$ are independent, the time-specific components appear in $\mathbf{b}_{\mathrm{GLS}}$, $\mathbf{e}_{i,\mathrm{BLUP}}$, and $\mathbf{a}_{i,\mathrm{BLUP}}$. Thus, the presence of $\{\boldsymbol{\lambda}_t\}$ influences the forecasts.

Example 8.3: Time-Varying Coefficients Suppose that the model is

$$y_{it} = \mathbf{x}'_{it}\boldsymbol{\beta}_t + \varepsilon_{it},$$

where $\{\boldsymbol{\beta}_t\}$ are i.i.d. We can rewrite this as

$$y_{it} = \mathbf{z}'_{\lambda,i,t}\boldsymbol{\lambda}_t + \mathbf{x}'_{it}\boldsymbol{\beta} + \varepsilon_{it},$$

where $\mathrm{E}\,\boldsymbol{\beta}_t = \boldsymbol{\beta}$, $\boldsymbol{\lambda}_t = \boldsymbol{\beta}_t - \boldsymbol{\beta}$, and $\mathbf{z}_{\lambda,i,t} = \mathbf{x}_{i,t}$. With this notation and Equation (8.16), the forecast of y_{i,T_i+L} is

$$\hat{y}_{i,T_i+L} = \mathbf{x}'_{i,T_i+L}\mathbf{b}_{\mathrm{GLS}}.$$

Example 8.4: Two-Way Error Components Model Consider the basic two-way model given in Equation (8.1), with $\{\lambda_t\}$ being i.i.d. Here, we have that $q = r = 1$ and $\mathbf{D} = \sigma_\alpha^2/\sigma^2$, $\mathbf{z}_{\alpha,i,T_i+L} = 1$, and $\mathbf{Z}_{\alpha,i} = \mathbf{1}_i$. Further, $\mathbf{Z}_{\lambda,i} = (\mathbf{I}_i : \mathbf{0}_{1i})$, where $\mathbf{0}_{1i}$ is a $T_i \times (T - T_i)$ matrix of zeros, and $\boldsymbol{\Sigma}_\lambda = (\sigma_\lambda^2/\sigma^2)\mathbf{I}_T$. Thus, from Equation (8.16), we have that the BLUP forecast of y_{i,T_i+L} is

$$\hat{y}_{i,T_i+L} = a_{i,\mathrm{BLUP}} + \mathbf{x}'_{i,T_i+L}\mathbf{b}_{\mathrm{GLS}}.$$

Here, from Equation (8.12), we have

$$\boldsymbol{\lambda}_{\mathrm{BLUP}} = \left(\sum_{i=1}^{n}\mathbf{Z}'_{\lambda,i}\mathbf{V}_{\alpha,i}^{-1}\mathbf{Z}_{\lambda,i} + \frac{\sigma^2}{\sigma_\lambda^2}\mathbf{I}_T\right)^{-1}\sum_{i=1}^{n}\mathbf{Z}'_{\lambda,i}\mathbf{V}_{\alpha,i}^{-1}\mathbf{e}_{i,\mathrm{GLS}},$$

where $\mathbf{Z}_{\lambda,i}$ as given before, $\mathbf{V}_{\alpha,i}^{-1} = \mathbf{I}_i - (\zeta_i/T_i)\mathbf{J}_i$, and $\zeta_i = T_i\sigma_\alpha^2/(\sigma^2 + T_i\sigma_\alpha^2)$. Further, Equation (8.16) yields.

$$a_{i,\text{BLUP}} = \zeta_i \left((\bar{y}_i - \bar{\mathbf{x}}_i'\mathbf{b}_{\text{GLS}}) - \frac{1}{T_i}\mathbf{Z}_{\lambda,i}\boldsymbol{\lambda}_{\text{BLUP}} \right).$$

For additional interpretation, we assume balanced data so that $T_i = T$; see Baltagi (2001E). To ease notation, recall that $\zeta = T\sigma_\alpha^2/(\sigma^2 + T\sigma_\alpha^2)$. Here, we have

$$\begin{aligned} \hat{y}_{i,T_i+L} &= \mathbf{x}_{i,T_i+L}'\,\mathbf{b}_{\text{GLS}} \\ &\quad + \zeta \left((\bar{y}_i - \bar{\mathbf{x}}_i'\mathbf{b}_{\text{GLS}}) - \frac{n(1-\zeta)\sigma_\lambda^2}{\sigma^2 + n(1-\zeta)\sigma_\lambda^2}(\bar{y} - \bar{\mathbf{x}}'\mathbf{b}_{\text{GLS}}) \right). \end{aligned}$$

Example 8.5: Random-Walk Model Through minor modifications, other temporal patterns of common, yet unobserved, components can be easily included. For this example, we assume that $r = 1$, $\{\lambda_t\}$ are i.i.d., so that the partial sum process $\{\lambda_1 + \lambda_2 + \cdots + \lambda_t\}$ is a random-walk process. Thus, the model is

$$y_{it} = \mathbf{z}_{\alpha,i,t}'\boldsymbol{\alpha}_i + \sum_{s=1}^{t}\lambda_s + \mathbf{x}_{it}'\boldsymbol{\beta} + \varepsilon_{it}, \quad t = 1, \ldots, T_i, \quad i = 1, \ldots, n. \tag{8.17}$$

Stacking over t, we can express this in matrix form as Equation (8.10), where the $T_i \times T$ matrix $\mathbf{Z}_{\lambda,i}$ is a lower triangular matrix of ones for the first T_i rows and zeros elsewhere. That is,

$$\mathbf{Z}_{\lambda,i} = \begin{pmatrix} 1 & 0 & 0 & \ldots & 0 & 0 & \cdots & 0 \\ 1 & 1 & 0 & \ldots & 0 & 0 & \cdots & 0 \\ 1 & 1 & 1 & \ldots & 0 & 0 & \cdots & 0 \\ \vdots & \vdots & \vdots & \ddots & 0 & 0 & \ldots & 0 \\ 1 & 1 & 1 & \ldots & 1 & 0 & \cdots & 0 \end{pmatrix}.$$

Thus, it can be shown that

$$\begin{aligned} \hat{y}_{i,T_i+L} &= \mathbf{x}_{i,T_i+L}'\mathbf{b}_{\text{GLS}} + \sum_{s=1}^{t}\lambda_{s,\text{BLUP}} + \mathbf{z}_{\alpha,i,T_i+L}'\boldsymbol{\alpha}_{i,\text{BLUP}} \\ &\quad + \text{Cov}(\varepsilon_{i,T_i+L}, \boldsymbol{\epsilon}_i)\left(\sigma^2\mathbf{R}_i\right)^{-1}\mathbf{e}_{i,\text{BLUP}}. \end{aligned}$$

8.5 Kalman Filter Approach

The Kalman (1960G) filter approach originated in time-series analysis. It is a technique for estimating parameters from complex, recursively specified systems. The essential idea is to use techniques from multivariate normal distributions to express the likelihood recursively, in an easily computable fashion. Then, parameters may be derived using maximum or restricted maximum likelihood. If this is your first exposure to Kalman filters, please skip ahead to the example in Section 8.6 and the introduction of the basic algorithm in Appendix D.

We now consider a class of models known as *state space models*. These models are well known in the time-series literature for their flexibility in describing broad categories of dynamic structures. As we will see, they can be readily fit using the Kalman fit algorithm. These models have been explored in the longitudinal data analysis literature extensively by Jones (1993S). We use recent modifications of this structure introduced by Tsimikas and Ledolter (1998S) for linear mixed-effects models. Specifically, we consider Equation (8.9), which, in the time-series literature, is called the *observation equation*. The time-specific quantities of Equation (8.9) are $\boldsymbol{\lambda}_t = (\lambda_{t1}, \ldots, \lambda_{tr})'$; this vector is our primary mechanism for specifying the dynamics. It is updated recursively through the *transition equation*,

$$\boldsymbol{\lambda}_t = \boldsymbol{\Phi}_{1t}\boldsymbol{\lambda}_{t-1} + \boldsymbol{\eta}_{1t}, \tag{8.18}$$

where $\{\boldsymbol{\eta}_{1t}\}$ are i.i.d., mean-zero, random vectors. With state space models, it is also possible to incorporate a dynamic error structure such as an *AR*(p) model. The *autoregressive of order* p(*AR*(p)) model for the disturbances $\{\varepsilon_{it}\}$ has the form

$$\varepsilon_{i,t} = \phi_1\varepsilon_{i,t-1} + \phi_2\varepsilon_{i,t-2} + \cdots + \phi_p\varepsilon_{i,t-p} + \zeta_{i,t}, \tag{8.19}$$

where $\{\zeta_{i,t}\}$ are initially assumed to be i.i.d., mean-zero, random variables. Harvey (1989S) illustrates the wide range of choices of dynamic error structures. Further, we shall see that state space models readily accommodate spatial correlations among the responses.

Both the linear mixed-effects models and the state space models are useful for forecasting. Because there is an underlying continuous stochastic process for the disturbances, both allow for unequally spaced (in time) observations. Furthermore, both accommodate missing data. Both classes of models can be represented as special cases of the linear mixed model. For state space models, the relationship to linear mixed models has been emphasized by Tsimikas and Ledolter (1994S, 1997S, 1998S).

Perhaps because of their longer history, the linear mixed-effects models are often easier to implement. These models are certainly adequate for data sets with shorter time dimensions. However, for longer time dimensions, the additional flexibility provided by the newer state space models leads to improved model fitting and forecasting. We first express the longitudinal data model in Equations (8.9) and (8.10) as a special case of a more general state space model. To this end, this section considers the transition equations, the set of observations available, and the measurement equation. It then describes how to calculate the likelihood associated with this general state space model.

The Kalman filter algorithm is a method for efficiently calculating the likelihood of complex time series using conditioning arguments. Appendix D introduces the general idea of the algorithm, as well as extensions to include fixed and random effects. This section presents only the computational aspects of the algorithm.

8.5.1 Transition Equations

We first collect the two sources of dynamic behavior, $\boldsymbol{\epsilon}$ and $\boldsymbol{\lambda}$, into a single transition equation. Equation (8.18) specifies the behavior of the latent, time-varying variable $\boldsymbol{\lambda}_t$. For the error $\boldsymbol{\epsilon}$, we assume that it is governed by a Markovian structure. Specifically, we use an *AR(p)* structure as specified in Equation (8.19).

To this end, define the $p \times 1$ vector $\boldsymbol{\xi}_{i,t} = (\varepsilon_{i,t}, \varepsilon_{i,t-1}, \ldots, \varepsilon_{i,t-p+1})'$ so that we may write

$$\boldsymbol{\xi}_{i,t} = \begin{pmatrix} \phi_1 & \phi_2 & \ldots & \phi_{p-1} & \phi_p \\ 1 & 0 & \ldots & 0 & 0 \\ 0 & 1 & \ldots & 0 & 0 \\ \vdots & \vdots & \ddots & \vdots & \vdots \\ 0 & 0 & \ldots & 1 & 0 \end{pmatrix} \boldsymbol{\xi}_{i,t-1} + \begin{pmatrix} \zeta_{i,t} \\ 0 \\ 0 \\ \vdots \\ 0 \end{pmatrix} = \boldsymbol{\Phi}_2 \boldsymbol{\xi}_{i,t-1} + \boldsymbol{\eta}_{2i,t}.$$

The first row is the *AR(p)* model in Equation (8.19). Stacking this over $i = 1, \ldots, n$ yields

$$\boldsymbol{\xi}_t = \begin{pmatrix} \boldsymbol{\xi}_{1,t} \\ \vdots \\ \boldsymbol{\xi}_{n,t} \end{pmatrix} = \begin{pmatrix} \boldsymbol{\Phi}_2 \boldsymbol{\xi}_{1,t-1} \\ \vdots \\ \boldsymbol{\Phi}_2 \boldsymbol{\xi}_{n,t-1} \end{pmatrix} + \begin{pmatrix} \boldsymbol{\eta}_{21,t} \\ \vdots \\ \boldsymbol{\eta}_{2n,t} \end{pmatrix} = (\mathbf{I}_n \otimes \boldsymbol{\Phi}_2)\, \boldsymbol{\xi}_{t-1} + \boldsymbol{\eta}_{2t},$$

where $\boldsymbol{\xi}_t$ is an $np \times 1$ vector, $\mathbf{I}_n$ is an $n \times n$ identity matrix, and $\otimes$ is a Kronecker (direct) product (see Appendix A.6). We assume that $\{\zeta_{i,t}\}$ are identically distributed with mean zero and variance σ^2. The spatial correlation matrix is defined as $\mathbf{H}_n = \mathrm{Var}(\zeta_{1,t}, \ldots, \zeta_{n,t})/\sigma^2$, for all t. We assume no cross-temporal spatial correlation so that $\mathrm{Cov}(\zeta_{i,s}, \zeta_{j,t}) = 0$ for $s \neq t$. Thus,

$$\mathrm{Var}\, \boldsymbol{\eta}_{2t} = \mathbf{H}_n \otimes \begin{pmatrix} \sigma^2 & \mathbf{0} \\ \mathbf{0} & \mathbf{0}_{p-1} \end{pmatrix},$$

where $\mathbf{0}_{p-1}$ is a $(p-1) \times (p-1)$ zero matrix. To initialize the recursion, we use $\zeta_{i,0} = 0$.

We may now collect the two sources of dynamic behavior, $\boldsymbol{\epsilon}$ and $\boldsymbol{\lambda}$, into a single transition equation. From Equations (8.18) and (8.19), we have

$$\begin{aligned} \boldsymbol{\delta}_t = \begin{pmatrix} \boldsymbol{\lambda}_t \\ \boldsymbol{\xi}_t \end{pmatrix} &= \begin{pmatrix} \boldsymbol{\Phi}_{1,t} \boldsymbol{\lambda}_{t-1} \\ (\mathbf{I}_n \otimes \boldsymbol{\Phi}_2)\, \boldsymbol{\xi}_{t-1} \end{pmatrix} + \begin{pmatrix} \boldsymbol{\eta}_{1t} \\ \boldsymbol{\eta}_{2t} \end{pmatrix} \\ &= \begin{pmatrix} \boldsymbol{\Phi}_{1,t} & \mathbf{0} \\ \mathbf{0} & (\mathbf{I}_n \otimes \boldsymbol{\Phi}_2) \end{pmatrix} \begin{pmatrix} \boldsymbol{\lambda}_{t-1} \\ \boldsymbol{\xi}_{t-1} \end{pmatrix} + \boldsymbol{\eta}_t = \mathbf{T}_t \boldsymbol{\delta}_{t-1} + \boldsymbol{\eta}_t. \end{aligned} \quad (8.20)$$

We assume that $\{\boldsymbol{\lambda}_t\}$ and $\{\boldsymbol{\xi}_t\}$ are independent stochastic processes and express the variance using

$$\begin{aligned} \mathbf{Q}_t = \mathrm{Var}\, \boldsymbol{\eta}_t &= \begin{pmatrix} \mathrm{Var}\, \boldsymbol{\eta}_{1t} & \mathbf{0} \\ \mathbf{0} & \mathrm{Var}\, \boldsymbol{\eta}_{2t} \end{pmatrix} \\ &= \sigma^2 \begin{pmatrix} \mathbf{Q}_{1t} & \mathbf{0} \\ \mathbf{0} & \left(\mathbf{H}_n \otimes \begin{pmatrix} 1 & \mathbf{0} \\ \mathbf{0} & \mathbf{0}_{p-1} \end{pmatrix} \right) \end{pmatrix} = \sigma^2 \mathbf{Q}_t^*. \end{aligned} \quad (8.21)$$

Finally, to initialize the recursion, we assume that $\boldsymbol{\delta}_0$ is a vector of parameters to be estimated.

8.5.2 Observation Set

To allow for unbalanced data, we use notation analogous to that introduced in Section 7.4.1. Specifically, let $\{i_1, \ldots, i_{n_t}\}$ denote the set of subjects that are available at time t, where $\{i_1, \ldots, i_{n_t}\} \subseteq \{1, 2, \ldots, n\}$. Further, define $\mathbf{M}_t$ to be the $n_t \times n$ design matrix that has a "one" in the i_jth column and zero otherwise, for $j = 1, \ldots, n_t$. With this design matrix, we have

$$\begin{pmatrix} i_1 \\ i_2 \\ \vdots \\ i_{n_t} \end{pmatrix} = \mathbf{M}_t \begin{pmatrix} 1 \\ 2 \\ \vdots \\ n \end{pmatrix}.$$

With this notation, we have

$$\begin{pmatrix} \boldsymbol{\alpha}_{i_1} \\ \boldsymbol{\alpha}_{i_2} \\ \vdots \\ \boldsymbol{\alpha}_{i_{n_t}} \end{pmatrix} = \left(\mathbf{M}_t \otimes \mathbf{I}_q\right) \begin{pmatrix} \boldsymbol{\alpha}_1 \\ \boldsymbol{\alpha}_2 \\ \vdots \\ \boldsymbol{\alpha}_n \end{pmatrix}. \tag{8.22}$$

Similarly, with $\varepsilon_{i,t} = (1 \quad 0 \quad \cdots \quad 0)\boldsymbol{\xi}_{i,t}$, we have

$$\begin{aligned} \begin{pmatrix} \varepsilon_{i_1,t} \\ \varepsilon_{i_2,t} \\ \vdots \\ \varepsilon_{i_{n_t},t} \end{pmatrix} &= \begin{pmatrix} (1 \quad 0 \quad \ldots \quad 0)\boldsymbol{\xi}_{i_1,t} \\ (1 \quad 0 \quad \ldots \quad 0)\boldsymbol{\xi}_{i_2,t} \\ \vdots \\ (1 \quad 0 \quad \ldots \quad 0)\boldsymbol{\xi}_{i_{n_t},t} \end{pmatrix} = (\mathbf{I}_{n_t} \otimes (1 \quad 0 \quad \cdots \quad 0)) \begin{pmatrix} \boldsymbol{\xi}_{i_1,t} \\ \boldsymbol{\xi}_{i_2,t} \\ \vdots \\ \boldsymbol{\xi}_{i_{n_t},t} \end{pmatrix} \\ &= (\mathbf{I}_{n_t} \otimes (1 \quad 0 \quad \cdots \quad 0))(\mathbf{M}_t \otimes \mathbf{I}_p) \begin{pmatrix} \boldsymbol{\xi}_{1,t} \\ \boldsymbol{\xi}_{2,t} \\ \vdots \\ \boldsymbol{\xi}_{n,t} \end{pmatrix} \\ &= (\mathbf{M}_t \otimes (1 \quad 0 \quad \cdots \quad 0))\boldsymbol{\xi}_t. \end{aligned} \tag{8.23}$$

8.5.3 Measurement Equations

With this observation set for the tth time period and Equation (8.9), we may write

$$\begin{aligned} \mathbf{y}_t = \begin{pmatrix} y_{i_1,t} \\ y_{i_2,t} \\ \vdots \\ y_{i_{n_t},t} \end{pmatrix} &= \begin{pmatrix} \mathbf{x}'_{i_1,t} \\ \mathbf{x}'_{i_2,t} \\ \vdots \\ \mathbf{x}'_{i_{n_t},t} \end{pmatrix} \boldsymbol{\beta} + \begin{pmatrix} \mathbf{z}'_{\alpha,i_1,t} & \mathbf{0} & \cdots & \mathbf{0} \\ \mathbf{0} & \mathbf{z}'_{\alpha,i_2,t} & \cdots & \mathbf{0} \\ \vdots & \vdots & \ddots & \vdots \\ \mathbf{0} & \mathbf{0} & \cdots & \mathbf{z}'_{\alpha,i_{n_t},t} \end{pmatrix} \begin{pmatrix} \boldsymbol{\alpha}_{i_1} \\ \boldsymbol{\alpha}_{i_2} \\ \vdots \\ \boldsymbol{\alpha}_{i_{n_t}} \end{pmatrix} \\ &+ \begin{pmatrix} \mathbf{z}'_{\lambda,i_1,t} \\ \mathbf{z}'_{\lambda,i_2,t} \\ \vdots \\ \mathbf{z}'_{\lambda,i_{n_t},t} \end{pmatrix} \boldsymbol{\lambda}_t + \begin{pmatrix} \varepsilon_{i_1,t} \\ \varepsilon_{i_2,t} \\ \vdots \\ \varepsilon_{i_{n_t},t} \end{pmatrix}. \end{aligned} \tag{8.24}$$

With Equations (8.22) and (8.23), we have

$$\begin{aligned} \mathbf{y}_t &= \mathbf{X}_t\boldsymbol{\beta} + \mathbf{Z}_{\alpha,t}\boldsymbol{\alpha} + \mathbf{Z}_{\lambda,t}\boldsymbol{\lambda}_t + \mathbf{W}_{1t}\boldsymbol{\xi}_t \\ &= \mathbf{X}_t\boldsymbol{\beta} + \mathbf{Z}_t\boldsymbol{\alpha} + \mathbf{W}_t\boldsymbol{\delta}_t, \end{aligned} \tag{8.25}$$

where

$$\mathbf{X}_t = \begin{pmatrix} \mathbf{x}'_{i_1,t} \\ \mathbf{x}'_{i_2,t} \\ \vdots \\ \mathbf{x}'_{i_{n_t},t} \end{pmatrix}, \qquad \mathbf{Z}_t = \mathbf{Z}_{\alpha,t} = \begin{pmatrix} \mathbf{z}'_{\alpha,i_1,t} & \mathbf{0} & \cdots & \mathbf{0} \\ \mathbf{0} & \mathbf{z}'_{\alpha,i_2,t} & \cdots & \mathbf{0} \\ \vdots & \vdots & \ddots & \vdots \\ \mathbf{0} & \mathbf{0} & \cdots & \mathbf{z}'_{\alpha,i_{n_t},t} \end{pmatrix} \left(\mathbf{M}_t \otimes \mathbf{I}_q\right),$$

$$\boldsymbol{\alpha} = \begin{pmatrix} \alpha_1 \\ \alpha_2 \\ \vdots \\ \alpha_n \end{pmatrix}, \qquad \mathbf{Z}_{\lambda,t} = \begin{pmatrix} \mathbf{z}'_{\lambda,i_1,t} \\ \mathbf{z}'_{\lambda,i_2,t} \\ \vdots \\ \mathbf{z}'_{\lambda,i_{n_t},t} \end{pmatrix}$$

and $\mathbf{W}_{1t} = (\mathbf{M}_t \otimes (1 \quad 0 \quad \cdots \quad 0))$. Thus, $\mathbf{W}_t = [\mathbf{Z}_{\lambda,t} : \mathbf{W}_{1t}]$, so that

$$\mathbf{W}_t \boldsymbol{\delta}_t = [\mathbf{Z}_{\lambda,t} \quad \mathbf{W}_{1t}] \begin{pmatrix} \boldsymbol{\lambda}_t \\ \boldsymbol{\xi}_t \end{pmatrix} = \mathbf{Z}_{\lambda,t} \boldsymbol{\lambda}_t + \mathbf{W}_{1t} \boldsymbol{\xi}_t.$$

Equation (8.25) collects the time t observations into a single expression. To complete the model specification, we assume that $\{\boldsymbol{\alpha}_i\}$ are i.i.d. with mean zero and $\mathbf{D} = \sigma^{-2} \operatorname{Var} \boldsymbol{\alpha}_i$. Thus, $\operatorname{Var} \boldsymbol{\alpha} = \sigma^2(\mathbf{I}_n \otimes \mathbf{D})$. Here, we write the variance of $\boldsymbol{\alpha}_i$ as a matrix times a constant σ^2 so that we may concentrate the constant out of the likelihood equations.

8.5.4 Initial Conditions

We first rewrite the measurement and observation equations so that the initial unobserved state vector, $\boldsymbol{\delta}_0$, is zero. To this end, define

$$\boldsymbol{\delta}_t^* = \boldsymbol{\delta}_t - \left(\prod_{r=1}^{t} \mathbf{T}_r\right) \boldsymbol{\delta}_0.$$

With these new variables, we may express Equations (8.25) and (8.20) as

$$\mathbf{y}_t = \mathbf{X}_t \boldsymbol{\beta} + \mathbf{W}_t \left(\prod_{r=1}^{t} \mathbf{T}_r\right) \boldsymbol{\delta}_0 + \mathbf{Z}_t \boldsymbol{\alpha} + \mathbf{W}_t \boldsymbol{\delta}_t^* + \boldsymbol{\epsilon}_t \tag{8.26}$$

and

$$\boldsymbol{\delta}_t^* = \mathbf{T}_t \boldsymbol{\delta}_{t-1}^* + \boldsymbol{\eta}_t, \tag{8.27}$$

respectively, where $\boldsymbol{\delta}_0^* = \mathbf{0}$.

With Equation (8.26), we may consider the initial state variable $\boldsymbol{\delta}_0$ to be fixed, random, or a combination of the two. With our assumptions of $\zeta_{i,0} = 0$

and $\boldsymbol{\lambda}_0$ as fixed, we may rewrite Equation (8.26) as

$$\mathbf{y}_t = \left(\mathbf{X}_t : \mathbf{W}_{1t}\left(\prod_{r=1}^{t} \boldsymbol{\Phi}_{1r}\right)\right)\begin{pmatrix}\boldsymbol{\beta}\\ \boldsymbol{\lambda}_0\end{pmatrix} + \mathbf{Z}_t\boldsymbol{\alpha} + \mathbf{W}_t\boldsymbol{\delta}_t^*. \tag{8.28}$$

Hence, the new vector of fixed parameters to be estimated is $(\boldsymbol{\beta}', \boldsymbol{\lambda}_0')'$ with corresponding $n_t \times (K + r)$ matrix of covariates $(\mathbf{X}_t : \mathbf{W}_{1t}(\prod_{r=1}^{t} \boldsymbol{\Phi}_{1r}))$. Thus, with this reparameterization, we henceforth consider the state space model with the assumption that $\boldsymbol{\delta}_0 = \mathbf{0}$.

8.5.5 The Kalman Filter Algorithm

We define the transformed variables $\mathbf{y}^*$, $\mathbf{X}^*$, and $\mathbf{Z}^*$ as follows. Recursively calculate

$$\mathbf{d}_{t+1/t}(\mathbf{y}) = \mathbf{T}_{t+1}\mathbf{d}_{t/t-1}(\mathbf{y}) + \mathbf{K}_t(\mathbf{y}_t - \mathbf{W}_t\mathbf{d}_{t/t-1}(\mathbf{y})), \tag{8.29a}$$

$$\mathbf{d}_{t+1/t}(\mathbf{X}) = \mathbf{T}_{t+1}\mathbf{d}_{t/t-1}(\mathbf{X}) + \mathbf{K}_t(\mathbf{X}_t - \mathbf{W}_t\mathbf{d}_{t/t-1}(\mathbf{X})), \tag{8.29b}$$

$$\mathbf{d}_{t+1/t}(\mathbf{Z}) = \mathbf{T}_{t+1}\mathbf{d}_{t/t-1}(\mathbf{Z}) + \mathbf{K}_t(\mathbf{Z}_t - \mathbf{W}_t\mathbf{d}_{t/t-1}(\mathbf{Z})) \tag{8.29c}$$

and

$$\mathbf{P}_{t+1/t} = \mathbf{T}_{t+1}\left(\mathbf{P}_{t/t-1} - \mathbf{P}_{t/t-1}\mathbf{W}_t'\mathbf{F}_t^{-1}\mathbf{W}_t\mathbf{P}_{t/t-1}\right)\mathbf{T}_{t+1}' + \mathbf{Q}_{t+1}, \tag{8.30a}$$

$$\mathbf{F}_{t+1} = \mathbf{W}_{t+1}\mathbf{P}_{t+1/t}\mathbf{W}_{t+1}', \tag{8.30b}$$

$$\mathbf{K}_{t+1} = \mathbf{T}_{t+2}\mathbf{P}_{t+1/t}\mathbf{W}_{t+1}'\mathbf{F}_{t+1}^{-1}. \tag{8.30c}$$

We begin the recursions in Equations (8.29a)–(8.29c) with $\mathbf{d}_{1/0}(\mathbf{y}) = \mathbf{0}$, $\mathbf{d}_{1/0}(\mathbf{X}) = \mathbf{0}$, and $\mathbf{d}_{1/0}(\mathbf{Z}) = \mathbf{0}$. Also, for Equation (8.30a), use $\mathbf{P}_{1/0} = \mathbf{Q}_1$. The tth components of each transformed variable are

$$\mathbf{y}_t^* = \mathbf{y}_t - \mathbf{W}_t\mathbf{d}_{t/t-1}(\mathbf{y}), \tag{8.31a}$$

$$\mathbf{X}_t^* = \mathbf{X}_t - \mathbf{W}_t\mathbf{d}_{t/t-1}(\mathbf{X}), \tag{8.31b}$$

$$\mathbf{Z}_t^* = \mathbf{Z}_t - \mathbf{W}_t\mathbf{d}_{t/t-1}(\mathbf{Z}). \tag{8.31c}$$

From Equations (8.29)–(8.31), note that the calculation of the transformed variables are unaffected by scale changes in $\{\mathbf{Q}_t\}$. Thus, using the sequence $\{\mathbf{Q}_t^*\}$ defined in Equation (8.21) in the Kalman filter algorithm yields the same transformed variables and rescaled conditional variances $\mathbf{F}_t^* = \sigma^{-2}\mathbf{F}_t$.

Likelihood Equations

To calculate parameter estimators and the likelihood, we use the following sums of squares:

$$\mathbf{S}_{\mathbf{XX},\mathbf{F}} = \sum_{t=1}^{T} \mathbf{X}_t^{*\prime} \mathbf{F}_t^{*-1} \mathbf{X}_t^*, \qquad \mathbf{S}_{\mathbf{XZ},\mathbf{F}} = \sum_{t=1}^{T} \mathbf{X}_t^{*\prime} \mathbf{F}_t^{*-1} \mathbf{Z}_t^*,$$

$$\mathbf{S}_{\mathbf{ZZ},\mathbf{F}} = \sum_{t=1}^{T} \mathbf{Z}_t^{*\prime} \mathbf{F}_t^{*-1} \mathbf{Z}_t^*, \qquad \mathbf{S}_{\mathbf{Xy},\mathbf{F}} = \sum_{t=1}^{T} \mathbf{X}_t^{*\prime} \mathbf{F}_t^{*-1} \mathbf{y}_t^*,$$

$$\mathbf{S}_{\mathbf{Zy},\mathbf{F}} = \sum_{t=1}^{T} \mathbf{Z}_t^{*\prime} \mathbf{F}_t^{*-1} \mathbf{y}_t^*,$$

and

$$\mathbf{S}_{\mathbf{yy},\mathbf{F}} = \sum_{t=1}^{T} \mathbf{y}_t^{*\prime} \mathbf{F}_t^{*-1} \mathbf{y}_t^*.$$

With this notation, the GLS estimator of $\boldsymbol{\beta}$ is

$$\mathbf{b}_{\mathrm{GLS}} = \left\{ \mathbf{S}_{\mathbf{XX},\mathbf{F}} - \mathbf{S}_{\mathbf{XZ},\mathbf{F}} \left(\mathbf{S}_{\mathbf{ZZ},\mathbf{F}} + \mathbf{I}_n \otimes \mathbf{D}^{-1} \right)^{-1} \mathbf{S}_{\mathbf{ZX},\mathbf{F}} \right\}^{-1} \\ \times \left\{ \mathbf{S}'_{\mathbf{Xy},\mathbf{F}} - \mathbf{S}'_{\mathbf{Zy},\mathbf{F}} \left(\mathbf{S}_{\mathbf{ZZ},\mathbf{F}} + \mathbf{I}_n \otimes \mathbf{D}^{-1} \right)^{-1} \mathbf{S}_{\mathbf{ZX},\mathbf{F}} \right\}'. \tag{8.32}$$

Let $\boldsymbol{\tau}$ denote the vector of the other variance components so that $(\sigma^2, \boldsymbol{\tau})$ represent all variance components. We may express the concentrated logarithmic likelihood as

$$L(\sigma^2, \boldsymbol{\tau}) \\ = -\frac{1}{2} \Bigg\{ N \ln 2\pi + N \ln \sigma^2 + \sigma^{-2}\, \mathit{Error\ SS} \\ + \sum_{t=1}^{T} \ln \det \mathbf{F}_t^* + n \ln \det \mathbf{D} + \ln \det \left(\mathbf{S}_{\mathbf{ZZ},\mathbf{F}} + \mathbf{I}_n \otimes \mathbf{D}^{-1} \right) \Bigg\}, \tag{8.33}$$

where

$$\mathit{Error\ SS} = \left(\mathbf{S}_{\mathbf{yy},\mathbf{F}} - \mathbf{S}'_{\mathbf{Zy},\mathbf{F}} \left(\mathbf{S}_{\mathbf{ZZ},\mathbf{F}} + \mathbf{I}_n \otimes \mathbf{D}^{-1} \right)^{-1} \mathbf{S}_{\mathbf{Zy},\mathbf{F}} \right) \\ - \left(\mathbf{S}'_{\mathbf{Xy},\mathbf{F}} - \mathbf{S}'_{\mathbf{Zy},\mathbf{F}} \left(\mathbf{S}_{\mathbf{ZZ},\mathbf{F}} + \mathbf{I}_n \otimes \mathbf{D}^{-1} \right)^{-1} \mathbf{S}_{\mathbf{ZX},\mathbf{F}} \right) \mathbf{b}_{\mathrm{GLS}}. \tag{8.34}$$

The restricted logarithmic likelihood is

$$L_{\mathrm{REML}}(\sigma^2, \boldsymbol{\tau}) = -\frac{1}{2} \Big\{ \ln \det \left(\mathbf{S}_{\mathbf{XX},\mathbf{F}} - \mathbf{S}_{\mathbf{XZ},\mathbf{F}} \left(\mathbf{S}_{\mathbf{ZZ},\mathbf{F}} + \mathbf{I}_n \otimes \mathbf{D}^{-1} \right)^{-1} \mathbf{S}_{\mathbf{ZX},\mathbf{F}} \right) \\ - K \ln \sigma^2 \Big\} + L(\sigma^2, \boldsymbol{\tau}), \tag{8.35}$$

up to an additive constant. Estimates of the variance components, σ^2 and $\boldsymbol{\tau}$, may be determined either by maximizing (8.33) or (8.35). This text uses (8.35), which yield the REML estimators. The REML estimator of σ^2 is

$$s^2_{\text{REML}} = Error\ SS/(N - K). \tag{8.36}$$

With Equation (8.35), the concentrated restricted log likelihood is

$$L_{\text{REML}}(\boldsymbol{\tau}) = -\frac{1}{2}\Big\{ \ln\det\left(\mathbf{S_{XX,F}} - \mathbf{S_{XZ,F}}\left(\mathbf{S_{ZZ,F}} + \mathbf{I}_n \otimes \mathbf{D}^{-1}\right)^{-1}\mathbf{S_{ZX,F}}\right) \\ - K \ln s^2_{\text{REML}}\Big\} + L\left(s^2_{\text{REML}}, \boldsymbol{\tau}\right). \tag{8.37}$$

Maximizing $L_{\text{REML}}(\boldsymbol{\tau})$ over $\boldsymbol{\tau}$ yields the REML estimator of $\boldsymbol{\tau}$.

8.6 Example: Capital Asset Pricing Model

The *capital asset pricing model* (CAPM) is a representation that is widely used in financial economics. An intuitively appealing idea, and one of the basic characteristics of the CAPM, is that there should be a relationship between the performance of a security and the performance of the market. One rationale is simply that if economic forces are such that the market improves, then those same forces should act upon an individual stock, suggesting that it also improve. We measure performance of a security through the return. To measure performance of the market, several market indices exist for each exchange. As an illustration, in the following we use the return from the "value-weighted" index of the market created by the Center for Research in Securities Prices (CRSP). The value-weighted index is defined by assuming a portfolio is created when investing an amount of money in proportion to the market value (at a certain date) of firms listed on the New York Stock Exchange, the American Stock Exchange, and the Nasdaq stock market.

Another rationale for a relationship between security and market returns comes from financial economics theory. This is the CAPM theory, attributed to Sharpe (1964O) and Lintner (1965O) and based on the portfolio diversification ideas of Markowitz (1952O). Other things being equal, investors would like to select a return with a high expected value and low standard deviation; the latter is a measure of riskiness. One of the desirable properties about using standard deviations as a measure of riskiness is that it is straightforward to calculate the standard deviation of a portfolio, a combination of securities. One only needs to know the standard deviation of each security and the correlations among securities. A notable security is a risk-free one, that is, a security that theoretically has a zero standard deviation. Investors often use a 30-day U.S. Treasury bill as

Table 8.2. *Summary statistics for market index and risk-free security based on sixty monthly observations (January 1995 to December 1999)*

Variable	Mean	Median	Minimum	Maximum	Standard deviation
VWRETD (Value weighted index)	2.091	2.946	−15.677	8.305	4.133
RISKFREE (Risk free)	0.408	0.415	0.296	0.483	0.035
VWFREE (Value weighted in excess of risk free)	1.684	2.517	−16.068	7.880	4.134

Source: Center for Research in Securities Prices.

an approximation of a risk-free security, arguing that the probability of default of the U.S. government within 30 days is negligible. Positing the existence of a risk-free asset and some other mild conditions, under the CAPM theory there exists an efficient frontier called the securities market line. This frontier specifies the minimum expected return that investors should demand for a specified level of risk. To estimate this line, we use the equation

$$y_{it} = \beta_{0i} + \beta_{1i} x_{mt} + \varepsilon_{it}, \tag{8.38}$$

where y is the security return in excess of the risk-free rate and x_m is the market return in excess of the risk-free rate. We interpret β_{1i} as a measure of the amount of the ith security's return that is attributed to the behavior of the market. According to the CAPM theory, the intercept β_{0i} is zero but we include it to study the robustness of the model.

To assess the empirical performance of the CAPM model, we study security returns from CRSP. We consider $n = 90$ firms from the insurance carriers that were listed on the CRSP files as of December 31, 1999. (The "insurance carriers" consist of those firms with standard industrial classification codes ranging from 6310 through 6331, inclusive.) For each firm, we used sixty months of data ranging from January 1995 through December 1999.

Table 8.2 summarizes the performance of the market through the return from the value-weighted index, VWRETD, and risk free instrument, RISKFREE. We also consider the difference between the two, VWFREE, and interpret this to be the return from the market in excess of the risk-free rate.

Table 8.3 summarizes the performance of individual securities through the monthly return, RET. These summary statistics are based on 5,400 monthly

Table 8.3. *Summary statistics for individual security returns based on 5,400 monthly observations (January 1995 to December 1999) taken from 90 firms*

Variable	Mean	Median	Minimum	Maximum	Standard deviation
RET (Individual security return)	1.052	0.745	−66.197	102.500	10.038
RETFREE return in excess of risk free)	0.645	0.340	−66.579	102.085	10.036

observations taken from 90 firms. The difference between the return and the corresponding risk-free instrument is RETFREE.

To examine the relationship between market and individual firm returns, a trellis plot is given in Figure 8.1. Here, only a subset of 18 randomly selected firms is presented; the subset allows one to see important patterns. Each panel in the figure represents a firm's experience; thus, the market returns (on the horizontal axis) are common to all firms. In particular, note the influential point on the left-hand side of each panel, corresponding to an August 1998 monthly return of -15%. So that this point would not dominate, a nonparametric line was fit for each panel. The lines superimposed show a positive relationship between the market and individual firm returns although the noise about each line is substantial.

Several fixed-effects models were fit using Equation (8.38) as a framework. Table 8.4 summarizes the fit of each model. Based on these fits, we will use the variable slopes with an $AR(1)$ error term model as the baseline for investigating time-varying coefficients.

For time-varying coefficients, we investigate models of the form

$$y_{it} = \beta_0 + \beta_{1,i,t} x_{mt} + \varepsilon_{it}, \tag{8.39}$$

where

$$\varepsilon_{it} = \rho_\varepsilon \varepsilon_{i,t-1} + \eta_{1,it} \tag{8.40}$$

and

$$\beta_{1,i,t} - \beta_{1,i} = \rho_\beta (\beta_{1,i,t-1} - \beta_{1,i}) + \eta_{2,it}. \tag{8.41}$$

Here, $\{\eta_{1,it}\}$ are i.i.d. noise terms. These are independent of $\{\eta_{2,it}\}$, which are mutually independent and identical for each firm i. For Equations (8.40) and (8.41), we assume that $\{\varepsilon_{it}\}$ and $\{\beta_{1,i,t}\}$ are stationary $AR(1)$ processes. The slope

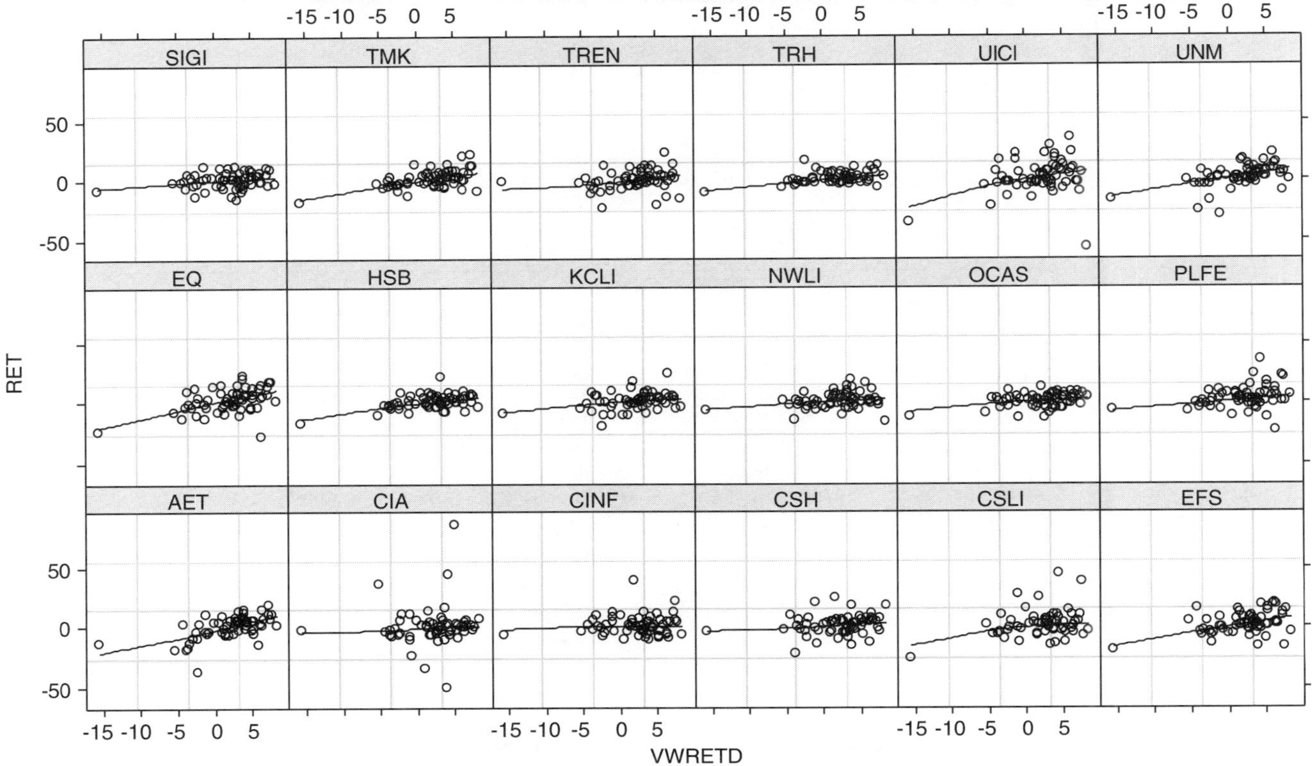

Figure 8.1. Trellis plot of returns versus market return. A random sample of 18 firms are plotted; each panel represents a firm. Within each panel, firm returns versus market returns are plotted. A nonparametric line is superimposed to provide a visual impression of the relationship between the market return and an individual firm's return.

Table 8.4. *Fixed effects models*

Summary measure	Homogeneous model	Variable intercepts model	Variable slopes model	Variable intercepts and slopes model	Variable slopes model with $AR(1)$ term
Residual standard deviation (s)	9.59	9.62	9.53	9.54	9.53
−2 ln likelihood	39,751.2	39,488.6	39,646.5	39,350.6	39,610.9
AIC	39,753.2	39,490.6	39,648.5	39,352.6	39,614.9
$AR(1)$ corr(ρ)					−0.084
t-statistic for ρ					−5.98

coefficient, $\beta_{1,i,t}$, is allowed to vary by both firm i and time t. We assume that each firm has its own stationary mean $\beta_{1,i}$ and variance $\mathrm{Var}\,\beta_{1,i,t}$. It is possible to investigate the model in Equations (8.39)–(8.41) for each firm i. However, by considering all firms simultaneously, we allow for efficient estimation of common parameters β_0, ρ_ε, ρ_β, and $\sigma^2 = \mathrm{Var}\,\varepsilon_{it}$.

To express this model formulation in the notation of Section 8.3, first define $\mathbf{j}_{n,i}$ to be an $n \times 1$ vector, with a one in the ith row and zeros elsewhere. Further, define

$$\boldsymbol{\beta} = \begin{pmatrix} \beta_0 \\ \beta_{1,1} \\ \vdots \\ \beta_{1,n} \end{pmatrix}, \qquad \mathbf{x}_{it} = \begin{pmatrix} 1 \\ \mathbf{j}_{n,i} x_{mt} \end{pmatrix},$$

$\mathbf{z}_{\lambda,it} = \mathbf{j}_{n,i} x_{mt}$, and

$$\boldsymbol{\lambda}_t = \begin{pmatrix} \beta_{1,1t} - \beta_{1,1} \\ \vdots \\ \beta_{1,nt} - \beta_{1,n} \end{pmatrix}.$$

Thus, with this notation, we have

$$y_{it} = \beta_0 + \beta_{1,i,t} x_{mt} + \varepsilon_{it} = \mathbf{z}'_{\lambda,i,t} \boldsymbol{\lambda}_t + \mathbf{x}'_{it} \boldsymbol{\beta} + \varepsilon_{it}.$$

This expresses the model as a special case of Equation (8.8), ignoring the time-invariant random-effects portion and using $r = n$ time-varying coefficients.

An important component of model estimation routines is $\mathrm{Var}\,\boldsymbol{\lambda} = \sigma^2 \boldsymbol{\Sigma}_\lambda$. Straightforward calculations show that this matrix may be expressed as $\mathrm{Var}\,\boldsymbol{\lambda} = \mathbf{R}_{AR}(\rho_\beta) \otimes \boldsymbol{\Sigma}_\beta$, where $\boldsymbol{\Sigma}_\beta = \sigma_\beta^2 \mathbf{I}_n$ and $\mathbf{R}_{AR}$ is defined in Section 8.2.2. Thus, this matrix is highly structured and easily invertible. However, it has dimension $nT \times nT$, which is large. Special routines must take advantage of the structure to make the estimation computationally feasible. The estimation procedure in Appendix 8A.2 assumes that r, the number of time-varying coefficients, is small. (See, for example, Equation (8A.5).) Thus, we look to the Kalman filter algorithm for this application.

To apply the Kalman filter algorithm, we use the following conventions. For the updating matrix for time-varying coefficients in Equation (8.18), we use $\mathbf{\Phi}_{1t} = \mathbf{I}_n \rho_\beta$. For the error structure in Equation (8.19), we use an $AR(1)$ structure so that $p = 1$ and $\mathbf{\Phi}_2 = \rho_\varepsilon$. Thus, we have

$$\boldsymbol{\delta}_t = \begin{pmatrix} \boldsymbol{\lambda}_t \\ \boldsymbol{\xi}_t \end{pmatrix} = \begin{pmatrix} \beta_{1,1t} - \beta_{1,1} \\ \vdots \\ \beta_{1,nt} - \beta_{1,n} \\ \varepsilon_{1t} \\ \vdots \\ \varepsilon_{nt} \end{pmatrix}$$

and

$$\mathbf{T}_t = \begin{pmatrix} \mathbf{I}_n \rho_\beta & \mathbf{0} \\ \mathbf{0} & \mathbf{I}_n \rho_\varepsilon \end{pmatrix} = \begin{pmatrix} \rho_\beta & 0 \\ 0 & \rho_\varepsilon \end{pmatrix} \otimes \mathbf{I}_n,$$

for the vector of time-varying parameters and updating matrix, respectively. As in Section 8.5.1, we assume that $\{\boldsymbol{\lambda}_t\}$ and $\{\boldsymbol{\xi}_t\}$ are independent stochastic processes and express the variance using

$$\mathbf{Q}_t = \begin{pmatrix} \mathrm{Var}\,\boldsymbol{\eta}_{1t} & \mathbf{0} \\ \mathbf{0} & \mathrm{Var}\,\boldsymbol{\eta}_{2t} \end{pmatrix} = \begin{pmatrix} (1-\rho_\beta^2)\sigma_\beta^2 \mathbf{I}_n & \mathbf{0} \\ \mathbf{0} & (1-\rho_\varepsilon^2)\sigma_\varepsilon^2 \mathbf{I}_n \end{pmatrix}.$$

To reduce the complexity, we assume that the initial vector is zero so that $\boldsymbol{\delta}_0 = \mathbf{0}$.

For the measurement equations, we have

$$\mathbf{Z}_{\lambda,t} = \begin{pmatrix} \mathbf{z}'_{\lambda,i_1,t} \\ \mathbf{z}'_{\lambda,i_2,t} \\ \vdots \\ \mathbf{z}'_{\lambda,i_{n_t},t} \end{pmatrix} = \begin{pmatrix} \mathbf{j}'_{n,i_1,t} x_{mt} \\ \mathbf{j}'_{n,i_2,t} x_{mt} \\ \vdots \\ \mathbf{j}'_{n,i_{n_t},t} x_{mt} \end{pmatrix} = \mathbf{M}_t x_{mt}$$

and

$$\mathbf{X}_t = \begin{pmatrix} \mathbf{x}'_{i_1,t} \\ \mathbf{x}'_{i_2,t} \\ \vdots \\ \mathbf{x}'_{i_{n_t},t} \end{pmatrix} = \begin{pmatrix} 1 & \mathbf{j}'_{n,i_1,t} x_{mt} \\ 1 & \mathbf{j}'_{n,i_2,t} x_{mt} \\ \vdots & \vdots \\ 1 & \mathbf{j}'_{n,i_{n_t},t} x_{mt} \end{pmatrix} = (\mathbf{1}_{n_t} \quad \mathbf{M}_t x_{mt}).$$

Further, we have $\mathbf{W}_{1t} = \mathbf{M}_t$ and thus, $\mathbf{W}_t = (\mathbf{M}_t x_{mt} : \mathbf{M}_t)$.

For parameter estimation, we have not specified any time-invariant random effects. Therefore, we need only use parts (a) and (b) of Equations (8.29) and (8.31), as well as all of Equation (8.30). To calculate parameter estimators and the likelihood, we use the following sums of squares: $\mathbf{S}_{\mathbf{XX},\mathbf{F}} = \sum_{t=1}^{T} \mathbf{X}_t^{*\prime} \mathbf{F}_t^{*-1} \mathbf{X}_t^*$,

Table 8.5. *Time-varying capital asset pricing models*

Parameter	σ	ρ_ε	ρ_β	σ_β
Model fit with ρ_ε parameter				
Estimate	9.527	−0.084	−0.186	0.864
Standard error	0.141	0.019	0.140	0.069
Model fit without ρ_ε parameter				
Estimate	9.527		−0.265	0.903
Standard error	0.141		0.116	0.068

$\mathbf{S}_{\mathbf{Xy,F}} = \sum_{t=1}^{T} \mathbf{X}_t^{*\prime} \mathbf{F}_t^{*-1} \mathbf{y}_t^*$, and $\mathbf{S}_{\mathbf{yy,F}} = \sum_{t=1}^{T} \mathbf{y}_t^{*\prime} \mathbf{F}_t^{*-1} \mathbf{y}_t^*$. With this notation, the GLS estimator of $\boldsymbol{\beta}$ is $\mathbf{b}_{\mathrm{GLS}} = \mathbf{S}_{\mathbf{XX,F}}^{-1} \mathbf{S}_{\mathbf{Xy,F}}$. We may express the concentrated logarithmic likelihood as

$$L(\sigma^2, \boldsymbol{\tau}) = -\frac{1}{2}\left\{ N \ln 2\pi + N \ln \sigma^2 + \sigma^{-2}\, \mathit{Error\ SS} + \sum_{t=1}^{T} \ln \det \mathbf{F}_t^* \right\},$$

where $\mathit{Error\ SS} = \mathbf{S}_{\mathbf{yy,F}} - \mathbf{S}'_{\mathbf{Xy,F}} \mathbf{b}_{\mathrm{GLS}}$. The restricted logarithmic likelihood is

$$L_{\mathrm{REML}}(\sigma^2, \boldsymbol{\tau}) = -\frac{1}{2}\left\{ \ln \det(\mathbf{S}_{\mathbf{XX,F}}) - K \ln \sigma^2 \right\} + L(\sigma^2, \boldsymbol{\tau})$$

up to an additive constant.

For prediction, we may again use the BLUPs introduced in Chapter 4 and extended in Section 8.4.3. Straightforward calculations show that the BLUP of $\beta_{1,i,t}$ is

$$\begin{aligned} b_{1,i,t,\mathrm{BLUP}} &= b_{1,i,t,\mathrm{GLS}} + \sigma_\beta^2 \left(\rho_\beta^{|t-1|} x_{m,1} \quad \ldots \quad \rho_\beta^{|t-T_i|} x_{m,T_i} \right) \\ &\quad \times (\operatorname{Var} \mathbf{y}_i)^{-1} (\mathbf{y}_i - b_{0,\mathrm{GLS}} \mathbf{1}_i - b_{1,i,t,\mathrm{GLS}} \mathbf{x}_m), \end{aligned} \tag{8.42}$$

where $\mathbf{x}_m = (x_{m,1} \quad \ldots \quad x_{m,T_i})'$, $\operatorname{Var} \mathbf{y}_i = \sigma_\beta^2 \mathbf{X}_m \mathbf{R}_{AR}(\rho_\beta) \mathbf{X}_m + \sigma_\varepsilon^2 \mathbf{R}_{AR}(\rho_\varepsilon)$, and $\mathbf{X}_m = \operatorname{diag}(x_{m,1} \quad \cdots \quad x_{m,T_i})'$.

Table 8.5 summarizes the fit of the time-varying CAPM, based on Equations (8.39)–(8.41) and the CRSP data. When fitting the model with both autoregressive processes in Equations (8.40) and (8.41), it can be difficult to separate the dynamic sources, thus flattening out the likelihood surface. When the likelihood surface is flat, it is difficult to obtain convergence of the likelihood maximization routine. Figure 8.2 shows that the likelihood function is less responsive to changes in the ρ_β parameter compared to those in the ρ_ε parameter.

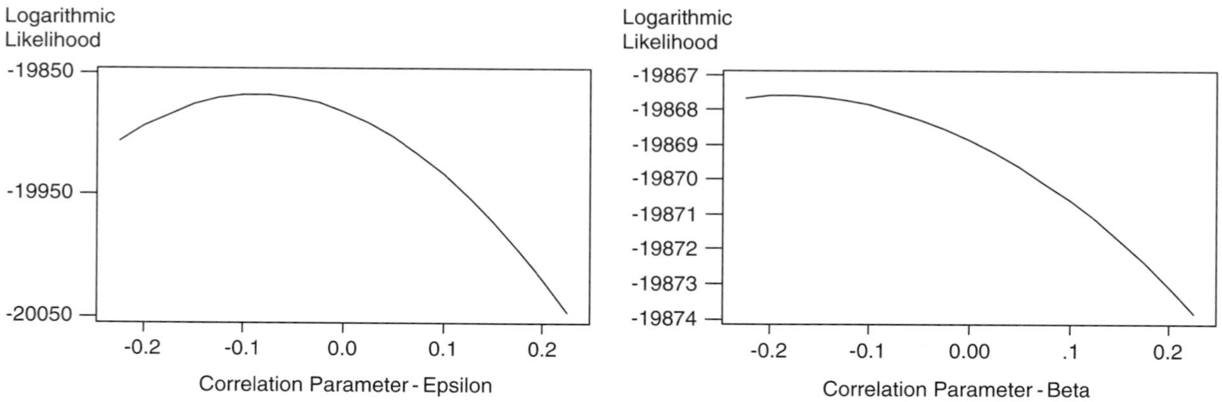

Figure 8.2. Logarithmic likelihood as a function of the correlation parameters. The left-hand panel shows the log likelihood as a function of ρ_ε, holding the other parameters fixed at their maximum likelihood values. The right-hand panel show log likelihood as a function of ρ_β. The likelihood surface is flatter in the direction of ρ_β than in the direction of ρ_ε.

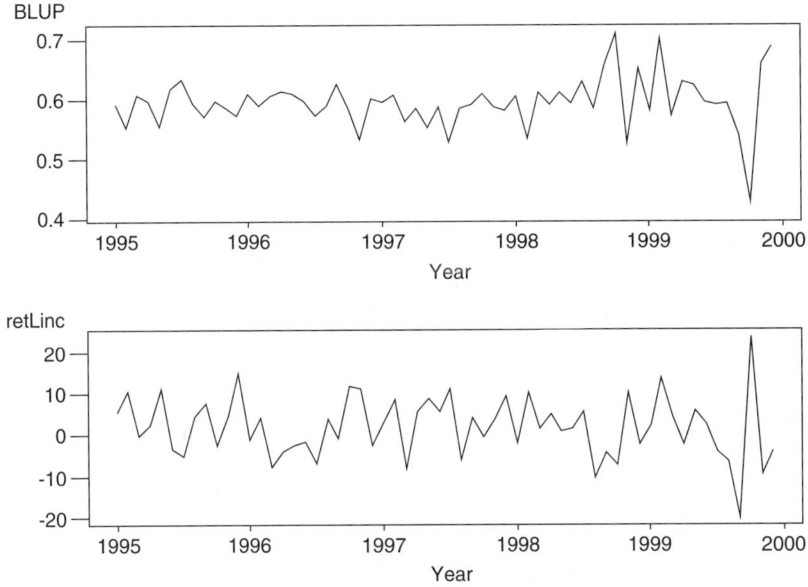

Figure 8.3. Time-series plot of BLUPs of the slope associated with the market returns and returns for the Lincoln National Corporation. The upper panel shows the BLUPs of the slopes. The lower panel shows the monthly returns.

Because of the interest in the changes of the slope parameter, the model was then refit without the correlation parameter for the noise process, ρ_ε. The likelihood surface was much steeper for this reduced model and the resulting standard errors are much sharper, as seen in Table 8.5. An alternative model would be to consider ρ_β equal to zero yet retain ρ_ε. We leave this as an exercise for the reader.

With the fitted model parameter estimates in Table 8.5, beta prediction and forecasting is possible. For illustration purposes, we calculated the predictions of the slope for each time point using Equation (8.42). Figure 8.2 summarizes these calculations for the Lincoln National Corporation. For reference, the GLS estimator of $\beta_{1,\text{LINCOLN}}$ for this time period is $b_{1,\text{LINCOLN}} = 0.599$.

The upper panel of Figure 8.3 shows the time series of the time-varying efficient predictors of the slope. The lower panel of Figure 8.3 shows the time series of Lincoln returns over the same time period. Here, we see the influence of the firm's returns on the efficient predictor of $\beta_{1,\text{LINCOLN}}$. For example, we see that the large drop in Lincoln's return for September of 1999 leads to a corresponding drop in the predictor of the slope.

Appendix 8A Inference for the Time-Varying Coefficient Model

8A.1 The Model

To allow for unbalanced data, recall the design matrix $\mathbf{M}_i$ specified in Equation (7.22). To allow for the observation set described in Section 7.4.1, we may use the matrix form of the linear mixed-effects model in Equation (8.9) with one exception. That exception is to expand the definition $\mathbf{Z}_{\lambda,i}$, a $T_i \times Tr$ matrix of explanatory variables. With the notation in Equation (7.22), we have

$$\begin{pmatrix} \boldsymbol{\lambda}_{t_1} \\ \boldsymbol{\lambda}_{t_2} \\ \vdots \\ \boldsymbol{\lambda}_{t_{T_i}} \end{pmatrix} = (\mathbf{M}_i \otimes \mathbf{I}_r) \begin{pmatrix} \boldsymbol{\lambda}_1 \\ \boldsymbol{\lambda}_2 \\ \vdots \\ \boldsymbol{\lambda}_T \end{pmatrix}. \tag{8A.1}$$

Thus, to complete the specification of Equation (8.9), we write

$$\begin{aligned} \mathbf{Z}_{\lambda,i} &= \begin{pmatrix} \mathbf{z}'_{\lambda,i,1} & \mathbf{0} & \cdots & \mathbf{0} \\ \mathbf{0} & \mathbf{z}'_{\lambda,i,2} & \cdots & \mathbf{0} \\ \vdots & \vdots & \ddots & \vdots \\ \mathbf{0} & \mathbf{0} & \cdots & \mathbf{z}'_{\lambda,i,T_i} \end{pmatrix} (\mathbf{M}_i \otimes \mathbf{I}_r) \\ &= \text{block diag}(\mathbf{z}'_{\lambda,i,1}, \ldots, \mathbf{z}'_{\lambda,i,T_i})\,(\mathbf{M}_i \otimes \mathbf{I}_r). \end{aligned} \tag{8A.2}$$

To express the model more compactly, we use the mixed linear model specification. Further, we also use the notation $\text{Var}\,\boldsymbol{\epsilon} = \sigma^2\mathbf{R} = \sigma^2 \times \text{block diag}(\mathbf{R}_1, \ldots, \mathbf{R}_n)$ and note that $\text{Var}\,\boldsymbol{\alpha} = \sigma^2(\mathbf{I}_n \otimes \mathbf{D})$. With this notation, we may express the variance–covariance matrix of $\mathbf{y}$ as $\text{Var}\,\mathbf{y} = \sigma^2\mathbf{V}$, where

$$\mathbf{V} = \mathbf{Z}_\alpha(\mathbf{I}_n \otimes \mathbf{D})\mathbf{Z}'_\alpha + \mathbf{Z}_\lambda \boldsymbol{\Sigma}_\lambda \mathbf{Z}'_\lambda + \mathbf{R}. \tag{8A.3}$$

8A.2 Estimation

For known variances, the usual GLS estimator of $\boldsymbol{\beta}$ is $\mathbf{b}_{\text{GLS}} = (\mathbf{X}'\mathbf{V}^{-1}\mathbf{X})^{-1}\mathbf{X}'\mathbf{V}^{-1}\mathbf{y}$ (where the scale parameter σ^2 drops out). To simplify calculations, we note that both $\mathbf{R}$ and $\mathbf{Z}_\alpha(\mathbf{I}_n \otimes \mathbf{D})\mathbf{Z}'_\alpha$ are block diagonal matrices and thus have readily computable inverses. Thus, we define

$$\mathbf{V}_\alpha = \mathbf{R} + \mathbf{Z}_\alpha(\mathbf{I}_n \otimes \mathbf{D})\mathbf{Z}'_\alpha = \text{block diag}(\mathbf{V}_{\alpha,1}, \ldots, \mathbf{V}_{\alpha,n}),$$

where $\mathbf{V}_{\alpha,i}$ is defined in Section 8.3.1. With this notation, we use Equation (A.4) of Appendix A.5 to write

$$\mathbf{V}^{-1} = (\mathbf{V}_\alpha + \mathbf{Z}_\lambda \boldsymbol{\Sigma}_\lambda \mathbf{Z}'_\lambda)^{-1} = \mathbf{V}_\alpha^{-1} - \mathbf{V}_\alpha^{-1}\mathbf{Z}_\lambda\left(\mathbf{Z}'_\lambda \mathbf{V}_\alpha^{-1}\mathbf{Z}_\lambda + \boldsymbol{\Sigma}_\lambda^{-1}\right)^{-1}\mathbf{Z}'_\lambda \mathbf{V}_\alpha^{-1}. \tag{8A.4}$$

In Equation (8A.4), only the block diagonal matrix $\mathbf{V}_\alpha$ and the $rT \times rT$ matrix $\mathbf{Z}'_\lambda \mathbf{V}_\alpha^{-1} \mathbf{Z}_\lambda + \boldsymbol{\Sigma}_\lambda^{-1}$ require inversion, not the $N \times N$ matrix $\mathbf{V}$.

Define the following sums of squares: $\mathbf{S_{XX}} = \sum_{i=1}^n \mathbf{X}'_i \mathbf{V}_{\alpha,i}^{-1} \mathbf{X}_i$, $\mathbf{S_{XZ}} = \sum_{i=1}^n \mathbf{X}'_i \mathbf{V}_{\alpha,i}^{-1} \mathbf{Z}_{\lambda,i}$, $\mathbf{S_{ZZ}} = \sum_{i=1}^n \mathbf{Z}'_{\lambda,i} \mathbf{V}_{\alpha,i}^{-1} \mathbf{Z}_{\lambda,i}$, $\mathbf{S_{Zy}} = \sum_{i=1}^n \mathbf{Z}'_{\lambda,i} \mathbf{V}_{\alpha,i}^{-1} \mathbf{y}_i$, and $\mathbf{S_{Xy}} = \sum_{i=1}^n \mathbf{X}'_i \mathbf{V}_{\alpha,i}^{-1} \mathbf{y}_i$. With this notation and Equation (8A.4), we may express the GLS estimator of $\boldsymbol{\beta}$ as

$$\mathbf{b}_{\mathrm{GLS}} = \left(\mathbf{S_{XX}} - \mathbf{S_{XZ}}\left(\mathbf{S_{ZZ}} + \boldsymbol{\Sigma}_\lambda^{-1}\right)^{-1} \mathbf{S}'_{\mathbf{XZ}}\right)^{-1} \left(\mathbf{S_{Xy}} - \mathbf{S_{XZ}}\left(\mathbf{S_{ZZ}} + \boldsymbol{\Sigma}_\lambda^{-1}\right)^{-1} \mathbf{S_{Zy}}\right). \tag{8A.5}$$

Likelihood Equations

We use the notation $\boldsymbol{\tau}$ to denote the remaining parameters so that $\{\sigma^2, \boldsymbol{\tau}\}$ represent the variance components. From standard normal theory (see Appendix B), the logarithmic likelihood is

$$L(\boldsymbol{\beta}, \sigma^2, \boldsymbol{\tau}) = -\frac{1}{2}\{N \ln 2\pi + N \ln \sigma^2 + \sigma^{-2}(\mathbf{y} - \mathbf{X}\boldsymbol{\beta})'\mathbf{V}^{-1}(\mathbf{y} - \mathbf{X}\boldsymbol{\beta}) + \ln \det \mathbf{V}\}. \tag{8A.6}$$

The corresponding restricted log likelihood is

$$L_{\mathrm{R}}(\boldsymbol{\beta}, \sigma^2, \boldsymbol{\tau}) = -\frac{1}{2}\{\ln \det(\mathbf{X}'\mathbf{V}^{-1}\mathbf{X}) - K \ln \sigma^2\} + L(\boldsymbol{\beta}, \sigma^2, \boldsymbol{\tau}) + \text{constant}. \tag{8A.7}$$

Either (8A.6) or (8A.7) can be maximized to determine an estimator of $\boldsymbol{\beta}$. The result is also the GLS estimator $\mathbf{b}_{\mathrm{GLS}}$, given in Equation (8A.5). Using $\mathbf{b}_{\mathrm{GLS}}$ for $\boldsymbol{\beta}$ in Equations (8A.6) and (8A.7) yields concentrated likelihoods. To determine the REML estimator of σ^2, we maximize $L_{\mathrm{R}}(\mathbf{b}_{\mathrm{GLS}}, \sigma^2, \boldsymbol{\tau})$ (holding $\boldsymbol{\tau}$ fixed), to get

$$s^2_{\mathrm{REML}} = (N - K)^{-1}(\mathbf{y} - \mathbf{X}\mathbf{b}_{\mathrm{GLS}})'\mathbf{V}^{-1}(\mathbf{y} - \mathbf{X}\mathbf{b}_{\mathrm{GLS}}). \tag{8A.8}$$

Thus, the log likelihood evaluated at these parameters is

$$L\left(\mathbf{b}_{\mathrm{GLS}}, s^2_{\mathrm{REML}}, \boldsymbol{\tau}\right) = -\frac{1}{2}\left\{N \ln 2\pi + N \ln s^2_{\mathrm{REML}} + N - K + \ln \det \mathbf{V}\right\}. \tag{8A.9}$$

The corresponding restricted log likelihood is

$$L_{\mathrm{R}} = -\frac{1}{2}\left\{\ln \det(\mathbf{X}'\mathbf{V}^{-1}\mathbf{X}) - K \ln s^2_{\mathrm{REML}}\right\} + L\left(\mathbf{b}_{\mathrm{GLS}}, s^2_{\mathrm{REML}}, \boldsymbol{\tau}\right) + \text{constant}. \tag{8A.10}$$

The likelihood expressions in Equations (8A.9) and (8A.10) are intuitively straightforward. However, because of the number of dimensions, they can be difficult to compute. We now provide alternative expressions that, although more complex in appearance, are simpler to compute. Using Equation (8A.4), we may express

$$\begin{aligned} Error\ SS = (N-K)s^2_{\text{REML}} &= \mathbf{y}\mathbf{V}^{-1}\mathbf{y} - \mathbf{y}'\mathbf{V}^{-1}\mathbf{X}\mathbf{b}_{\text{GLS}} \\ &= \mathbf{S}_{\mathbf{yy}} - \mathbf{S}'_{\mathbf{Zy}}\left(\mathbf{S}_{\mathbf{ZZ}} + \boldsymbol{\Sigma}_\lambda^{-1}\right)^{-1}\mathbf{S}_{\mathbf{Zy}} - \mathbf{S}'_{\mathbf{Xy}}\mathbf{b}_{\text{GLS}} \\ &\quad + \mathbf{S}'_{\mathbf{Zy}}\left(\mathbf{S}_{\mathbf{ZZ}} + \boldsymbol{\Sigma}_\lambda^{-1}\right)^{-1}\mathbf{S}'_{\mathbf{XZ}}\mathbf{b}_{\text{GLS}}. \end{aligned} \tag{8A.11}$$

From Equation (A.5) of Appendix A.5, we have

$$\ln\det\mathbf{V} = \ln\det\mathbf{V}_\alpha + \ln\det\boldsymbol{\Sigma}_\lambda + \ln\det\left(\mathbf{Z}'_\lambda\mathbf{V}_\alpha^{-1}\mathbf{Z}_\lambda + \boldsymbol{\Sigma}_\lambda^{-1}\right). \tag{8A.12}$$

Thus, the logarithmic likelihood evaluated at these parameters is

$$\begin{aligned} L\left(\mathbf{b}_{\text{GLS}}, s^2_{\text{REML}}, \boldsymbol{\tau}\right) = -\frac{1}{2}\{&N\ln 2\pi + N\ln s^2_{\text{REML}} + N - K + \mathbf{S}_{\mathbf{V}} + \ln\det\boldsymbol{\Sigma}_\lambda \\ &+ \ln\det\left(\mathbf{S}_{\mathbf{ZZ}} + \boldsymbol{\Sigma}_\lambda^{-1}\right)\}, \end{aligned} \tag{8A.13}$$

where $\mathbf{S}_{\mathbf{V}} = \sum_{i=1}^n \ln\det\mathbf{V}_{\alpha,i}$. The corresponding restricted log likelihood is

$$\begin{aligned} L_{\text{R}}(\boldsymbol{\tau}) = -\frac{1}{2}\{&\ln\det\left(\mathbf{S}_{\mathbf{XX}} - \mathbf{S}_{\mathbf{XZ}}\left(\mathbf{S}_{\mathbf{ZZ}} + \boldsymbol{\Sigma}_\lambda^{-1}\right)^{-1}\mathbf{S}'_{\mathbf{XZ}}\right) - K\ln s^2_{\text{REML}}\} \\ &+ L\left(\mathbf{b}_{\text{GLS}}, s^2_{\text{REML}}, \boldsymbol{\tau}\right) + \text{constant}. \end{aligned} \tag{8A.14}$$

Maximizing $L_{\text{R}}(\boldsymbol{\tau})$ over $\boldsymbol{\tau}$ yields the REML estimator of $\boldsymbol{\tau}$, say, $\boldsymbol{\tau}_{\text{REML}}$.

8A.3 Prediction

To derive the BLUP of $\boldsymbol{\lambda}$, we let $\mathbf{c}_\lambda$ be an arbitrary vector of constants and set $w = \mathbf{c}'_\lambda\boldsymbol{\lambda}$. With this choice, we have $\text{E}w = 0$. Using Equation (4.7), we have

$$\begin{aligned} \mathbf{c}'_\lambda\boldsymbol{\lambda}_{\text{BLUP}} &= \sigma^{-2}\mathbf{c}'_\lambda\text{Cov}(\boldsymbol{\lambda}, \mathbf{y})\mathbf{V}^{-1}(\mathbf{y} - \mathbf{X}\mathbf{b}_{\text{GLS}}) \\ &= \sigma^{-2}\mathbf{c}'_\lambda\text{Cov}(\boldsymbol{\lambda}, \mathbf{Z}_\lambda\boldsymbol{\lambda})\mathbf{V}^{-1}(\mathbf{y} - \mathbf{X}\,\mathbf{b}_{\text{GLS}}). \end{aligned}$$

With Wald's device, this yields $\boldsymbol{\lambda}_{\text{BLUP}} = \boldsymbol{\Sigma}_\lambda\mathbf{Z}'_\lambda\mathbf{V}^{-1}(\mathbf{y} - \mathbf{X}\mathbf{b}_{\text{GLS}})$. Further, using Equation (8A.4), we have

$$\begin{aligned} \boldsymbol{\Sigma}_\lambda\mathbf{Z}'_\lambda\mathbf{V}^{-1} &= \boldsymbol{\Sigma}_\lambda\mathbf{Z}'_\lambda\left(\mathbf{V}_\alpha^{-1} - \mathbf{V}_\alpha^{-1}\mathbf{Z}_\lambda\left(\mathbf{Z}'_\lambda\mathbf{V}_\alpha^{-1}\mathbf{Z}_\lambda + \boldsymbol{\Sigma}_\lambda^{-1}\right)^{-1}\mathbf{Z}'_\lambda\mathbf{V}_\alpha^{-1}\right) \\ &= \boldsymbol{\Sigma}_\lambda\left(\mathbf{I} - \mathbf{S}_{\mathbf{ZZ}}\left(\mathbf{S}_{\mathbf{ZZ}} + \boldsymbol{\Sigma}_\lambda^{-1}\right)^{-1}\right)\mathbf{Z}'_\lambda\mathbf{V}_\alpha^{-1} \\ &= \boldsymbol{\Sigma}_\lambda\left(\left(\mathbf{S}_{\mathbf{ZZ}} + \boldsymbol{\Sigma}_\lambda^{-1}\right) - \mathbf{S}_{\mathbf{ZZ}}\right)\left(\mathbf{S}_{\mathbf{ZZ}} + \boldsymbol{\Sigma}_\lambda^{-1}\right)^{-1}\mathbf{Z}'_\lambda\mathbf{V}_\alpha^{-1} \\ &= \left(\mathbf{S}_{\mathbf{ZZ}} + \boldsymbol{\Sigma}_\lambda^{-1}\right)^{-1}\mathbf{Z}'_\lambda\mathbf{V}_\alpha^{-1}. \end{aligned}$$

Thus,

$$\begin{aligned} \boldsymbol{\lambda}_{\text{BLUP}} &= \left(\mathbf{S}_{\mathbf{ZZ}} + \boldsymbol{\Sigma}_\lambda^{-1}\right)^{-1}\mathbf{Z}'_\lambda\mathbf{V}_\alpha^{-1}(\mathbf{y} - \mathbf{X}\mathbf{b}_{\text{GLS}}) \\ &= \left(\mathbf{S}_{\mathbf{ZZ}} + \boldsymbol{\Sigma}_\lambda^{-1}\right)^{-1}(\mathbf{S}_{\mathbf{Zy}} - \mathbf{S}_{\mathbf{ZX}}\mathbf{b}_{\text{GLS}}). \end{aligned} \tag{8A.15}$$

To simplify this expression, we recall the vector of residuals $\mathbf{e}_{i,\mathrm{GLS}} = \mathbf{y}_i - \mathbf{X}_i \mathbf{b}_{\mathrm{GLS}}$. This yields

$$\begin{aligned}\boldsymbol{\lambda}_{\mathrm{BLUP}} &= \left(\mathbf{S}_{\mathbf{ZZ}} + \boldsymbol{\Sigma}_{\lambda}^{-1}\right)^{-1} \sum_{i=1}^{n} \mathbf{Z}'_{\lambda,i} \mathbf{V}_{\alpha,i}^{-1} \left(\mathbf{y}_i - \mathbf{X}_i \mathbf{b}_{\mathrm{GLS}}\right) \\ &= \left(\mathbf{S}_{\mathbf{ZZ}} + \boldsymbol{\Sigma}_{\lambda}^{-1}\right)^{-1} \sum_{i=1}^{n} \mathbf{Z}'_{\lambda,i} \mathbf{V}_{\alpha,i}^{-1} \mathbf{e}_{i,\mathrm{GLS}},\end{aligned}$$

as in Equation (8.12).

We now consider predicting a linear combination of residuals, $w = \mathbf{c}'_{\epsilon} \boldsymbol{\epsilon}_i$, where $\mathbf{c}_{\varepsilon}$ is a vector of constants. With this choice, we have $\mathrm{E}w = 0$. Straightforward calculations show that

$$\mathrm{Cov}(w, \mathbf{y}_j) = \begin{cases} \mathbf{c}'_{\varepsilon} \sigma^2 \mathbf{R}_i & \text{for} \quad j = i, \\ \mathbf{0} & \text{for} \quad j \neq i. \end{cases}$$

Using this, Equations (4.7), (8A.4), and (8A.15) yield

$$\begin{aligned}\mathbf{c}'_{\varepsilon} \mathbf{e}_{i,\mathrm{BLUP}} &= \sigma^{-2} \mathrm{Cov}(\mathbf{c}'_{\varepsilon} \boldsymbol{\epsilon}_i, \mathbf{y}) \mathbf{V}^{-1} (\mathbf{y} - \mathbf{X} \mathbf{b}_{\mathrm{GLS}}) \\ &= \sigma^{-2} (\mathrm{Cov}(\mathbf{c}'_{\varepsilon} \boldsymbol{\epsilon}_i, \mathbf{y}) \mathbf{V}_{\alpha}^{-1} (\mathbf{y} - \mathbf{X} \mathbf{b}_{\mathrm{GLS}}) \\ &\quad - \mathrm{Cov}(\mathbf{c}'_{\varepsilon} \boldsymbol{\epsilon}_i, \mathbf{y}) \mathbf{V}_{\alpha}^{-1} \mathbf{Z}_{\lambda} (\mathbf{S}_{\mathbf{ZZ}} + \boldsymbol{\Sigma}_{\lambda}^{-1})^{-1} (\mathbf{S}_{\mathbf{Zy}} - \mathbf{S}_{\mathbf{ZX}} \mathbf{b}_{\mathrm{GLS}})) \\ &= \mathbf{c}'_{\varepsilon} \mathbf{R}_i \mathbf{V}_{\alpha,i}^{-1} \big((\mathbf{y}_i - \mathbf{X}_i \mathbf{b}_{\mathrm{GLS}}) - \mathbf{Z}_{\lambda,i} \left(\mathbf{S}_{\mathbf{ZZ}} + \boldsymbol{\Sigma}_{\lambda}^{-1}\right)^{-1} \\ &\quad \times (\mathbf{S}_{\mathbf{Zy}} - \mathbf{S}_{\mathbf{ZX}} \mathbf{b}_{\mathrm{GLS}}) \big) \\ &= \mathbf{c}'_{\varepsilon} \mathbf{R}_i \mathbf{V}_{\alpha,i}^{-1} (\mathbf{e}_{i,\mathrm{GLS}} - \mathbf{Z}_{\lambda,i} \boldsymbol{\lambda}_{\mathrm{BLUP}}).\end{aligned}$$

This yields Equation (8.13).

Similarly, we derive the BLUP of $\boldsymbol{\alpha}_i$. Let $\mathbf{c}_{\alpha}$ be an arbitrary vector of constants and set $w = \mathbf{c}'_{\alpha} \boldsymbol{\alpha}_i$. For this choice of w, we have $\mathrm{E}w = 0$. Further, we have

$$\mathrm{Cov}(\mathbf{c}'_{\alpha} \boldsymbol{\alpha}_i, \mathbf{y}_j) = \begin{cases} \sigma^2 \mathbf{c}'_{\alpha} \mathbf{D} \mathbf{Z}'_{\alpha,i} & \text{for} \quad j = i, \\ \mathbf{0} & \text{for} \quad j \neq i. \end{cases}$$

Using this, Equations (8A.4) and (4.7) yield

$$\begin{aligned}\mathbf{c}'_{\alpha} \boldsymbol{\alpha}_{i,\mathrm{BLUP}} &= \sigma^{-2} \mathrm{Cov}(\mathbf{c}'_{\alpha} \boldsymbol{\alpha}_i, \mathbf{y}) \mathbf{V}^{-1} (\mathbf{y} - \mathbf{X} \mathbf{b}_{\mathrm{GLS}}) \\ &= \sigma^{-2} \big(\mathrm{Cov}(\mathbf{c}'_{\alpha} \boldsymbol{\alpha}_i, \mathbf{y}) \mathbf{V}_{\alpha}^{-1} (\mathbf{y} - \mathbf{X} \mathbf{b}_{\mathrm{GLS}}) \\ &\quad - \mathrm{Cov}(\mathbf{c}'_{\alpha} \boldsymbol{\alpha}_i, \mathbf{y}) \mathbf{V}_{\alpha}^{-1} \mathbf{Z}_{\lambda} \left(\mathbf{S}_{\mathbf{ZZ}} + \boldsymbol{\Sigma}_{\lambda}^{-1}\right)^{-1} (\mathbf{S}_{\mathbf{Zy}} - \mathbf{S}_{\mathbf{ZX}} \mathbf{b}_{\mathrm{GLS}})\big) \\ &= \mathbf{c}'_{\alpha} \mathbf{D} \mathbf{Z}'_{\alpha,i} \mathbf{V}_{\alpha,i}^{-1} (\mathbf{e}_{i,\mathrm{GLS}} - \mathbf{Z}_{\lambda,i} \boldsymbol{\lambda}_{\mathrm{BLUP}}).\end{aligned}$$

Using Wald's device, we have the BLUP of $\boldsymbol{\alpha}_i$, given in Equation (8.14).

Forecasting

First note, from the calculation of BLUPs in Equation (4.7), that the BLUP projection is linear. That is, consider estimating the sum of two random variables, $w_1 + w_2$. Then, it is immediate that

$$\text{BLUP}(w_1 + w_2) = \text{BLUP}(w_1) + \text{BLUP}(w_2).$$

With this and Equation (8.9), we have

$$\begin{aligned}\hat{y}_{i,T_i+L} &= \text{BLUP}(y_{i,T_i+L}) = \text{BLUP}(\mathbf{z}'_{\alpha,i,T_i+L}\boldsymbol{\alpha}_i) + \text{BLUP}(\mathbf{z}'_{\lambda,i,T_i+L}\boldsymbol{\lambda}_{T_i+L}) \\ &\quad + \text{BLUP}(\mathbf{x}'_{i,T_i+L}\boldsymbol{\beta}) + \text{BLUP}(\varepsilon_{i,T_i+L}) \\ &= \mathbf{z}'_{\alpha,i,T_i+L}\mathbf{a}_{i,\text{BLUP}} + \mathbf{z}'_{\lambda,i,T_i+L}\text{BLUP}(\boldsymbol{\lambda}_{T_i+L}) \\ &\quad + \mathbf{x}'_{i,T_i+L}\mathbf{b}_{\text{GLS}} + \text{BLUP}(\varepsilon_{i,T_i+L}).\end{aligned}$$

From Equation (4.7) and the expression of $\boldsymbol{\lambda}_{\text{BLUP}}$, we have

$$\begin{aligned}\text{BLUP}(\boldsymbol{\lambda}_{T_i+L}) &= \sigma^{-2}\text{Cov}(\boldsymbol{\lambda}_{T_i+L}, \mathbf{y})\mathbf{V}^{-1}(\mathbf{y} - \mathbf{X}\,\mathbf{b}_{\text{GLS}}) \\ &= \sigma^{-2}\text{Cov}(\boldsymbol{\lambda}_{T_i+L}, \boldsymbol{\lambda})\mathbf{Z}'_{\lambda}\mathbf{V}^{-1}(\mathbf{y} - \mathbf{X}\,\mathbf{b}_{\text{GLS}}) \\ &= \sigma^{-2}\text{Cov}(\boldsymbol{\lambda}_{T_i+L}, \boldsymbol{\lambda})\boldsymbol{\Sigma}_{\lambda}^{-1}\boldsymbol{\lambda}_{\text{BLUP}}.\end{aligned}$$

From Equation (4.7) and the calculation of $\mathbf{e}_{i,\text{BLUP}}$, we have

$$\begin{aligned}&\text{BLUP}(\varepsilon_{i,T_i+L}) \\ &\quad = \sigma^{-2}\text{Cov}(\varepsilon_{i,T_i+L}, \mathbf{y})\mathbf{V}^{-1}(\mathbf{y} - \mathbf{X}\,\mathbf{b}_{\text{GLS}}) = \sigma^{-2}\text{Cov}(\varepsilon_{i,T_i+L}, \boldsymbol{\epsilon}_i)\mathbf{V}_{\alpha,i}^{-1} \\ &\qquad \times \left((\mathbf{y}_i - \mathbf{X}_i\mathbf{b}_{\text{GLS}}) - \mathbf{Z}_{\lambda,i}\left(\mathbf{S}_{\mathbf{ZZ}} + \boldsymbol{\Sigma}_{\lambda}^{-1}\right)^{-1}(\mathbf{S}_{\mathbf{Zy}} - \mathbf{S}_{\mathbf{ZX}}\mathbf{b}_{\text{GLS}})\right) \\ &\quad = \sigma^{-2}\text{Cov}(\varepsilon_{i,T_i+L}, \boldsymbol{\epsilon}_i)\mathbf{V}_{\alpha,i}^{-1}(\mathbf{e}_{i,\text{GLS}} - \mathbf{Z}_{\lambda,i}\boldsymbol{\lambda}_{\text{BLUP}}) \\ &\quad = \sigma^{-2}\text{Cov}(\varepsilon_{i,T_i+L}, \boldsymbol{\epsilon}_i)\mathbf{R}_i^{-1}\mathbf{e}_{i,\text{BLUP}}.\end{aligned}$$

Thus, the BLUP forecast of y_{i,T_i+L} is

$$\begin{aligned}\hat{y}_{i,T_i+L} &= \mathbf{x}'_{i,T_i+L}\mathbf{b}_{\text{GLS}} + \sigma^{-2}\mathbf{z}'_{\lambda,i,T_i+L}\text{Cov}(\boldsymbol{\lambda}_{T_i+L}, \boldsymbol{\lambda})\boldsymbol{\Sigma}_{\lambda}^{-1}\boldsymbol{\lambda}_{\text{BLUP}} \\ &\quad + \mathbf{z}'_{\alpha,i,T_i+L}\boldsymbol{\alpha}_{i,\text{BLUP}} + \sigma^{-2}\text{Cov}(\varepsilon_{i,T_i+L}, \boldsymbol{\epsilon}_i)\mathbf{R}_i^{-1}\mathbf{e}_{i,\text{BLUP}},\end{aligned}$$

as in Equation (8.16).

For forecasting, we wish to predict $w = y_{i,T_i+L}$, given in Equation (8.11). It is easy to see that

$$\begin{aligned}\text{Var}\, y_{i,T_i+L} &= \sigma^2\mathbf{z}'_{\alpha,i,T_i+L}\mathbf{D}\mathbf{z}_{\alpha,i,T_i+L} \\ &\quad + \mathbf{z}'_{\lambda,i,T_i+L}(\text{Var}\,\boldsymbol{\lambda}_{T_i+L})\,\mathbf{z}_{\lambda,i,T_i+L} + \text{Var}\,\varepsilon_{i,T_i+L}. \qquad (8\text{A}.16)\end{aligned}$$

To calculate the variance of the forecast error, we use Equation (4.9). First, note that

$$\mathbf{X}'\mathbf{V}^{-1}\mathbf{X} = \mathbf{S}_{\mathbf{XX}} - \mathbf{S}_{\mathbf{XZ}}\left(\mathbf{S}_{\mathbf{ZZ}} + \boldsymbol{\Sigma}_{\lambda}^{-1}\right)^{-1}\mathbf{S}'_{\mathbf{XZ}}. \tag{8A.17}$$

Next, we have

$$\begin{aligned}\operatorname{Cov}(y_{i,T_i+L}, \mathbf{y})\mathbf{V}^{-1}\mathbf{X} = {} & \sum_{j=1}^{n} \operatorname{Cov}(y_{i,T_i+L}, \mathbf{y}_j)\mathbf{V}_{\alpha,j}^{-1}\mathbf{X}_j \\ & - \left(\sum_{j=1}^{n} \operatorname{Cov}(y_{i,T_i+L}, \mathbf{y}_j)\mathbf{V}_{\alpha,j}^{-1}\mathbf{Z}_{\lambda,j}\right) \\ & \times \left(\mathbf{S}_{\mathbf{ZZ}} + \boldsymbol{\Sigma}_{\lambda}^{-1}\right)^{-1}\mathbf{S}_{\mathbf{ZX}}.\end{aligned} \tag{8A.18}$$

Similarly, we have

$$\begin{aligned}& \operatorname{Cov}(y_{i,T_i+L}, \mathbf{y})\mathbf{V}^{-1}\operatorname{Cov}(y_{i,T_i+L}, \mathbf{y}) \\ & \quad = \sum_{j=1}^{n} \operatorname{Cov}(y_{i,T_i+L}, \mathbf{y}_j)\mathbf{V}_{\alpha,j}^{-1}\operatorname{Cov}(y_{i,T_i+L}, \mathbf{y}_j) \\ & \qquad - \left(\sum_{j=1}^{n} \operatorname{Cov}(y_{i,T_i+L}, \mathbf{y}_j)\mathbf{V}_{\alpha,j}^{-1}\mathbf{Z}_{\lambda,j}\right)\left(\mathbf{S}_{\mathbf{ZZ}} + \boldsymbol{\Sigma}_{\lambda}^{-1}\right)^{-1} \\ & \qquad \times \left(\sum_{j=1}^{n} \operatorname{Cov}(y_{i,T_i+L}, \mathbf{y}_j)\mathbf{V}_{\alpha,j}^{-1}\mathbf{Z}_{\lambda,j}\right)'.\end{aligned} \tag{8A.19}$$

Thus, using Equation (4.8), we obtain the variance of the forecast error:

$$\begin{aligned}\operatorname{Var}(\hat{y}_{i,T_i+L} - y_{i,T_i+L}) = {} & (\mathbf{x}'_{i,T_i+L} - \operatorname{Cov}(y_{i,T_i+L}, \mathbf{y})\mathbf{V}^{-1}\mathbf{X})(\mathbf{X}'\mathbf{V}^{-1}\mathbf{X})^{-1} \\ & \times(\mathbf{x}'_{i,T_i+L} - \operatorname{Cov}(y_{i,T_i+L}, \mathbf{y})\mathbf{V}^{-1}\mathbf{X})' \\ & - \operatorname{Cov}(y_{i,T_i+L}, \mathbf{y})\mathbf{V}^{-1}\operatorname{Cov}(y_{i,T_i+L}, \mathbf{y})' + \operatorname{Var} y_{i,T_i+L},\end{aligned} \tag{8A.20}$$

where $\operatorname{Cov}(y_{i,T_i+L}, \mathbf{y})\mathbf{V}^{-1}\mathbf{X}$ is specified in Equation (8A.18), $\mathbf{X}'\mathbf{V}^{-1}\mathbf{X}$ is specified in Equation (8A.17), $\operatorname{Var} y_{i,T_i+L}$ is specified in Equation (8A.15), and $\operatorname{Cov}(y_{i,T_i+L}, \mathbf{y})\mathbf{V}^{-1}\operatorname{Cov}(y_{i,T_i+L}, \mathbf{y})'$ is specified in Equation (8A.19).

9
Binary Dependent Variables

Abstract. This chapter considers situations where the response of interest, y, takes on values 0 or 1, a *binary* dependent variable. To illustrate, one could use y to indicate whether or not a subject possesses an attribute or to indicate a choice made, for example, whether or not a taxpayer employs a professional tax preparer to file income tax returns.

Regression models that describe the behavior of binary dependent variables are more complex than linear regression models. Thus, Section 9.1 reviews basic modeling and inferential techniques without the heterogeneity components (so-called *homogeneous models*). Sections 9.2 and 9.3 include heterogeneity components by describing random- and fixed-effects models. Section 9.4 introduces a broader class of models known as *marginal models*, which can be estimated using a moment-based procedure known as *generalized estimating equations*.

9.1 Homogeneous Models

To introduce some of the complexities encountered with binary dependent variables, denote the probability that the response equals 1 by $p_{it} = \text{Prob}(y_{it} = 1)$. Then, we may interpret the mean response to be the probability that the response equals 1; that is, $\text{E}y_{it} = 0 \times \text{Prob}(y_{it} = 0) + 1 \times \text{Prob}(y_{it} = 1) = p_{it}$. Further, straightforward calculations show that the variance is related to the mean through the expression $\text{Var}\, y_{it} = p_{it}(1 - p_{it})$.

Linear Probability Models

Without heterogeneity terms, we begin by considering a linear model of the form

$$y_{it} = \mathbf{x}_{it}'\beta + \varepsilon_{it}, \tag{9.1}$$

known as a *linear probability model*. Assuming $\mathrm{E}\varepsilon_{it} = 0$, we have that $\mathrm{E}y_{it} = p_{it} = \mathbf{x}_{it}'\boldsymbol{\beta}$ and $\mathrm{Var}\, y_{it} = \mathbf{x}_{it}'\boldsymbol{\beta}(1 - \mathbf{x}_{it}'\boldsymbol{\beta})$. Linear probability models are widely applied because of the ease of parameter interpretations. For large data sets, the computational simplicity of OLS estimators is attractive when compared to some complex alternative nonlinear models introduced in this chapter. Further, OLS estimators for $\boldsymbol{\beta}$ have desirable properties. It is straightforward to check that they are consistent and asymptotically normal under mild conditions on the explanatory variables $\{\mathbf{x}_{it}\}$.

However, linear probability models have several drawbacks that are serious for many applications. These drawbacks include the following:

- The expected response is a probability and thus must vary between 0 and 1. However, the linear combination, $\mathbf{x}_{it}'\boldsymbol{\beta}$, can vary between negative and positive infinity. This mismatch implies, for example, that fitted values may be unreasonable.
- Linear models assume homoscedasticity (constant variance), yet the variance of the response depends on the mean that varies over observations. The problem of such fluctuating variance is known as *heteroscedasticity*.
- The response must be either a 0 or 1 although the regression models typically regard distribution of the error term as continuous. This mismatch implies, for example, that the usual residual analysis in regression modeling is meaningless.

To handle the heteroscedasticity problem, a (two-stage) weighted least-squares procedure is possible. That is, in the first stage, one uses ordinary least squares to compute estimates of $\boldsymbol{\beta}$. With this estimate, an estimated variance for each subject can be computed using the relation $\mathrm{Var}\, y_{it} = \mathbf{x}_{it}'\boldsymbol{\beta}(1 - \mathbf{x}_{it}'\boldsymbol{\beta})$. At the second stage, a weighted least squares is performed using the inverse of the estimated variances as weights to arrive at new estimates of $\boldsymbol{\beta}$. It is possible to iterate this procedure, although studies have shown that there are few advantages in doing so (see Carroll and Rupert, 1988G). Alternatively, one can use OLS estimators of $\boldsymbol{\beta}$ with standard errors that are robust to heteroscedasticity.

9.1.1 Logistic and Probit Regression Models

Using Nonlinear Functions of Explanatory Variables

To circumvent the drawbacks of linear probability models, we consider alternative models in which we express the expectation of the response as a function of explanatory variables, $p_{it} = \pi(\mathbf{x}_{it}'\beta) = \text{Prob}(y_{it} = 1 \mid \mathbf{x}_{it})$. We focus on two special cases of the function $\pi(.)$:

- $\pi(z) = 1/(1+e^{-z}) = e^z/(e^z+1)$, the *logit case*, and
- $\pi(z) = \Phi(z)$, the *probit case*.

Here, $\Phi(.)$ is the standard normal distribution function. Note that the choice of the identity function (a special kind of linear function), $\pi(z) = z$, yields the linear probability model. Thus, we focus on nonlinear choices of π. The inverse of the function, π^{-1}, specifies the form of the probability that is linear in the explanatory variables; that is, $\pi^{-1}(p_{it}) = \mathbf{x}_{it}'\beta$. In Chapter 10, we will refer to this inverse as the *link* function.

These two functions are similar in that they are almost linearly related over the interval $0.1 \leq \pi \leq 0.9$ (see McCullagh and Nelder, 1989G, p. 109). This similarity means that it will be difficult to distinguish between the two specifications with most data sets. Thus, to a large extent, the function choice depends on the preferences of the analyst.

Threshold Interpretation

Both the logit and probit cases can be justified by appealing to the following "threshold" interpretation of the model. To this end, suppose that there exists an underlying linear model, $y_{it}^* = \mathbf{x}_{it}'\beta + \varepsilon_{it}^*$. Here, we do not observe the response y_{it}^* yet interpret it to be the "propensity" to possess a characteristic. For example, we might think about the speed of a horse as a measure of its propensity to win a race. Under the threshold interpretation, we do not observe the propensity but we do observe when the propensity crosses a threshold. It is customary to assume that this threshold is 0, for simplicity. Thus, we observe

$$y_{it} = \begin{cases} 0 & y_{it}^* \leq 0, \\ 1 & y_{it}^* > 0. \end{cases}$$

To see how the logit case can be derived from the threshold model, we assume a logit distribution function for the disturbances, so that

$$\text{Prob}(\varepsilon_{it}^* \leq a) = \frac{1}{1+\exp(-a)}.$$

Because the logit distribution is symmetric about zero, we have that $\text{Prob}(\varepsilon_{it}^* \leq a) = \text{Prob}(-\varepsilon_{it}^* \leq a)$. Thus,

$$\begin{aligned} p_{it} &= \text{Prob}(y_{it} = 1) = \text{Prob}(y_{it}^* > 0) = \text{Prob}(\varepsilon_{it}^* \leq \mathbf{x}_{it}'\beta) \\ &= \frac{1}{1+\exp(-\mathbf{x}_{it}'\beta)} = \pi(\mathbf{x}_{it}'\beta). \end{aligned}$$

This establishes the threshold interpretation for the logit case. The development for the probit case is similar and is omitted.

Random-Utility Interpretation

Both the logit and probit cases can also be justified by appealing to the following "random utility" interpretation of the model. In economic applications, we think of an individual as selecting between two choices. To illustrate, in Section 9.1.3 we will consider whether or not a taxpayer chooses to employ a professional tax preparer to assist in filing an income tax return. Here, preferences among choices are indexed by an unobserved utility function; individuals select the choice that provides the greater utility.

For the ith individual at the tth time period, we use the notation u_{it} for this utility. We model utility as a function of an underlying value plus random noise, that is, $U_{itj} = u_{it}(V_{itj} + \varepsilon_{itj})$, where j may be 0 or 1, corresponding to the choice. To illustrate, we assume that the individual chooses the category corresponding to $j = 1$ if $U_{it1} > U_{it0}$ and denote this choice as $y_{it} = 1$. Assuming that u_{it} is a strictly increasing function, we have

$$\begin{aligned}\text{Prob}(y_{it} = 1) &= \text{Prob}(U_{it0} < U_{it1}) = \text{Prob}\left(u_{it}(V_{it0} + \varepsilon_{it0}) < u_{it}(V_{it1} + \varepsilon_{it1})\right) \\ &= \text{Prob}\left(\varepsilon_{it0} - \varepsilon_{it1} < V_{it1} - V_{it0}\right).\end{aligned}$$

To parameterize the problem, assume that the value function is an unknown linear combination of explanatory variables. Specifically, we take $V_{it0} = 0$ and $V_{it1} = \mathbf{x}_{it}'\boldsymbol{\beta}$. We may take the difference in the errors, $\varepsilon_{it0} - \varepsilon_{it1}$, to be normal or logistic, corresponding to the probit and logit cases, respectively. In Section 11.1, we will show that the logistic distribution is satisfied if the errors are assumed to have an extreme-value, or Gumbel, distribution. In Section 9.1.3, linear combinations of taxpayer characteristics will allow us to model the choice of using a professional tax preparer. The analysis allows for taxpayer preferences to vary by subject and over time.

Example 9.1: Job Security Valletta (1999E) studied declining job security using the PSID (Panel Survey of Income Dynamics) database (see Appendix F). We consider here one of the regressions presented by Valetta, based on a sample of male household heads that consists of $N = 24{,}168$ observations over the years 1976–1992, inclusive. The PSID survey records reasons why men left their most recent employment, including plant closures, "quitting" and changing jobs for other reasons. However, Valetta focused on dismissals ("laid off" or "fired") because involuntary separations are associated with job

Table 9.1. *Dismissal probit regression estimates*

Variable	Parameter estimate	Standard error
Tenure	−0.084	0.010
Time trend	−0.002	0.005
Tenure* (time trend)	0.003	0.001
Change in logarithmic sector employment	0.094	0.057
Tenure* (change in logarithmic sector employment)	−0.020	0.009
−2 log likelihood	7,027.8	
Pseudo-R^2	0.097	

insecurity. Chapter 11 will expand this discussion to consider the other sources of job turnover.

Table 9.1 presents a probit regression model run by Valetta (1999E), using dismissals as the dependent variable. In addition to the explanatory variables listed in Table 9.1, other variables controlled for consisted of education, marital status, number of children, race, years of full-time work experience and its square, union membership, government employment, logarithmic wage, the U.S. employment rate, and location as measured through the Metropolitan Statistical Area residence. In Table 9.1, tenure is years employed at the current firm. Sector employment was measured by examining Consumer Price Survey employment in 387 sectors of the economy, based on 43 industry categories and 9 regions of the country.

On the one hand, the tenure coefficient reveals that more experienced workers are less likely to be dismissed. On the other hand, the coefficient associated with the interaction between tenure and time trend reveals an increasing dismissal rate for experienced workers.

The interpretation of the sector employment coefficients is also of interest. With an average tenure of about 7.8 years in the sample, we see that low-tenure men are relatively unaffected by changes in sector employment. However, for more experienced men, there is an increasing probability of dismissal associated with sectors of the economy where growth declines.

Valetta also fit a random-effects model, which will be described in Section 9.2; the results were qualitatively similar to those presented here.

Logistic Regression

An advantage of the logit case is that it permits closed-form expressions, unlike the normal distribution function. *Logistic regression* is another phrase used to describe the logit case.

Using $p = \pi(z)$, the inverse of π can be calculated as $z = \pi^{-1}(p) = \ln(p/(1-p))$. To simplify future presentations, we define $\text{logit}(p) = \ln(p/(1-p))$ to be the *logit function*. With a logistic regression model, we represent the linear combination of explanatory variables as the logit of the success probability; that is, $\mathbf{x}_{it}'\boldsymbol{\beta} = \text{logit}(p_{it})$.

Odds Ratio Interpretation

When the response y is binary, knowing only the probability of $y = 1$, p, summarizes the distribution. In some applications, a simple transformation of p has an important interpretation. The lead example of this is the *odds ratio*, given by $p/(1-p)$. For example, suppose that y indicates whether or not a horse wins a race; that is, $y = 1$ if the horse wins and $y = 0$ if the horse does not. Interpret p to be the probability of the horse winning the race and, as an example, suppose that $p = 0.25$. Then, the *odds* of the horse winning the race is $0.25/(1.00 - 0.25) = 0.3333$. We might say that the odds of winning are 0.3333 to 1, or one to three. Equivalently, we can say that the probability of not winning is $1 - p = 0.75$. Thus, the odds of the horse not winning is $0.75/(1 - 0.75) = 3$. We interpret this to mean the odds against the horse are three to one.

Odds have a useful interpretation from a betting standpoint. Suppose that we are playing a fair game and that we place a bet of \$1 with odds of one to three. If the horse wins, then we get our \$1 back plus winnings of \$3. If the horse loses, then we lose our bet of \$1. It is a fair game in the sense that the expected value of the game is zero because we win \$3 with probability $p = 0.25$ and lose \$1 with probability $1 - p = 0.75$. From an economic standpoint, the odds provide the important numbers (bet of \$1 and winnings of \$3), not the probabilities. Of course, if we know p, then we can always calculate the odds. Similarly, if we know the odds, we can always calculate the probability p.

The logit is the logarithmic odds function, also known as the *log odds*.

Logistic Regression Parameter Interpretation

To interpret the regression coefficients in the logistic regression model, $\boldsymbol{\beta} = (\beta_1, \beta_2, \ldots, \beta_K)'$, we begin by assuming that the jth explanatory variable, x_{itj}, is either 0 or 1. Then, with the notation $\mathbf{x}_{it} = (x_{it1} \;\cdots\; x_{itj} \;\cdots\; x_{itK})'$, we may interpret

$$\begin{aligned}
\beta_j &= (x_{it1} \;\cdots\; 1 \;\cdots\; x_{itK})'\boldsymbol{\beta} - (x_{it1} \;\cdots\; 0 \;\cdots\; x_{itK})'\boldsymbol{\beta} \\
&= \ln\left(\frac{\text{Prob}(y_{it} = 1 \mid x_{itj} = 1)}{1 - \text{Prob}(y_{it} = 1 \mid x_{itj} = 1)}\right) - \ln\left(\frac{\text{Prob}(y_{it} = 1 \mid x_{itj} = 0)}{1 - \text{Prob}(y_{it} = 1 \mid x_{itj} = 0)}\right).
\end{aligned}$$

Thus,

$$e^{\beta_j} = \frac{\mathrm{Prob}(y_{it} = 1 \mid x_{itj} = 1)/(1 - \mathrm{Prob}(y_{it} = 1 \mid x_{itj} = 1))}{\mathrm{Prob}(y_{it} = 1 \mid x_{itj} = 0)/(1 - \mathrm{Prob}(y_{it} = 1 \mid x_{itj} = 0))}.$$

We note that the numerator of this expression is the odds when $x_{itj} = 1$, whereas the denominator is the odds when $x_{itj} = 0$. Thus, we can say that the odds when $x_{itj} = 1$ are $\exp(\beta_j)$ times as large as the odds when $x_{itj} = 0$. For example, if $\beta_j = 0.693$, then $\exp(\beta_j) = 2$. From this, we say that the odds (for $y = 1$) are twice as great for $x_j = 1$ as those for $x_j = 0$.

Similarly, assuming that the jth explanatory variable is continuous (differentiable), we have

$$\beta_j = \frac{\partial}{\partial x_{itj}} \mathbf{x}_{it}'\beta = \frac{\partial}{\partial x_{itj}} \ln\left(\frac{\mathrm{Prob}(y_{it} = 1 \mid x_{itj})}{1 - \mathrm{Prob}(y_{it} = 1 \mid x_{itj})}\right)$$

$$= \frac{\frac{\partial}{\partial x_{itj}}(\mathrm{Prob}(y_{it} = 1 \mid x_{itj})/(1 - \mathrm{Prob}(y_{it} = 1 \mid x_{itj})))}{\mathrm{Prob}(y_{it} = 1 \mid x_{itj})/(1 - \mathrm{Prob}(y_{it} = 1 \mid x_{itj}))}.$$

Thus, we may interpret β_j as the proportional change in the odds ratio, known as an *elasticity* in economics.

9.1.2 Inference for Logistic and Probit Regression Models

Parameter Estimation

The customary method of estimation for homogenous models is maximum likelihood, described in further detail in Appendix C. To provide intuition, we outline the ideas in the context of binary dependent variable regression models.

The *likelihood* is the observed value of the density or mass function. For a single observation, the likelihood is

$$\begin{cases} 1 - p_{it} & \text{if } y_{it} = 0, \\ p_{it} & \text{if } y_{it} = 1. \end{cases}$$

The objective of maximum likelihood estimation is to find the parameter values that produce the largest likelihood. Finding the maximum of the logarithmic function typically yields the same solution as finding the maximum of the corresponding function. Because it is generally computationally simpler, we consider the logarithmic (log) likelihood, written as

$$\begin{cases} \ln(1 - p_{it}) & \text{if } y_{it} = 0, \\ \ln p_{it} & \text{if } y_{it} = 1. \end{cases}$$

More compactly, the log likelihood of a single observation is

$$y_{it} \ln \pi(\mathbf{x}_{it}'\beta) + (1 - y_{it}) \ln(1 - \pi(\mathbf{x}_{it}'\beta)),$$

where $p_{it} = \pi(\mathbf{x}_{it}'\beta)$. Assuming independence among observations, the likelihood of the data set is a product of likelihoods of each observation. Thus, taking logarithms, the log likelihood of the data set is the sum of log likelihoods of single observations. The log likelihood of the data set is

$$L(\beta) = \sum_{it} \{y_{it} \ln \pi(\mathbf{x}_{it}'\beta) + (1 - y_{it}) \ln(1 - \pi(\mathbf{x}_{it}'\beta))\}, \tag{9.2}$$

where the sum ranges over $\{t = 1, \ldots, T_i, i = 1, \ldots, n\}$. The (log) likelihood is viewed as a function of the parameters, with the data held fixed. In contrast, the joint probability mass (density) function is viewed as a function of the realized data, with the parameters held fixed.

The method of maximum likelihood means finding the values of β that maximize the log likelihood. The customary method of finding the maximum is taking partial derivatives with respect to the parameters of interest and finding roots of these equations. In this case, taking partial derivatives with respect to β yields the *score equations*

$$\frac{\partial}{\partial \beta} L(\beta) = \sum_{it} \mathbf{x}_{it}(y_{it} - \pi(\mathbf{x}_{it}'\beta)) \frac{\pi'(\mathbf{x}_{it}'\beta)}{\pi(\mathbf{x}_{it}'\beta)(1 - \pi(\mathbf{x}_{it}'\beta))} = \mathbf{0}. \tag{9.3}$$

The solution of these equations, say $\mathbf{b}_{\mathrm{MLE}}$, is the MLE. For example, for the logit function, the score equations in Equation (9.3) reduce to

$$\sum_{it} \mathbf{x}_{it}(y_{it} - \pi(\mathbf{x}_{it}'\beta)) = \mathbf{0}, \tag{9.4}$$

where $\pi(z) = (1 + \exp(-z))^{-1}$. We note that the solution depends on the responses y_{it} only through the statistics $\Sigma_{it}\mathbf{x}_{it} y_{it}$. This property, known as *sufficiency*, will be important in Section 9.3.

An alternative expression for the score equations in Equation (9.3) is

$$\sum_{it} \frac{\partial}{\partial \beta}(\mathrm{E} y_{it})(\mathrm{Var}\ y_{it})^{-1} (y_{it} - \mathrm{E} y_{it}) = \mathbf{0}, \tag{9.5}$$

where $\mathrm{E} y_{it} = \pi(\mathbf{x}_{it}'\beta)$,

$$\frac{\partial}{\partial \beta} \mathrm{E} y_{it} = \mathbf{x}_{it} \pi'(\mathbf{x}_{it}'\beta),$$

and $\mathrm{Var}\ y_{it} = \pi(\mathbf{x}_{it}'\beta)(1 - \pi(\mathbf{x}_{it}'\beta))$. The expression in Equation (9.5) is an example of a *generalized estimating equation*, which will be introduced formally in Section 9.4.

An estimator of the asymptotic variance of $\boldsymbol{\beta}$ may be calculated by taking partial derivatives of the score equations. Specifically, the term

$$-\frac{\partial^2}{\partial \boldsymbol{\beta} \partial \boldsymbol{\beta}'} L(\boldsymbol{\beta}) \bigg|_{\boldsymbol{\beta}=\mathbf{b}_{\text{MLE}}}$$

is the *information matrix* evaluated at $\mathbf{b}_{\text{MLE}}$. As an illustration, using the logit function, straightforward calculations show that the information matrix is

$$\sum_{it} \mathbf{x}_{it}\mathbf{x}_{it}' \pi(\mathbf{x}_{it}'\boldsymbol{\beta})(1 - \pi(\mathbf{x}_{it}'\boldsymbol{\beta})).$$

The square root of the jth diagonal element of this matrix yields the standard error for the jth row of $\mathbf{b}_{j,\text{MLE}}$, which we denote as $se(\mathbf{b}_{j,\text{MLE}})$.

To assess the overall model fit, it is customary to cite *likelihood ratio test statistics* in nonlinear regression models. To test the overall model adequacy $\text{H}_0 : \boldsymbol{\beta} = \mathbf{0}$, we use the statistic

$$LRT = 2 \times (L(\mathbf{b}_{\text{MLE}}) - L_0),$$

where L_0 is the maximized log likelihood with only an intercept term. Under the null hypothesis H_0, this statistic has a chi-square distribution with K degrees of freedom. Appendix C.8 describes likelihood ratio test statistics in greater detail.

As described in Appendix C.9, measures of goodness of fit are difficult to interpret in nonlinear models. One measure is the *max-scaled* R^2, defined as $R^2_{\text{ms}} = R^2 / R^2_{\text{max}}$, where

$$R^2 = 1 - \left(\frac{\exp(L_0/N)}{\exp(L(\mathbf{b}_{\text{MLE}})/N)} \right)^2$$

and $R^2_{\text{max}} = 1 - \exp(L_0/N)^2$. Here, L_0/N represents the average value of this log likelihood.

9.1.3 Example: Income Tax Payments and Tax Preparers

To illustrate the methods described in this section, we return to the income tax payments example introduced in Section 3.2. For this chapter, we now will use the demographic and economic characteristics of a taxpayer described in Table 3.1 to model the choice of using a professional tax preparer, denoted by PREP. The one exception is that we will not consider the variable LNTAX, the tax paid in logarithmic dollars. Although tax paid is clearly related to the

Table 9.2. *Averages of binary variables by level of PREP*

PREP	Number	MS	HH	AGE	EMP
0	671	0.542	0.106	0.072	0.092
1	619	0.709	0.066	0.165	0.212

Table 9.3. *Summary statistics for other variables by level of PREP*

Variable	PREP	Mean	Median	Minimum	Maximum	Standard deviation
DEPEND	0	2.267	2	0	6	1.301
	1	2.585	2	0	6	1.358
LNTPI	0	9.732	9.921	−0.128	12.043	1.089
	1	10.059	10.178	−0.092	13.222	1.220
MR	0	21.987	21	0	50	11.168
	1	25.188	25	0	50	11.536

Table 9.4. *Counts of taxpayers by levels of PREP and EMP*

		EMP		
		0	1	Total
PREP	0	609	62	671
	1	488	131	619
Total		1,097	193	1,290

choice of a professional tax preparer, it is not clear whether this can serve as an "independent" explanatory variable. In econometric terminology, this is considered an *endogenous* variable (see Chapter 6).

Many summary statistics of the data were discussed in Sections 3.2 and 7.2. Tables 9.2 and 9.3 show additional statistics, by level of PREP. Table 9.2 shows that those taxpayers using a professional tax preparer (PREP = 1) were more likely to be married, not the head of a household, age 65 and over, and self-employed. Table 9.3 shows that those taxpayers using a professional tax preparer had more dependents, had a larger income, and were in a higher tax bracket.

Table 9.4 provides additional information about the relation between EMP and PREP. For example, for those self-employed individuals (EMP = 1),

Display 9.1 Selected SAS Output

```
The LOGISTIC Procedure
                              Model Information
        Response Variable               PREP
        Number of Response Levels       2
        Number of Observations          1290
        Link Function                   Logit
        Optimization Technique          Fisher's scoring
                            Model Fit Statistics
                                                    Intercept
                                      Intercept       and
                Criterion               Only        Covariates
                AIC                   1788.223        1726.981
                SC                    1793.385        1747.630
                -2 Log L              1786.223        1718.981
         R-Square     0.0508     Max-rescaled R-Square     0.0678
                Testing Global Null Hypothesis: BETA=0
        Test                        Chi-Square        DF      Pr > ChiSq
        Likelihood Ratio               67.2422         3          <.0001
        Score                          65.0775         3          <.0001
        Wald                           60.5549         3          <.0001
               Analysis of Maximum Likelihood Estimates
                                     Standard
Parameter     DF     Estimate           Error      Chi-Square      Pr > ChiSq
Intercept      1      -2.3447          0.7754          9.1430          0.0025
LNTPI          1       0.1881          0.0940          4.0017          0.0455
MR             1       0.0108         0.00884          1.4964          0.2212
EMP            1       1.0091          0.1693         35.5329          <.0001
                         Odds Ratio Estimates
                         Point            95% Wald
               Effect    Estimate       Confidence Limits
               LNTPI        1.207       1.004         1.451
               MR           1.011       0.994         1.029
               EMP          2.743       1.969         3.822
```

67.9% (= 131/193) of the time they chose to use a tax preparer compared to 44.5% (= 488/1,097) for those not self-employed. Put another way, the odds of self-employed individuals using a preparer are $2.11(=0.679/(1-0.679))$ compared to $0.801(=0.445/(1-0.445))$ for those not self-employed.

Display 9.1 shows a fitted logistic regression model, using LNTPI, MR, and EMP as explanatory variables. The calculations were done using SAS PROC LOGISTIC. To interpret this output, we first note that the likelihood ratio test statistic for checking model adequacy is

$$LRT = 67.2422 = 2 \times L(\mathbf{b}_{\text{MLE}}) - 2 \times L_0 = 1786.223 - 1718.981.$$

Compared to a chi-square with $K = 3$ degrees of freedom, this indicates that at least one of the variables LNTPI, MR, and EMP is a statistically significant predictors of PREP. Additional model fit statistics, including *Akaikie's information criterion* and *Schwarz's criterion* are described in Appendix C.9.

We interpret the R^2 statistic to mean that there is substantial information regarding PREP that is not explained by LNTPI, MR, and EMP. It is useful to confirm the calculation of this statistic, which is

$$\begin{aligned} R^2 &= 1 - \frac{\exp(2 \times L_0/N)}{\exp(2 \times L(\mathbf{b}_{\mathrm{MLE}})/N)} \\ &= 1 - \frac{\exp(2 \times 1786.223/1290)}{\exp(2 \times 1718.981/1290)} = 0.05079. \end{aligned}$$

For parameter interpretation, we note that the coefficient associated with EMP is $b_{\mathrm{EMP}} = 1.0091$. Thus, we interpret the odds associated with this estimator, $\exp(1.0091) = 2.743$, to mean that self-employed taxpayers (EMP $= 1$) are 2.743 times more likely to employ a professional tax preparer compared to taxpayers that are not self-employed.

9.2 Random-Effects Models

This section introduces models that use random effects to accommodate heterogeneity. Section 9.3 follows up with the corresponding fixed-effects formulation. In contrast, in the linear models portion of the text, we first introduced fixed effects (in Chapter 2) followed by random effects (in Chapter 3). The consistency between these seemingly different approaches is that the text approaches data modeling from an applications orientation. Specifically, for estimation and ease of explanation to users, typically the fixed-effects formulation is simpler than the random-effects alternative in linear models. This is because fixed-effects models are simply special cases of analysis of covariance models, representations that are familiar from applied regression analysis. In contrast, in nonlinear cases such as models with binary dependent variables, random-effects models are simpler than corresponding fixed-effects alternatives. Here, this is in part computational because random-effects summary statistics are easier to calculate. Further, as we will see in Section 9.3, standard estimation routines, such as maximum likelihood, yield fixed-effects estimators that do not have the usual desirable asymptotic properties. Thus, the fixed-effects formulation requires specialized estimators that can be cumbersome to compute and explain to users.

As in Section 9.1, we expressed the probability of a response equal to one as a function of linear combinations of explanatory variables. To accommodate heterogeneity, we incorporate subject-specific variables of the form $\mathrm{Prob}(y_{it} = 1 \mid \alpha_i) = \pi(\alpha_i + \mathbf{x}_{it}'\boldsymbol{\beta})$. Here, the subject-specific effects account only for the intercepts and do not include other variables. Chapter 10 will introduce extensions to variable-slope models. We assume that $\{\alpha_i\}$ are random effects in this section.

To motivate the random-effects formulation, we may assume the two-stage sampling scheme introduced in Section 3.1.

Stage 1. Draw a random sample of n subjects from a population. The subject-specific parameter α_i is associated with the ith subject.
Stage 2. Conditional on α_i, draw realizations of $\{y_{it}, \mathbf{x}_{it}\}$, for $t = 1, \ldots, T_i$ for the ith subject.

In the first stage, one draws subject-specific effects $\{\alpha_i\}$ from a population. In the second stage, for each subject i, one draws a random sample of T_i responses y_{it}, for $t = 1, \ldots, T_i$, and also observes the explanatory variables $\{\mathbf{x}_{it}\}$.

Random-Effects Likelihood

To develop the likelihood, first note that from the second sampling stage, conditional on α_i, the likelihood for the ith subject at the tth observation is

$$p(y_{it}; \boldsymbol{\beta} \mid \alpha_i) = \begin{cases} \pi(\alpha_i + \mathbf{x}_{it}'\boldsymbol{\beta}) & \text{if } y_{it} = 1, \\ 1 - \pi(\alpha_i + \mathbf{x}_{it}'\boldsymbol{\beta}) & \text{if } y_{it} = 0. \end{cases}$$

We summarize this as

$$p(y_{it}; \boldsymbol{\beta} \mid \alpha_i) = \pi(\alpha_i + \mathbf{x}_{it}'\boldsymbol{\beta})^{y_{it}}(1 - \pi(\alpha_i + \mathbf{x}_{it}'\boldsymbol{\beta}))^{1-y_{it}}.$$

Because of the independence among responses for a subject conditional on α_i, the conditional likelihood for the ith subject is

$$p(\mathbf{y}_i; \boldsymbol{\beta} \mid \alpha_i) = \prod_{t=1}^{T_i} \pi(\alpha_i + \mathbf{x}_{it}'\boldsymbol{\beta})^{y_{it}}(1 - \pi(\alpha_i + \mathbf{x}_{it}'\boldsymbol{\beta}))^{1-y_{it}}.$$

Taking expectations over α_i yields the unconditional likelihood. Thus, the (unconditional) likelihood for the ith subject is

$$p(\mathbf{y}_i; \boldsymbol{\beta}, \tau) = \int \left\{ \prod_{t=1}^{T_i} \pi(a + \mathbf{x}_{it}'\boldsymbol{\beta})^{y_{it}}(1 - \pi(a + \mathbf{x}_{it}'\boldsymbol{\beta}))^{1-y_{it}} \right\} dF_\alpha(a). \quad (9.6)$$

In Equation (9.6), τ is a parameter of the distribution of α_i, $F_\alpha(.)$. Although not necessary, it is customary to use a normal distribution for $F_\alpha(.)$. In this case, τ represents the variance of this mean-zero distribution. With a specification for F_α, the log likelihood for the data set is

$$L(\boldsymbol{\beta}, \tau) = \sum_{i=1}^{n} \ln \mathrm{p}(\mathbf{y}_i, \boldsymbol{\beta}, \tau).$$

To determine the MLEs, one maximizes the log likelihood $L(\boldsymbol{\beta}, \tau)$ as a function of $\boldsymbol{\beta}$ and τ. Closed-form analytical solutions for this maximization problem do not exist in general, although numerical solutions are feasible with high-speed computers. The MLEs can be determined by solving for the roots of the $K + 1$ score equations

$$\frac{\partial}{\partial \boldsymbol{\beta}} L(\boldsymbol{\beta}, \tau) = \mathbf{0}$$

and

$$\frac{\partial}{\partial \tau} L(\boldsymbol{\beta}, \tau) = 0.$$

Further, asymptotic variances can be computed by taking the matrix of second derivatives of the log likelihood, known as the *information matrix*. Appendix C provides additional details.

There are two commonly used specifications of the conditional distribution in the random-effects model:

- A logit model for the conditional distribution of a response. That is,

$$\text{Prob}(y_{it} = 1 \mid \alpha_i) = \pi(\alpha_i + \mathbf{x}'_{it}\boldsymbol{\beta}) = \frac{1}{1 + \exp(-(\alpha_i + \mathbf{x}'_{it}\boldsymbol{\beta}))}.$$

- A probit model for the conditional distribution of a response. That is, $\text{Prob}(y_{it} = 1 \mid \alpha_i) = \Phi(\alpha_i + \mathbf{x}'_{it}\boldsymbol{\beta})$, where Φ is the standard normal distribution function.

There are no important advantages or disadvantages when choosing the conditional probability π to be either a logit or a probit. The likelihood involves roughly the same amount of work to evaluate and maximize, although the logit function is slightly easier to evaluate than the standard normal distribution function. The probit model can be easier to interpret because unconditional probabilities can be expressed in terms of the standard normal distribution function. That is, assuming normality for α_i, we have

$$\text{Prob}(y_{it} = 1) = \text{E}\Phi(\alpha_i + \mathbf{x}'_{it}\boldsymbol{\beta}) = \Phi\left(\frac{\mathbf{x}'_{it}\boldsymbol{\beta}}{\sqrt{1 + \tau}}\right).$$

Example: Income Tax Payments and Tax Preparers (continued) To see how a random-effects dependent-variable model works with a data set, we return to the Section 9.1.3 example. Display 9.2 shows a fitted model, using LNTPI, MR, and EMP as explanatory variables. The calculations were done using the SAS procedure NLMIXED. This procedure uses a numerical integration technique for mixed-effect models, called *adaptive Gaussian quadrature*

Display 9.2 Selected SAS Output

```
                    The NLMIXED Procedure
                       Specifications
Dependent Variable                       PREP
Distribution for Dependent Variable      Binomial
Distribution for Random Effects          Normal
Optimization Technique                   Dual Quasi-Newton
Integration Method                       Adaptive Gaussian Quadrature
                       Fit Statistics
          -2 Log Likelihood                  1719.7
          AIC (smaller is better)            1729.7
          AICC (smaller is better)           1729.8
          BIC (smaller is better)            1755.5
             Correlation Matrix of Parameter Estimates
```

Row	Parameter	int	bLNTPI	bMR	bEMP	sigma
1	int	1.0000	-0.9995	-0.9900	-0.9986	-0.9988
2	bLNTPI	-0.9995	1.0000	0.9852	0.9969	0.9971
3	bMR	-0.9900	0.9852	1.0000	0.9940	0.9944
4	bEMP	-0.9986	0.9969	0.9940	1.0000	0.9997
5	sigma	-0.9988	0.9971	0.9944	0.9997	1.0000

(see Pinheiro and Bates, 2000S, for a description). Display 9.2 shows that this random-effects specification is not a desirable model for this data set. By conditioning on the random effects, the parameter estimates turn out to be highly correlated with one another.

Multilevel Model Extensions

We saw that there were a sufficient number of applications to devote an entire chapter (5) to a multilevel framework for linear models. However, extensions to nonlinear models such as binary dependent-variable models are only now coming into regular use in the applied sciences. Here we present the development of three-level extensions of probit and logit models due to Gibbons and Hedeker (1997B).

We use a set of notation similar to that developed in Section 5.1.2. Let there be $i = 1, \ldots, n$ subjects (schools, for example), $j = 1, \ldots, J_i$ clusters within each subject (classrooms, for example), and $t = 1, \ldots, T_{ij}$ observations within each cluster (over time, or students with a classroom). Combining the three levels, Gibbons and Hedeker considered

$$y_{ijt}^* = \alpha_i + \mathbf{z}_{ijt}'\boldsymbol{\alpha}_{ij} + \mathbf{x}_{ijt}'\boldsymbol{\beta} + \varepsilon_{ijt}, \tag{9.7}$$

where α_i represents a level-three heterogeneity term, $\boldsymbol{\alpha}_{ij}$ represents a level-two vector of heterogeneity terms, and ε_{ijt} represents the level-one disturbance term. All three terms have zero mean; the means of each level are already accounted for in $\mathbf{x}_{ijt}'\boldsymbol{\beta}$. The left-hand variable of Equation (9.7) is latent; we actually observe y_{ijt}, which is a binary variable, corresponding to whether the latent variable y_{ijt}^* crosses a threshold. The conditional probability of this event is

$$\text{Prob}(y_{ijt} = 1 \mid \alpha_i, \boldsymbol{\alpha}_{ij}) = \pi(\alpha_i + \mathbf{z}_{ijt}'\boldsymbol{\alpha}_{ij} + \mathbf{x}_{ijt}'\boldsymbol{\beta}),$$

where $\pi(.)$ is either logit or probit. To complete the model specification, we assume that $\{\alpha_i\}$ and $\{\boldsymbol{\alpha}_{ij}\}$ are each i.i.d. and independent of one another.

The model parameters can be estimated via maximum likelihood. As pointed out by Gibbons and Hedeker, the main computational technique is to take advantage of the independence that we typically assume among levels in multilevel modeling. That is, defining $\mathbf{y}_{ij} = (y_{ij1}, \ldots, y_{ijT_{ij}})'$, the conditional probability mass function is

$$p(\mathbf{y}_{ij}; \boldsymbol{\beta} \mid \alpha_i, \boldsymbol{\alpha}_{ij})$$
$$= \prod_{t=1}^{T_{ij}} \pi(\alpha_i + \mathbf{z}'_{ijt}\boldsymbol{\alpha}_{ij} + \mathbf{x}'_{ijt}\boldsymbol{\beta})^{y_{ijt}} (1 - \pi(\alpha_i + \mathbf{z}'_{ijt}\boldsymbol{\alpha}_{ij} + \mathbf{x}'_{ijt}\boldsymbol{\beta}))^{1-y_{ijt}}.$$

Integrating over the level-two heterogeneity effects, we have

$$p(\mathbf{y}_{ij}; \boldsymbol{\beta}, \boldsymbol{\Sigma}_2 \mid \alpha_i) = \int_{\mathbf{a}} p(\mathbf{y}_{ij}; \boldsymbol{\beta} \mid \alpha_i, \mathbf{a}) dF_{\alpha,2}(\mathbf{a}), \qquad (9.8)$$

where $F_{\alpha,2}(\cdot)$ is the distribution function of $\{\boldsymbol{\alpha}_{ij}\}$ and $\boldsymbol{\Sigma}_2$ are the parameters associated with it. Following Gibbons and Hedeker, we assume that $\boldsymbol{\alpha}_{ij}$ is normally distributed with mean zero and variance–covariance $\boldsymbol{\Sigma}_2$.

Integrating over the level-three heterogeneity effects, we have

$$p\left(\mathbf{y}_{i1}, \ldots, \mathbf{y}_{iJ_i}; \boldsymbol{\beta}, \boldsymbol{\Sigma}_2, \sigma_3^2\right) = \int_a \prod_{j=1}^{J_i} p(\mathbf{y}_{ij}; \boldsymbol{\beta}, \boldsymbol{\Sigma}_2 \mid a) d\mathrm{F}_{\alpha,3}(a), \qquad (9.9)$$

where $F_{\alpha,3}(.)$ is the distribution function of $\{\alpha_i\}$ and σ_3^2 is the parameter associated with it (typically, normal). With Equations (9.8) and (9.9), the log likelihood is

$$L\left(\boldsymbol{\beta}, \boldsymbol{\Sigma}_2, \sigma_3^2\right) = \sum_{i=1}^{n} \ln p\left(\mathbf{y}_{i1}, \ldots, \mathbf{y}_{iJ_i}; \boldsymbol{\beta}, \boldsymbol{\Sigma}_2, \sigma_3^2\right). \qquad (9.10)$$

Computation of the log likelihood requires numerical integration. However, the integral in Equation (9.9) is only one-dimensional and the integral in Equation (9.8) depends on the dimension of $\{\boldsymbol{\alpha}_{ij}\}$, say q, which is typically only 1 or 2. This is in contrast to a more direct approach that combines the heterogeneity terms $(\alpha_i, \boldsymbol{\alpha}'_{i1}, \ldots, \boldsymbol{\alpha}'_{iJ_i})'$. The dimension of this vector is $(1 + qJ_i) \times 1$; using this directly in Equation (9.6) is much more computationally intense.

Example 9.2: Family Smoking Prevention To illustrate their procedures, Gibbons and Hedeker (1997B) considered data from the Television School and Family Smoking Prevention and Cessation Project. In their report of this

Table 9.5. *Tobacco and health scale postintervention performance empirical probabilities*

Social resistance classroom	Television-based curriculum	Count	Knowledgeable[a]	
			Yes	No
No	No	421	41.6	58.4
No	Yes	416	48.3	51.7
Yes	No	380	63.2	36.8
Yes	Yes	383	60.3	39.7
Total		1,600	52.9	47.1

[a] In percent.

study, Gibbons and Hedeker considered 1,600 seventh-grade students, from 135 classrooms within 28 schools. The data set was unbalanced; there were between 1 and 13 classrooms from each school and between 2 and 28 students from each classroom. The schools were randomly assigned to one of four study conditions:

- a social resistance classroom in which a school-based curriculum was used to promote tobacco use prevention and cessation,
- a television-based curriculum,
- a combination of social resistance and television-based curricula, and
- no treatments (control).

A tobacco and health scale was used to classify each student as knowledgeable or not, both before and after the intervention.

Table 9.5 provides empirical probabilities of students' knowledge after the interventions, by type of intervention. This table suggests that social resistance classroom curricula are effective in promoting tobacco use prevention and awareness.

Gibbons and Hedeker estimated both logit and probit models using type of intervention and performance of the preintervention test as explanatory variables. They considered a model with random effects as well as a model with classroom as the second level and school as the third level (as well as two two-level models for robustness purposes). For both models, the social resistance classroom curriculum was statistically significant. However, they also found that the model without random effects indicated that the television-based intervention was statistically significant whereas the three-level model did not reveal such a strong effect.

9.3 Fixed-Effects Models

As in Section 9.1, we express the probability of the response being a one as a nonlinear function of linear combinations of explanatory variables. To accommodate heterogeneity, we incorporate subject-specific variables of the form

$$p_{it} = \pi(\alpha_i + \mathbf{x}_{it}'\boldsymbol{\beta}).$$

Here, the subject-specific effects account only for the intercepts and do not include other variables. Extensions to variable-slope models are possible but, as we will see, even variable-intercept models are difficult to estimate. We assume that $\{\alpha_i\}$ are fixed effects in this section.

Maximum Likelihood Estimation

Similar to Equation (9.2), the log likelihood of the data set is

$$L = \sum_{it} \{y_{it} \ln \pi(\alpha_i + \mathbf{x}_{it}'\boldsymbol{\beta}) + (1 - y_{it}) \ln(1 - \pi(\alpha_i + \mathbf{x}_{it}'\boldsymbol{\beta}))\}. \quad (9.11)$$

This log likelihood can be maximized to yield maximum likelihood estimators of α_i and $\boldsymbol{\beta}$, denoted as $a_{i,\text{MLE}}$ and $\mathbf{b}_{\text{MLE}}$, respectively. Note that there are $n + K$ parameters to be estimated simultaneously. As in Section 9.1, we consider the logit specification of π, so that

$$p_{it} = \pi(\alpha_i + \mathbf{x}_{it}'\boldsymbol{\beta}) = \frac{1}{1 + \exp(-(\alpha_i + \mathbf{x}_{it}'\boldsymbol{\beta}))}. \quad (9.12)$$

Because $\ln(\pi(x)/(1 - \pi(x)) = x$, we have that the log likelihood in Equation (9.11) is

$$\begin{aligned} L &= \sum_{it} \left\{ \ln(1 - \pi(\alpha_i + \mathbf{x}_{it}'\boldsymbol{\beta})) + y_{it} \ln \frac{\pi(\alpha_i + \mathbf{x}_{it}'\boldsymbol{\beta})}{1 - \pi(\alpha_i + \mathbf{x}_{it}'\boldsymbol{\beta})} \right\} \\ &= \sum_{it} \left\{ \ln(1 - \pi(\alpha_i + \mathbf{x}_{it}'\boldsymbol{\beta})) + y_{it}(\alpha_i + \mathbf{x}_{it}'\boldsymbol{\beta}) \right\}. \end{aligned} \quad (9.13)$$

Straightforward calculations show that the score equations are

$$\frac{\partial L}{\partial \alpha_i} = \sum_t (y_{it} - \pi(\alpha_i + \mathbf{x}_{it}'\boldsymbol{\beta})), \qquad \frac{\partial L}{\partial \boldsymbol{\beta}} = \sum_{it} \mathbf{x}_{it}(y_{it} - \pi(\alpha_i + \mathbf{x}_{it}'\boldsymbol{\beta})). \quad (9.14)$$

Finding the roots of these equations yields our MLEs.

Example: Income Tax Payments and Tax Preparers (continued) To see how maximum likelihood works with a data set, we return to the Section 9.1.3 example. For this data set, we have $n = 258$ taxpayers and consider $K = 3$ explanatory variables, LNTPI, MR, and EMP. Fitting this model yields $-2 \times$

(log likelihood) = 416.024 and $R^2 = 0.6543$. According to standard likelihood ratio tests, the additional intercept terms are highly statistically significant. That is, the likelihood ratio test statistic for assessing the null hypothesis $H_0: \alpha_1 = \cdots = \alpha_{258}$ is

$$LRT = 1{,}718.981 - 416.024 = 1{,}302.957.$$

The null hypothesis is rejected based on a comparison of this statistic with a chi-square distribution with 258 degrees of freedom.

Unfortunately, the previous analysis is based on approximations that are known to be unreliable. The difficulty is that, as the subject size n tends to infinity, the number of parameters also tends to infinity. It turns out that our ability to estimate $\boldsymbol{\beta}$ is corrupted by our inability to estimate consistently the subject-specific effects $\{\alpha_i\}$. In contrast, in the linear case, MLEs are equivalent to the least-squares estimators that are consistent. The least-squares procedure "sweeps out" intercept estimates when producing estimates of $\boldsymbol{\beta}$. This is not the case in nonlinear regression models.

To get a better feel for the types of things that can go wrong, suppose that we have no explanatory variables. In this case, from Equations (9.14), the root of the score equation is

$$\frac{\partial L}{\partial \alpha_i} = \sum_t \left\{ -\frac{\exp(\alpha_i)}{1+\exp(\alpha_i)} + y_{it} \right\} = 0.$$

The solution $a_{i,\mathrm{MLE}}$ is

$$\bar{y}_i = \frac{\exp(a_{i,\mathrm{MLE}})}{1+\exp(a_{i,\mathrm{MLE}})}$$

or

$$a_{i,\mathrm{MLE}} = \mathrm{logit}(\bar{y}_i).$$

Thus, if $\bar{y}_i = 1$, then $a_{i,\mathrm{MLE}} = \infty$, and if $\bar{y}_i = 0$, then $a_{i,\mathrm{MLE}} = -\infty$. Thus, intercept estimators are unreliable in these circumstances. An examination of the score functions in Equations (9.14) shows that similar phenomena also occur even in the presence of explanatory variables. To illustrate, in Example 9.1 we have 97 taxpayers who do not use a professional tax preparer in any of the five years under considerations ($\bar{y}_i = 0$) whereas 89 taxpayers always use a tax preparer ($\bar{y}_i = 1$).

Even when the intercept estimators are finite, MLEs of global parameters $\boldsymbol{\beta}$ are inconsistent in fixed-effects binary dependent-variable models. See Illustration 9.1.

Illustration 9.1: Inconsistency of Maximum Likelihood Estimates (Hsiao, 2002E) Now, as a special case, let $T_i = 2$, $K = 1$, and $x_{i1} = 0$ and $x_{i2} = 1$. Using the score equations of (9.14), Appendix 9A shows how to calculate directly the MLE of β, b_{MLE}, for this special case. Further, Appendix 9A.1 argues that the probability limit of b_{MLE} is 2β. Hence, it is an inconsistent estimator of β.

Conditional Maximum Likelihood Estimation

To circumvent the problem of the intercept estimators corrupting the estimator of $\boldsymbol{\beta}$, we use the *conditional maximum likelihood estimator*. This estimation technique is due to Chamberlain (1980E) in the context of fixed-effects binary dependent-variable models. We consider the logit specification of π as in Equation (9.13). With this specification, $\Sigma_t y_{it}$ becomes a *sufficient* statistic for α_i. The idea of sufficiency is reviewed in Appendix 10A.2. In this context, it means that, if we condition on $\Sigma_t y_{it}$, then the distribution of the responses will not depend on α_i.

Illustration 9.2: Sufficiency Continuing with the setup of Illustration 9.1, we now illustrate how to separate the intercept from the slope effects. Here, we only assume that $T_i = 2$, not that $K = 1$.

To show that the conditional distribution of the responses does not depend on α_i, begin by supposing that *sum* $(= \Sigma_t y_{it} = y_{i1} + y_{i2})$ equals either 0 or 2. Consider three cases. For the first case, assume that *sum* equals 0. Then, the conditional distribution of the responses is $\text{Prob}(y_{i1} = 0, y_{i2} = 0 \mid y_{i1} + y_{i2} = sum) = 1$; this clearly does not depend on α_i. For the second case, assume that *sum* equals 2. Then, the conditional distribution of the responses is $\text{Prob}(y_{i1} = 1, y_{i2} = 1 \mid y_{i1} + y_{i2} = sum) = 1$, which also does not depend on α_i. Stated another way, if $\bar{y}_i$ is either 0 or 1, then the statistic that is being conditioned on determines all the responses, resulting in no contribution to a conditional likelihood.

Now consider the third case, where the *sum* equals 1. Basic probability calculations establish that

$$\begin{aligned}\text{Prob}\,(y_{i1} + y_{i2} = 1) &= \text{Prob}\,(y_{i1} = 0)\,\text{Prob}\,(y_{i2} = 1) \\ &\quad + \text{Prob}\,(y_{i1} = 1)\,\text{Prob}\,(y_{i2} = 0) \\ &= \frac{\exp(\alpha_i + \mathbf{x}_{i1}'\boldsymbol{\beta}) + \exp(\alpha_i + \mathbf{x}_{i2}'\boldsymbol{\beta})}{(1 + \exp(\alpha_i + \mathbf{x}_{i1}'\boldsymbol{\beta}))(1 + \exp(\alpha_i + \mathbf{x}_{i2}'\boldsymbol{\beta}))}.\end{aligned}$$

Thus, if *sum* equals 1, then

$$\begin{aligned}\text{Prob}\,(y_{i1} = 0,\, y_{i2} = 1 \mid y_{i1} + y_{i2} = 1) &= \frac{\text{Prob}\,(y_{i1} = 0)\,\text{Prob}\,(y_{i2} = 1)}{\text{Prob}\,(y_{i1} + y_{i2} = 1)} \\ &= \frac{\exp(\alpha_i + \mathbf{x}_{i2}'\boldsymbol{\beta})}{\exp(\alpha_i + \mathbf{x}_{i1}'\boldsymbol{\beta}) + \exp(\alpha_i + \mathbf{x}_{i2}'\boldsymbol{\beta})} \\ &= \frac{\exp(\mathbf{x}_{i2}'\boldsymbol{\beta})}{\exp(\mathbf{x}_{i1}'\boldsymbol{\beta}) + \exp(\mathbf{x}_{i2}'\boldsymbol{\beta})}.\end{aligned}$$

Thus, the conditional distribution of the responses does not depend on α_i. We also note that, if an explanatory variable x_j is constant in time ($\mathbf{x}_{i2,j} = \mathbf{x}_{i1,j}$), then the corresponding parameter β_j disappears from the conditional likelihood.

Conditional Likelihood Estimation

To define the conditional likelihood, let S_i be the random variable representing $\Sigma_t y_{it}$ and let sum_i be the realization of $\Sigma_t y_{it}$. With this notation, the conditional likelihood of the data set becomes

$$\prod_{i=1}^{n} \left\{ \frac{1}{\text{Prob}(S_i = sum_i)} \prod_{t=1}^{T_i} p_{it}^{y_{it}} (1 - p_{it})^{1-y_{it}} \right\}.$$

Note that the ratio within the curly braces equals one when sum_i equals 0 or T_i. Taking the log of the function and then finding values of $\boldsymbol{\beta}$ that maximize it yields $\mathbf{b}_{\text{CMLE}}$, the conditional maximum likelihood estimator. We remark that this process can be computationally intensive because, the distribution of S_i is complicated and is difficult to compute for moderate-size data sets with T more than 10. Appendix 9A.2 provides details.

Illustration 9.3: Conditional Maximum Likelihood Estimator To see that the conditional maximum likelihood estimator is consistent in a case where the maximum likelihood is not, we continue with Illustration 9.1. As argued in Illustration 9.2, we need only be concerned with the case $y_{i1} + y_{i2} = 1$. The conditional likelihood is

$$\begin{aligned}&\prod_{i=1}^{n} \left\{ \frac{1}{\text{Prob}(y_{i1} + y_{i2} = 1)} \prod_{t=1}^{2} p_{it}^{y_{it}} (1 - p_{it})^{1-y_{it}} \right\} \\ &\quad = \prod_{i=1}^{n} \frac{y_{i1} \exp(\mathbf{x}_{i1}'\boldsymbol{\beta}) + y_{i2} \exp(\mathbf{x}_{i2}'\boldsymbol{\beta})}{\exp(\mathbf{x}_{i1}'\boldsymbol{\beta}) + \exp(\mathbf{x}_{i2}'\boldsymbol{\beta})}. \qquad (9.15)\end{aligned}$$

As in Example 9.2, take $K = 1$ and $x_{i1} = 0$ and $x_{i2} = 1$. Then, by taking the derivative with respect to β of the log of the conditional likelihood and setting this equal to zero, one can determine explicitly the conditional maximum likelihood estimator, denoted as b_{CMLE}. Straightforward limit theory shows this to be a consistent estimator of β. Appendix 9A.1 provides details.

We should note that the conditional maximum likelihood estimation for the logit model differs from the *conditional logit model*, which we will introduce in Section 11.1.

9.4 Marginal Models and GEE

For *marginal models*, we require only the specification of the first two moments, specifically the mean and variance of a response, as well as the covariances among responses. This entails much less information than the entire distribution, as required by the likelihood-based approaches in Sections 9.2 and 9.3. Of course, if the entire distribution is assumed known, then we can always calculate the first two moments. Thus, the estimation techniques applicable to marginal models can also be used when the entire distribution is specified.

Marginal models are estimated using a special type of moment estimation known as the *generalized estimating equations*, or *GEE*, approach in the biological sciences. In the social sciences, this approach is part of the *generalized method of moments*, or *GMM*. For the applications that we have in mind, it is most useful to develop the estimation approach using the GEE notation. However, analysts should keep in mind that this estimator is really just another type of GMM estimator.

To describe GEE estimators, one must specify a mean, variance, and covariance structure. To illustrate the development, we begin by assuming that the Section 9.2 random-effects model is valid and we wish to estimate parameters of this distribution. The general GEE procedure is described in Appendix C.6 and will be further developed in Chapter 10.

GEE Estimators for the Random-Effects Binary Dependent-Variable Model

From Section 9.2, we have that the conditional first moment of the response is $\mathrm{E}(y_{it} \mid \alpha_i) = \mathrm{Prob}(y_{it} = 1 \mid \alpha_i) = \pi(\alpha_i + \mathbf{x}_{it}'\boldsymbol{\beta})$. Thus, the mean may be expressed as

$$\mu_{it} = \mu_{it}(\boldsymbol{\beta}, \tau) = \int \pi(a + \mathbf{x}_{it}'\boldsymbol{\beta}) dF_{\alpha}(a). \tag{9.16}$$

Recall that τ is a parameter of the distribution function $F_{\alpha}(.)$. For example, if $F_{\alpha}(.)$ represents a normal distribution, then τ represents the variance.

Occasionally, it is useful to use the notation $\mu_{it}(\boldsymbol{\beta}, \boldsymbol{\tau})$ to remind ourselves that the mean function μ_{it} depends on the parameters $\boldsymbol{\beta}$ and $\boldsymbol{\tau}$. Let $\boldsymbol{\mu}_i = (\mu_{i1}, \ldots, \mu_{iT_i})'$ denote the $T_i \times 1$ vector of means.

For this model, straightforward calculations show that the variance can be expressed as Var $y_{it} = \mu_{it}(1 - \mu_{it})$. For the covariances, for $r \neq s$, we have

$$\begin{aligned}\mathrm{Cov}(y_{ir}, y_{is}) &= \mathrm{E}(y_{ir} y_{is}) - \mu_{ir}\mu_{is} \\ &= \mathrm{E}(\mathrm{E}(y_{ir} y_{is} \mid \alpha_i)) - \mu_{ir}\mu_{is} \\ &= \mathrm{E}\pi(\alpha_i + \mathbf{x}_{ir}'\boldsymbol{\beta})\pi(\alpha_i + \mathbf{x}_{is}'\boldsymbol{\beta}) - \mu_{ir}\mu_{is} \\ &= \int \pi(a + \mathbf{x}_{ir}'\boldsymbol{\beta})\pi(a + \mathbf{x}_{is}'\boldsymbol{\beta})dF_\alpha(a) - \mu_{ir}\mu_{is}. \qquad (9.17)\end{aligned}$$

Let $\mathbf{V}_i = \mathbf{V}_i(\boldsymbol{\beta}, \boldsymbol{\tau})$ be the usual $T_i \times T_i$ variance–covariance matrix for the ith subject; that is, the tth diagonal element of $\mathbf{V}_i$ is Var y_{it}, whereas for nondiagonal elements, the rth row and sth column of $\mathbf{V}_i$ is given by $\mathrm{Cov}(y_{ir}, y_{is})$.

For GEE, we also require derivatives of certain moments. For the mean function, from Equation (9.16), we will use

$$\frac{\partial}{\partial \boldsymbol{\beta}}\mu_{it} = \mathbf{x}_{it}\int \pi'(a + \mathbf{x}_{it}'\boldsymbol{\beta})dF_\alpha(a).$$

As is customary in applied data analysis, this calculation assumes a sufficient amount of regularity of the distribution function $F_\alpha(.)$ so that we may interchange the order of differentiation and integration. In general, we will use the notation

$$\mathbf{G}_\mu(\boldsymbol{\beta}, \boldsymbol{\tau}) = \left(\frac{\partial \mu_{i1}}{\partial \boldsymbol{\beta}} \quad \cdots \quad \frac{\partial \mu_{iT_i}}{\partial \boldsymbol{\beta}}\right),$$

a $K \times T_i$ matrix of derivatives.

GEE Estimation Procedure

The GEE estimators are computed according to the following general recursion. Begin with initial estimators of $(\boldsymbol{\beta}, \boldsymbol{\tau})$, say $(\mathbf{b}_{0,\mathrm{EE}}, \boldsymbol{\tau}_{0,\mathrm{EE}})$. Typically, initial estimators $\mathbf{b}_{0,\mathrm{EE}}$ are calculated assuming zero covariances among responses. Initial estimators $\boldsymbol{\tau}_{0,\mathrm{EE}}$ are computed using residuals based on the $\mathbf{b}_{0,\mathrm{EE}}$ estimate. Then, at the $(n + 1)$st stage, recursively, perform the following:

1. Use $\boldsymbol{\tau}_{n,\mathrm{EE}}$ and the solution of the equation

$$\mathbf{0}_K = \sum_{i=1}^{n} \mathbf{G}_\mu(\mathbf{b}, \boldsymbol{\tau}_{n,\mathrm{EE}})(\mathbf{V}_i(\mathbf{b}, \boldsymbol{\tau}_{n,\mathrm{EE}}))^{-1}(\mathbf{y}_i - \boldsymbol{\mu}_i(\mathbf{b}, \boldsymbol{\tau}_{n,\mathrm{EE}})) \qquad (9.18)$$

to determine an updated estimator of $\boldsymbol{\beta}$, say $\mathbf{b}_{n+1,\mathrm{EE}}$.

2. Use the residuals $\{y_{it} - \mu_{it}(\mathbf{b}_{n+1,\text{EE}}, \tau_{n,\text{EE}})\}$ to determine an updated estimator of τ, say $\tau_{n+1,\text{EE}}$.
3. Repeat steps 1 and 2 until convergence is obtained.

Let $\mathbf{b}_{\text{EE}}$ and τ_{EE} denote the resulting estimators of $\boldsymbol{\beta}$ and τ. Under broad conditions, $\mathbf{b}_{\text{EE}}$ is consistent and asymptotically normal with asymptotic variance

$$\left(\sum_{i=1}^{n} \mathbf{G}_{\mu}(\mathbf{b}_{\text{EE}}, \tau_{\text{EE}}) \left(\mathbf{V}_i(\mathbf{b}_{\text{EE}}, \tau_{\text{EE}})\right)^{-1} \mathbf{G}_{\mu}(\mathbf{b}_{\text{EE}}, \tau_{\text{EE}})'\right)^{-1}. \tag{9.19}$$

The solution $\mathbf{b}_{\text{EE}}$ in Equation (9.18) can be computed quickly using iterated reweighted least squares, a procedure described in Appendix C.3. However, the specified estimation procedure remains tedious because it relies on the numerical integration computations in calculating μ_{it} in Equation (9.16) and $\text{Cov}(y_{ir}, y_{is})$ in Equation (9.17). Now, in Section 9.2, we saw that numerical integration in (9.16) could be avoided by specifying normal distributions for π and F_α, resulting in $\mu_{it} = \Phi(\mathbf{x}_{it}'\boldsymbol{\beta}/\sqrt{1+\tau})$. However, even with this specification, we would still require numerical integration to calculate $\text{Cov}(y_{ir}, y_{is})$ in Equation (9.17). A single numerical integration is straightforward in the modern-day computing environment. However, evaluation of $\mathbf{V}_i$ would require $T_i(T_i - 1)/2$ numerical integrations for the covariance terms. Thus, each evaluation of Equation (9.18) would require $\Sigma_i\{T_i(T_i - 1)/2\}$ numerical integrations (which is $258 \times 5 \times (5-1)/2 = 2{,}580$ for the Section 9.1.3 example). Many evaluations of Equation (9.18) would be required prior to successful convergence of the recursive procedure. In summary, this approach is often computationally prohibitive.

To reduce these computational complexities, the focus of *marginal models* is the representation of the first two moments directly, with or without reference to underlying probability distributions. By focusing directly on the first two moments, we may keep the specification simple and computationally feasible.

To illustrate, we may choose to specify the mean function as $\mu_{it} = \Phi(\mathbf{x}_{it}'\boldsymbol{\beta})$. This is certainly plausible under the random-effects binary dependent-variable model. For the variance function, we consider $\text{Var}\, y_{it} = \phi\mu_{it}(1 - \mu_{it})$, where ϕ is an overdispersion parameter that we may either assume to be 1 or to be estimated from the data. Finally, it is customary in the literature to specify correlations in lieu of covariances. Use the notation $\text{Corr}(y_{ir}, y_{is})$ to denote the correlation between y_{ir} and y_{is}. For example, it is common to use the *exchangeable* correlation structure specified as

$$\text{Corr}(y_{ir}, y_{is}) = \begin{cases} 1 & \text{for} \quad r = s, \\ \rho & \text{for} \quad r \neq s. \end{cases}$$

Table 9.6. *Comparison of GEE estimators*

	Exchangeable working correlation			Unstructured working correlation		
Parameter	Estimate	Empirical standard error	Model-based standard error	Estimate	Empirical standard error	Model-based standard error
Intercept	−0.9684	0.7010	0.5185	0.1884	0.6369	0.3512
LNTPI	0.0764	0.0801	0.0594	−0.0522	0.0754	0.0395
MR	0.0024	0.0083	0.0066	0.0099	0.0076	0.0052
EMP	0.5096	0.2676	0.1714	0.2797	0.1901	0.1419

Here, the motivation is that the latent variable α_i is common to all observations within a subject, thus inducing a common correlation. For this illustration, the parameters $\boldsymbol{\tau} = (\phi, \rho)'$ constitute the variance components.

Estimation may then proceed as described in the recursion beginning with Equation (9.18). However, as with linear models, the second moments may be misspecified. For this reason, the correlation specification is commonly known as a *working correlation*. For linear models, weighted least squares provides estimators with desirable properties. Although not optimal compared to generalized least squares, weighted least-squares estimators are typically consistent and asymptotically normal. In the same fashion, GEE estimators based on working correlations have desirable properties, even when the correlation structure is not perfectly specified.

However, if the correlation structure is not valid, then the asymptotic standard errors provided through the asymptotic variance in Equation (9.18) are not valid. Instead, *empirical standard errors* may be calculated using the following estimator of the asymptotic variance of $\mathbf{b}_{\mathrm{EE}}$:

$$\left(\sum_{i=1}^{n} \mathbf{G}_{\mu} \mathbf{V}_i^{-1} \mathbf{G}_{\mu}'\right)^{-1} \left(\sum_{i=1}^{n} \mathbf{G}_{\mu} \mathbf{V}_i^{-1} (\mathbf{y}_i - \boldsymbol{\mu}_i)(\mathbf{y}_i - \boldsymbol{\mu}_i)' \mathbf{V}_i^{-1} \mathbf{G}_{\mu}'\right) \left(\sum_{i=1}^{n} \mathbf{G}_{\mu} \mathbf{V}_i^{-1} \mathbf{G}_{\mu}'\right)^{-1}. \tag{9.20}$$

Specifically, the standard error of the jth component of $\mathbf{b}_{\mathrm{EE}}$, $se(\mathbf{b}_{j,\mathrm{EE}})$, is defined to be the square root of the jth diagonal element of the variance–covariance matrix in (9.20).

Example: Income Tax Payments and Tax Preparers (continued) To see how a marginal model works with a data set, we return to the Section 9.1.3 example. Table 9.6 shows the fit of two models, each using LNTPI, MR, and EMP as explanatory variables. The calculations were done using the SAS

Table 9.7. *Estimate of unstructured correlation matrix*

	Time = 1	Time = 2	Time = 3	Time = 4	Time = 5
Time = 1	1.0000	0.8663	0.7072	0.6048	0.4360
Time = 2	0.8663	1.0000	0.8408	0.7398	0.5723
Time = 3	0.7072	0.8408	1.0000	0.9032	0.7376
Time = 4	0.6048	0.7398	0.9032	1.0000	0.8577
Time = 5	0.4360	0.5723	0.7376	0.8577	1.0000

procedure GENMOD. For the first model, the exchangeable working correlation structure was used. Parameter estimates, as well as empirical standard errors from (9.20) and model-based standard errors from Equation (9.19), appear in Table 9.6. The estimated correlation parameter works out to be $\hat{\rho} = 0.712$. For the second model, an unstructured working correlation matrix was used. (Table 2.4 provides an example of an unstructured covariance matrix.) Table 9.7 provides these estimated correlations.

We may interpret the ratio of the estimate to standard error as a t-statistic and use this to assess the statistical significance of a variable. Examining Table 9.6, we see that LNTPI and MR are not statistically significant, using either type of standard error or correlation structure. The variable EMP ranges from being strongly statistically significant, for the case with model-based standard errors and an exchangeable working correlation, to being not statistically significant, for the case of empirical standard errors and an unstructured working correlation. Overall, the GEE estimates of the marginal model provide dramatically different results when compared to either the Section 9.1 homogeneous model or the Section 9.2 random-effects results.

Further Reading

More extensive introductions to (homogeneous) binary dependent-variable models are available in Agresti (2002G) and Hosmer and Lemeshow (2000G). For an econometrics perspective, see Cameron and Trivedi (1998E).

For discussions of binary dependent-variable models with endogenous explanatory variables, see Wooldridge (2002E) and Arellano and Honoré (2001E).

For models of binary dependent variables with random intercepts, MLEs can be computed using numerical integration techniques to approximate the likelihood. McCulloch and Searle (2001G) discuss numerical integration for mixed-effects models. Pinheiro and Bates (2000S) describe the adaptive Gaussian quadrature method used in SAS PROC NLMIXED.

Appendix 9A Likelihood Calculations

9A.1 Consistency of Likelihood Estimators

Illustration 9.1: Inconsistency of Maximum-Likelihood Estimates (continued) Recall that $T_i = 2$, $K = 1$, and $x_{i1} = 0$ and $x_{i2} = 1$. Thus, from Equation (9.14), we have

$$\frac{\partial L}{\partial \alpha_i} = y_{i1} + y_{i2} - \frac{\mathrm{e}^{\alpha_i}}{1+\mathrm{e}^{\alpha_i}} - \frac{\mathrm{e}^{\alpha_i+\beta}}{1+\mathrm{e}^{\alpha_i+\beta}} = 0 \tag{9A.1}$$

and

$$\frac{\partial L}{\partial \beta} = \sum_i \left\{ y_{i2} - \frac{\mathrm{e}^{\alpha_i+\beta}}{1+\mathrm{e}^{\alpha_i+\beta}} \right\} = 0. \tag{9A.2}$$

From Equation (9A.1), it is easy to see that if $y_{i1} + y_{i2} = 0$ then $a_{i,\mathrm{MLE}} = -\infty$. Further, if $y_{i1} + y_{i2} = 2$, then $a_{i,\mathrm{MLE}} = \infty$. For both cases, the contribution to the sum in Equation (9A.2) is zero. Thus, we consider the case $y_{i1} + y_{i2} = 1$ and let d_i be the indicator variable for the case that $y_{i1} + y_{i2} = 1$. In this case, we have that $a_{i,\mathrm{MLE}} = -b_{\mathrm{MLE}}/2$ from Equation (9A.1). Putting this into Equation (9A.2) yields

$$\begin{aligned}\sum_i d_i y_{i2} &= \sum_i d_i \frac{\exp(a_{i,\mathrm{MLE}} + b_{\mathrm{MLE}})}{1+\exp(a_{i,\mathrm{MLE}} + b_{\mathrm{MLE}})} = \sum_i d_i \frac{\exp(b_{\mathrm{MLE}}/2)}{1+\exp(b_{\mathrm{MLE}}/2)} \\ &= n_1 \frac{\exp(b_{\mathrm{MLE}}/2)}{1+\exp(b_{\mathrm{MLE}}/2)},\end{aligned}$$

where $n_1 = \Sigma_i d_i$ is the number of subjects where $y_{i1} + y_{i2} = 1$. Thus, with the notation $\bar{y}_2^+ = (1/n_1)\sum_i d_i y_{i2}$, we have $b_{\mathrm{MLE}} = 2\ln(\bar{y}_2^+/(1-\bar{y}_2^+))$.

To establish the inconsistency, straightforward weak law of large numbers can be used to show that the probability limit of $\bar{y}_2^+$ is $e^{\beta}/(1+e^{\beta})$. Thus, the probability limit of b_{MLE} is 2β, and hence it is an inconsistent estimator of β.

Illustration 9.3: Conditional Maximum Likelihood Estimator (continued) Recall that $K = 1, x_{i1} = 0$, and $x_{i2} = 1$. Then, from Equation (9.15), the conditional likelihood is

$$\prod_{i=1}^{n} \left\{ \frac{y_{i1} + y_{i2}\exp(\beta)}{1+\exp(\beta)} \right\}^{d_i} = \prod_{i=1}^{n} \left\{ \frac{\exp(\beta y_{i2})}{1+\exp(\beta)} \right\}^{d_i},$$

because $y_{i1} + y_{i2} = 1$ and d_i is a variable to indicate $y_{i1} + y_{i2} = 1$. Thus, the conditional log likelihood is $L_c(\beta) = \sum_{i=1}^{n} d_i \{\beta y_{i2} - \ln(1 + e^{\beta})\}$. To find the conditional maximum likelihood estimator, we have

$$\frac{\partial L_c(\beta)}{\partial \beta} = \sum_{i=1}^{n} d_i \left\{ y_{i2} - \frac{\partial}{\partial \beta} \ln(1 + e^{\beta}) \right\} = \sum_{i=1}^{n} d_i \left\{ y_{i2} - \frac{e^{\beta}}{1 + e^{\beta}} \right\} = 0.$$

The root of this is

$$b_{\text{CMLE}} = \ln \frac{\bar{y}_2^+}{1 - \bar{y}_2^+}.$$

In Example 9.2, we used the fact that the probability limit of $\bar{y}_2^+$ is $e^{\beta}/(1 + e^{\beta})$. Thus, the probability limit of b_{CMLE} is β and hence is consistent.

9A.2 Computing Conditional Maximum Likelihood Estimators

Computing the Distribution of Sums of Nonidentically, Independently Distributed Bernoulli Random Variables

We begin by presenting an algorithm for the computation of the distribution of sums of nonidentically, independently distributed Bernoulli random variables. Thus, we take y_{it} to be independent Bernoulli random variables with $\text{Prob}(y_{it} = 1) = \pi(\mathbf{x}_{it}'\boldsymbol{\beta})$. For convenience, we use the logit form of π. Define the sum random variable $S_{iT} = y_{i1} + y_{i2} + \cdots + y_{iT}$. We wish to evaluate $\text{Prob}(S_{iT} = s)$. For notational convenience, we omit the i subscript on T.

We first note that it is straightforward to compute

$$\text{Prob}(S_{iT} = 0) = \prod_{t=1}^{T} \{1 - \pi(\mathbf{x}_{it}'\boldsymbol{\beta})\}$$

using a logit form for π. Continuing, we have

$$\begin{aligned}\text{Prob}(S_{iT} = 1) &= \sum_{t=1}^{T} \pi(\mathbf{x}_{it}'\boldsymbol{\beta}) \prod_{r=1, r\neq t}^{T} \{1 - \pi(\mathbf{x}_{ir}'\boldsymbol{\beta})\} \\ &= \sum_{t=1}^{T} \frac{\pi(\mathbf{x}_{it}'\boldsymbol{\beta})}{1 - \pi(\mathbf{x}_{it}'\boldsymbol{\beta})} \prod_{r=1}^{T} \{1 - \pi(\mathbf{x}_{ir}'\boldsymbol{\beta})\}.\end{aligned}$$

Using a logit form for π, we have $\pi(z)/(1 - \pi(z)) = e^z$. Thus, with this notation, we have

$$\text{Prob}(S_{iT} = 1) = \text{Prob}(S_{iT} = 0) \left(\sum_{t=1}^{T} \exp(\mathbf{x}_{it}'\boldsymbol{\beta}) \right).$$

Let $\{j_1, j_2, \ldots, j_s\}$ be a subset of $\{1, 2, \ldots, T\}$ and $\Sigma_{s,T}$ be the sum over all such subsets. Thus, for the next step in the iteration, we have

$$\begin{aligned}\text{Prob}(S_{iT} = 2) &= \sum_{2,T} \pi(\mathbf{x}'_{i,j_1}\boldsymbol{\beta})\pi(\mathbf{x}'_{i,j_2}\boldsymbol{\beta}) \prod_{r=1, r\neq j_1, r\neq j_2}^{T} \{1 - \pi(\mathbf{x}'_{ir}\boldsymbol{\beta})\} \\ &= \text{Prob}(S_{iT} = 0) \sum_{2,T} \frac{\pi(\mathbf{x}'_{i,j_1}\boldsymbol{\beta})}{1 - \pi(\mathbf{x}'_{i,j_1}\boldsymbol{\beta})} \frac{\pi(\mathbf{x}'_{i,j_2}\boldsymbol{\beta})}{1 - \pi(\mathbf{x}'_{i,j_2}\boldsymbol{\beta})}.\end{aligned}$$

Continuing this, in general we have

$$\text{Prob}(S_{iT} = s) = \text{Prob}(S_{iT} = 0) \sum_{s,T} \frac{\pi(\mathbf{x}'_{i,j_1}\boldsymbol{\beta})}{1 - \pi(\mathbf{x}'_{i,j_1}\boldsymbol{\beta})} \cdots \frac{\pi(\mathbf{x}'_{i,j_s}\boldsymbol{\beta})}{1 - \pi(\mathbf{x}'_{i,j_s}\boldsymbol{\beta})}.$$

Using a logit form for π, we may express the distribution as

$$\text{Prob}(S_{iT} = s) = \text{Prob}(S_{iT} = 0) \sum_{s,T} \exp((\mathbf{x}_{i,j_1} + \cdots + \mathbf{x}_{i,j_s})'\boldsymbol{\beta}). \qquad (9A.3)$$

Thus, even with the logit form for π, we see the difficulty in computing $\text{Prob}(S_{it} = s)$ is that it involves the sum over $\binom{T}{s}$ quantities in $\Sigma_{s,T}$. This expresses the distribution in terms of $\text{Prob}(S_{it} = 0)$. It is also possible to derive a similar expression in terms of $\text{Prob}(S_{it} = T)$; this alternative expression is more computationally useful than Equation (9A.3) for evaluation of the distribution at large values of s.

Computing the Conditional Maximum Likelihood Estimator

From Section 9.3, the logarithmic conditional likelihood is

$$\begin{aligned}\ln CL = \sum_{i=1}^{n} \Bigg\{ \left(\sum_{t=1}^{T_i} y_{it} \ln \pi(\mathbf{x}'_{it}\boldsymbol{\beta}) + (1 - y_{it}) \ln(1 - \pi(\mathbf{x}'_{it}\boldsymbol{\beta})) \right) \\ - \ln \text{Prob}(S_{iT_i} = sum_{iT_i}) \Bigg\},\end{aligned}$$

where we have taken α_i to be zero, without loss of generality. As remarked in Section 9.3, when summing over all subjects $i = 1, \ldots, n$, we need not consider those subjects where $\sum_{t=1}^{T_i} y_{it}$ equal 0 or T_i because the conditional likelihood is identically equal to one.

To find those values of $\boldsymbol{\beta}$ that maximize ln CL, one could use the Newton–Raphson recursive algorithm (see Appendix C.2). To this end, we require the vector of partial derivatives

$$\frac{\partial}{\partial \boldsymbol{\beta}} \ln CL = \sum_{i=1}^{n} \left\{ \left(\sum_{t=1}^{T_i} \mathbf{x}_{it}(y_{it} - \pi(\mathbf{x}_{it}'\boldsymbol{\beta})) \right) - \frac{\partial}{\partial \boldsymbol{\beta}} \ln \text{Prob}(S_{iT_i} = sum_{iT_i}) \right\}.$$

The Newton–Raphson algorithm also requires the matrix of second derivatives, but computational considerations of this matrix are similar to the ones for the first derivative and are omitted.

From the form of the vector of partial derivatives, we see that the main task is to compute the gradient of $\ln \text{Prob}(S_{iT} = s)$. Using a logit form for π and Equation (9A.3) (dropping the i subscript on T), we have

$$\begin{aligned}
&\frac{\partial}{\partial \boldsymbol{\beta}} \ln \text{Prob}(S_{iT} = s) \\
&\quad = \frac{\partial}{\partial \boldsymbol{\beta}} \ln \text{Prob}(S_{iT} = 0) + \frac{\partial}{\partial \boldsymbol{\beta}} \ln \sum_{s,T} \exp((\mathbf{x}_{i,j_1} + \cdots + \mathbf{x}_{i,j_s})'\boldsymbol{\beta}) \\
&\quad = \frac{\sum_{s,T} (\mathbf{x}_{i,j_1} + \cdots + \mathbf{x}_{i,j_s}) \exp((\mathbf{x}_{i,j_1} + \cdots + \mathbf{x}_{i,j_s})'\boldsymbol{\beta})}{\sum_{s,T} \exp\{(\mathbf{x}_{i,j_1} + \cdots + \mathbf{x}_{i,j_s})'\boldsymbol{\beta}\}} \\
&\qquad - \sum_{t=1}^{T} \frac{\mathbf{x}_{it} \exp(\mathbf{x}_{it}'\boldsymbol{\beta})}{1 + \exp(\mathbf{x}_{it}'\boldsymbol{\beta})}.
\end{aligned}$$

As with the probability in Equation (9A.3), this is easy to compute for values of s that are close to 0 or T. However, in general the calculation requires summing over $\binom{T}{s}$ quantities in $\Sigma_{s,T}$, both in the numerator and denominator. Moreover, this is required for each subject i at each stage of the likelihood maximization process. Thus, the calculation of conditional likelihood estimators becomes burdensome for large values of T.

Exercises and Extensions

Section 9.1

9.1 *Threshold interpretation of the probit regression model* Consider an underlying linear model, $y_{it}^* = \mathbf{x}_{it}'\boldsymbol{\beta} + \varepsilon_{it}^*$, where ε_{it}^* is normally distributed with mean zero and variance σ^2. Define

$$y_{it} = \begin{cases} 0 & y_{it}^* \le 0, \\ 1 & y_{it}^* > 0. \end{cases}$$

Show that $p_{it} = \text{Prob}(y_{it} = 1) = \Phi(\mathbf{x}'_{it}\beta/\sigma)$, where Φ is the standard normal distribution function.

9.2 Random-utility interpretation of the logistic regression model Under the random-utility interpretation, an individual with utility $U_{itj} = u_{it}(V_{itj} + \varepsilon_{itj})$, where j may be 0 or 1, selects a category corresponding to $j = 1$ with probability

$$\begin{aligned} p_{it} &= \text{Prob}(y_{it} = 1) = \text{Prob}(U_{it0} < U_{it1}) \\ &= \text{Prob}(u_{it}(V_{it0} + \varepsilon_{it0}) < u_{it}(V_{it1} + \varepsilon_{it1})) \\ &= \text{Prob}(\varepsilon_{it0} - \varepsilon_{it1} < V_{it1} - V_{it0}). \end{aligned}$$

Suppose that the errors are from an extreme value distribution of the form

$$\text{Prob}(\varepsilon_{itj} < a) = \exp(-e^{-a}).$$

Show that the choice probability p_{it} has a logit form. That is, show that

$$p_{it} = \frac{1}{1 + \exp(-\mathbf{x}'_{it}\beta)}.$$

9.3 Marginal distribution of the probit random-effects model Consider a normal model for the conditional distribution of a response, that is, $\text{Prob}(y_{it} = 1 \mid \alpha_i) = \Phi(\alpha_i + \mathbf{x}'_{it}\beta)$, where Φ is the standard normal distribution function. Assume further that α_i is normally distributed with mean zero and variance τ. Show that

$$p_{it} = \Phi\left(\frac{\mathbf{x}'_{it}\beta}{\sqrt{1+\tau}}\right).$$

9.4 Choice of yogurt brands These data are known as *scanner data* because they are obtained from optical scanning of grocery purchases at checkout. The subjects consist of $n = 100$ households in Springfield, Missouri. The response of interest is the type of yogurt purchased. For this exercise, we consider only the brand Yoplait or another choice. The households were monitored over a two-year period with the number of purchases ranging from 4 to 185; the total number of purchases is $N = 2{,}412$. More extensive motivation is provided in Section 11.1.

The two marketing variables of interest are price and features. We use two price variables for this study, PY, the price of Yoplait, and PRICEOTHER, the lowest price of the other brands. For features, these are binary variables, defined to be one if there was a newspaper feature advertising the brand at time of purchase and zero otherwise. We use two feature variables, FY, the features associated with Yoplait, and FEATOTHER, a variable to indicate if any other brand has a feature displayed.

a. *Basic summary statistics*
   Create the basic summary statistics for Yoplait and the four explanatory variables.
   i. What are the odds of purchasing Yoplait?
   ii. Determine the odds ratio when assessing whether or not Yoplait is featured. Interpret your result.
b. *Logistic regression models*
   Run a logistic regression model with the two explanatory variables, PY and FY. Further, run a second logistic regression model with four explanatory variables, PY, FY, PRICEOTHER, and FEATOTHER. Compare these two models and say which you prefer. Justify your choice by appealing to standard statistical hypothesis tests.
c. *Random-effects model*
   Fit a random-effects model with four explanatory variables. Interpret the regression coefficients of this model.
d. *GEE model*
   Fit a GEE model with four explanatory variables.
   i. Give a brief description of the theory behind this model.
   ii. Compare the part (d)(i) description to the random-effects model in part (c). In particular, how does each model address the heterogeneity?
   iii. For this data set, describe whether or not your model choice is robust to any important interpretations about the regression coefficients.

10
Generalized Linear Models

Abstract. This chapter extends the linear model introduced in Chapters 1–8 and the binary dependent-variable model in Chapter 9 to the *generalized linear model* formulation. Generalized linear models (GLMs) represent an important class of nonlinear regression models that have found extensive use in practice. In addition to the normal and Bernoulli distributions, these models include the binomial, Poisson, and Gamma families as distributions for dependent variables.

Section 10.1 begins this chapter with a review of homogeneous GLMs, models that do not incorporate heterogeneity. The Section 10.2 example reinforces this review. Section 10.3 then describes marginal models and generalized estimating equations, a widely applied framework for incorporating heterogeneity. Then, Sections 10.4 and 10.5 allow for heterogeneity by modeling subject-specific quantities as random and fixed effects, respectively. Section 10.6 ties together fixed and random effects under the umbrella of Bayesian inference.

10.1 Homogeneous Models

This section introduces the generalized linear model (GLM) due to Nelder and Wedderburn (1972G); a more extensive treatment may be found in the classic work by McCullagh and Nelder (1989G). The GLM framework generalizes linear models in the following sense. Linear model theory provides a platform for choosing appropriate linear combinations of explanatory variables to predict a response. In Chapter 9, we saw how to use nonlinear functions of these linear combinations to provide better predictors, at least for responses with Bernoulli (binary) outcomes. With GLMs, we widen the class of distributions to allow us to handle other types of nonnormal outcomes. This broad class includes as special cases the normal, Bernoulli, and Poisson distributions. As we will see, the normal distribution corresponds to linear models, and the Bernoulli to the Chapter 9 binary response models. To motivate our discussion of GLMs, we focus on the Poisson distribution; this distribution allows us to readily model count data. Our treatment follows the structure introduced in Chapter 9; we first introduce the *homogenous* version of the GLM framework, that is, models without subject-specific parameters nor terms to account for serial correlation. Subsequent sections will introduce techniques for handling heterogeneity in longitudinal and panel data.

This section begins with an introduction of the response distribution, then shows how to link the distribution's parameters to regression variables, and then describes estimation principles.

10.1.1 Linear Exponential Families of Distributions

This chapter considers the *linear exponential family* of the form

$$p(y, \theta, \phi) = \exp\left(\frac{y\theta - b(\theta)}{\phi} + S(y, \phi)\right), \qquad (10.1)$$

where y is a dependent variable and θ is the parameter of interest. The quantity ϕ is a scale parameter, which we often assume to be known. The term $b(\theta)$ depends only on the parameter θ, not the dependent variable. The statistic $S(y, \phi)$ is a function of the dependent variable and the scale parameter, not the parameter θ.

The dependent variable y may be discrete, continuous, or a mixture. Thus, $p(.)$ may be interpreted to be a density or mass function, depending on the application. Table 10A.1 in Appendix 10A provides several examples, including the normal, binomial, and Poisson distributions. To illustrate, consider a normal distribution with a probability density function of the form

$$\begin{aligned} f(y, \mu, \sigma^2) &= \frac{1}{\sqrt{2\pi}\sigma} \exp\left(-\frac{(y-\mu)^2}{2\sigma^2}\right) \\ &= \exp\left(\frac{(y\mu - \mu^2/2)}{\sigma^2} - \frac{y^2}{2\sigma^2} - \frac{1}{2}\ln(2\pi\sigma^2)\right). \end{aligned}$$

With the choices $\theta = \mu, \phi = \sigma^2, b(\theta) = \theta^2/2$, and $S(y, \phi) = -y^2/(2\phi) - \ln(2\pi\phi)/2$, we see that the normal probability density function can be expressed as in Equation (10.1).

For the function in Equation (10.1), some straightforward calculations show that

- $\mathrm{E}y = b'(\theta)$ and
- $\mathrm{Var}\, y = \phi b''(\theta)$.

For reference, these calculations appear in Appendix 10A.1. For example, in the context of the normal distribution case here, it is easy to check that $\mathrm{E}y = b'(\theta) = \theta = \mu$ and $\mathrm{Var}\, y = \sigma^2 b''(\mu) = \sigma^2$, as anticipated.

10.1.2 Link Functions

In regression modeling situations, the distribution of y_{it} varies by observation through the subscripts *it*. Specifically, we allow the distribution's parameters to vary by observation through the notation θ_{it} and ϕ_{it}. For our applications, the variation of the scale parameter is due to known weight factors. Specifically, when the scale parameter varies by observation, it is according to $\phi_{it} = \phi / w_{it}$, that is, a constant divided by a known weight w_{it}. With the relation $\mathrm{Var}\, y_{it} = \phi_{it} b''(\theta) = \phi b''(\theta)/w_{it}$, we have that a larger weight implies a smaller variance, other things being equal.

In regression situations, we wish to understand the impact of a linear combination of explanatory variables on the distribution. In the GLM context, it is customary to call $\eta_{it} = \mathbf{x}_{it}'\boldsymbol{\beta}$ the *systematic component* of the model. This systematic component is related to the mean through the expression

$$\eta_{it} = g(\mu_{it}), \tag{10.2}$$

where $g(.)$ is known as the *link function*. As we saw in the previous subsection, we can express the mean of y_{it} as $\mathrm{E}y_{it} = \mu_{it} = b'(\theta_{it})$. Thus, Equation (10.2) serves to "link" the systematic component to the parameter θ_{it}. It is possible to use the identity function for $g(.)$ so that $\mu_{it} = \mathbf{x}_{it}'\boldsymbol{\beta}$. Indeed, this is the usual case in linear regression. However, linear combinations of explanatory variables, $\mathbf{x}_{it}'\boldsymbol{\beta}$, may vary between negative and positive infinity whereas means are often restricted to a smaller range. For example, Poisson means vary between zero and infinity. The link function serves to map the domain of the mean function onto the whole real line.

Special Case: Links for the Bernoulli Distribution Bernoulli means are probabilities and thus vary between zero and one. For this case, it is useful to choose a link function that maps the unit interval (0, 1) onto the whole real line. The following are three important examples of link functions for the Bernoulli distribution:

Table 10.1. *Mean functions and canonical links for selected distributions*

Distribution	Mean function ($b'(\theta)$)	Canonical link ($g(\theta)$)
Normal	θ	θ
Bernoulli	$e^{\theta}/(1+e^{\theta})$	logit(θ)
Poisson	e^{θ}	$\ln\theta$
Gamma	$-1/\theta$	$-\theta$

- Logit: $g(\mu) = \text{logit}(\mu) = \ln(\mu/(1-\mu))$.
- Probit: $g(\mu) = \Phi^{-1}(\mu)$, where Φ^{-1} is the inverse of the standard normal distribution function.
- Complementary log–log: $g(\mu) = \ln(-\ln(1-\mu))$.

This example demonstrates that there may be several link functions that are suitable for a particular distribution. To help with the selection, an intuitively appealing case occurs when the systematic component equals the parameter of interest ($\eta = \theta$). To see this, first recall that $\eta = g(\mu)$ and $\mu = b'(\theta)$, dropping the *it* subscripts for the moment. Then, it is easy to see that if $g^{-1} = b'$, then $\eta = g(b'(\theta)) = \theta$. The choice of g that is the inverse of b' is called the *canonical* link.

Table 10.1 shows the mean function and corresponding canonical link for several important distributions.

10.1.3 Estimation

This section presents maximum likelihood, the customary form of estimation. To provide intuition, we begin with the simpler case of canonical links and then extend the results to more general links.

Maximum Likelihood Estimation for Canonical Links

From Equation (10.1) and the independence among observations, the log likelihood is

$$\ln\, p(\mathbf{y}) = \sum_{it} \left\{ \frac{y_{it}\theta_{it} - b(\theta_{it})}{\phi_{it}} + S(y_{it}, \phi_{it}) \right\}. \tag{10.3}$$

Recall that, for canonical links, we have equality between the distribution's parameter and the systematic component, so that $\theta_{it} = \eta_{it} = \mathbf{x}_{it}'\boldsymbol{\beta}$. Thus, the log likelihood is

$$\ln p(\mathbf{y}) = \sum_{it} \left\{ \frac{y_{it}\mathbf{x}_{it}'\boldsymbol{\beta} - b(\mathbf{x}_{it}'\boldsymbol{\beta})}{\phi_{it}} + S(y_{it}, \phi_{it}) \right\}. \tag{10.4}$$

Taking the partial derivative with respect to $\boldsymbol{\beta}$ yields the score function

$$\frac{\partial}{\partial \boldsymbol{\beta}} \ln p(\mathbf{y}) = \sum_{it} \left\{ \mathbf{x}_{it} \frac{y_{it} - b'(\mathbf{x}_{it}'\boldsymbol{\beta})}{\phi_{it}} \right\}.$$

Because $\mu_{it} = b'(\theta_{it}) = b'(\mathbf{x}_{it}'\boldsymbol{\beta})$ and $\phi_{it} = \phi / w_{it}$, we can solve for the MLEs of $\boldsymbol{\beta}, \mathbf{b}_{\mathrm{MLE}}$, through the "normal equations"

$$\mathbf{0} = \sum_{it} w_{it}\mathbf{x}_{it} \left(y_{it} - \mu_{it} \right). \tag{10.5}$$

One reason for the widespread use of GLM methods is that the MLEs can be computed quickly through a technique known as *iterated reweighted least squares*, described in Appendix C.3.

Note that, like ordinary linear regression normal equations, we do not need to consider estimation of the variance scale parameter ϕ at this stage. That is, we can first compute $\mathbf{b}_{\mathrm{MLE}}$ and then estimate ϕ.

Maximum Likelihood Estimation for General Links

For general links, we no longer assume the relationship $\theta_{it} = \mathbf{x}_{it}'\boldsymbol{\beta}$ but assume that $\boldsymbol{\beta}$ is related to θ_{it} through the relations $\mu_{it} = b'(\theta_{it})$ and $\mathbf{x}_{it}'\boldsymbol{\beta} = g(\mu_{it})$. Using Equation (10.3), we have that the jth element of the score function is

$$\frac{\partial}{\partial \beta_j} \ln p(\mathbf{y}) = \sum_{it} \left\{ \frac{\partial \theta_{it}}{\partial \beta_j} \frac{y_{it} - \mu_{it}}{\phi_{it}} \right\},$$

because $b'(\theta_{it}) = \mu_{it}$. Now, use the chain rule and the relation $\mathrm{Var}\, y_{it} = \phi_{it} b''(\theta_{it})$ to get

$$\frac{\partial \mu_{it}}{\partial \beta_j} = \frac{\partial b'(\theta_{it})}{\partial \beta_j} = b''(\theta_{it}) \frac{\partial \theta_{it}}{\partial \beta_j} = \frac{\mathrm{Var}\, y_{it}}{\phi_{it}} \frac{\partial \theta_{it}}{\partial \beta_j}.$$

Thus, we have

$$\frac{\partial \theta_{it}}{\partial \beta_j} \frac{1}{\phi_{it}} = \frac{\partial \mu_{it}}{\partial \beta_j} \frac{1}{\mathrm{Var}\, y_{it}}.$$

This yields

$$\frac{\partial}{\partial \beta_j} \ln p(\mathbf{y}) = \sum_{it} \left\{ \frac{\partial \mu_{it}}{\partial \beta_j} \left(\mathrm{Var}\, y_{it} \right)^{-1} \left(y_{it} - \mu_{it} \right) \right\},$$

which is the GEE form. This is the topic of Section 10.3.

Standard Errors for Regression Coefficient Estimators

As described in Appendix C.2, MLEs are consistent and asymptotically normally distributed under broad conditions. To illustrate, consider the MLE determined from Equation (10.4) using the canonical link. The asymptotic variance–covariance matrix of $\mathbf{b}_{\text{MLE}}$ is the inverse of the information matrix that, from Equation (C.4), is

$$\sum_{it} b''(\mathbf{x}_{it}'\mathbf{b}_{\text{MLE}}) \frac{\mathbf{x}_{it}\mathbf{x}_{it}'}{\phi_{it}}. \tag{10.6}$$

The square root of the jth diagonal element of the inverse of this matrix yields the standard error for $\mathbf{b}_{j,\text{MLE}}$, which we denote as $se(\mathbf{b}_{j,\text{MLE}})$. Extensions to general links are similar.

Overdispersion

An important feature of several members of the linear exponential family of distributions, such as the Bernoulli and the Poisson distributions, is that the variance is determined by the mean. In contrast, the normal distribution has a separate parameter for the variance, or dispersion. When fitting models to data with binary or count-dependent variables, it is common to observe that the variance exceeds that anticipated by the fit of the mean parameters. This phenomenon is known as *overdispersion*. Several alternative probabilistic models may be available to explain this phenomenon, depending on the application at hand. See Section 10.4 for an example and McCullagh and Nelder (1989G) for a more detailed inventory.

Although arriving at a satisfactory probabilistic model is the most desirable route, in many situations analysts are content to postulate an approximate model through the relation

$$\text{Var } y_{it} = \sigma^2 \phi b''(\mathbf{x}_{it}'\beta)/w_{it}.$$

The scale parameter ϕ is specified through the choice of the distribution whereas the scale parameter σ^2 allows for extra variability. For example, Table 10A.1 shows that, by specifying either the Bernoulli or Poisson distribution, we have $\phi = 1$. Although the scale parameter σ^2 allows for extra variability, it may also accommodate situations in which the variability is smaller than specified by the distributional form (although this situation is less common). Finally, note that for some distributions such as the normal distribution, the extra term is already incorporated in the ϕ parameter and thus serves no useful purpose.

When the additional scale parameter σ^2 is included, it is customary to estimate it by Pearson's chi-square statistic divided by the error degrees of freedom;

that is,

$$\hat{\sigma}^2 = \frac{1}{N-K}\sum_{it} w_{it} \frac{(y_{it} - b'(\mathbf{x}'_{it}\mathbf{b}_{\text{MLE}}))^2}{\phi b''(\mathbf{x}'_{it}\mathbf{b}_{\text{MLE}})}.$$

10.2 Example: Tort Filings

There is a widespread belief that, in the United States, contentious parties have become increasingly willing to go to the judicial system to settle disputes. This is particularly true when one party is from the insurance industry, an industry designed to spread risk among individuals. Litigation in the insurance industry arises from two types of disagreement among parties, breach of faith and tort. A breach of faith is a failure by a party to the contract to perform according to its terms. A tort action is a civil wrong, other than breach of contract, for which the court will provide a remedy in the form of action for damages. A civil wrong may include malice, wantonness, oppression, or capricious behavior by a party. Generally, large damages can be collected for tort actions because the award may be large enough to "sting" the guilty party. Because large insurance companies are viewed as having "deep pockets," these awards can be quite large.

The response that we consider is the number of filings, NUMFILE, of tort actions against insurance companies, y. For each of six years, 1984–1989, the data were obtained from 19 states with two observations unavailable for a total of $N = 112$ observations. The issue is to understand ways in which state legal, economic, and demographic characteristics affect the number of filings. Table 10.2 describes these characteristics. More extensive motivation is provided in Lee (1994O) and Lee, Browne, and Schmit (1994O).

Tables 10.3 and 10.4 summarize the state legal, economic, and demographic characteristics. To illustrate, in Table 10.3 we see that 23.2% of the 112 state-year observations were under limits (caps) on noneconomic reform. Those observations not under limits on noneconomic reforms had a larger average number of filings. The correlations in Table 10.4 show that several of the economic and demographic variables appear to be related to the number of filings. In particular, we note that the number of filings is highly related to the state population.

Figure 10.1 is a multiple time-series plot of the number of filings. The state with the largest number of filings is California. This plot shows the state-level heterogeneity of filings.

When data represent counts such as the number of tort filings, it is customary to consider the Poisson model to represent the distribution of responses. From mathematical statistics theory, it is known that the sums of

Table 10.2. *State characteristics*

Dependent variable	
NUMFILE	Number of filings of tort actions against insurance companies
State legal characteristics	
JSLIAB	An indicator of joint and several liability reform
COLLRULE	An indicator of collateral source reform
CAPS	An indicator of caps on noneconomic reform
PUNITIVE	An indicator of limits of punitive damage
State economic and demographic characteristics	
POP	The state population, in millions
POPLAWYR	The population per lawyer
VEHCMILE	Number of automobile-miles per mile of road, in thousands
POPDENSY	Number of people per ten square miles of land
WCMPMAX	Maximum workers' compensation weekly benefit
URBAN	Percentage of population living in urban areas
UNEMPLOY	State unemployment rate, in percentages

Source: Lee (1994O).

Table 10.3. *Averages with explanatory binary variables*

	Explanatory variable			
	JSLIAB	COLLRULE	CAPS	PUNITIVE
Average explanatory variable	0.491	0.304	0.232	0.321
Average NUMFILE				
When explanatory variable = 0	15,530	20,727	24,682	17,693
When explanatory variable = 1	25,967	20,027	6,727	26,469

Table 10.4. *Summary statistics for other variables*

Variable	Mean	Median	Minimum	Maximum	Standard deviation	Correlation with NUMFILE
NUMFILE	20514	9085	512	137455	29039	1.0
POP	6.7	3.4	0.5	29.0	7.2	0.902
POPLAWYR	377.3	382.5	211.0	537.0	75.7	−0.378
VEHCMILE	654.8	510.5	63.0	1899.0	515.4	0.518
POPDENSY	168.2	63.9	0.9	1043.0	243.9	0.368
WCMPMAX	350.0	319.0	203.0	1140.0	151.7	−0.265
URBAN	69.4	78.9	18.9	100.0	24.8	0.550
UNEMPLOY	6.2	6.0	2.6	10.8	1.6	0.008

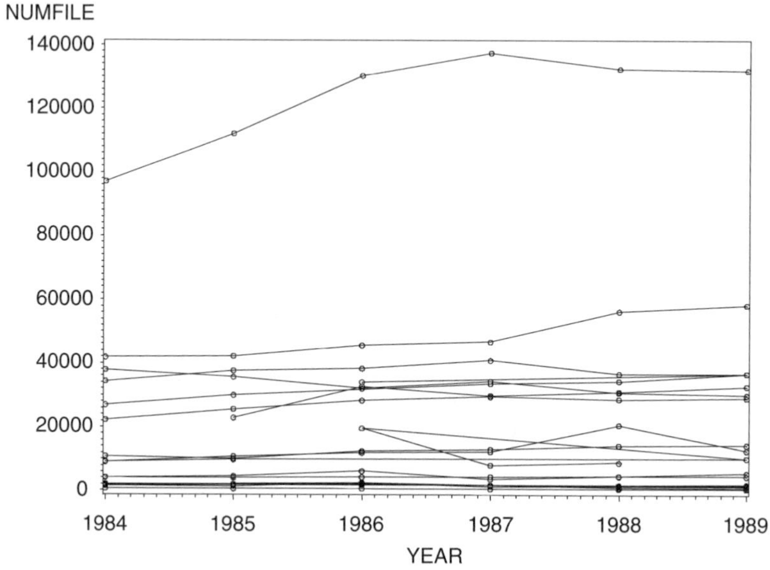

Figure 10.1. Multiple time-series plot of NUMFILE.

independent Poisson random variables have a Poisson distribution. Thus, if $\lambda_1, \ldots, \lambda_n$ represent independent Poisson random variables, each with parameter $\mathrm{E}\lambda_i = \mu_i$, then $\lambda_1 + \cdots + \lambda_n$ is Poisson distributed with parameter $n\bar{\mu}$, where $\bar{\mu} = (\mu_1 + \cdots + \mu_n)/n$. We assume that y_{it} has a Poisson distribution with parameter $\mathrm{POP}_{it} \exp(\mathbf{x}_{it}'\boldsymbol{\beta})$, where POP_{it} is the population of the ith state at time t. To account for this "known" relationship with population, we assume that the (natural) logarithmic population is one of the explanatory variables, yet has a known regression coefficient equal to one. In GLM terminology, such a variable is known as an *offset*. Thus, our Poisson parameter for y_{it} is

$$\begin{aligned}\exp(\ln \mathrm{POP}_{it} + x_{it,1}\beta_1 + \cdots + x_{it,K}\beta_K) &= \exp(\ln \mathrm{POP}_{it} + \mathbf{x}_{it}'\boldsymbol{\beta}) \\ &= \mathrm{POP}_{it}\exp(\mathbf{x}_{it}'\boldsymbol{\beta}).\end{aligned}$$

An alternative approach is to use the *average* number of tort filings as the response and assume approximate normality. This was the approach taken by Lee et al. (1994O); the reader has an opportunity to practice this approach in Exercises 2.18 and 3.12. Note that, in our Poisson model, the expectation of the average response is $\mathrm{E}\,(y_{it}/\mathrm{POP}_{it}) = \exp(\mathbf{x}_{it}'\boldsymbol{\beta})$, whereas the variance is $\mathrm{Var}\,(y_{it}/\mathrm{POP}_{it}) = \exp(\mathbf{x}_{it}'\boldsymbol{\beta})/\mathrm{POP}_{it}$. Thus, to make these two approaches compatible, one must use weighted regression, using estimated reciprocal variances as the weights.

Table 10.5 summarizes the fit of three Poisson models. With the basic homogeneous Poisson model, all explanatory variables turn out to be statistically

Table 10.5. *Tort filings model coefficient estimates based on $N = 112$ observations from $n = 19$ states and $T = 6$ years with logarithmic population used as an offset*

	Homogeneous Poisson model		Model with estimated scale parameter		Model with scale parameter and time categorical variable	
Variable	Parameter estimate	p-values	Parameter estimate	p-values	Parameter estimate	p-values
Intercept	−7.943	<.0001	−7.943	<.0001		
POPLAWYR/1000	2.163	<.0001	2.163	0.0002	2.123	0.0004
VEHCMILE/1000	0.862	<.0001	0.862	<.0001	0.856	<.0001
POPDENSY/1000	0.392	<.0001	0.392	0.0038	0.384	0.0067
WCMPMAX/1000	−0.802	<.0001	−0.802	0.1226	−0.826	0.1523
URBAN/1000	0.892	<.0001	0.892	0.8187	0.977	0.8059
UNEMPLOY	0.087	<.0001	0.087	0.0005	0.086	0.0024
JSLIAB	0.177	<.0001	0.177	0.0292	0.130	0.2705
COLLRULE	−0.030	<.0001	−0.030	0.7444	−0.023	0.8053
CAPS	−0.032	<.0001	−0.032	0.7457	−0.056	0.6008
PUNITIVE	0.030	<.0001	0.030	0.6623	0.053	0.4986
Scale	1.000		35.857		36.383	
Deviance	118,309.0		118,309.0		115,496.4	
Pearson chi-square	129,855.7		129,855.7		127,073.9	

significant, as evidenced by the small p-values. However, the Poisson model assumes that the variance equals the mean; this is often a restrictive assumption for empirical work. Thus, to account for potential overdispersion, Table 10.5 also summarizes a homogenous Poisson model with an estimated scale parameter. Table 10.5 emphasizes that, although the regression coefficient estimates do not change with the introduction of the scale parameter, estimated standard errors and thus p-values do change. Many variables, such as CAPS, turn out to be statistically insignificant predictors of the number of filings when a more flexible model for the variance is introduced. Subsequent sections will introduce models of the state-level heterogeneity. Although not as important for this data set, it is still easy to examine temporal heterogeneity in this context through year binary (dummy) variables. The goodness-of-fit statistics – the deviance and Pearson chi-square – favor including the time categorical variable. (See Appendix C.8 for definitions of these goodness-of-fit statistics.)

10.3 Marginal Models and GEE

Marginal models underpin a widely applied framework for handling heterogeneity in longitudinal data. As introduced in Section 9.4, marginal models are *semiparametric*. Specifically, we model the first and second moments as functions of unknown parameters without specifying a full parametric distribution for the responses. For these models, parameters can be estimated using the GEE extension of the method of moment estimation technique.

Marginal Models

To reintroduce marginal models, again use $\boldsymbol{\beta}$ to be a $K \times 1$ vector of parameters associated with the explanatory variables $\mathbf{x}_{it}$. Further, let $\boldsymbol{\tau}$ be a $q \times 1$ vector of variance component parameters. Assume that the mean of each response is

$$\mathrm{E} y_{it} = \mu_{it} = \mu_{it}(\boldsymbol{\beta}, \boldsymbol{\tau}),$$

where μ_{it} is a known function. Further, denote the variance of each response as

$$\mathrm{Var}\, y_{it} = v_{it} = v_{it}(\boldsymbol{\beta}, \boldsymbol{\tau}),$$

where v_{it} is a known function.

To complete the second moment specification, we need assumptions on the covariances among responses. We assume that observations among different subjects are independent and thus have zero covariance. For observations within a subject, we assume that the correlation between two observations within the same subject is a known function. Specifically, we will denote

$$\text{corr}(y_{ir}, y_{is}) = \rho_{rs}(\mu_{ir}, \mu_{is}, \boldsymbol{\tau}),$$

where $\rho_{rs}(.)$ is a known function. As in Chapter 3, marginal models use correlations among observations within a subject to represent heterogeneity, that is, the tendency for observations from the same subject to be related.

Special Case: Generalized Linear Model with Canonical Links As we saw in Section 10.1, in the context of the generalized linear model with canonical links, we have $\mu_{it} = \text{E}y_{it} = b'(\mathbf{x}'_{it}\boldsymbol{\beta})$ and $v_{it} = \text{Var}\, y_{it} = \phi b''(\mathbf{x}'_{it}\boldsymbol{\beta})/w_{it}$, where $b(.)$ is a known function that depends on the choice of the distribution of responses. There is only one variance component, so $q = 1$ and $\boldsymbol{\tau} = \phi$. For the cases of independent observations among subjects discussed in Section 10.1, the correlation function ρ_{rs} is 1 if $r = s$ and 0 otherwise. For this example, note that the mean function depends on $\boldsymbol{\beta}$ but does not depend on the variance component $\boldsymbol{\tau}$.

Special Case: Error-Components Model For the error-components model introduced in Section 3.1, we may use a generalized linear model with a normal distribution for responses, so that $b(\theta) = \theta^2/2$. Thus, from the previous example, we have $\mu_{it} = \text{E}y_{it} = \mathbf{x}'_{it}\boldsymbol{\beta}$ and $v_{it} = \text{Var}\, y_{it} = \phi = \sigma_\alpha^2 + \sigma^2$. Unlike the previous example, observations within subjects are not independent but have an *exchangeable* correlation structure. Specifically, in Section 3.1 we saw that, for $r \neq s$, we have

$$\text{corr}(y_{ir}, y_{is}) = \rho = \frac{\sigma_\alpha^2}{\sigma_\alpha^2 + \sigma^2}.$$

With this specification, the vector of variance components is $\boldsymbol{\tau} = (\phi, \sigma_\alpha^2)'$, where $\phi = \sigma_\alpha^2 + \sigma^2$.

In Section 10.1, the parametric form of the first two moments was a consequence of the assumption of exponential family distribution. With marginal models, the parametric form is now a basic assumption. We now give matrix notation to express these assumptions more compactly. To this end, define $\boldsymbol{\mu}_i(\boldsymbol{\beta}, \boldsymbol{\tau}) = \boldsymbol{\mu}_i = (\mu_{i1}\ \mu_{i2}\ \cdots\ \mu_{iT_i})'$ to be the vector of means for the ith subject. Let $\mathbf{V}_i$ be the $T_i \times T_i$ variance covariance matrix, $\text{Var}\, \mathbf{y}_i$, where the (r, s)th element of $\mathbf{V}_i$ is given by

$$\text{Cov}(y_{ir}, y_{is}) = \text{corr}(y_{ir}, y_{is})\sqrt{v_{ir}v_{is}}.$$

As in Section 9.4, for estimation purposes we will also require the $K \times T_i$ gradient matrix,

$$\mathbf{G}_\mu(\boldsymbol{\beta}, \tau) = \left(\frac{\partial \mu_{i1}}{\partial \boldsymbol{\beta}} \cdots \frac{\partial \mu_{iT_i}}{\partial \boldsymbol{\beta}} \right). \tag{10.7}$$

Generalized Estimating Equations

Marginal models provide sufficient structure to allow for use of GEE. As with generalized least squares, we first describe this estimation method assuming the variance component parameters are known. We then describe extensions to incorporate variance component estimation. The general GEE procedure is described in Appendix C.6; recall that an introduction was provided in Section 9.4.

Assuming that the variance components in $\boldsymbol{\tau}$ are known, the GEE estimator of $\boldsymbol{\beta}$ is the solution of

$$\mathbf{0}_K = \sum_{i=1}^{n} \mathbf{G}_\mu(\mathbf{b}) \mathbf{V}_i^{-1}(\mathbf{b})(\mathbf{y}_i - \boldsymbol{\mu}_i(\mathbf{b})). \tag{10.8}$$

We will denote this solution as $\mathbf{b}_{\mathrm{EE}}$. Under mild regularity conditions, this estimator is consistent and asymptotically normal with variance–covariance matrix

$$\operatorname{Var} \mathbf{b}_{\mathrm{EE}} = \left(\sum_{i=1}^{n} \mathbf{G}_\mu \mathbf{V}_i^{-1}(\mathbf{b}) \mathbf{G}_\mu' \right)^{-1}. \tag{10.9}$$

Special Case: Generalized Linear Model with Canonical Links (continued) Because $\mu_{it} = b'(\mathbf{x}_{it}'\boldsymbol{\beta})$, it is easy to see that

$$\frac{\partial \mu_{it}}{\partial \boldsymbol{\beta}} = \mathbf{x}_{it} b''(\mathbf{x}_{it}'\boldsymbol{\beta}) = w_{it} \mathbf{x}_{it} \frac{v_{it}}{\phi}.$$

Thus,

$$\mathbf{G}_\mu(\boldsymbol{\beta}, \boldsymbol{\tau}) = \frac{1}{\phi} (\, w_{i1} \mathbf{x}_{i1} v_{i1} \ \cdots \ w_{iT_i} \mathbf{x}_{iT_i} v_{iT_i} \,)$$

is our $K \times T_i$ matrix of derivatives. Assuming independence among observations within a subject we have $\mathbf{V}_i = \mathrm{diag}(v_{i1}, \ldots, v_{iT_i})$. Thus, we may express the generalized estimating equations as

$$\mathbf{0}_K = \sum_{i=1}^{n} \mathbf{G}_\mu(\mathbf{b}) \mathbf{V}_i^{-1}(\mathbf{b})(\mathbf{y}_i - \boldsymbol{\mu}_i(\mathbf{b})) = \frac{1}{\phi} \sum_{i=1}^{n} \sum_{t=1}^{T_i} w_{it} \mathbf{x}_{it} (y_{it} - \mu_{it}(\mathbf{b})).$$

This yields the same solution as the maximum likelihood "normal equations" in Equation (10.5). Thus, the GEE estimators are equal to the MLEs for this special case. Note that this solution does not depend on knowledge of the variance component, ϕ.

GEEs with Unknown Variance Components

To determine GEE estimates, the first task is to determine an initial estimator of $\boldsymbol{\beta}$, say $\mathbf{b}_{0,\mathrm{EE}}$. To illustrate, one might use the GLM with independence among observations, as before, to get an initial estimator. Next, we use the initial estimator $\mathbf{b}_{0,\mathrm{EE}}$ to compute residuals and determine an initial estimator of the variance components, say $\boldsymbol{\tau}_{0,\mathrm{EE}}$. Then, at the $(n+1)$st stage, recursively, perform the following:

1. Use $\boldsymbol{\tau}_{n,\mathrm{EE}}$ and the solution of the equation
$$\mathbf{0}_K = \sum_{i=1}^{n} \mathbf{G}_\mu(\mathbf{b}, \boldsymbol{\tau}_{n,\mathrm{EE}}) \mathbf{V}_i^{-1}(\mathbf{b}, \boldsymbol{\tau}_{n,\mathrm{EE}})(\mathbf{y}_i - \boldsymbol{\mu}_i(\mathbf{b}, \boldsymbol{\tau}_{n,\mathrm{EE}}))$$
to determine an updated estimator of $\boldsymbol{\beta}$, say $\mathbf{b}_{n+1,\mathrm{EE}}$.
2. Use the residuals $\{y_{it} - \mu_{it}(\mathbf{b}_{n+1,\mathrm{EE}}, \boldsymbol{\tau}_{n,\mathrm{EE}})\}$ to determine an updated estimator of $\boldsymbol{\tau}$, say $\boldsymbol{\tau}_{n+1,\mathrm{EE}}$.
3. Repeat steps 1 and 2 until convergence is obtained.

Alternatively, for the first stage, we may update estimators using a one-step procedure such as a Newton–Raphson iteration. As another example, note that the statistical package SAS uses a Fisher scoring type update of the form

$$\mathbf{b}_{n+1,\mathrm{EE}} = \mathbf{b}_{n,\mathrm{EE}} + \left(\sum_{i=1}^{n} \mathbf{G}_\mu(\mathbf{b}_{n,\mathrm{EE}}) \mathbf{V}_i^{-1}(\mathbf{b}_{n,\mathrm{EE}}) \mathbf{G}'_\mu(\mathbf{b}_{n,\mathrm{EE}}) \right)^{-1} \times \left(\sum_{i=1}^{n} \mathbf{G}_\mu(\mathbf{b}_{n,\mathrm{EE}}) \mathbf{V}_i^{-1}(\mathbf{b}_{n,\mathrm{EE}})(\mathbf{y}_i - \boldsymbol{\mu}_i(\mathbf{b}_{n,\mathrm{EE}})) \right)$$

For GEE in more complex problems with unknown variance components, Prentice (1988B) suggests using a second estimating equation of the form

$$\sum_i \left(\frac{\partial \mathrm{E}\mathbf{y}_i^*}{\partial \boldsymbol{\tau}} \right) \mathbf{W}_i^{-1}(\mathbf{y}_i^* - \mathrm{E}\mathbf{y}_i^*) = \mathbf{0},$$

where $\mathbf{y}_i^*$ is a vector of squares and cross-products of observations within a subject of the form

$$\mathbf{y}_i^* = \left(y_{i1}y_{i2} \; y_{i1}y_{i3} \; \cdots \; y_{i1}y_{iT_i} \; \cdots \; y_{i1}^2 \; y_{i2}^2 \; \cdots \; y_{iT_i}^2 \right)'.$$

For the variance of $\mathbf{y}_i^*$, Diggle et al. (2002S) suggest using the identity matrix for $\mathbf{W}_i$ with most discrete data. However, for binary responses, they note that the last T_i observations are redundant because $y_{it} = y_{it}^2$. These should be ignored; Diggle et al. recommend using

$$\mathbf{W}_i = \text{diag}(\,\text{Var}(y_{i1}y_{i2}) \cdots \text{Var}(y_{i,T_i-1}y_{iT_i})).$$

Robust Estimation of Standard Errors

As discussed in Sections 2.5.3 and 3.4 for linear models, one must be concerned with unsuspected serial correlation and heteroscedasticity. This is particularly true with marginal models, where the specification is based only on the first two moments and not a full probability distribution. Because the specification of the correlation structure may be suspect, these are often called *working correlations*. Standard errors that rely on the correlation structure specification from the relation in Equation (10.9) are known as *model-based* standard errors. In contrast, *empirical* standard errors may be calculated using the following estimator of the asymptotic variance of $\mathbf{b}_{\text{EE}}$:

$$\left(\sum_{i=1}^{n} \mathbf{G}_{\mu}\mathbf{V}_i^{-1}\mathbf{G}_{\mu}'\right)^{-1} \left(\sum_{i=1}^{n} \mathbf{G}_{\mu}\mathbf{V}_i^{-1}(\mathbf{y}_i - \boldsymbol{\mu}_i)(\mathbf{y}_i - \boldsymbol{\mu}_i)'\mathbf{V}_i^{-1}\mathbf{G}_{\mu}'\right) \times \left(\sum_{i=1}^{n} \mathbf{G}_{\mu}\mathbf{V}_i^{-1}\mathbf{G}_{\mu}'\right)^{-1}. \tag{10.10}$$

Specifically, the standard error of the jth component of $\mathbf{b}_{\text{EE}}$, $se(\mathbf{b}_{j,\text{EE}})$, is defined to be the square root of the jth diagonal element of the variance–covariance matrix in (10.10).

Example: Tort Filings (continued) To illustrate marginal models and GEE estimators, we return to the Section 10.2 tort filing example. The fitted models, which appear in Table 10.6, were estimated using the SAS procedure GENMOD. Assuming an independent working correlation, we arrive at the same parameter estimators as in Table 10.5, under the homogenous Poisson model with an estimated scale parameter. Further, although not displayed, the fits provide the same model-based standard errors. Compared to the empirical standard errors, we see that almost all empirical standard errors are larger than the corresponding model-based standard errors (the exception being CAPS).

To test the robustness of this model fit, we fit the same model with an *AR* (1) working correlation. This model fit also appears in Table 10.6. Because we use the working correlation as a weight matrix, many of the coefficients have different values than the corresponding fit with the independent working correlation matrix. Note that the asterisk indicates when an estimate exceeds

Table 10.6. *Comparison of GEE estimators*[a]

	Independent working correlation			*AR*(1) working correlation		
Parameter	Estimate[b]	Empirical standard error	Model-based standard error	Estimate[b]	Empirical standard error	Model-based standard error
Intercept	−7.943*	0.612	0.435	−7.840*	0.870	0.806
POPLAWYR/1000	2.163	1.101	0.589	2.231	1.306	0.996
VEHCMILE/1000	0.862*	0.265	0.120	0.748*	0.166	0.180
POPDENSY/1000	0.392*	0.175	0.135	0.400*	0.181	0.223
WCMPMAX/1000	−0.802	0.895	0.519	−0.764	0.664	0.506
URBAN/1000	0.892	5.367	3.891	3.508	7.251	7.130
UNEMPLOY	0.087*	0.042	0.025	0.048*	0.018	0.021
JSLIAB	0.177*	0.089	0.081	0.139*	0.020	0.049
COLLRULE	−0.030	0.120	0.091	−0.014	0.079	0.065
CAPS	−0.032	0.098	0.098	0.142*	0.068	0.066
PUNITIVE	0.030	0.125	0.068	−0.043	0.054	0.049
Scale		35.857			35.857	
AR(1) coefficient					0.854	

[a] All models use an estimated scale parameter. Logarithmic population is used as an offset.
[b] The asterisk (*) indicates that the estimate is more than twice the empirical standard error, in absolute value.

twice the empirical standard error; this gives a rough idea of the statistical significance of the variable. Under the $AR(1)$ working correlation, variables that are statistically significant using empirical standard errors are also statistically significant using model-based standard errors. (Again, CAPS is a borderline exception.) Comparing the independent to the $AR(1)$ working correlation results, we see that VEHCMILE, UNEMPLOY, and JSLIAB are consistently statistically significant whereas POPLAWYR, POPDENSY, WCMPMAX, URBAN, COLLRULE, and PUNITIVE are consistently statistically insignificant. However, judging statistical significance depends on the model selection for CAPS.

10.4 Random-Effects Models

An important method for accounting for heterogeneity in longitudinal and panel data is through a random-effects formulation. As in Sections 3.1 and 3.3.1 for linear models, this model is easy to introduce and interpret in the following hierarchical fashion:

Stage 1. Subject effects $\{\alpha_i\}$ are a random sample from a distribution that is known up to a vector of parameters.

Stage 2. Conditional on $\{\alpha_i\}$, the responses $\{y_{i1}\ y_{i2} \cdots y_{iT_i}\}$ are a random sample from a GLM with systematic component $\eta_{it} = \mathbf{z}'_{it}\boldsymbol{\alpha}_i + \mathbf{x}'_{it}\boldsymbol{\beta}$.

Additional motivation and sampling issues regarding random effects were introduced in Chapter 3 for linear models; these also pertain to generalized linear models.

Thus, the random-effects model introduced here is a generalization of the following:

- The Chapter 3 linear random-effects model. Here, we used a normal distribution for the GLM component.
- The Section 9.2 binary dependent-variables model with random effects. Here, we used a Bernoulli distribution for the GLM component. (In Section 9.2, we also focused on the case $z_{it} = 1$.)

Assuming a distribution for the heterogeneity components α_i, typically requires fewer parameters to be estimated when compared to the fixed-effects models that will be described in Section 10.5. The maximum likelihood method of estimation is available although computationally intensive. To use this method, the customary practice is to assume normally distributed random effects. Other estimation methods are also available in the literature. For example, the EM (for expectation–maximization) algorithm for estimation is described in Diggle et al. (2002S) and McCulloch and Searle (2001G).

Random-Effects Likelihood

To develop the likelihood, we first note that from the second sampling stage, conditional on $\boldsymbol{\alpha}_i$, the likelihood for the ith subject at the tth observation is

$$p(y_{it};\boldsymbol{\beta}\mid\boldsymbol{\alpha}_i)=\exp\left(\frac{y_{it}\theta_{it}-b(\theta_{it})}{\phi}+S(y_{it},\phi)\right),$$

where $b'(\theta_{it})=\mathrm{E}(y_{it}\mid\boldsymbol{\alpha}_i)$ and $\eta_{it}=\mathbf{z}_{it}'\boldsymbol{\alpha}_i+\mathbf{x}_{it}'\boldsymbol{\beta}=g(\mathrm{E}(y_{it}\mid\boldsymbol{\alpha}_i))$. Because of the independence among responses within a subject conditional on $\boldsymbol{\alpha}_i$, the conditional likelihood for the ith subject is

$$p(\mathbf{y}_i;\boldsymbol{\beta}\mid\boldsymbol{\alpha}_i)=\exp\left\{\sum_t\left(\frac{y_{it}\theta_{it}-b(\theta_{it})}{\phi}+S(y_{it},\phi)\right)\right\}.$$

Taking expectations over $\boldsymbol{\alpha}_i$ yields the (unconditional) likelihood. To see this explicitly, we use the canonical link so that $\theta_{it}=\eta_{it}$. Thus, the (unconditional) likelihood for the ith subject is

$$p(\mathbf{y}_i;\boldsymbol{\beta})=\exp\left\{\sum_t S(y_{it},\phi)\right\} \times\int\exp\left\{\sum_t\left(\frac{y_{it}(\mathbf{z}_{it}'\mathbf{a}+\mathbf{x}_{it}'\boldsymbol{\beta})-b(\mathbf{z}_{it}'\mathbf{a}+\mathbf{x}_{it}'\boldsymbol{\beta})}{\phi}\right)\right\}dF_\alpha(\mathbf{a}), \tag{10.11}$$

where $F_\alpha(.)$ represents the distribution of $\boldsymbol{\alpha}_i$, which we will assume to be multivariate normal with mean zero and variance–covariance matrix $\boldsymbol{\Sigma}_\alpha$. Consistent with the notation in Chapter 3, let $\boldsymbol{\tau}$ denote the vector of parameters associated with scaling in stage 2, ϕ, and the stage-1 parameters in the matrix $\mathbf{D}$. With this notation, we may write the total log likelihood as

$$\ln p(\mathbf{y},\boldsymbol{\beta},\boldsymbol{\tau})=\sum_i\ln p(\mathbf{y}_i,\boldsymbol{\beta}).$$

From Equation (10.11), we see that evaluating the log likelihood requires numerical integration and thus it is more difficult to compute than likelihoods for homogeneous models.

Special Case: Poisson Distribution To illustrate, assume $q=1$ and $z_{it}=1$ so that only intercepts α_i vary by subject. Assuming a Poisson distribution for the conditional responses, we have $\phi=1$, $b(a)=e^a$, and $S(y,\phi)=-\ln(y!)$.

Thus, from Equation (10.11), the log likelihood for the ith subject is

$$\begin{aligned}
\ln\ p(\mathbf{y}_i, \boldsymbol{\beta}) &= -\sum_t \ln(y_{it}!) \\
&\quad + \ln\left(\int \exp\left\{\sum_t (y_{it}(a + \mathbf{x}_{it}'\boldsymbol{\beta}) - \exp(a + \mathbf{x}_{it}'\boldsymbol{\beta}))\right\} f_\alpha(a)da\right) \\
&= -\sum_t \ln(y_{it}!) + \sum_t y_{it}\mathbf{x}_{it}'\boldsymbol{\beta} \\
&\quad + \ln\left(\int \exp\left\{\sum_t (y_{it}a - \exp(a + \mathbf{x}_{it}'\boldsymbol{\beta}))\right\} f_\alpha(a)da\right),
\end{aligned}$$

where $f_\alpha(.)$ is the probability density function of α_i. As before, evaluating and maximizing the log likelihood requires numerical integration.

Serial Correlation and Overdispersion

The random-effects GLM introduces terms that account for heterogeneity in the data. However, the model also introduces certain forms of serial correlation and overdispersion that may or may not be evident in the data. Recall that in Chapter 3 we saw that permitting subject-specific effects, α_i, to be random induced serial correlation in the responses $\{y_{i1}\ y_{i2}\ \cdots\ y_{iT_i}\}$. This is also true for the nonlinear GLMs, as shown in the following example.

Special Case: Poisson Distribution (continued) Recall that, for a canonical link, we have $\mathrm{E}(y_{it} \mid \alpha_i) = b'(\theta_{it}) = b'(\eta_{it}) = b'(\alpha_i + \mathbf{x}_{it}'\boldsymbol{\beta})$. For the Poisson distribution, we have $b'(a) = e^a$ so that

$$\mu_{it} = \mathrm{E}y_{it} = \mathrm{E}(\mathrm{E}(y_{it} \mid \alpha_i)) = \mathrm{E}b'(\alpha_i + \mathbf{x}_{it}'\boldsymbol{\beta}) = \exp(\mathbf{x}_{it}'\boldsymbol{\beta})\mathrm{E}e^\alpha,$$

where we have dropped the subscript on α_i because the distribution is identical over i.

To see the serial correlation, we examine the covariance between two observations, for example, y_{i1} and y_{i2}. By the conditional independence, we have

$$\begin{aligned}
\mathrm{Cov}\,(y_{i1}, y_{i2}) &= \mathrm{E}(\mathrm{Cov}(y_{i1}, y_{i2}) \mid \alpha_i) + \mathrm{Cov}(\mathrm{E}(y_{i1} \mid \alpha_i), \mathrm{E}(y_{i2} \mid \alpha_i)) \\
&= \mathrm{Cov}(b'(\alpha_i + \mathbf{x}_{i1}'\boldsymbol{\beta}), b'(\alpha_i + \mathbf{x}_{i2}'\boldsymbol{\beta})) \\
&= \mathrm{Cov}(e^\alpha \exp(\mathbf{x}_{i1}'\boldsymbol{\beta}), e^\alpha \exp(\mathbf{x}_{i2}'\boldsymbol{\beta})) \\
&= \exp((\mathbf{x}_{i1} + \mathbf{x}_{i2})'\boldsymbol{\beta})\ \mathrm{Var}\, e^\alpha.
\end{aligned}$$

This covariance is always nonnegative, indicating that we can anticipate positive serial correlation using this model.

Table 10.7. *Tort filings model coefficient estimates (random effects) with logarithmic population used as an offset*

	Homogeneous model with estimated scale parameter		Random-effects model	
Variable	Parameter estimate	*p*-values	Parameter estimate	*p*-values
Intercept	−7.943	<.0001	−2.753	<.0001
POPLAWYR/1000	2.163	0.0002	−2.694	<.0001
VEHCMILE/1000	0.862	<.0001	−0.183	0.0004
POPDENSY/1000	0.392	0.0038	9.547	<.0001
WCMPMAX/1000	−0.802	0.1226	−1.900	<.0001
URBAN/1000	0.892	0.8187	−47.820	<.0001
UNEMPLOY	0.087	0.0005	0.035	<.0001
JSLIAB .	0.177	0.0292	0.546	0.3695
COLLRULE	−0.030	0.7444	−1.031	0.1984
CAPS	−0.032	0.7457	0.391	0.5598
PUNITIVE	0.030	0.6623	0.110	0.8921
State variance			2.711	
−2 log likelihood	119,576		15,623	

Similar calculations show that

$$\begin{aligned}
\text{Var}\, y_{it} &= \text{E}(\text{Var}\,(y_{it} \mid \alpha_i)) + \text{Var}\,(\text{E}(y_{it} \mid \alpha_i)) \\
&= \text{E}\phi b''(\alpha_i + \mathbf{x}_{it}'\boldsymbol{\beta}) + \text{Var}\,(b'(\alpha_i + \mathbf{x}_{it}'\boldsymbol{\beta})) \\
&= \text{E}e^{\alpha} \exp(\mathbf{x}_{i1}'\boldsymbol{\beta}) + \text{Var}\,(\exp(\alpha_i + \mathbf{x}_{it}'\boldsymbol{\beta})) \\
&= \mu_{it} + \exp(2\mathbf{x}_{it}'\boldsymbol{\beta})\text{Var}\, e^{\alpha}.
\end{aligned}$$

Thus, the variance always exceeds the mean. Compared to the usual Poisson models, which require equality between the mean and the variance, the random-effects specification induces a larger variance. This is a specific example of overdispersion.

Example: Tort Filings (continued) To illustrate the random-effects estimators, we return to the Section 10.2 tort filing example. Table 10.7 summarizes a random-effects model that was fit using the SAS statistical procedure NLMIXED. For comparison, the Table 10.5 fits from the homogeneous Poisson model with an estimated scale parameter are included in Table 10.7. The random-effects model assumes a conditional Poisson-distributed response, with a scalar homogeneity parameter that has a normal distribution.

Computational Considerations

As is evident from Equation (10.11), maximum likelihood estimation of regression coefficients requires one or more q-dimensional numerical integrations for each subject and each iteration of an optimization routine. As we have seen in our Chapter 9 and 10 examples, this computational complexity is manageable for random intercept models where $q = 1$. According to McCulloch and Searle (2001G), this direct method is also available for applications with $q = 2$ or 3; however, for higher order models, such as with crossed-random effects, alternative approaches are necessary.

We have already mentioned the EM algorithm in Chapter 9 as one alternative; see McCulloch and Searle (2001G) or Diggle et al. (2002S) for more details. Another alternative is to use simulation techniques. McCulloch and Searle (2001G) summarize a Monte Carlo Newton–Raphson approach for approximating the score function, a simulated maximum likelihood approach for approximating the integrated likelihood, and a stochastic approximation method for a more efficient and sequential approach of simulation.

The most widely used set of alternatives are based on Taylor-series expansions, generally about the link function or the integrated likelihood. There are several justifications for this set of alternatives. One is that a Taylor series is used to produce adjusted variables that follow an approximate linear (mixed-effects) model. (Appendix C.3 describes this adjustment in the linear case.) Another justification is that these methods are determined through a "penalized" quasi-likelihood function, where there is a so-called penalty term for the random effects. This set of alternatives is the basis for the SAS macro GLM800.sas and, for example, the S-plus (a statistical package) procedure *nlme* (for nonlinear mixed effects). The disadvantage of this set of alternatives is that they do not work well for distributions that are far from normality, such as Bernoulli distributions (Lin and Breslow, 1996S). The advantage is that the approximation procedures work well even for a relatively large number of random effects. We refer the reader to McCulloch and Searle (2001G) for further discussion.

We also note that generalized linear models can be expressed as special cases of *nonlinear regression models*. Here, by "nonlinear," we mean that the regression function need not be a linear function of the predictors but can be expressed as a nonlinear function of the form $f(\mathbf{x}_{it}, \boldsymbol{\beta})$. The exponential family of distributions provides a special case of nonlinear regression functions. This is relevant to computing considerations because many computational routines have been developed for nonlinear regression and can be used directly in the GLM context. This is also true of extensions to mixed-effects models, as suggested by our reference to the *nmle* procedure. For discussions of nonlinear regression models with random effects, we refer the reader to Davidian and Giltinan (1995S), Vonesh and Chinchilli (1997S), and Pinheiro and Bates (2000S).

10.5 Fixed-Effects Models

As we have seen, another method for accounting for heterogeneity in longitudinal and panel data is through a fixed-effects formulation. Specifically, to incorporate heterogeneity, we allow for a systematic component of the form $\eta_{it} = \mathbf{z}_{it}'\boldsymbol{\alpha}_i + \mathbf{x}_{it}'\boldsymbol{\beta}$. Thus, this section extends the Section 9.3 discussion where we only permitted varying intercepts α_i. However, many of the same computational difficulties arise. Specifically, it will turn out that MLEs are generally inconsistent. Conditional maximum likelihood estimators are consistent but difficult to compute for many distributional forms. The two exceptions are the normal and Poisson families, where we show that it is easy to compute conditional maximum likelihood estimators. Because of the computational difficulties, we restrict consideration in this section to canonical links.

10.5.1 Maximum Likelihood Estimation for Canonical Links

Assume a canonical link so that $\theta_{it} = \eta_{it} = \mathbf{z}_{it}'\boldsymbol{\alpha}_i + \mathbf{x}_{it}'\boldsymbol{\beta}$. Thus, with Equation (10.3), we have the log likelihood

$$\ln p(\mathbf{y}) = \sum_{it} \left\{ \frac{y_{it}(\mathbf{z}_{it}'\boldsymbol{\alpha}_i + \mathbf{x}_{it}'\boldsymbol{\beta}) - b(\mathbf{z}_{it}'\boldsymbol{\alpha}_i + \mathbf{x}_{it}'\boldsymbol{\beta})}{\phi} + S(y_{it}, \phi) \right\}. \quad (10.12)$$

To determine maximum likelihood estimators of $\boldsymbol{\alpha}_i$ and $\boldsymbol{\beta}$, we take derivatives of $\ln p(\mathbf{y})$, set the derivatives equal to zero, and solve for the roots of these equations.

Taking the partial derivative with respect to $\boldsymbol{\alpha}_i$ yields

$$\mathbf{0} = \sum_{t} \left\{ \frac{\mathbf{z}_{it}(y_{it} - b'(\mathbf{z}_{it}'\boldsymbol{\alpha}_i + \mathbf{x}_{it}'\boldsymbol{\beta}))}{\phi} \right\} = \sum_{t} \left\{ \frac{\mathbf{z}_{it}\,(y_{it} - \mu_{it})}{\phi} \right\},$$

because $\mu_{it} = b'(\theta_{it}) = b'(\mathbf{z}_{it}'\boldsymbol{\alpha}_i + \mathbf{x}_{it}'\boldsymbol{\beta})$. Taking the partial derivative with respect to $\boldsymbol{\beta}$ yields

$$\mathbf{0} = \sum_{it} \left\{ \frac{\mathbf{x}_{it}(y_{it} - b'(\mathbf{z}_{it}'\boldsymbol{\alpha}_i + \mathbf{x}_{it}'\boldsymbol{\beta}))}{\phi} \right\} = \sum_{it} \left\{ \frac{\mathbf{x}_{it}\,(y_{it} - \mu_{it})}{\phi} \right\}.$$

Thus, we can solve for the MLEs of $\boldsymbol{\alpha}_i$ and $\boldsymbol{\beta}$ through the "normal equations"

$$\mathbf{0} = \Sigma_t \mathbf{z}_{it}(y_{it} - \mu_{it}), \qquad \mathbf{0} = \Sigma_{it} \mathbf{x}_{it}(y_{it} - \mu_{it}). \quad (10.13)$$

Table 10.8. *Tort filings model coefficient estimates (fixed effects)*[a]

	Homogeneous model		Model with state categorical variable		Model with state and time categorical variables	
Variable	Parameter estimate	p-values	Parameter estimate	p-values	Parameter estimate	p-values
Intercept	−7.943	<.0001				
POPLAWYR/1000	2.163	0.0002	0.788	0.5893	−0.428	0.7869
VEHCMILE/1000	0.862	<.0001	0.093	0.7465	0.303	0.3140
POPDENSY/1000	0.392	0.0038	4.351	0.2565	3.123	0.4385
WCMPMAX/1000	−0.802	0.1226	0.546	0.3791	1.150	0.0805
URBAN/1000	0.892	0.8187	−33.941	0.3567	−31.910	0.4080
UNEMPLOY	0.087	0.0005	0.028	0.1784	0.014	0.5002
JSLIAB	0.177	0.0292	0.131	0.0065	0.120	0.0592
COLLRULE	−0.030	0.7444	−0.024	0.6853	−0.016	0.7734
CAPS	−0.032	0.7457	0.079	0.2053	0.040	0.5264
PUNITIVE	0.030	0.6623	−0.022	0.6377	0.039	0.4719
Scale	35.857		16.779		16.315	
Deviance	118,309.0		22,463.4		19,834.2	
Pearson chi-square	129,855.7		23,366.1		20,763.0	

[a] All models have an estimated scale parameter. Logarithmic population is used as an offset.

This is a special case of the method of moments. Unfortunately, as we have seen in Section 9.3, this procedure may produce inconsistent estimates of $\boldsymbol{\beta}$. The difficulty is that the number of parameter estimators, $q \times n + K$, grows with the number of subjects, n. Thus, the usual asymptotic theorems that ensure our distributional approximations are no longer valid.

Example 10.1: Tort Filings (continued) To illustrate the fixed-effects estimators, we return to the Section 10.2 tort filing example. Table 10.8 summarizes the fit of the fixed-effects model. For comparison, the Table 10.5 fits from the homogeneous Poisson model with an estimated scale parameter are included in Table 10.8.

10.5.2 Conditional Maximum Likelihood Estimation for Canonical Links

Using Equation (10.12), we see that certain parameters depend on the responses through certain summary statistics. Specifically, using the *factorization theorem* described in Appendix 10A.2, we have that the statistics $\Sigma_t y_{it}\mathbf{z}_{it}$ are sufficient for $\boldsymbol{\alpha}_i$ and that the statistics $\Sigma_{it} y_{it}\mathbf{x}_{it}$ are sufficient for $\boldsymbol{\beta}$. This convenient property of canonical links is not available for other choices of links. Recall that sufficiency means that the distribution of the responses will not depend on a parameter when conditioned by the corresponding sufficient statistic. Specifically, we now consider the likelihood of the data conditional on $\{\Sigma_t y_{it}\mathbf{z}_{it}, i = 1, \ldots, n\}$, so that the conditional likelihood will not depend on $\{\boldsymbol{\alpha}_i\}$. By maximizing this conditional likelihood, we achieve consistent estimators of $\boldsymbol{\beta}$.

To this end, let S_i be the random variable representing $\Sigma_t \mathbf{z}_{it} y_{it}$ and let sum_i be the realization of $\Sigma_t \mathbf{z}_{it} y_{it}$. The conditional likelihood of the responses is

$$\prod_{i=1}^{n} \left\{ \frac{p(y_{i1}, \boldsymbol{\alpha}_i, \boldsymbol{\beta}) p(y_{i2}, \boldsymbol{\alpha}_i, \boldsymbol{\beta}) \cdots p(y_{iT_i}, \boldsymbol{\alpha}_i, \boldsymbol{\beta})}{p_{S_i}(sum_i)} \right\}, \qquad (10.14)$$

where $p_{S_i}(sum_i)$ is the probability density (or mass) function of S_i evaluated at sum_i. This likelihood does not depend on $\{\boldsymbol{\alpha}_i\}$, only on $\boldsymbol{\beta}$. Thus, when evaluating it, we can take $\boldsymbol{\alpha}_i$ to be a zero vector without loss of generality. Under broad conditions, maximizing Equation (10.14) with respect to $\boldsymbol{\beta}$ yields root-n consistent estimators; see, for example, McCullagh and Nelder (1989G). Still, as in Section 9.3, for most parametric families, it is difficult to compute the distribution of S_i. Clearly, the normal distribution is one exception, because if the responses are normal, then the distribution of S_i is also normal. The following subsection describes another important application where the computation is feasible under conditions likely to be encountered in applied data analysis.

10.5.3 Poisson Distribution

This subsection considers Poisson-distributed data, a widely used distribution for count responses. To illustrate, we consider the canonical link with $q = 1$ and $z_{it} = 1$, so that $\theta_{it} = \alpha_i + \mathbf{x}'_{it}\boldsymbol{\beta}$. The Poisson distribution is given by

$$p(y_{it}, \alpha_i, \boldsymbol{\beta}) = \frac{(\mu_{it})^{y_{it}} \exp(-\mu_{it})}{y_{it}!}$$

with $\mu_{it} = b'(\theta_{it}) = \exp(\alpha_i + \mathbf{x}'_{it}\boldsymbol{\beta})$. Conversely, because $\ln \mu_{it} = \alpha_i + \mathbf{x}'_{it}\boldsymbol{\beta}$, we have that the logarithmic mean is a linear combination of explanatory variables. This is the basis of the so-called log-linear model.

As noted in Appendix 10A.2, $\Sigma_t y_{it}$ is a sufficient statistic for α_i. Further, it is easy to check that the distribution of $\Sigma_t y_{it}$ turns out to be Poisson, with mean $\Sigma_t \exp(\alpha_i + \mathbf{x}'_{it}\boldsymbol{\beta})$. In the subsequent development, we will use the ratio of means,

$$\pi_{it}(\boldsymbol{\beta}) = \frac{\mu_{it}}{\sum_t \mu_{it}} = \frac{\exp(\mathbf{x}'_{it}\boldsymbol{\beta})}{\sum_t \exp(\mathbf{x}'_{it}\boldsymbol{\beta})}, \tag{10.15}$$

which does not depend on α_i.

Now, using Equation (10.14), the conditional likelihood for the ith subject is

$$\begin{aligned}
&\frac{p(y_{i1}, \alpha_i, \boldsymbol{\beta}) p(y_{i2}, \alpha_i, \boldsymbol{\beta}) \cdots p(y_{iT_i}, \alpha_i, \boldsymbol{\beta})}{\text{Prob}\left(S_i = \sum_t y_{it}\right)} \\
&\quad = \frac{(\mu_{i1})^{y_{i1}} \cdots (\mu_{iT_i})^{y_{iT_i}} \exp(-(\mu_{i1} + \cdots + \mu_{iT_i}))}{y_{i1}! \cdots y_{iT_i}!} \Bigg/ \\
&\qquad \frac{\left(\sum_t \mu_{it}\right)^{(\sum_t y_{it})} \exp\left(-\sum_t \mu_{it}\right)}{\left(\sum_t y_{it}\right)!} \\
&\quad = \frac{\left(\sum_t y_{it}\right)!}{y_{i1}! \cdots y_{iT_i}!} \prod_t \pi_{it}(\boldsymbol{\beta})^{y_{it}},
\end{aligned}$$

where $\pi_{it}(\boldsymbol{\beta})$ is given in Equation (10.15). This is a *multinomial* distribution. Thus, the joint distribution of $\{y_{i1} \cdots y_{iT_i}\}$ given $\sum_t y_{it}$ has a multinomial distribution.

Using Equation (10.14), the conditional likelihood is

$$CL = \prod_{i=1}^{n} \left\{ \frac{\left(\sum_t y_{it}\right)!}{y_{i1}! \cdots y_{iT_i}!} \prod_t \pi_{it}(\boldsymbol{\beta})^{y_{it}} \right\}.$$

Taking partial derivatives of the log conditional likelihood yields

$$\frac{\partial}{\partial \boldsymbol{\beta}} \ln CL = \sum_{it} y_{it} \frac{\partial}{\partial \boldsymbol{\beta}} \ln \pi_{it}(\boldsymbol{\beta}) = \sum_{it} y_{it} \left(\mathbf{x}_{it} - \sum_r \mathbf{x}_{ir} \pi_{ir}(\boldsymbol{\beta}) \right).$$

Thus, the conditional maximum likelihood estimate, $\mathbf{b}_{\text{CMLE}}$, is the solution of

$$\sum_{i=1}^{n} \sum_{t=1}^{T_i} y_{it} \left(\mathbf{x}_{it} - \sum_{r=1}^{T_i} \mathbf{x}_{ir} \pi_{ir}\ (\mathbf{b}_{\text{CMLE}}) \right) = \mathbf{0}.$$

Example 10.1: Air Quality Regulation Becker and Henderson (2000E) investigated the effects of air quality regulations on firm decisions concerning plant locations and births (plant start-ups) in four major polluting industries: industrial organic chemicals, metal containers, miscellaneous plastic products, and wood furniture. They focused on air quality regulation of ground-level ozone, a major component of smog. With the 1977 amendments to the Clean Air Act, each county in the United States is classified as being either in or out of attainment of the national standards for each of several pollutants, including ozone. This binary variable is the primary variable of interest. Becker and Henderson examined the birth of plants in a county, using attainment status and several control variables, to evaluate hypotheses concerning the location shift from nonattainment to attainment areas. They used plant and industry data from the Longitudinal Research Database, developed by the U.S. Census Bureau, to derive the number of new physical plants, births, by county for $T = 6$ five-year time periods, from 1963 to 1992. The control variables consist of time dummies, county manufacturing wages, and county scale, as measured by all other manufacturing employment in the county.

Becker and Henderson employed a conditional likelihood Poisson model; in this way, they could allow for fixed effects that account for time-constant unmeasured county effects. They found that nonattainment status reduces the expected number of births in a county by 25–45%, depending on the industry. This was a statistically significant finding that corroborated a main research hypothesis, that there has been a shift in births from nonattainment to attainment counties.

10.6 Bayesian Inference

In Section 4.5, Bayesian inference was introduced in the context of the normal linear model. This section provides an introduction in a more general context that is suitable for handling generalized linear models. Only the highlights will be presented; we refer to Gelman et al. (2004G) for a more detailed treatment.

Begin by recalling *Bayes' rule*

$$p(parameters \mid data) = \frac{p(data \mid parameters) \times p(parameters)}{p(data)},$$

where the terms are defined as follows:

- *p(parameters)* is the distribution of the parameters, known as the *prior* distribution.
- *p(data | parameters)* is the sampling distribution. In a frequentist context, it is used for making inferences about the parameters and is known as the *likelihood*.
- *p(parameters | data)* is the distribution of the parameters having observed the data, known as the *posterior* distribution.
- *p(data)* is the marginal distribution of the data. It is generally obtained by integrating (or summing) the joint distribution of data and parameters over parameter values. This is often the difficult step in Bayesian inference.

In a regression context, we have two types of data, the response variable $\mathbf{y}$ and the set of explanatory variables $\mathbf{X}$. Let $\boldsymbol{\theta}$ and $\boldsymbol{\psi}$ denote the sets of parameters that describe the sampling distributions of $\mathbf{y}$ and $\mathbf{X}$, respectively. Moreover, assume that $\boldsymbol{\theta}$ and $\boldsymbol{\psi}$ are independent. Then, using Bayes' rule, the posterior distribution is

$$\begin{aligned}
p(\boldsymbol{\theta}, \boldsymbol{\psi} \mid \mathbf{y}, \mathbf{X}) &= \frac{p(\mathbf{y}, \mathbf{X} \mid \boldsymbol{\theta}, \boldsymbol{\psi}) \times p(\boldsymbol{\theta}, \boldsymbol{\psi})}{p(\mathbf{y}, \mathbf{X})} \\
&= \frac{p(\mathbf{y} \mid \mathbf{X}, \boldsymbol{\theta}, \boldsymbol{\psi}) \times p(\boldsymbol{\theta}) \times p(\mathbf{X} \mid \boldsymbol{\theta}, \boldsymbol{\psi}) \times p(\boldsymbol{\psi})}{p(\mathbf{y}, \mathbf{X})} \\
&\propto \{p(\mathbf{y} \mid \mathbf{X}, \boldsymbol{\theta}) \times p(\boldsymbol{\theta})\} \times \{p(\mathbf{X} \mid \boldsymbol{\psi}) \times p(\boldsymbol{\psi})\} \\
&\propto p(\boldsymbol{\theta} \mid \mathbf{y}, \mathbf{X}) \times p(\boldsymbol{\psi} \mid \mathbf{X}).
\end{aligned}$$

Here, the symbol "$\propto$" means "is proportional to." Thus, the joint posterior distribution of the parameters can be factored into two pieces, one for the responses and one for the explanatory variables. Assuming no dependencies between $\boldsymbol{\theta}$ and $\boldsymbol{\psi}$, there is no loss of information in the traditional regression setting by ignoring the distributions associated with the explanatory variables. By the "traditional regression setting," we mean that one essentially treats the explanatory variables as nonstochastic.

Most statistical inference can be accomplished readily having computed the posterior. With this entire distribution, summarizing likely values of the parameters through confidence intervals or unlikely values through hypothesis tests is straightforward. Bayesian methods are also especially suitable for forecasting. In the regression context, suppose we wish to summarize the distribution of a set of new responses, $\mathbf{y}_{\text{new}}$, given new explanatory variables, $\mathbf{X}_{\text{new}}$, and previously observed data $\mathbf{y}$ and $\mathbf{X}$. This distribution, $p(\mathbf{y}_{\text{new}} \mid \mathbf{X}_{\text{new}}, \mathbf{y}, \mathbf{X})$, is a type of *predictive distribution*. We have that

$$\begin{aligned} p(\mathbf{y}_{\text{new}} \mid \mathbf{y}, \mathbf{X}, \mathbf{X}_{\text{new}}) &= \int p(\mathbf{y}_{\text{new}}, \boldsymbol{\theta} \mid \mathbf{y}, \mathbf{X}, \mathbf{X}_{\text{new}})\, d\boldsymbol{\theta} \\ &= \int p(\mathbf{y}_{\text{new}} \mid \boldsymbol{\theta}, \mathbf{y}, \mathbf{X}, \mathbf{X}_{\text{new}}) p(\boldsymbol{\theta} \mid \mathbf{y}, \mathbf{X}, \mathbf{X}_{\text{new}})\, d\boldsymbol{\theta}. \end{aligned}$$

This assumes that the parameters $\boldsymbol{\theta}$ are continuous. Here, $p(\mathbf{y}_{\text{new}} \mid \boldsymbol{\theta}, \mathbf{y}, \mathbf{X}, \mathbf{X}_{\text{new}})$ is the sampling distribution of $\mathbf{y}_{\text{new}}$ and $p(\boldsymbol{\theta} \mid \mathbf{y}, \mathbf{X}, \mathbf{X}_{\text{new}}) = p(\boldsymbol{\theta} \mid \mathbf{y}, \mathbf{X})$ is the posterior distribution (assuming that values of the new explanatory variables are independent of $\boldsymbol{\theta}$). Thus, the predictive distribution can be computed as a weighted average of the sampling distribution, where the weights are given by the posterior.

A difficult aspect of Bayesian inference can be the assignment of priors. Classical assignments of priors are generally either *noninformative* or *conjugate*. Noninformative priors are distributions that are designed to interject the least amount of information possible. Two important types of noninformative priors are *uniform* (also known as *flat*) *priors* and *Jeffreys priors*. A uniform prior is simply a constant value; thus, no value is more likely than any other. A drawback of this type of prior is that it is not invariant under transformation of the parameters. To illustrate, consider the normal linear regression model, so that $\mathbf{y} \mid \mathbf{X}, \boldsymbol{\beta}, \sigma^2 \sim N(\mathbf{X}\boldsymbol{\beta}, \sigma^2\mathbf{I})$. A widely used noninformative prior is a flat prior on $(\boldsymbol{\beta}, \log \sigma^2)$, so that the joint distribution of $(\boldsymbol{\beta}, \sigma^2)$ turns out to be proportional to σ^{-2}. Thus, although uniform in $\log \sigma^2$, this prior gives heavier weight to small values of σ^2. A Jeffrey's prior is one that is invariant under transformation. Jeffrey's priors are complex in the case of multidimensional parameters and thus we will not consider them further here.

Conjugacy of a prior is actually a property that depends on both the prior as well as sampling distribution. When the prior and posterior distributions come from the same family of distributions, then the prior is known as a conjugate prior. Appendix 10A.3 gives several examples of the more commonly used conjugate priors.

For longitudinal and panel data, it is convenient to formulate Bayesian models in three stages: one for the parameters, one for the data, and one stage for the latent variables used to represent the heterogeneity. Thus, extending Section 7.3.1, we have the following:

Stage 0 (*Prior distribution*). Draw a realization of a set of parameters from a population. The parameters consist of regression coefficients $\boldsymbol{\beta}$ and variance components $\boldsymbol{\tau}$.

Stage 1 (*Heterogeneity-effects distribution*). Conditional on the parameters from stage 0, draw a random sample of n subjects from a population. The vector of subject-specific effects $\boldsymbol{\alpha}_i$ is associated with the ith subject.

Stage 2 (*Conditional sampling distribution*). Conditional on $\boldsymbol{\alpha}_i$, $\boldsymbol{\beta}$, and $\boldsymbol{\tau}$, draw realizations of $\{y_{it}, \mathbf{z}_{it}, \mathbf{x}_{it}\}$, for $t = 1, \ldots, T_i$ for the ith subject. Summarize these draws as $\{\mathbf{y}_i, \mathbf{Z}_i, \mathbf{X}_i\}$.

A common method of analysis is to combine stages 0 and 1 and to treat $\boldsymbol{\beta}^* = (\boldsymbol{\beta}', \boldsymbol{\alpha}_1, \ldots, \boldsymbol{\alpha}_n)'$ as the regression parameters of interest. Also common is to use a normal prior distribution for this set of regression parameters. This is the conjugate prior when the sampling distribution is normal. When the sampling distribution is from the GLM family (but not normal), there is no general recipe for conjugate priors. A normal prior is useful because it is flexible and computationally convenient. To be specific, we consider the following:

Stage 1* (*Prior distribution*). Assume that

$$\boldsymbol{\beta}^* = \begin{pmatrix} \boldsymbol{\beta} \\ \boldsymbol{\alpha}_1 \\ \vdots \\ \boldsymbol{\alpha}_n \end{pmatrix} \sim N\left(\begin{pmatrix} \mathrm{E}\boldsymbol{\beta} \\ \mathbf{0}_n \end{pmatrix}, \begin{pmatrix} \boldsymbol{\Sigma}_{\beta} & \mathbf{0} \\ \mathbf{0} & \mathbf{D} \otimes \mathbf{I}_n \end{pmatrix} \right).$$

Thus, both $\boldsymbol{\beta}$ and $(\alpha_1, \ldots, \alpha_n)$ are normally distributed.

Stage 2 (*Conditional sampling distribution*). Conditional on $\boldsymbol{\beta}^*$ and $\{\mathbf{Z}, \mathbf{X}\}$, $\{y_{it}\}$ are independent and the distribution of y_{it} is from a GLM family with parameter $\eta_{it} = \mathbf{z}_{it}'\boldsymbol{\alpha}_i + \mathbf{x}_{it}'\boldsymbol{\beta}$.

For some types of GLM families (such as the normal), an additional scale parameter, typically included in the prior distribution specification, is used.

With this specification, in principle one simply applies Bayes' rule to determine the posterior distribution of β^* given the data $\{\mathbf{y}, \mathbf{X}, \mathbf{Z}\}$. However, as a practical matter, this is difficult to do without conjugate priors. Specifically, to compute the marginal distribution of the data, one must use numerical integration to remove parameter distributions; this is computationally intensive for many problems of interest. To circumvent this difficulty, modern Bayesian analysis regularly employs simulation techniques known as Markov Chain Monte Carlo (MCMC) methods and an especially important special case, the Gibbs sampler. MCMC methods produce simulated values of the posterior distribution and are available in many statistical packages, including the shareware favored by many Bayesian analysts, BUGS/WINBUGS (available at http://www.mrc-bsu.cam.ac.uk/bugs/). There are many specialized treatments that discuss the theory and applications of this approach; we refer the reader to Gelman et al. (2004G), Gill (2002G), and Congdon (2003G).

For some applications such as prediction, the interest is in the full joint posterior distribution of the global regression parameters and the heterogeneity effects, $\boldsymbol{\beta}, \boldsymbol{\alpha}_1, \ldots, \boldsymbol{\alpha}_n$. For other applications, the interest is in the posterior distribution of the global regression coefficients, $\boldsymbol{\beta}$. In this case, one integrates out the heterogeneity effects from the joint posterior distribution of $\boldsymbol{\beta}^*$.

Example 10.2: Respiratory Infections Zeger and Karim (1991S) introduced Bayesian inference for GLM longitudinal data. Specifically, Zeger and Karim considered the stage 2 GLM conditional sampling distribution and, for the heterogeneity distribution, assumed that $\boldsymbol{\alpha}_i$ are conditionally i.i.d. $N(\mathbf{0}, \mathbf{D})$. They allowed for a general form for the prior distribution.

To illustrate this setup, Zeger and Karim examined infectious disease data on $n = 250$ Indonesian children who were observed up to $T = 6$ consecutive quarters, for a total of $N = 1{,}200$ observations. The goal was to assess determinants of a binary variable that indicated the presence of a respiratory disease. They used normally distributed random intercepts ($q = 1$ and $z_{it} = 1$) with a logistic conditional sampling distribution.

Example 10.3: Patents Chib, Greenberg, and Winkelman (1998E) discussed parameterization of Bayesian Poisson models. They considered a conditional Poisson model with a canonical link so that $y_{it} \mid \boldsymbol{\alpha}_i, \boldsymbol{\beta} \sim$ Poisson (θ_{it}), where $\theta_{it} = \mathrm{E}(y_{it} \mid \boldsymbol{\alpha}_i, \boldsymbol{\beta}) = \mathbf{z}_{it}'\boldsymbol{\alpha}_i + \mathbf{x}_{it}'\boldsymbol{\beta}$. Unlike as in prior work, they did not assume that the heterogeneity effects have zero mean but instead used $\boldsymbol{\alpha}_i \sim N(\boldsymbol{\eta}_\alpha, \mathbf{D})$. With this specification, they did not use the usual convention of assuming that $\mathbf{z}_{it}$ is a subset of $\mathbf{x}_{it}$. This, they argued, leads to more stable convergence algorithms when computing posterior distributions. To complete the specification,

Chib et al. used the following prior distributions: $\boldsymbol{\beta} \sim N(\mu_\beta, \boldsymbol{\Sigma}_\beta)$, $\boldsymbol{\eta}_\alpha \sim N(\boldsymbol{\eta}_0, \boldsymbol{\Sigma}_\eta)$, and $\mathbf{D}^{-1} \sim$ Wishart. The Wishart distribution is a multivariate extension of the chi-square distribution; see, for example, Anderson (1958G).

To illustrate, Chib et al. used patent data first considered by Hausman, Hall, and Griliches (1984E). These data include the number of patents received by $n = 642$ firms over $T = 5$ years, 1975–1979. The explanatory variables included the logarithm of research and development expenditures, as well as their one-, two-, and three-year lags, and time dummy variables. Chib et al. used a variable intercept and a variable slope for the logarithmic research and development expenditures.

Further Reading

McCullagh and Nelder (1989G) and Fahrmeir and Tutz (2001S) provide more extensive introductions to generalized linear models.

Conditional likelihood testing in connection with exponential families was suggested by Rasch (1961EP) and asymptotic properties were developed by Andersen (1970G), motivated in part by the "presence of infinitely many nuisance parameters" of Neyman and Scott (1948E). Panel data conditional likelihood estimation was introduced by Chamberlain (1980E) for binary logit models and by Hausman et al. (1984E) for Poisson (as well as negative binomial) count models.

Three excellent sources for further discussions of nonlinear mixed-effects models are Davidian and Giltinan (1995S), Vonesh and Chinchilli (1997S), and Pinheiro and Bates (2000S).

For additional discussions on computing aspects, we refer to McCulloch and Searle (2001G) and Pinheiro and Bates (2000S).

Appendix 10A Exponential Families of Distributions

The distribution of the random variable y may be discrete, continuous, or a mixture. Thus, $p(.)$ in Equation (10.1) may be interpreted to be a density or mass function, depending on the application. Table 10A.1 provides several examples, including the normal, binomial and Poisson distributions.

10A.1 Moment-Generating Function

To assess the moments of exponential families, it is convenient to work with the moment-generating function. For simplicity, we assume that the random

Table 10A.1. *Selected distributions of the one-parameter exponential family*

Distribution	Parameters	Density or mass function	Components	Ey	Var y
General	θ, ϕ	$\exp\left(\frac{y\theta - b(\theta)}{\phi} + S(y, \phi)\right)$	$\theta, \phi, b(\theta), S(y, \phi)$	$b'(\theta)$	$b''(\theta)\phi$
Normal	μ, σ^2	$\frac{1}{\sigma\sqrt{2\pi}} \exp\left(-\frac{(y-\mu)^2}{2\sigma^2}\right)$	$\mu, \sigma^2, \frac{\theta^2}{2}, -\left(\frac{y^2}{2\phi} + \frac{\ln(2\pi\phi)}{2}\right)$	$\theta = \mu$	$\phi = \sigma^2$
Binomial	π	$\binom{n}{y} \pi^y (1-\pi)^{n-y}$	$\ln\left(\frac{\pi}{1-\pi}\right), 1, n\ln(1+e^\theta), \ln\binom{n}{y}$	$n\frac{e^\theta}{1+e^\theta} = n\pi$	$n\frac{e^\theta}{(1+e^\theta)^2} = n\pi(1-\pi)$
Poisson	λ	$\frac{\lambda^y \exp(-\lambda)}{y!}$	$\ln\lambda, 1, e^\theta, -\ln(y!)$	$e^\theta = \lambda$	$e^\theta = \lambda$
Gamma	α, β	$\frac{\beta^\alpha}{\Gamma(\alpha)} y^{\alpha-1} \exp(-y\beta)$	$-\frac{\beta}{\alpha}, \frac{1}{\alpha}, -\ln(-\theta),$ $-\phi^{-1}\ln\phi - \ln(\Gamma(\phi^{-1})) + (\phi^{-1} - 1)\ln y$	$-\frac{1}{\theta} = \frac{\alpha}{\beta}$	$\frac{\phi}{\theta^2} = \frac{\alpha}{\beta^2}$

variable y is continuous. Define the moment-generating function

$$\begin{aligned}
M(s) = \mathrm{E}e^{sy} &= \int \exp\left(sy + \frac{y\theta - b(\theta)}{\phi} + S(y,\phi)\right)dy \\
&= \exp\left(\frac{b(\theta + s\phi) - b(\theta)}{\phi}\right) \\
&\quad \times \int \exp\left(\frac{y(\theta + s\phi) - b(\theta + s\phi)}{\phi} + S(y,\phi)\right)dy \\
&= \exp\left(\frac{b(\theta + s\phi) - b(\theta)}{\phi}\right) \int \exp\left(\frac{y\theta^* - b(\theta^*)}{\phi} + S(y,\phi)\right)dy \\
&= \exp\left(\frac{b(\theta + s\phi) - b(\theta)}{\phi}\right).
\end{aligned}$$

With this expression, we can generate the moments. Thus, for the mean, we have

$$\begin{aligned}
\mathrm{E}y = M'(0) &= \frac{\partial}{\partial s} \exp\left(\frac{b(\theta + s\phi) - b(\theta)}{\phi}\right)\Bigg|_{s=0} \\
&= \left[b'(\theta + s\phi)\exp\left(\frac{b(\theta + s\phi) - b(\theta)}{\phi}\right)\right]_{s=0} = b'(\theta).
\end{aligned}$$

Similarly, for the second moment, we have

$$\begin{aligned}
M''(s) &= \frac{\partial}{\partial s}\left[b'(\theta + s\phi)\exp\left(\frac{b(\theta + s\phi) - b(\theta)}{\phi}\right)\right] \\
&= \phi b''(\theta + s\phi)\exp\left(\frac{b(\theta + s\phi) - b(\theta)}{\phi}\right) \\
&\quad + (b'(\theta + s\phi))^2 \exp\left(\frac{b(\theta + s\phi) - b(\theta)}{\phi}\right).
\end{aligned}$$

This yields $\mathrm{E}y^2 = M''(0) = \phi b''(\theta) + (b'(\theta))^2$ and $\mathrm{Var}\, y = \phi b''(\theta)$.

10A.2 Sufficiency

In complex situations, it is convenient to be able to decompose the likelihood into several pieces that can be analyzed separately. To accomplish this, the concept of sufficiency is useful. A statistic $T(y_1, \ldots, y_n) = T(\mathbf{y})$ is *sufficient* for a parameter θ if the distribution of $y_1, \ldots, y_n$ conditional on $T(\mathbf{y})$ does not depend on θ.

When checking whether or not a statistic is sufficient, an important result is the *factorization theorem*. This result indicates, under certain regularity conditions, that a statistic $T(\mathbf{y})$ is sufficient for θ if and only if the density (mass) function of $\mathbf{y}$ can be decomposed into the product of two components:

$$p(\mathbf{y}, \theta) = p_1(T(\mathbf{y}), \theta) p_2(\mathbf{y}). \qquad (10A.1)$$

Here, the first portion, p_1, may depend on θ but depends on the data only through the sufficient statistic $T(\mathbf{y})$. The second portion, p_2, may depend on the data but does not depend on the parameter θ. See, for example, Bickel and Doksum (1977G).

For example, if $\{y_1, \ldots, y_n\}$ are independent and follow the distribution in Equation (10.1), then the joint distribution is

$$p(\mathbf{y}, \theta, \phi) = \exp\left(\frac{\theta \left(\sum_{i=1}^n y_i\right) - nb(\theta)}{\phi} + \sum_{i=1}^n S(y_i, \phi) \right).$$

Thus, with $p_2(\mathbf{y}) = \exp(\sum_{i=1}^n S(y_i, \phi))$ and

$$p_1(T(\mathbf{y}), \theta) = \exp\left(\frac{\theta T(\mathbf{y}) - nb(\theta)}{\phi} \right),$$

the statistic $T(\mathbf{y}) = \sum_{i=1}^n y_i$ is sufficient for θ.

10A.3 Conjugate Distributions

Assume that the parameter θ is random with distribution $\pi(\theta, \boldsymbol{\tau})$, where $\boldsymbol{\tau}$ is a vector of parameters that describes the distribution of θ. In Bayesian models, the distribution π is known as the *prior* and reflects our belief or information about θ. The likelihood $p(\mathbf{y}, \theta) = p(\mathbf{y} \mid \theta)$ is a probability conditional on θ. The distribution of θ with knowledge of the random variables, $\pi(\theta, \boldsymbol{\tau} \mid \mathbf{y})$, is called the *posterior* distribution. For a given likelihood distribution, priors and posteriors that come from the same parametric family are known as *conjugate* families of distributions.

For a linear exponential likelihood, there exists a natural conjugate family. For the likelihood in Equation (10.1), define the prior distribution

$$\pi(\theta, \boldsymbol{\tau}) = C \exp(\theta a_1(\boldsymbol{\tau}) - b(\theta) a_2(\boldsymbol{\tau})), \qquad (10A.2)$$

where C is a normalizing constant and $a_1(\boldsymbol{\tau})$ and $a_2(\boldsymbol{\tau})$ are functions of the parameters $\boldsymbol{\tau}$. The joint distribution of $\mathbf{y}$ and θ is given by $p(\mathbf{y}, \theta) = p(\mathbf{y} \mid \theta) \pi(\theta, \boldsymbol{\tau})$. Using Bayes' rule, the posterior distribution is

$$\pi(\theta, \boldsymbol{\tau} \mid y) = C_1 \exp\left(\left(a_1(\boldsymbol{\tau}) + \frac{y}{\phi} \right) \theta - b(\theta) \left(a_2(\boldsymbol{\tau}) + \frac{1}{\phi} \right) \right),$$

where C_1 is a normalizing constant. Thus, we see that $\pi(\theta, \boldsymbol{\tau} \mid \mathbf{y})$ has the same form as $\pi(\theta, \boldsymbol{\tau})$.

Special Case 10A.1: Normal–Normal Model Consider a normal likelihood in Equation (10.1) so that $b(\theta) = \theta^2/2$. Thus, with Equation (10A.2), we have

$$\pi(\theta, \boldsymbol{\tau}) = C \exp\left(a_1(\boldsymbol{\tau})\theta - \frac{\theta^2}{2} a_2(\boldsymbol{\tau}) \right)$$
$$= C_1(\boldsymbol{\tau}) \exp\left(-\frac{a_2(\boldsymbol{\tau})}{2} \left(\theta - \frac{a_1(\boldsymbol{\tau})}{a_2(\boldsymbol{\tau})} \right)^2 \right).$$

Thus, the prior distribution of θ is normal with mean $a_1(\boldsymbol{\tau})/a_2(\boldsymbol{\tau})$ and variance $(a_2(\boldsymbol{\tau}))^{-1}$. The posterior distribution of θ given y is normal with mean $(a_1(\boldsymbol{\tau}) + y/\phi)/(a_2(\boldsymbol{\tau}) + \phi^{-1})$ and variance $(a_2(\boldsymbol{\tau}) + \phi^{-1})^{-1}$.

Special Case 10A.2: Binomial-Beta Model Consider a binomial likelihood in Equation (10.1) so that $b(\theta) = n \ln(1 + e^\theta)$. Thus, with Equation (10A.2), we have

$$\pi(\theta, \boldsymbol{\tau}) = C \exp\left(a_1(\boldsymbol{\tau})\theta - n a_2(\boldsymbol{\tau}) \ln(1 + e^\theta)\right)$$
$$= C \left(\frac{e^\theta}{1 + e^\theta} \right)^{a_1(\boldsymbol{\tau})} (1 + e^\theta)^{-n a_2(\boldsymbol{\tau}) + a_1(\boldsymbol{\tau})}.$$

Thus, we have that the prior of logit (θ) is a beta distribution with parameters $a_1(\boldsymbol{\tau})$ and $n a_2(\boldsymbol{\tau}) - a_1(\boldsymbol{\tau})$. The posterior of logit (θ) is a beta distribution with parameters $a_1(\boldsymbol{\tau}) + y/\phi$ and $n(a_2(\boldsymbol{\tau}) + \phi^{-1}) - (a_1(\boldsymbol{\tau}) + y/\phi)$.

Special Case 10A.3: Poisson–Gamma Model Consider a Poisson likelihood in Equation (10.1) so that $b(\theta) = e^\theta$. Thus, with Equation (10.A.2), we have

$$\pi(\theta, \boldsymbol{\tau}) = C \exp\left(a_1(\boldsymbol{\tau})\theta - a_2(\boldsymbol{\tau})e^\theta\right) = C(e^\theta)^{a_1(\boldsymbol{\tau})} \exp\left(-e^\theta a_2(\boldsymbol{\tau})\right).$$

Thus, we have that the prior of e^θ is a Gamma distribution with parameters $a_1(\boldsymbol{\tau})$ and $a_2(\boldsymbol{\tau})$. The posterior of e^θ is a Gamma distribution with parameters $a_1(\boldsymbol{\tau}) + y/\phi$ and $a_2(\boldsymbol{\tau}) + \phi^{-1}$.

10A.4 Marginal Distributions

In many conjugate families, the marginal distribution of the random variable y turns out to be from a well-known parametric family. To see this, consider the prior distribution in Equation (10.A.2). Because densities integrate (sum, in the discrete case) to 1, we may express the constant C as

$$C^{-1} = (C(a_1, a_2))^{-1} = \int \exp(\theta\, a_1 - b(\theta) a_2) d\theta,$$

a function of the parameters $a_1 = a_1(\boldsymbol{\tau})$ and $a_2 = a_2(\boldsymbol{\tau})$. With this and Equation (10A.2), the marginal distribution of y is

$$
\begin{aligned}
g(y) &= \int p(y,\theta)\pi(\theta)d\theta \\
&= C(a_1,a_2)\exp\left(S(y,\phi)\right)\int \exp\left(\theta\left(\frac{y}{\phi}+a_1\right)-b(\theta)\left(\frac{1}{\phi}+a_2\right)\right)d\theta \\
&= \frac{C(a_1,a_2)\exp\left(S(y,\phi)\right)}{C\left(a_1+\frac{y}{\phi},a_2+\frac{1}{\phi}\right)}. \qquad (10A.3)
\end{aligned}
$$

It is of interest to consider several special cases.

Special Case 10A.1: Normal–Normal Model (continued) Using Table 10A.1, we have

$$\exp\left(S(y,\phi)\right) = \frac{1}{\sqrt{2\pi\phi}}\exp\left(-\frac{y^2}{2\phi}\right).$$

Straightforward calculations show that

$$C(a_1,a_2) = C_1\exp\left(-\frac{a_1^2}{2a_2}\right) = \frac{\sqrt{a_2}}{\sqrt{2\pi}}\exp\left(-\frac{a_1^2}{2a_2}\right).$$

Thus, from Equation (10A.3), the marginal distribution is

$$
\begin{aligned}
g(y) &= \frac{\sqrt{a_2/(2\pi)}\exp\left(-\frac{a_1^2}{2a_2}\right)\sqrt{1/(2\pi\phi)}\exp\left(-\frac{y^2}{2\phi}\right)}{\sqrt{\left(a_2+\phi^{-1}\right)/(2\pi)}\exp\left(-\frac{(a_1+y/\phi)^2}{2(a_2+\phi^{-1})}\right)} \\
&= \frac{1}{\sqrt{2\pi\left(\phi+a_2^{-1}\right)}}\exp\left(-\frac{1}{2}\left(\frac{a_1^2}{a_2}+\frac{y^2}{\phi}-\frac{(a_1+y/\phi)^2}{a_2+\phi^{-1}}\right)\right) \\
&= \frac{1}{\sqrt{2\pi\left(\phi+a_2^{-1}\right)}}\exp\left(-\frac{(y-a_1/a_2)^2}{2\left(a_2+\phi^{-1}\right)}\right).
\end{aligned}
$$

Thus, y is normally distributed with mean a_1/a_2 and variance $\phi + a_2^{-1}$.

Special Case 10A.2: Binomial–Beta Model (continued) Using Table 10A.1, we have $\exp\left(S(y,\phi)\right) = \binom{n}{y}$ and $\phi = 1$. Further, straightforward calculations show that

$$C(a_1,a_2) = \frac{\Gamma(na_2)}{\Gamma(a_1)\Gamma(na_2-a_1)} = \frac{1}{\mathrm{B}(a_1,na_2-a_1)},$$

where $B(.,.)$ is the Beta function. Thus, from Equation (10A.3), we have

$$g(y) = \binom{n}{y} \frac{B(a_1 + y, n(a_2 + 1) - (a_1 + y))}{B(a_1, na_2 - a_1)}.$$

This is called the *beta-binomial distribution.*

Special Case 10A.3: Poisson–Gamma Model (continued) Using Table 10A.1, we have $\exp(S(y, \phi)) = 1/y! = 1/\Gamma(y+1)$ and $\phi = 1$. Further, straightforward calculations show that $C(a_1, a_2) = (a_2)^{a_1}/\Gamma(a_1)$. Thus, from Equation (10A.3), we have

$$g(y) = \frac{\frac{(a_2)^{a_1}}{\Gamma(a_1)} \frac{1}{\Gamma(y+1)}}{\frac{(a_2+1)^{a_1+y}}{\Gamma(a_1+y)}} = \frac{\Gamma(a_1 + y)(a_2)^{a_1}}{\Gamma(a_1)\Gamma(y+1)(a_2+1)^{a_1+y}}$$
$$= \frac{\Gamma(a_1+y)}{\Gamma(a_1)\Gamma(y+1)} \left(\frac{a_2}{a_2+1}\right)^{a_1} \left(\frac{1}{a_2+1}\right)^{y}.$$

This is a negative binomial distribution.

Exercises and Extensions

10.1 Poisson and binomials Assume that y_i has a Poisson distribution with mean parameter $\mu_i, i = 1, 2$, and suppose that y_1 is independent of y_2.

a. Show that $y_1 + y_2$ has a Poisson distribution with mean parameter $\mu_1 + \mu_2$.
b. Show that y_1, given $y_1 + y_2$, has a binomial distribution. Determine the parameters of the binomial distribution.
c. Suppose that you tell the computer to evaluate the likelihood of (y_1, y_2) assuming that y_1 is independent of y_2, with each having a Poisson distribution with mean parameters μ_1 and $\mu_2 = 1 - \mu_1$.
   i. Evaluate the likelihood of (y_1, y_2).
   ii. However, the data are such that y_1 is binary and $y_2 = 1 - y_1$. Evaluate the likelihood that the computer is evaluating.
   iii. Show that this equals the likelihood assuming y_1 has a Bernoulli distribution with mean μ_1, up to a proportionality constant. Determine the constant.
d. Consider a longitudinal data set $\{y_{it}, \mathbf{x}_{it}\}$, where y_{it} are binary with mean $p_{it} = \pi(\mathbf{x}_{it}'\beta)$, where $\pi(.)$ is known. You would like to evaluate the likelihood but can only do so in terms of a package using Poisson distributions. Using the part (c) result, explain how to do so.

11
Categorical Dependent Variables and Survival Models

Abstract. Extending Chapter 9, this chapter considers dependent variables having more than two possible categorical alternatives. As in Chapter 9, we often interpret a categorical variable to be an attribute possessed or choice made by an individual, household, or firm. By allowing more than two alternatives, we substantially broaden the scope of applications to complex social science problems of interest.

The pedagogic approach of this chapter follows the pattern established in earlier chapters; we begin with homogeneous models in Section 11.1 followed by the Section 11.2 model that incorporates random effects, thus providing a heterogeneity component. Section 11.3 introduces an alternative method for accommodating time patterns through *transition*, or *Markov, models*. Although transition models are applicable in the Chapter 10 generalized linear models, they are particularly useful in the context of categorical dependent variables. Many repeated applications of the idea of transitioning give rise to survival models, another important class of longitudinal models. Section 11.4 develops this link.

11.1 Homogeneous Models

We now consider a response that is an unordered categorical variable. We assume that the dependent variable y may take on values $1, 2, \ldots, c$, corresponding to c categories. We first introduce the *homogenous* version so that this section does not use any subject-specific parameters nor introduce terms to account for serial correlation.

In many social science applications, the response categories correspond to attributes possessed or choices made by individuals, households, or firms. Some applications of categorical dependent variable models include the following:

- employment choice (e.g., Valletta, 1999E),
- mode of transportation choice, such as the classic work by McFadden (1978E),
- choice of political party affiliation (e.g., Brader and Tucker, 2001O), and
- marketing brand choice (e.g., Jain, Vilcassion, and Chintagunta, 1994O).

Example 11.1: Political Party Affiliation Brader and Tucker (2001O) studied the choice made by Russian voters of political parties in the 1995–1996 elections. Their interest was in assessing effects of meaningful political party attachments during Russia's transition to democracy. They examined a $T = 3$ wave survey of $n = 2{,}841$ respondents, taken (1) three to four weeks before the 1995 parliamentary elections, (2) immediately after the parliamentary elections, and (3) after the 1996 presidential elections. This survey design was modeled on the American National Election Studies (see Appendix F).

The dependent variable was the political party voted for, consisting of $c = 10$ parties including the Liberal Democratic Party of Russia, Communist Party of the Russian Federation, Our Home is Russia, and others. The independent variables included social characteristics such as education, gender, religion, nationality, age, and location (urban versus rural), economic characteristics such as income and employment status, and political attitudinal characteristics such as attitudes toward market transitions and privatization.

11.1.1 Statistical Inference

For an observation from subject i at time t, denote the probability of choosing the jth category as $\pi_{it,j} = \text{Prob}(y_{it} = j)$, so that $\pi_{it,1} + \cdots + \pi_{it,c} = 1$. In the homogeneous framework, we assume that observations are independent of one another.

In general, we will model these probabilities as a (known) function of parameters and use maximum likelihood estimation for statistical inference. We use the notation y_{itj} to be an indicator of the event $y_{it} = j$. The likelihood for the ith subject at the tth time point is

$$\prod_{j=1}^{c} (\pi_{it,j})^{y_{it,j}} = \begin{cases} \pi_{it,1} & \text{if } y_{it} = 1, \\ \pi_{it,2} & \text{if } y_{it} = 2, \\ \vdots & \quad\vdots \\ \pi_{it,c} & \text{if } y_{it} = c. \end{cases}$$

Thus, assuming independence among observations, the total log likelihood is

$$L = \sum_{it} \sum_{j=1}^{c} y_{it,j} \ln \pi_{it,j}.$$

With this framework, standard maximum likelihood estimation is available. Thus, our main task is to specify an appropriate form for π.

11.1.2 Generalized Logit

Like standard linear regression, generalized logit models employ linear combinations of explanatory variables of the form

$$V_{it,j} = \mathbf{x}_{it}' \boldsymbol{\beta}_j. \tag{11.1}$$

Because the dependent variables are not numerical, we cannot model the response y as a linear combination of explanatory variables plus an error. Instead we use the probabilities

$$\text{Prob}(y_{it} = j) = \pi_{it,j} = \frac{\exp(V_{it,j})}{\sum_{k=1}^{c} \exp(V_{it,k})}, \quad j = 1, 2, \ldots, c. \tag{11.2}$$

Note here that $\boldsymbol{\beta}_j$ is the corresponding vector of parameters that may depend on the alternative, or choice, whereas the explanatory variables $\mathbf{x}_{it}$ do not. So that probabilities sum to one, a convenient normalization for this model is $\boldsymbol{\beta}_c = 0$. With this normalization and the special case of $c = 2$, the generalized logit reduces to the logit model introduced in Section 9.1.

Parameter Interpretations

We now describe an interpretation of coefficients in generalized logit models, similar to that of Section 9.1.1 for the logistic model. From Equations (11.1) and (11.2), we have

$$\ln \left(\frac{\text{Prob}(y_{it} = j)}{\text{Prob}(y_{it} = c)} \right) = V_{it,j} - V_{it,c} = \mathbf{x}_{it}' \boldsymbol{\beta}_j.$$

The left-hand side of this equation is interpreted to be the logarithmic odds of choosing choice j compared to choice c. Thus, as in Section 9.1.1, we may interpret β_j as the proportional change in the odds ratio.

Generalized logits have an interesting nested structure, which we will explore briefly in Section 11.1.5. That is, it is easy to check that conditional on not choosing the first category, the form of $\text{Prob}(y_{it} = j \mid y_{it} \neq 1)$ has a generalized logit form in Equation (11.2). Further, if j and h are different alternatives, we note that

$$\begin{aligned}\text{Prob}(y_{it} = j \mid \{y_{it} = j \quad \text{or} \quad y_{it} = h\}) &= \frac{\text{Prob}(y_{it} = j)}{\text{Prob}(y_{it} = j) + \text{Prob}(y_{it} = h)} \\ &= \frac{\exp(V_{it,j})}{\exp(V_{it,j}) + \exp(V_{it,h})} \\ &= \frac{1}{1 + \exp(\mathbf{x}'_{it}(\boldsymbol{\beta}_h - \boldsymbol{\beta}_j))}.\end{aligned}$$

This has a logit form that was introduced in Section 9.1.1.

Special Case: Intercept-Only Model To develop intuition, we now consider the model with only intercepts. Thus, let $\mathbf{x}_{it} = 1$ and $\boldsymbol{\beta}_j = \beta_{0,j} = \alpha_j$. With the convention $\alpha_c = 0$, we have

$$\text{Prob}(y_{it} = j) = \pi_{it,j} = \frac{e^{\alpha_j}}{e^{\alpha_1} + e^{\alpha_2} + \cdots + e^{\alpha_{c-1}} + 1}$$

and

$$\ln\left(\frac{\text{Prob}(y_{it} = j)}{\text{Prob}(y_{it} = c)}\right) = \alpha_j.$$

From the second relation, we may interpret the jth intercept α_j to be the logarithmic odds of choosing alternative j compared to alternative c.

Example 11.2: Job Security This is a continuation of Example 9.1 on the determinants of job turnover, based on the work of Valetta (1999E). The Chapter 9 analysis of these data considered only the binary dependent variable dismissal, the motivation being that this is the main source of job insecurity. Valetta (1999E) also presented results from a generalized logit model, his primary motivation being that the economic theory describing turnover implies that other reasons for leaving a job may affect dismissal probabilities.

For the generalized logit model, the response variable has $c = 5$ categories: dismissal, left job because of plant closures, quit, changed jobs for other reasons, and no change in employment. The latter category is the omitted one in Table 11.1. The explanatory variables of the generalized logit are the same as the probit regression described in Example 9.1; the estimates summarized in Example 9.1 are reproduced here for convenience.

Table 11.1 shows that turnover declines as tenure increases. To illustrate, consider a typical man in the 1992 sample where we have time = 16 and focus on dismissal probabilities. For this value of time, the coefficient associated with

Table 11.1. *Turnover generalized logit and probit regression estimates*[a]

Variable[c]	Probit regression model	Generalized logit model[b]			
		Dismissed	Plant closed	Other reason	Quit
Tenure	−0.084 (0.010)	−0.221 (0.025)	−0.086 (0.019)	−0.068 (0.020)	−0.127 (0.012)
Time trend	−0.002 (0.005)	−0.008 (0.011)	−0.024 (0.016)	0.011 (0.013)	−0.022 (0.007)
Tenure × (time trend)	0.003 (0.001)	0.008 (0.002)	0.004 (0.001)	−0.005 (0.002)	0.006 (0.001)
Change in logarithmic sector employment	0.094 (0.057)	0.286 (0.123)	0.459 (0.189)	−0.022 (0.158)	0.333 (0.082)
Tenure × (change in logarithmic sector employment)	−0.020 (0.009)	−0.061 (0.023)	−0.053 (0.025)	−0.005 (0.025)	−0.027 (0.012)

[a] Standard errors in parentheses.
[b] Omitted category is no change in employment.
[c] Other variables controlled for consist of education, marital status, number of children, race, years of full-time work experience and its square, union membership, government employment, logarithmic wage, the U.S. employment rate, and location.

tenure for dismissal is $-0.221 + 16 \times (0.008) = -0.093$ (owing to the interaction term). From this, we interpret an additional year of tenure to imply that the dismissal probability is $\exp(-0.093) = 91\%$ of what it would be otherwise, representing a decline of 9%.

Table 11.1 also shows that the generalized coefficients associated with dismissal are similar to the probit fits. The standard errors are also qualitatively similar, although they are higher for the generalized logits than for the probit model. In particular, we again see that the coefficient associated with the interaction between tenure and time trend reveals an increasing dismissal rate for experienced workers. The same is true for the rate of quitting.

11.1.3 Multinomial (Conditional) Logit

Similar to Equation (11.1), an alternative linear combination of explanatory variables is

$$V_{it,j} = \mathbf{x}'_{it,j}\boldsymbol{\beta}, \tag{11.3}$$

where $\mathbf{x}_{it,j}$ is a vector of explanatory variables that depends on the jth alternative, whereas the parameters $\boldsymbol{\beta}$ do not. Using this expression in Equation (11.2) is the basis for the *multinomial logit model*, also known as the *conditional logit model* (McFadden, 1974E). With this specification, the total log likelihood becomes

$$L(\boldsymbol{\beta}) = \sum_{it} \sum_{j=1}^{c} y_{it,j} \ln \pi_{it,j} = \sum_{it} \left\{ \sum_{j=1}^{c} y_{it,j} \mathbf{x}_{it,j}' \boldsymbol{\beta} - \ln \left(\sum_{k=1}^{c} \exp(\mathbf{x}_{it,k}' \boldsymbol{\beta}) \right) \right\}. \tag{11.4}$$

This straightforward expression for the likelihood enables maximum likelihood inference to be easily performed.

The generalized logit model is a special case of the multinomial logit model. To see this, consider explanatory variables $\mathbf{x}_{it}$ and parameters $\boldsymbol{\beta}_j$, each of dimension $K \times 1$. Define

$$\mathbf{x}_{it,j} = \begin{pmatrix} \mathbf{0} \\ \vdots \\ \mathbf{0} \\ \mathbf{x}_{it} \\ \mathbf{0} \\ \vdots \\ \mathbf{0} \end{pmatrix}$$

and

$$\boldsymbol{\beta} = \begin{pmatrix} \boldsymbol{\beta}_1 \\ \boldsymbol{\beta}_2 \\ \vdots \\ \boldsymbol{\beta}_c \end{pmatrix}.$$

Specifically, $\mathbf{x}_{it,j}$ is defined as $j-1$ zero vectors (each of dimension $K \times 1$), followed by $\mathbf{x}_{it}$ and then followed by $c-j$ zero vectors. With this specification, we have $\mathbf{x}_{it,j}'\boldsymbol{\beta} = \mathbf{x}_{it}'\boldsymbol{\beta}_j$. Thus, a statistical package that performs multinomial logit estimation can also perform generalized logit estimation through the appropriate coding of explanatory variables and parameters. Another consequence of this connection is that some authors use the descriptor multinomial logit when referring to the Section 11.1.2 generalized logit model.

Moreover, through similar coding schemes, multinomial logit models can also handle linear combinations of the form

$$V_{it,j} = \mathbf{x}_{it,1,j}'\boldsymbol{\beta} + \mathbf{x}_{it,2}'\boldsymbol{\beta}_j. \tag{11.5}$$

Here, $\mathbf{x}_{it,1,j}$ are explanatory variables that depend on the alternative whereas $\mathbf{x}_{it,2}$ do not. Similarly, $\boldsymbol{\beta}_j$ are parameters that depend on the alternative whereas $\boldsymbol{\beta}$ do not. This type of linear combination is the basis of a *mixed logit model*.

As with conditional logits, it is customary to choose one set of parameters as the baseline and specify $\boldsymbol{\beta}_c = \mathbf{0}$ to avoid redundancies.

To interpret parameters for the multinomial logit model, we may compare alternatives h and k using Equations (11.2) and (11.3) to get

$$\ln\left[\frac{\text{Prob}(y_{it} = h)}{\text{Prob}(y_{it} = k)}\right] = \left(\mathbf{x}_{it,h} - \mathbf{x}_{it,k}\right)' \boldsymbol{\beta}. \tag{11.6}$$

Thus, we may interpret β_j as the proportional change in the odds ratio, where the change is the value of the jth explanatory variable, moving from the kth to the hth alternative.

With Equation (11.2), note that $\pi_{it,1}/\pi_{it,2} = \exp(V_{it,1})/\exp(V_{it,2})$. This ratio does not depend on the underlying values of the other alternatives, $V_{it,j}$, for $j = 3, \ldots, c$. This feature, called the *independence of irrelevant alternatives*, can be a drawback of the multinomial logit model for some applications.

Example 11.3: Choice of Yogurt Brands We now consider a marketing data set introduced by Jain et al. (1994O) and further analyzed by Chen and Kuo (2001S). These data, obtained from A. C. Nielsen, are known as *scanner data* because they are obtained from optical scanning of grocery purchases at check-out. The subjects consist of $n = 100$ households in Springfield, Missouri. The response of interest is the type of yogurt purchased, consisting of four brands: Yoplait, Dannon, Weight Watchers, and Hiland. The households were monitored over a two-year period with the number of purchases ranging from 4 to 185; the total number of purchases is $N = 2{,}412$.

The two marketing variables of interest are PRICE and FEATURES. For the brand purchased, PRICE is recorded as price paid, that is, the shelf price net of the value of coupons redeemed. For other brands, PRICE is the shelf price. FEATURES is a binary variable, defined to be one if there was a newspaper feature advertising the brand at the time of purchase and zero otherwise. Note that the explanatory variables vary by alternative, suggesting the use of a multinomial (conditional) logit model.

Tables 11.2 and 11.3 summarize some important aspects of the data. Table 11.2 shows that Yoplait was the most frequently selected (33.9%) type of yogurt in our sample whereas Hiland was the least frequently selected (2.9%). Yoplait was also the most heavily advertised, appearing in newspaper advertisements 5.6% of the time that the brand was chosen. Table 11.3 shows that Yoplait was also the most expensive, costing 10.7 cents per ounce, on average. Table 11.3 also shows that there are several prices that were far below the average, suggesting some potential influential observations.

Table 11.2. *Summary statistics by choice of yogurt*

Summary statistics	Yoplait	Dannon	Weight Watchers	Hiland	Totals
Number of choices	818	970	553	71	2,412
Number of choices, in percent	33.9	40.2	22.9	2.9	100.0
Feature averages, in percent	5.6	3.8	3.8	3.7	4.4

Table 11.3. *Summary statistics for prices*

Variable	Mean	Median	Minimum	Maximum	Standard deviation
Yoplait	0.107	0.108	0.003	0.193	0.019
Dannon	0.082	0.086	0.019	0.111	0.011
Weight Watchers	0.079	0.079	0.004	0.104	0.008
Hiland	0.054	0.054	0.025	0.086	0.008

A multinomial logit model was fit to the data, using the following specification for the systematic component:

$$V_{it,j} = \alpha_j + \beta_1 PRICE_{it,j} + \beta_2 FEATURE_{it,j},$$

using Hiland as the omitted alternative. The results are summarized in Table 11.4. Here, we see that each parameter is statistically significantly different from zero. Thus, the parameter estimates may be useful when predicting the probability of choosing a brand of yogurt. Moreover, in a marketing context, the coefficients have important substantive interpretations. Specifically, we interpret the coefficient associated with FEATURES to suggest that a consumer is $\exp(0.491) = 1.634$ times more likely to purchase a product that is featured in a newspaper ad compared to one that is not. For the PRICE coefficient, a one-cent decrease in price suggests that a consumer is $\exp(0.3666) = 1.443$ times more likely to purchase a brand of yogurt.

11.1.4 Random-Utility Interpretation

In economic applications, we think of an individual choosing among c categories where preferences among categories are determined by an unobserved utility function. For the ith individual at the tth time period, use $U_{it,j}$ for the utility of

Table 11.4. *Yogurt multinomial logit model estimates*

Variable	Parameter estimate	t-statistic
Yoplait	4.450	23.78
Dannon	3.716	25.55
Weight Watchers	3.074	21.15
FEATURES	0.491	4.09
PRICE	−36.658	−15.04
−2 log likelihood	10,138	
AIC	10,148	

the jth choice. To illustrate, assume that the individual chooses the first category if $U_{it,1} > U_{it,j}$ for $j = 2, \ldots, c$ and denote this choice as $y_{it} = 1$. We model utility as an underlying value plus random noise, that is, $U_{it,j} = V_{it,j} + \varepsilon_{it,j}$, where $V_{it,j}$ is specified in Equation (11.4). The noise variable is assumed to have an extreme-value distribution of the form

$$F(a) = \text{Prob}(\varepsilon_{it,j} \leq a) = \exp(e^{-a}).$$

This form is computationally convenient. Omitting the observation-level subscripts $\{it\}$ for the moment, we have

$$\begin{aligned}
\text{Prob}(y = 1) &= \text{Prob}(U_1 > U_j \text{ for } j = 2, \ldots, c) \\
&= \text{Prob}(\varepsilon_j < \varepsilon_1 + V_1 - V_j \text{ for } j = 2, \ldots, c) \\
&= \text{E}\{\text{Prob}(\varepsilon_j < \varepsilon_1 + V_1 - V_j \text{ for } j = 2, \ldots, c \mid \varepsilon_1)\} \\
&= \text{E}\{F(\varepsilon_1 + V_1 - V_2) \cdots F(\varepsilon_1 + V_1 - V_c)\} \\
&= \text{E}\exp[\exp(-(\varepsilon_1 + V_1 - V_2)) + \cdots + \exp(-(\varepsilon_1 + V_1 - V_c))] \\
&= \text{E}\exp[k_V \exp(-\varepsilon_1)],
\end{aligned}$$

where $k_V = \sum_{j=2}^{c} \exp(V_j - V_1)$. Now, it is a simple exercise in calculus to show, with the distribution function given here, that $\text{E}\exp[k_V \exp(-\varepsilon_1)] = 1/(k_V + 1)$. Thus, we have

$$\text{Prob}(y = 1) = \frac{1}{1 + \sum_{j=2}^{c} \exp(V_j - V_1)} = \frac{\exp(V_1)}{\sum_{j=1}^{c} \exp(V_j)}.$$

Because this argument is valid for all alternatives $j = 1, 2, \ldots, c$, the random-utility representation yields the multinomial logit model.

11.1.5 Nested Logit

To mitigate the problem of independence of irrelevant alternatives, we now introduce a type of hierarchical model known as a *nested logit model*. To interpret the nested logit model, in the first stage one chooses an alternative (say the first type) with probability

$$\pi_{it,1} = \text{Prob}(y_{it} = 1) = \frac{\exp(V_{it,1})}{\exp(V_{it,1}) + \left(\sum_{k=2}^{c} \exp(V_{it,k}/\rho)\right)^{\rho}}. \tag{11.7}$$

Then, conditional on not choosing the first alternative, the probability of choosing any one of the other alternatives follows a multinomial logit model with probabilities

$$\frac{\pi_{it,j}}{1-\pi_{it,1}} = \text{Prob}(y_{it} = j \mid y_{it} \neq 1) = \frac{\exp(V_{it,j}/\rho)}{\sum_{k=2}^{c} \exp(V_{it,k}/\rho)}, \quad j = 2, \ldots, c. \tag{11.8}$$

In Equations (11.7) and (11.8), the parameter ρ measures the association among the choices $j = 2, \ldots, c$. The value of $\rho = 1$ reduces to the multinomial logit model that we interpret to mean independence of irrelevant alternatives. We also interpret $\text{Prob}(y_{it} = 1)$ to be a weighted average of values from the first choice and the others. Conditional on not choosing the first category, the form of $\text{Prob}(y_{it} = j \mid y_{it} \neq 1)$ has the same form as the multinomial logit.

The advantage of the nested logit is that it generalizes the multinomial logit model in such a way that we no longer have the problem of independence of irrelevant alternatives. A disadvantage, pointed out by McFadden (1981E), is that only one choice is observed; thus, we do not know which category belongs in the first stage of the nesting without additional theory regarding choice behavior. Nonetheless, the nested logit generalizes the multinomial logit by allowing alternative "dependence" structures. That is, one may view the nested logit as a robust alternative to the multinomial logit and examine each one of the categories in the first stage of the nesting.

11.1.6 Generalized Extreme-Value Distribution

The nested logit model can also be given a random-utility interpretation. To this end, return to the random-utility model but assume that the choices are related through a dependence within the error structure. McFadden (1978E) introduced the *generalized extreme-value distribution* of the form

$$F(a_1, \ldots, a_c) = \exp\{-G(e^{-a_1}, \ldots, e^{-a_c})\}.$$

Under regularity conditions on G, McFadden (1978E) showed that this yields

$$\begin{aligned}\text{Prob}(y = j) &= \text{Prob}(U_j > U_k \text{ for } k = 1, \ldots, c, k \neq j)\\ &= \frac{e^{V_j} G_j(e^{-V_1}, \ldots, e^{-V_c})}{G(e^{-V_1}, \ldots, e^{-V_c})},\end{aligned}$$

where

$$G_j(x_1, \ldots, x_c) = \frac{\partial}{\partial x_j} G(x_1, \ldots, x_c)$$

is the jth partial derivative of G.

Special Cases

1. Let $G(x_1, \ldots, x_c) = x_1 + \cdots + x_c$. In this case, $G_j = 1$ and

$$\text{Prob}(y = j) = \frac{\exp(V_j)}{\sum_{k=1}^{c} \exp(V_k)}.$$

   This is the multinomial logit case.
2. Let

$$G(x_1, \ldots, x_c) = x_1 + \left(\sum_{k=2}^{c} x_k^{1/\rho} \right)^{\rho}.$$

   In this case, $G_1 = 1$ and

$$\text{Prob}(y = 1) = \frac{\exp(V_1)}{\exp(V_1) + \left(\sum_{k=2}^{c} \exp(V_k/\rho)\right)^{\rho}}.$$

   Additional calculations show that

$$\text{Prob}(y = j \mid y \neq 1) = \frac{\exp(V_j/\rho)}{\sum_{k=2}^{c} \exp(V_k/\rho)}.$$

   This is the nested logit case.

Thus, the generalized extreme-value distribution provides a framework that encompasses the multinomial and conditional logit models. Amemiya (1985E) provides background on more complex nested models that utilize the generalized extreme-value distribution.

11.2 Multinomial Logit Models with Random Effects

Repeated observations from an individual tend to be similar; in the case of categorical choices, this means that individuals tend to make the same choices from one observation to the next. This section models that similarity through a common heterogeneity term. To this end, we augment our systematic component with a heterogeneity term and, similarly to Equation (11.4), consider linear combinations of explanatory variables of the form

$$V_{it,j} = \mathbf{z}'_{it,j}\boldsymbol{\alpha}_i + \mathbf{x}'_{it,j}\boldsymbol{\beta}. \tag{11.9}$$

As before, $\boldsymbol{\alpha}_i$ represents the heterogeneity term that is subject-specific. The form of Equation (11.9) is quite general and includes many applications of interest. However, to develop intuition, we focus on the special case

$$V_{it,j} = \alpha_{ij} + \mathbf{x}'_{it,j}\boldsymbol{\beta}. \tag{11.10}$$

Here, intercepts vary by individual and alternative but are common over time.

With this specification for the systematic component, the conditional (on the heterogeneity) probability that the ith subject at time t chooses the jth alternative is

$$\pi_{it,j}\left(\boldsymbol{\alpha}_i\right) = \frac{\exp(V_{it,j})}{\sum_{k=1}^{c}\exp(V_{it,k})} = \frac{\exp(\alpha_{ij} + \mathbf{x}'_{it,j}\boldsymbol{\beta})}{\sum_{k=1}^{c}\exp(\alpha_{ik} + \mathbf{x}'_{it,k}\boldsymbol{\beta})}, \qquad j = 1, 2, \ldots, c, \tag{11.11}$$

where we now denote the set of heterogeneity terms as $\boldsymbol{\alpha}_i = (\alpha_{i1}, \ldots, \alpha_{ic})'$. From the form of this equation, we see that a heterogeneity term that is constant over alternatives j does not affect the conditional probability. To avoid parameter redundancies, a convenient normalization is to take $\alpha_{ic} = 0$.

For statistical inference, we begin with likelihood equations. Similar to the development in Section 11.1.1, the conditional likelihood for the ith subject is

$$L\left(\mathbf{y}_i \mid \boldsymbol{\alpha}_i\right) = \prod_{t=1}^{T_i}\prod_{j=1}^{c}\left(\pi_{it,j}(\boldsymbol{\alpha}_i)\right)^{y_{it,j}} = \prod_{t=1}^{T_i}\left\{\frac{\exp\left(\sum_{j=1}^{c} y_{it,j}(\alpha_{ij} + \mathbf{x}'_{it,j}\boldsymbol{\beta})\right)}{\sum_{k=1}^{c}\exp(\alpha_{ik} + \mathbf{x}'_{it,k}\boldsymbol{\beta})}\right\}. \tag{11.12}$$

We assume that $\{\boldsymbol{\alpha}_i\}$ is i.i.d. with distribution function G_α, which is typically taken to be multivariate normal. With this convention, the (unconditional) likelihood for the ith subject is

$$L\left(\mathbf{y}_i\right) = \int L\left(\mathbf{y}_i \mid \mathbf{a}\right) dG_\alpha(\mathbf{a}).$$

Assuming independence among subjects, the total log likelihood is $L = \sum_i \ln L(\mathbf{y}_i)$.

Relation with Nonlinear Random-Effects Poisson Model

With this framework, standard maximum likelihood estimation is available. However, for applied work, there are relatively few statistical packages available for estimating multinomial logit models with random effects. As an alternative, one can look to properties of the multinomial distribution and link it to other distributions. Chen and Kuo (2001S) recently surveyed these linkages in the context of random-effects models and we now present one link to a nonlinear Poisson model. Statistical packages for nonlinear Poisson models are readily available; with this link, they can be used to estimate parameters of the multinomial logit model with random effects.

To this end, an analyst would instruct a statistical package to "assume" that the binary random variables $y_{it,j}$ are Poisson distributed with conditional means $\pi_{it,j}$ and, conditional on the heterogeneity terms, are independent. This is a nonlinear Poisson model because, from Section 10.5.3, a linear Poisson model takes the logarithmic (conditional) mean to be a linear function of explanatory variables. In contrast, from Equation (11.11), $\log \pi_{it,j}$ is a nonlinear function. Of course, this "assumption" is not valid. Binary random variables have only two outcomes and thus cannot have a Poisson distribution. Moreover, the binary variables must sum to one (that is, $\sum_j y_{it,j} = 1$) and thus are not even conditionally independent. Nonetheless, with this "assumption" and the Poisson distribution (reviewed in Section 10.5.3), the conditional likelihood interpreted by the statistical package is

$$\begin{aligned} L(\mathbf{y}_i \mid \alpha_i) &= \prod_{t=1}^{T_i} \prod_{j=1}^{c} \frac{(\pi_{it,j}(\alpha_i))^{y_{it,j}} \exp(-\pi_{it,j}(\alpha_i))}{y_{it,j}!} \\ &= \prod_{t=1}^{T_i} \left\{ \prod_{j=1}^{c} (\pi_{it,j}(\alpha_i))^{y_{it,j}} \right\} e^{-1}. \end{aligned}$$

Up to the constant, this is the same conditional likelihood as in Equation (11.12) (see Exercise 10.1). Thus, a statistical package that performs nonlinear Poisson models with random effects can be used to get maximum likelihood estimates for the multinomial logit model with random effects. See Chen and Kuo (2001S) for a related algorithm based on a linear Poisson model with random effects.

Example 11.3: Choice of Yogurt Brands (continued) To illustrate, we used a multinomial logit model with random effects on the yogurt data introduced in Example 11.1. Following Chen and Kuo (2001S), random intercepts for Yoplait,

Table 11.5. *Yogurt multinomial logit model estimates*

	Without random effects		With random effects	
Variable	Parameter estimate	t-statistic	Parameter estimate	t-statistic
Yoplait	4.450	23.78	5.622	7.29
Dannon	3.716	25.55	4.772	6.55
Weight Watchers	3.074	21.15	1.896	2.09
FEATURES	0.491	4.09	0.849	4.53
PRICE	−36.658	−15.04	−44.406	−11.08
−2 log likelihood	10,138		7,301.4	
AIC	10,148		7,323.4	

Dannon, and Weight Watchers were assumed to follow a multivariate normal distribution with an unstructured covariance matrix. Table 11.5 shows results from fitting this model, based on the nonlinear Poisson model link and using SAS PROC NLMIXED. Here, we see that the coefficients for FEATURES and PRICE are qualitatively similar to the model without random effects, reproduced for convenience from Table 11.4. They are qualitatively similar in the sense that they have the same sign and same degree of statistical significance. Overall, the *AIC* statistic suggests that the model with random effects is the preferred model.

As with binary dependent variables, conditional maximum likelihood estimators have been proposed; see, for example, Conaway (1989S). Appendix 11A provides a brief introduction to these alternative estimators.

11.3 Transition (Markov) Models

Another way of accounting for heterogeneity is to trace the development of a dependent variable over time and represent the distribution of its current value as a function of its history. To this end, define H_{it} to be the history of the ith subject up to time t. For example, if the explanatory variables are assumed to be nonstochastic, then we might use $H_{it} = \{y_{i1}, \ldots, y_{i,t-1}\}$. With this information set, we may partition the likelihood for the ith subject as

$$L(\mathbf{y}_i) = f(y_{i1}) \prod_{t=2}^{T_i} f(y_{it} \mid H_{it}), \tag{11.13}$$

where $f(y_{it} \mid H_{it})$ is the conditional distribution of y_{it} given its history and $f(y_{i1})$ is the marginal distribution of y_{i1}. For example, one type of application is through a conditional GLM of the form

$$f(y_{it} \mid H_{it}) = \exp\left(\frac{y_{it}\theta_{it} - b(\theta_{it})}{\phi} + S(y_{it}, \phi)\right),$$

where $\mathrm{E}(y_{it} \mid H_{it}) = b'(\theta_{it})$ and $\mathrm{Var}(y_{it} \mid H_{it}) = \phi b''(\theta_{it})$. Assuming a canonical link, for the systematic component, one could use

$$\theta_{it} = g\left(\mathrm{E}(y_{it} \mid H_{it})\right) = \mathbf{x}_{it}'\beta + \sum_j \varphi_j y_{i,t-j}.$$

See Diggle et al. (2002S, Chapter 10) for further applications of and references to general transition GLMs. We focus on categorical responses.

Unordered Categorical Response

To simplify our discussion of unordered categorical responses, we also assume discrete unit time intervals. To begin, we consider Markov models of order 1. Thus, the history H_{it} need only contain $y_{i,t-1}$. More formally, we assume that

$$\begin{aligned}\pi_{it,jk} &= \mathrm{Prob}(y_{it} = k \mid y_{i,t-1} = j) \\ &= \mathrm{Prob}(y_{it} = k \mid \{y_{i,t-1} = j, y_{i,t-2}, \ldots, y_{i,1}\}).\end{aligned}$$

That is, given the information in $y_{i,t-1}$, there is no additional information content in $\{y_{i,t-2}, \ldots, y_{i,1}\}$ about the distribution of y_{it}.

Without covariate information, it is customary to organize the set of *transition probabilities* $\pi_{it,jk}$ as a matrix of the form

$$\mathbf{\Pi}_{it} = \begin{pmatrix} \pi_{it,11} & \pi_{it,12} & \cdots & \pi_{it,1c} \\ \pi_{it,21} & \pi_{it,22} & \cdots & \pi_{it,2c} \\ \vdots & \vdots & \ddots & \vdots \\ \pi_{it,c1} & \pi_{it,c2} & \cdots & \pi_{it,cc} \end{pmatrix}.$$

Here, each row sums to one. With covariate information and an initial state distribution $\mathrm{Prob}(y_{i1})$, one can trace the history of the process knowing only the *transition matrix* $\mathbf{\Pi}_{it}$. We call the row identifier, j, the *state of origin* and the column identifier, k, the *destination state*.

For complex transition models, it can be useful to graphically summarize the set of feasible transitions under consideration. To illustrate, Figure 11.1 summarizes an employee retirement system with $c = 4$ categories, where

- 1 denotes active continuation in the pension plan,
- 2 denotes retirement from the pension plan,
- 3 denotes withdrawal from the pension plan, and
- 4 denotes death.

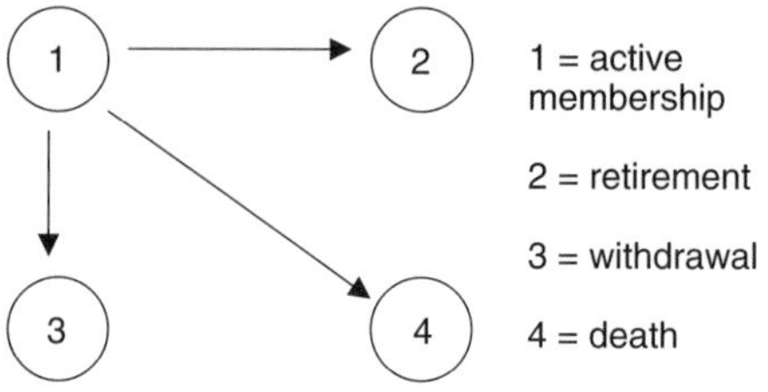

Figure 11.1. Graphical summary of a transition model for a hypothetical employment retirement system.

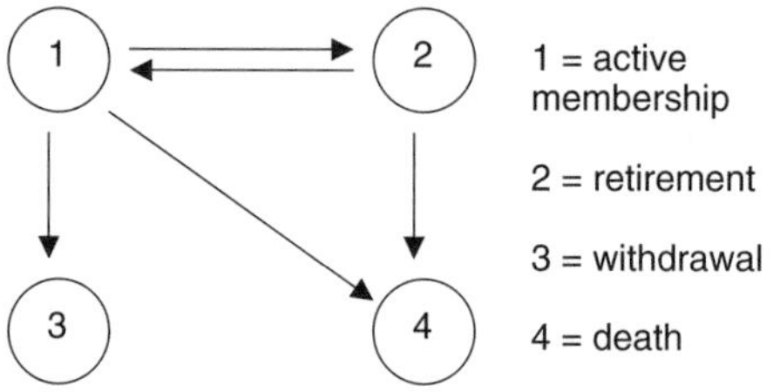

Figure 11.2. A modified transition model for an employment retirement system.

For this system, the circles represent the nodes of the graph and correspond to the response categories. The arrows, or arcs, indicate the modes of possible transitions. This graph indicates that movement from state 1 to states 1, 2, 3, or 4 is possible, so that we would assume $\pi_{1j} \geq 0$, for $j = 1, 2, 3, 4$. However, once an individual is in states 2, 3, or 4, it is not possible to move from those states (known as *absorbing states*). Thus, we use $\pi_{jj} = 1$ for $j = 2, 3, 4$ and $\pi_{jk} = 0$ for $j = 2, 3, 4$ and $k \neq j$. Note that although death is certainly possible (and eventually certain) for those in retirement, we assume $\pi_{24} = 0$ with the understanding that the plan has paid pension benefits at retirement and needs no longer be concerned with additional transitions after exiting the plan, regardless of the reason. This assumption is often convenient because it is difficult to track individuals having left active membership in a benefit plan.

For another example, consider the modification summarized in Figure 11.2. Here, we see that retirees are now permitted to reenter the workforce so that π_{21} may be positive. Moreover, now the transition from retirement to death is also explicitly accounted for so that $\pi_{24} \geq 0$. This may be of interest in a system that pays retirement benefits as long as a retiree lives. We refer to Haberman and Pitacco (1999O) for many additional examples of Markov transition models that are of interest in employee benefit and other types of actuarial systems.

We can parameterize the problem by choosing a multinomial logit, one for each state of origin. Thus, we use

$$\pi_{it,jk} = \frac{\exp(V_{it,jk})}{\sum_{h=1}^{c} \exp(V_{it,jh})}, \quad j,k = 1,2,\ldots,c, \tag{11.14}$$

where the systematic component $V_{it,jk}$ is given by

$$V_{it,jk} = \mathbf{x}'_{it,jk}\boldsymbol{\beta}_j. \tag{11.15}$$

As discussed in the context of employment retirement systems, in a given problem one assumes that a certain subset of transition probabilities are zero, thus constraining the estimation of $\boldsymbol{\beta}_j$.

For estimation, we may proceed as in Section 11.1. Define

$$y_{it,jk} = \begin{cases} 1 & \text{if } y_{it} = k \text{ and } y_{i,t-1} = j, \\ 0 & \text{otherwise.} \end{cases}$$

With this notation, the conditional likelihood is

$$f(y_{it} \mid y_{i,t-1}) = \prod_{j=1}^{c} \prod_{k=1}^{c} (\pi_{it,jk})^{y_{it,jk}}. \tag{11.16}$$

Here, in the case that $\pi_{it,jk} = 0$ (by assumption), we have that $y_{it,jk} = 0$ and use the convention that $0^0 = 1$.

To simplify matters, we assume that the initial state distribution, $\text{Prob}(y_{i1})$, is described by a different set of parameters than the transition distribution, $f(y_{it} \mid y_{i,t-1})$. Thus, to estimate this latter set of parameters, one only needs to maximize the *partial log likelihood*

$$L_{\text{P}} = \sum_i \sum_{t=2}^{T_i} \ln f(y_{it} \mid y_{i,t-1}), \tag{11.17}$$

where $f(y_{it} \mid y_{i,t-1})$ is specified in Equation (11.16). In some cases, the interesting aspect of the problem is the transition. In this case, one loses little by focusing on the partial likelihood. In other cases, the interesting aspect is the state, such as the proportion of retirements at a certain age. Here, a representation for the initial state distribution takes on greater importance.

In Equation (11.15), we specified separate components for each alternative. Assuming no implicit relationship among the components, this specification yields a particularly simple analysis. That is, we may write the partial log likelihood as

$$L_{\text{P}} = \sum_{j=1}^{c} L_{\text{P},j}(\boldsymbol{\beta}_j),$$

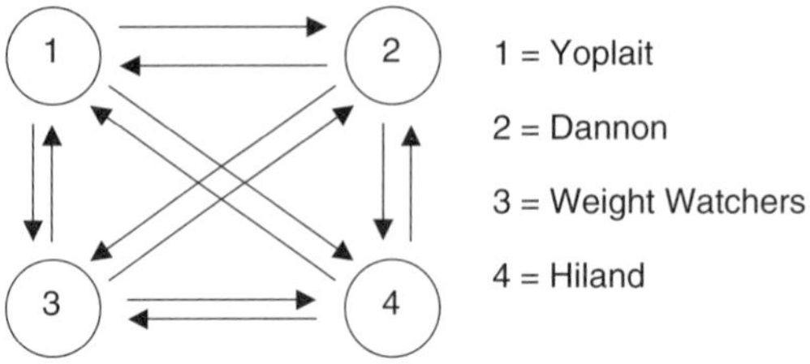

Figure 11.3. A transition model for yogurt brand choice.

where, from Equations 11.14–11.16, we have

$$\begin{aligned} L_{\mathrm{P},j}(\boldsymbol{\beta}_j) &= \sum_i \sum_{t=2}^{T_i} \ln f(y_{it} \mid y_{i,t-1} = j) \\ &= \sum_i \sum_{t=2}^{T_i} \left\{ \sum_{k=1}^{c} y_{it,jk} \mathbf{x}'_{it,jk} \boldsymbol{\beta}_j - \ln \left(\sum_{k=1}^{c} \exp(\mathbf{x}'_{it,jk} \boldsymbol{\beta}_j) \right) \right\}, \end{aligned}$$

as in Equation (11.4). Thus, we can split up the data according to each (lagged) choice and determine MLEs for each alternative, in isolation of the others.

Example 11.3: Choice of Yogurt Brands (continued) To illustrate, we return to the Yogurt data set. We now explicitly model the transitions between brand choices, as denoted in Figure 11.3. Here, purchases of yogurt occur intermittently over a two-year period; the data are not observed at discrete time intervals. By ignoring the length of time between purchases, we are using what is sometimes referred to as "operational time." In effect, we are assuming that one's most recent choice of a brand of yogurt has the same effect on the current choice, regardless of whether the prior choice was one day or one month ago. This assumption suggests future refinements to the transition approach in modeling yogurt choice.

Tables 11.6a and 11.6b show the relation between the current and most recent choice of yogurt brands. Here, we call the most recent choice the "origin state" and the current choice the "destination state." Table 11.6a shows that there are only 2,312 observations under consideration; this is because initial values from each of 100 subjects are not available for the transition analysis. For most observation pairs, the current choice of the brand of yogurt is the same as that chosen most recently, exhibiting "brand loyalty." Other observation pairs can be described as "switchers." Brand loyalty and switching behavior are more apparent in Table 11.6b, where we rescale counts by row totals to

Table 11.6a. *Yogurt transition counts*

	Destination state				
Origin state	Yoplait	Dannon	Weight Watchers	Hiland	Total
Yoplait	654	65	41	17	777
Dannon	71	822	19	16	928
Weight Watchers	44	18	473	5	540
Hiland	14	17	6	30	67
Total	783	922	539	68	2,312

Table 11.6b. *Yogurt transition empirical probabilities (in percent)*

	Destination state				
Origin state	Yoplait	Dannon	Weight Watchers	Hiland	Total
Yoplait	84.2	8.4	5.3	2.2	100.0
Dannon	7.7	88.6	2.0	1.7	100.0
Weight Watchers	8.1	3.3	87.6	0.9	100.0
Hiland	20.9	25.4	9.0	44.8	100.0

give (rough) empirical transition probabilities. Here, we see that customers of Yoplait, Dannon, and Weight Watchers exhibit more brand loyalty compared to those of Hiland, who are more prone to switching.

Of course, Tables 11.6a and 11.6b do not account for changing aspects of price and features. In contrast, these explanatory variables are captured in the multinomial logit fit, displayed in Table 11.7. Table 11.8 shows that purchase probabilities for customers of Dannon, Weight Watchers, and Hiland are more responsive to a newspaper ad than Yoplait customers. Moreover, compared to the other three brands, Hiland customers are not price sensitive in that changes in PRICE have relatively little impact on the purchase probability (nor is it even statistically significant).

Table 11.6b suggests that prior purchase information is important when estimating purchase probabilities. To test this, it is straightforward to use a likelihood ratio test of the null hypothesis $H_0 : \boldsymbol{\beta}_j = \boldsymbol{\beta}$, that is, the case where the components do not vary by origin state. Table 11.7 shows that the total (minus two times the partial) log likelihood is $2{,}379.8 + \cdots + 281.5 = 6{,}850.3$. Estimation of this model under the null hypothesis yields a corresponding value of 9,741.2. Thus, the likelihood ratio test statistic is $LRT = 2{,}890.9$. There

Table 11.7. *Yogurt transition model estimates*

	State of origin							
	Yoplait		Dannon		Weight Watchers		Hiland	
Variable	Estimate	*t*-stat	Estimate	*t*-stat	Estimate	*t*-stat	Estimate	*t*-stat
Yoplait	5.952	12.75	4.125	9.43	4.266	6.83	0.215	0.32
Dannon	2.529	7.56	5.458	16.45	2.401	4.35	0.210	0.42
Weight Watchers	1.986	5.81	1.522	3.91	5.699	11.19	−1.105	−1.93
FEATURES	0.593	2.07	0.907	2.89	0.913	2.39	1.820	3.27
PRICE	−41.257	−6.28	−48.989	−8.01	−37.412	−5.09	−13.840	−1.21
−2 log likelihood	2,397.8		2,608.8		1,562.2		281.5	

Table 11.8. *Tax preparers transition empirical probabilities*

	Destination state (%)	
Origin state	PREP = 0	PREP = 1
PREP = 0 (Count = 546)	89.4	10.6
PREP = 1 (Count = 486)	8.4	91.6

are fifteen degrees of freedom for this test statistic. Thus, this provides strong evidence for rejecting the null hypothesis, corroborating the intuition that the most recent type of purchase has a strong influence on the current brand choice.

Example 11.3 on the choice of yogurt brands illustrated an application of a conditional (multinomial) logit model where the explanatory variables FEATURES and PRICE depend on the alternative. To provide an example where this is not so, we return to Example 9.1.3 on the choice of a professional tax preparer. Here, financial and demographic characteristics do not depend on the alternative and so we apply a straightforward logit model.

Example 11.4: Income Tax Payments and Tax Preparers This is a continuation of the Section 9.1.3 example on the choice of whether or not a tax filer will use a professional tax preparer. Our Chapter 9 analysis of these data indicated strong similarities in responses within a subject. In fact, our Section 9.3 fixed-effects analysis indicated that 97 tax filers never used a preparer in the five years under consideration and 89 always did, out of 258 subjects. We now model these time transitions explicitly in lieu of using a time-constant latent variable α_i to account for this similarity.

Table 11.8 shows the relationship between the current and most recent choice of preparer. Although we began with 1,290 observations, 258 initial observations are not available for the transition analysis, reducing our sample size to $1{,}290 - 258 = 1{,}032$. Table 11.8 strongly suggests that the most recent choice is an important predictor of the current choice.

Table 11.9 provides a more formal assessment with a fit of a logit transition model. To assess whether or not the transition aspect is an important piece of the model, we can use a likelihood ratio test of the null hypothesis H_0 : $\boldsymbol{\beta}_j = \boldsymbol{\beta}$, that is, the case where the coefficients do not vary by origin state. Table 11.9 shows that the total (minus two times the partial) log likelihood is $361.5 + 264.6 = 626.1$. Estimation of this model under the null hypothesis yields a corresponding value of 1,380.3. Thus, the likelihood ratio test statistic

Table 11.9. *Tax preparers transition model estimates*

	State of origin			
	PREP = 0		PREP = 1	
Variable	Estimate	*t*-statistic	Estimate	*t*-statistic
Intercept	−10.704	−3.06	0.208	0.18
LNTPI	1.024	2.50	0.104	0.73
MR	−0.072	−2.37	0.047	2.25
EMP	0.352	0.85	0.750	1.56
−2 log likelihood	361.5		264.6	

is $LRT = 754.2$. There are four degrees of freedom for this test statistic. Thus, this provides strong evidence for rejecting the null hypothesis, corroborating the intuition that the most recent choice is an important predictor of the current choice.

To interpret the regression coefficients in Table 11.9, we use the summary statistics in Section 9.1.3 to describe a "typical" tax filer and assume that LNTPI = 10, MR = 23, and EMP = 0. If this tax filer had not previously chosen to use a preparer, the estimated systematic component is $V = -10.704 + 1.024(10) - 0.072(23) + 0.352(0) = -2.12$. Thus, the estimated probability of choosing to use a preparer is $\exp(-2.12)/(1 + \exp(-2.12)) = 0.107$. Similar calculations show that, if this tax filer had chosen to use a preparer, then the estimated probability is 0.911. These calculations are in accord with the estimates in Table 11.8 that do not account for the explanatory variables. This illustration points out the importance of the intercept in determining these estimated probabilities.

Higher Order Markov Models

There are strong serial relationships in the taxpayer data and these may not be completely captured by simply looking at the most recent choice. For example, it may be that a tax filer who uses a preparer for two consecutive periods has a substantially different choice probability than a comparable filer who does not use a preparer in one period but elects to use a preparer in the subsequent period. It is customary in Markov modeling to simply expand the state space to handle such higher order time relationships. To this end, we may define a new categorical response, $y_{it}^* = \{y_{it}, y_{i,t-1}\}$. With this new response, the order-1 transition probability, $f(y_{it}^* \mid y_{i,t-1}^*)$, is equivalent to an order-2 transition probability of the original response, $f(y_{it} \mid y_{i,t-1}, y_{i,t-2})$. This is because the conditioning events are the same, $y_{i,t-1}^* = \{y_{i,t-1}, y_{i,t-2}\}$, and because $y_{i,t-1}$ is

Table 11.10. *Tax preparers order-2 transition empirical probabilities*

Origin state			Destination state (%)	
Lag PREP	Lag 2 PREP	Count	PREP = 0	PREP = 1
0	0	376	89.1	10.9
0	1	28	67.9	32.1
1	0	43	25.6	74.4
1	1	327	6.1	93.9

completely determined by the conditioning event $y^*_{i,t-1}$. Expansions to higher orders can be readily accomplished in a similar fashion.

To simplify the exposition, we consider only binary outcomes so that $c = 2$. Examining the transition probability, we are now conditioning on four states, $y^*_{i,t-1} = \{y_{i,t-1}, y_{i,t-2}\} = \{(0, 0), (0, 1), (1, 0), (1, 1)\}$. As before, one can split up the (partial) likelihood into four components, one for each state. Alternatively, one can write the logit model as

$$\text{Prob}(y_{it} \mid y_{i,t-1}, y_{i,t-2}) = \text{logit}(V_{it})$$

with

$$\begin{aligned} V_{it} &= \mathbf{x}'_{it,1}\beta_1 I(y_{i,t-1} = 0, y_{i,t-2} = 0) + \mathbf{x}'_{it,2}\beta_2 I(y_{i,t-1} = 1, y_{i,t-2} = 0) \\ &\quad + \mathbf{x}'_{it,3}\beta_3 I(y_{i,t-1} = 0, y_{i,t-2} = 1) + \mathbf{x}'_{it,4}\beta_4 I(y_{i,t-1} = 1, y_{i,t-2} = 1), \end{aligned}$$

where $I(.)$ is the indicator function of a set. The advantage of running the model in this fashion, instead of splitting it up into four distinct components, is that one can test directly the equality of parameters and consider a reduced parameter set by combining them. The advantage of the alternative approach is computational convenience; one performs a maximization procedure over a smaller data set and a smaller set of parameters, albeit several times.

Example 11.4: Income Tax Payments and Tax Preparers (continued) To investigate the usefulness of a second-order component, $y_{i,t-2}$, in the transition model, we begin in Table 11.10 with empirical transition probabilities. Table 11.10 suggests that there are important differences in the transition probabilities for each lag-1 origin state ($y_{i,t-1}$ = Lag PREP) between levels of the lag-2 origin state ($y_{i,t-2}$ = Lag 2 PREP).

Table 11.11 provides a more formal analysis by incorporating potential explanatory variables. The total (minus two times the partial) log likelihood is 469.4. Estimation of this model under the null hypothesis yields a

Table 11.11. *Tax preparers order-2 transition model estimates*

	State of origin							
	Lag PREP = 0, Lag 2 PREP = 0		Lag PREP = 0, Lag 2 PREP = 1		Lag PREP = 1, Lag 2 PREP = 0		Lag PREP = 1, Lag 2 PREP = 1	
Variable	Estimate	*t*-stat	Estimate	*t*-stat	Estimate	*t*-stat	Estimate	*t*-stat
Intercept	−9.866	−2.30	−7.331	−0.81	1.629	0.25	−0.251	−0.19
LNTPI	0.923	1.84	0.675	0.63	−0.210	−0.27	0.197	1.21
MR	−0.066	−1.79	−0.001	−0.01	0.065	0.89	0.040	1.42
EMP	0.406	0.84	0.050	0.04	NA	NA	1.406	1.69
−2 log likelihood	254.2		33.4		42.7		139.1	

corresponding value of 1,067.7. Thus, the likelihood ratio test statistic is $LRT = 567.3$. There are twelve degrees of freedom for this test statistic. Thus, this provides strong evidence for rejecting the null hypothesis. With this data set, estimation of the model incorporating lag-1 differences yields total (partial) (minus two times) log likelihood of 490.2. Thus, the likelihood ratio test statistic is $LRT = 20.8$. With eight degrees of freedom, comparing this test statistic to a chi-square distribution yields a p-value of 0.0077. Thus, the lag-2 component is a statistically significant contribution to the model.

Just as one can incorporate higher order lags into a Markov structure, it is also possible to bring in the time spent in a state. This may be of interest in a model of health states, where we might wish to accommodate the time spent in a "healthy" state or an "at-risk" state. This phenomenon is known as "lagged duration dependence." Similarly, the transition probabilities may depend on the number of prior occurrences of an event, known as "occurrence dependence." For example, when modeling employment, we may wish to allow transition probabilities to depend on the number of previous employment spells. For further considerations of these and other specialized transition models, see Lancaster (1990E) and Haberman and Pitacco (1999O).

11.4 Survival Models

Categorical data transition models, where one models the probability of movement from one state to another, are closely related to *survival models*. In survival models, the dependent variable is the time until an event of interest. The classic example is time until death (the complement of death being survival). Survival models are now widely applied in many scientific disciplines; other examples of events of interest include the onset of Alzheimer's disease (biomedical), time until bankruptcy (economics), and time until divorce (sociology).

Like the data studied elsewhere in this text, survival data are *longitudinal*. The cross-sectional aspect typically consists of multiple subjects, such as persons or firms, under study. There may be only one measurement on each subject, but the measurement is taken with respect to time. This combination of cross-sectional and temporal aspects gives survival data their longitudinal flavor. Because of the importance of survival models, it is not uncommon for researchers to equate the phrase "longitudinal data" with survival data.

Some events of interest, such as bankruptcy or divorce, may not happen for a specific subject. It is common that an event of interest may not have yet occurred within the study period so that the data are (*right*) *censored*. Thus, the complete observation times may not be available owing to the design of the study. Moreover, firms may merge or be acquired by other firms and subjects may move from a geographical area, leaving the study. Thus, the data may be incomplete because events that are extraneous to the research question under consideration; this is known as *random censoring*. Censoring is a regular feature of survival data; large values of a dependent variable are more difficult to observe than small values, other things being equal. In Section 7.4 we introduced mechanisms and models for handling incomplete data. For the repeated cross-sectional data described in this text, models for incompleteness have become available only relatively recently (although researchers have long been aware of these issues, focusing on attrition). In contrast, models of incompleteness have been historically important and one of the distinguishing features of survival data.

Some survival models can be written in terms of the Section 11.3 transition models. To illustrate, suppose that Y_i is the time until an event of interest and, for simplicity, assume that it is a discrete positive integer. From knowledge of Y_i, we may define y_{it} to be one if $Y_i = t$ and zero otherwise. With this notation, we may write the likelihood

$$\begin{aligned}\text{Prob}(Y_i = n) &= \text{Prob}(y_{i1} = 0, \ldots, y_{i,n-1} = 0, y_{in} = 1) \\ &= \text{Prob}(y_{i1} = 0) \left(\prod_{t=2}^{n-1} \text{Prob}(y_{it} = 0 \mid y_{i,t-1} = 0) \right) \\ &\quad \times \text{Prob}(y_{in} = 1 \mid y_{i,n-1} = 0),\end{aligned}$$

in terms of transition probabilities $\text{Prob}(y_{it} \mid y_{i,t-1})$ and the initial state distribution $\text{Prob}(y_{i1})$. Note that in Section 11.3 we considered n to be the nonrandom number of time units under consideration whereas here it is a realized value of a random variable.

Example 11.5: Time until Bankruptcy Shumway (2001O) examined the time to bankruptcy for 3,182 firms listed on the Compustat Industrial File and the CRSP Daily Stock Return File for the New York Stock Exchange over the period 1962–1992. Several explanatory financial variables were examined, including working capital to total assets, retained to total assets, earnings before interest and taxes to total assets, market equity to total liabilities, sales to total assets, net income to total assets, total liabilities to total assets, and current assets to current liabilities. The data set included 300 bankruptcies from 39,745 firm-years.

See also Kim et al. (1995O) for a similar study on insurance insolvencies.

Survival models are frequently expressed in terms of continuous dependent random variables. To summarize the distribution of Y, define the *hazard function* as

$$h(t) = \frac{\textit{probability density function}}{\textit{survival function}}$$
$$= \frac{-\frac{\partial}{\partial t}\text{Prob}\,(Y > t)}{\text{Prob}\,(Y > t)} = -\frac{\partial}{\partial t}\ln \text{Prob}\,(Y > t)\,,$$

the "instantaneous" probability of failure, conditional on survivorship up to time t. This is also known as the *force of mortality* in actuarial science, as well as the *failure rate* in engineering. A related quantity of interest is the *cumulative hazard function,*

$$H(t) = \int_0^t h(s)ds.$$

This quantity can also be expressed as minus the log survival function, and conversely, $\text{Prob}(Y > t) = \exp(-H(t))$.

Survival models regularly allow for "noninformative" censoring. Thus, define δ to be an indicator function for right-censoring, that is,

$$\delta = \begin{cases} 1 & \text{if } Y \text{ is censored,} \\ 0 & \text{otherwise.} \end{cases}$$

Then, the likelihood of a realization of (Y, δ), say (y, d), can be expressed in terms of the hazard function and cumulative hazard as

$$\begin{cases} \text{Prob}\,(Y > y) & \text{if } Y \text{ is censored,} \\ -\frac{\partial}{\partial y}\,\text{Prob}\,(Y > y) & \text{otherwise} \end{cases}$$
$$= (\text{Prob}\,(Y > y))^d\,(h(y)\,\text{Prob}\,(Y > y))^{1-d}$$
$$= h(y)^{1-d}\exp\,(-H(y))\,.$$

There are two common methods for introducing regression explanatory variables: the *accelerated failure time model* and the *Cox proportional hazard model.* Under the former, one essentially assumes a linear model in the logarithmic time to failure. We refer to any standard treatment of survival models for more discussion of this mechanism. Under the latter, one assumes that the hazard function can be written as the product of some "baseline" hazard and a function of a linear combination of explanatory variables. To illustrate, we use

$$h(t) = h_0(t)\exp(\mathbf{x}_i'\boldsymbol{\beta}), \qquad (11.18)$$

where $h_0(t)$ is the baseline hazard. This is known as a proportional hazards model because if one takes the ratio of hazard functions for any two sets of covariates, say $\mathbf{x}_1$ and $\mathbf{x}_2$, one gets

$$\frac{h(t \mid \mathbf{x}_1)}{h(t \mid \mathbf{x}_2)} = \frac{h_0(t)\exp(\mathbf{x}_1'\boldsymbol{\beta})}{h_0(t)\exp(\mathbf{x}_2'\boldsymbol{\beta})} = \exp((\mathbf{x}_1 - \mathbf{x}_2)'\boldsymbol{\beta}).$$

Note that the ratio is independent of time t.

To express the likelihood function for the Cox model, let H_0 be the cumulative hazard function associated with the baseline hazard function h_0. Let $(Y_1, \delta_1), \ldots, (Y_n, \delta_n)$ be independent and assume that Y_i follows a Cox proportional hazard model with regressors $\mathbf{x}_i$. Then the likelihood is

$$\begin{aligned} L(\boldsymbol{\beta}, h_0) &= \prod_{i=1}^{n} h(Y_i)^{1-\delta_i} \exp\left(-H(Y_i)\right) \\ &= \prod_{i=1}^{n} (h_0(Y_i)\exp(\mathbf{x}_i'\boldsymbol{\beta}))^{1-\delta_i} \exp(-H_0(Y_i)\exp(\mathbf{x}_i'\boldsymbol{\beta})). \end{aligned}$$

Maximizing this in terms of h_0 yields the *partial likelihood*

$$L_{\mathrm{P}}(\boldsymbol{\beta}) = \prod_{i=1}^{n} \left(\frac{\exp(\mathbf{x}_i'\boldsymbol{\beta})}{\sum_{j \in R(Y_i)} \exp(\mathbf{x}_j'\boldsymbol{\beta})} \right)^{1-\delta_i}, \tag{11.19}$$

where $R(t)$ is the set of all $\{Y_1, \ldots, Y_n\}$ such that $Y_i \geq t$, that is, the set of all subjects still under study at time t.

From Equation (11.19), we see that inference for the regression coefficients depends only on the ranks of the dependent variables $\{Y_1, \ldots, Y_n\}$, not their actual values. Moreover, Equation (11.19) suggests (correctly) that large-sample distribution theory has properties similar to the usual desirable (fully) parametric theory. This is mildly surprising because the proportional hazards model is *semiparametric*; in Equation (11.18) the hazard function has a fully parametric component, $\exp(\mathbf{x}_i'\boldsymbol{\beta})$, but it also contains a nonparametric baseline hazard, $h_0(t)$. In general, nonparametric models are more flexible than parametric counterparts for model fitting but result in less desirable large-sample properties (specifically, slower rates of convergence to an asymptotic distribution).

An important feature of the proportional hazards model is that it can readily handle time-dependent covariates of the form $\mathbf{x}_i(t)$. In this case, one can write the partial likelihood as

$$L_{\mathrm{P}}(\boldsymbol{\beta}) = \prod_{i=1}^{n} \left(\frac{\exp(\mathbf{x}_i'(Y_i)\boldsymbol{\beta})}{\sum_{j \in R(Y_i)} \exp(\mathbf{x}_j'(Y_i)\boldsymbol{\beta})} \right)^{1-\delta_i}.$$

Maximization of this likelihood is somewhat complex but can be readily accomplished with modern statistical software.

In summary, there is a large overlap between survival models and the longitudinal and panel data models considered in this text. Survival models are concerned with dependent variables that indicate time until an event of interest whereas the focus of longitudinal and panel data models is broader. Because they concern time, survival models use conditioning arguments extensively in model specification and estimation. Also because of the time element, survival models heavily involve censoring and truncation of variables (it is often more difficult to observe large values of a time variable, other things being equal). Like longitudinal and panel data models, survival models address repeated observations on a subject. Unlike longitudinal and panel data models, survival models also address repeated occurrences of an event (such as marriage). To track events over time, survival models may be expressed in terms of stochastic processes. This formulation allows one to model many complex data patterns of interest. There are many excellent applied introductions to survival modeling; see, for example, Klein and Moeschberger (1997B) and Singer and Willet (2003EP). For a more technical treatment, see Hougaard (2000B).

Appendix 11A Conditional Likelihood Estimation for Multinomial Logit Models with Heterogeneity Terms

To estimate the parameters $\boldsymbol{\beta}$ in the presence of the heterogeneity terms α_{ij}, we may again look to conditional likelihood estimation. The idea is to condition the likelihood using sufficient statistics, which were introduced in Appendix 10A.2.

From Equations (11.4) and (11.11), the log likelihood for the ith subject is

$$\begin{aligned}\ln L(\mathbf{y}_i \mid \alpha_i) &= \sum_t \sum_{j=1}^{c} y_{itj} \ln \pi_{itj} = \sum_t \left\{ \sum_{j=1}^{c-1} y_{itj} \ln \frac{\pi_{itj}}{\pi_{itc}} + \ln \pi_{itc} \right\} \\ &= \sum_t \left\{ \sum_{j=1}^{c-1} y_{itj}(\alpha_{ij} + (\mathbf{x}_{it,j} - \mathbf{x}_{it,c})'\boldsymbol{\beta}) + \ln \pi_{itc} \right\},\end{aligned}$$

because $\ln \pi_{itj}/\pi_{itc} = V_{itj} - V_{itc} = \alpha_{ij} + (\mathbf{x}_{itj} - \mathbf{x}_{itc})'\boldsymbol{\beta}$. Thus, using the factorization theorem in Appendix 10A.2, $\Sigma_t y_{itj}$ is sufficient for α_{ij}. We interpret $\Sigma_t y_{itj}$ to be the number of choices of alternative j in T_i time periods.

To calculate the conditional likelihood, we let S_{ij} be the random variable representing $\Sigma_t y_{itj}$ and let sum_{ij} be the realization of $\Sigma_t y_{itj}$. With this, the distribution of the sufficient statistic is

$$\text{Prob}(S_{ij} = sum_{ij}) = \sum_{B_i} \prod_{j=1}^{c} (\pi_{itj})^{y_{itj}},$$

where B_i is the sum over all sets of the form $\{y_{itj} : \Sigma_t y_{itj} = sum_{ij})$. By sufficiency, we may take $\alpha_{ij} = 0$ without loss of generality. Thus, the conditional likelihood of the ith subject is

$$\frac{L(\mathbf{y}_i \mid \alpha_i)}{\text{Prob}(S_{ij} = sum_{ij})} = \frac{\exp\left(\sum_t \left(\sum_{j=1}^{c-1} y_{itj}(\mathbf{x}_{it,j} - \mathbf{x}_{it,c})'\boldsymbol{\beta} + \ln \pi_{itc}\right)\right)}{\sum_{B_i} \prod_{j=1}^{c} (\pi_{itj})^{y_{itj}}}$$
$$= \frac{\exp\left(\sum_t \left(\sum_{j=1}^{c-1} y_{itj}(\mathbf{x}_{it,j} - \mathbf{x}_{it,c})'\boldsymbol{\beta} + \ln \pi_{itc}\right)\right)}{\sum_{B_i} \exp\left(\sum_t \left(\sum_{j=1}^{c-1} y_{itj}(\mathbf{x}_{it,j} - \mathbf{x}_{it,c})'\boldsymbol{\beta} + \ln \pi_{itc}\right)\right)}.$$

As in Appendix 9A.2, this can be maximized in $\boldsymbol{\beta}$. However, the process is computationally intensive.

Appendix A
Elements of Matrix Algebra

A.1 Basic Terminology

- *Matrix* – a rectangular array of numbers arranged in rows and columns (the plural of matrix is matrices).
- *Dimension* of a matrix – the number of rows and columns of a matrix.
- Consider a matrix $\mathbf{A}$ that has dimension $m \times k$. Let a_{ij} be the symbol for the number in the ith row and jth column of $\mathbf{A}$. In general, we work with matrices of the form

$$\mathbf{A} = \begin{pmatrix} a_{11} & a_{12} & \cdots & a_{1k} \\ a_{21} & a_{22} & \cdots & a_{2k} \\ \vdots & \vdots & \ddots & \vdots \\ a_{m1} & a_{m2} & \cdots & a_{mk} \end{pmatrix}.$$

- *Vector* – a matrix containing only one column ($k = 1$); also called a *column vector.*
- *Row vector* – a matrix containing only one row ($m = 1$).
- *Transpose* – defined by interchanging the rows and columns of a matrix. The transpose of $\mathbf{A}$ is denoted by $\mathbf{A}'$ (or $\mathbf{A}^{\mathrm{T}}$). Thus, if $\mathbf{A}$ has dimension $m \times k$, then $\mathbf{A}'$ has dimension $k \times m$.
- *Square matrix* – a matrix where the number of rows equals the number of columns; that is, $m = k$.
- *Diagonal element* – the number in the rth row and column of a square matrix,$r = 1, 2, \ldots$.
- *Diagonal matrix* – a square matrix where all nondiagonal elements are equal to zero.
- *Identity matrix* – a diagonal matrix where all the diagonal elements are equal to one; denoted by $\mathbf{I}$.
- *Symmetric matrix* – a square matrix $\mathbf{A}$ such that the matrix remains unchanged if we interchange the roles of the rows and columns; that is, if $\mathbf{A} = \mathbf{A}'$. Note that a diagonal matrix is a symmetric matrix.
- *Gradient vector* – a vector of partial derivatives. If $f(.)$ is a function of the vector $\mathbf{x} = (x_1, \ldots, x_m)'$, then the gradient vector is $\partial f(\mathbf{x})/\partial \mathbf{x}$. The ith row of the gradient vector is $\partial f(\mathbf{x})/\partial x_i$.

- *Hessian matrix* – a matrix of second derivatives. If $f(.)$ is a function of the vector $\mathbf{x} = (x_1, \ldots, x_m)'$, then the Hessian matrix is $\partial^2 f(\mathbf{x})/\partial\mathbf{x}\partial\mathbf{x}'$. The element in the ith row and jth column of the Hessian matrix is $\partial^2 f(\mathbf{x})/\partial x_i \partial x_j$.

A.2 Basic Operations

- *Scalar multiplication.* Let c be a real number, called a *scalar*. Multiplying a scalar c by a matrix $\mathbf{A}$ is denoted by $c\,\mathbf{A}$ and defined by

$$c\mathbf{A} = \begin{pmatrix} ca_{11} & ca_{12} & \ldots & ca_{1k} \\ ca_{21} & ca_{22} & \ldots & ca_{2k} \\ \vdots & \vdots & \ddots & \vdots \\ ca_{m1} & ca_{m2} & \ldots & ca_{mk} \end{pmatrix}.$$

- *Matrix addition and subtraction.* Let $\mathbf{A}$ and $\mathbf{B}$ be matrices, each with dimension $m \times k$. Use a_{ij} and b_{ij} to denote the numbers in the ith row and jth column of $\mathbf{A}$ and $\mathbf{B}$, respectively. Then, the matrix $\mathbf{C} = \mathbf{A} + \mathbf{B}$ is defined to be the matrix with the number $(a_{ij} + b_{ij})$ to denote the number in the ith row and jth column. Similarly, the matrix $\mathbf{C} = \mathbf{A} - \mathbf{B}$ is defined to be the matrix with the number $(a_{ij} - b_{ij})$ to denote the numbers in the ith row and jth column.
- *Matrix multiplication.* If $\mathbf{A}$ is a matrix of dimension $m \times c$ and $\mathbf{B}$ is a matrix of dimension $c \times k$, then $\mathbf{C} = \mathbf{AB}$ is a matrix of dimension $m \times k$. The number in the ith row and jth column of $\mathbf{C}$ is $\sum_{s=1}^{c} a_{is} b_{sj}$.
- *Determinant.* The determinant is a function of a square matrix, denoted by det($\mathbf{A}$), or $|\mathbf{A}|$. For a 1×1 matrix, the determinant is det($\mathbf{A}$) = a_{11}. To define determinants for larger matrices, let $\mathbf{A}_{rs}$ be the $(m-1) \times (m-1)$ submatrix of $\mathbf{A}$ defined be removing the rth row and sth column. Recursively, define $\det(\mathbf{A}) = \sum_{s=1}^{m} (-1)^{r+s} a_{rs} \det(\mathbf{A}_{rs})$, for any $r = 1, \ldots, m$. For example, for $m = 2$, we have $\det(\mathbf{A}) = a_{11}a_{22} - a_{12}a_{21}$.
- *Matrix inverse.* In matrix algebra, there is no concept of division. Instead, we extend the concept of reciprocals of real numbers. To begin, suppose that $\mathbf{A}$ is a square matrix of dimension $m \times m$ such that det($\mathbf{A}$) $\neq 0$. Further, let $\mathbf{I}$ be the $m \times m$ identity matrix. If there exists a $m \times m$ matrix $\mathbf{B}$ such that $\mathbf{AB} = \mathbf{I} = \mathbf{BA}$, then $\mathbf{B}$ is called the *inverse* of $\mathbf{A}$ and is written $\mathbf{B} = \mathbf{A}^{-1}$.

A.3 Additional Definitions

- *Linearly dependent vectors* – a set of vectors $\mathbf{c}_1, \ldots, \mathbf{c}_k$ is said to be linearly dependent if one of the vectors in the set can be written as a linear combination of the others.
- *Linearly independent vectors* – a set of vectors $\mathbf{c}_1, \ldots, \mathbf{c}_k$ is said to be linearly independent if they are not linearly dependent. Specifically, a set of vectors $\mathbf{c}_1, \ldots, \mathbf{c}_k$ is said to be linearly independent if and only if the only solution of the equation $x_1\mathbf{c}_1 + \cdots + x_k\mathbf{c}_k = 0$ is $x_1 = \cdots = x_k = 0$.

- *Rank of a matrix* – the largest number of linearly independent columns (or rows) of a matrix.
- *Singular matrix* – a square matrix $\mathbf{A}$ such that $\det(\mathbf{A}) = 0$.
- *Nonsingular matrix* – a square matrix $\mathbf{A}$ such that $\det(\mathbf{A}) \neq 0$.
- *Positive definite matrix* – a symmetric square matrix $\mathbf{A}$ such that $\mathbf{x}'\mathbf{A}\mathbf{x} > 0$ for $\mathbf{x} \neq \mathbf{0}$.
- *Nonnegative definite matrix* – a symmetric square matrix $\mathbf{A}$ such that $\mathbf{x}'\mathbf{A}\mathbf{x} \geq 0$ for $\mathbf{x} \neq \mathbf{0}$.
- *Orthogonal* – two matrices $\mathbf{A}$ and $\mathbf{B}$ are orthogonal if $\mathbf{A}'\mathbf{B} = \mathbf{0}$, a zero matrix.
- *Idempotent* – a square matrix such that $\mathbf{AA} = \mathbf{A}$.
- *Trace* – the sum of all diagonal elements of a square matrix.
- *Eigenvalues* – the solutions of the nth-degree polynomial $\det(\mathbf{A} - \lambda\mathbf{I}) = 0$; also known as *characteristic roots* and *latent roots*.
- *Eigenvector* – a vector $\mathbf{x}$ such that $\mathbf{Ax} = \lambda\mathbf{x}$, where λ is an eigenvalue of $\mathbf{A}$; also known as a *characteristic vector* and *latent vector*.
- The *generalized inverse* of a matrix $\mathbf{A}$ is a matrix $\mathbf{B}$ such that $\mathbf{ABA} = \mathbf{A}$. We use the notation $\mathbf{A}^-$ to denote the generalized inverse of $\mathbf{A}$. In the case that $\mathbf{A}$ is invertible, then $\mathbf{A}^-$ is unique and equals $\mathbf{A}^{-1}$. Although there are several definitions of generalized inverses, this definition suffices for our purposes. See Searle (1987G) for further discussion of alternative definitions of generalized inverses.

A.4 Matrix Decompositions

- Let $\mathbf{A}$ be an $m \times m$ symmetric matrix. Then $\mathbf{A}$ has m pairs of eigenvalues and eigenvectors $(\lambda_1, \mathbf{e}_1), \ldots, (\lambda_m, \mathbf{e}_m)$. The eigenvectors can be chosen to have unit length, $\mathbf{e}_j'\mathbf{e}_j = 1$, and to be orthogonal to one another, $\mathbf{e}_i'\mathbf{e}_j = 0$, for $i \neq j$. The eigenvectors are unique unless two or more eigenvalues are equal.
- *Spectral decomposition*. Let $\mathbf{A}$ be an $m \times m$ symmetric matrix. Then we can write $\mathbf{A} = \lambda_1\mathbf{e}_1\mathbf{e}_1' + \cdots + \lambda_m\mathbf{e}_m\mathbf{e}_m'$, where the eigenvectors have unit length and are mutually orthogonal.
- Suppose that $\mathbf{A}$ is positive definite. Then, each eigenvalue is positive. Similarly, if $\mathbf{A}$ is nonnegative definite, then each eigenvalue is nonnegative.
- Using the spectral decomposition, we may write an $m \times m$ symmetric matrix $\mathbf{A}$ as $\mathbf{A} = \mathbf{P}\boldsymbol{\Lambda}\mathbf{P}'$, where $\mathbf{P} = [\mathbf{e}_1 : \ldots : \mathbf{e}_m]$ and $\boldsymbol{\Lambda} = \mathrm{diag}(\lambda_1, \ldots, \lambda_m)$.
- *Square-root matrix*. For a nonnegative definite matrix $\mathbf{A}$, we may define the square-root matrix as $\mathbf{A}^{1/2} = \mathbf{P}\boldsymbol{\Lambda}^{1/2}\mathbf{P}'$, where $\boldsymbol{\Lambda}^{1/2} = \mathrm{diag}(\lambda_1^{1/2}, \ldots, \lambda_m^{1/2})$. The matrix $\mathbf{A}^{1/2}$ is symmetric and is such that $\mathbf{A}^{1/2}\mathbf{A}^{1/2} = \mathbf{A}$.
- *Matrix power*. For a positive definite matrix $\mathbf{A}$, we may define $\mathbf{A}^c = \mathbf{P}\boldsymbol{\Lambda}^c\mathbf{P}'$, where $\boldsymbol{\Lambda}^c = \mathrm{diag}(\lambda_1^c, \ldots, \lambda_m^c)$, for any scalar c.
- *Cholesky factorization*. Suppose that $\mathbf{A}$ is positive definite. Then, we may write $\mathbf{A} = \mathbf{LL}'$, where $\mathbf{L}$ is a lower triangular matrix ($l_{ij} = 0$ for $i < j$). Further, $\mathbf{L}$ is invertible so that $\mathbf{A}^{-1} = (\mathbf{L}')^{-1}\mathbf{L}^{-1}$. $\mathbf{L}$ is known as the *Cholesky square-root matrix*.
- Let $\mathbf{A}$ be an $m \times k$ matrix. Then, there exists an $m \times m$ orthogonal matrix $\mathbf{U}$ and a $k \times k$ orthogonal matrix $\mathbf{V}$ such that $\mathbf{A} = \mathbf{U}\boldsymbol{\Lambda}\mathbf{V}'$. Here, $\boldsymbol{\Lambda}$ is an $m \times k$ matrix such that the (i, i) entry of $\boldsymbol{\Lambda}$ is $\lambda_i \geq 0$ for $i = 1, \ldots, \min(m, k)$ and the other entries are zero. The elements λ_i are called the *singular values* of $\mathbf{A}$.

- *Singular value decomposition*. For $\lambda_i > 0$, let $\mathbf{u}_i$ and $\mathbf{v}_i$ be the corresponding columns of $\mathbf{U}$ and $\mathbf{V}$, respectively. The *singular value decomposition* of $\mathbf{A}$ is $\mathbf{A} = \sum_{i=1}^{r} \lambda_i \mathbf{u}_i \mathbf{v}_i'$, where r is the rank of $\mathbf{A}$.
- *QR decomposition*. Let $\mathbf{A}$ be an $m \times k$ matrix, with $m \geq k$ of rank k. Then, there exists an $m \times m$ orthogonal matrix $\mathbf{Q}$ and a $k \times k$ upper triangular matrix $\mathbf{R}$ such that

$$\mathbf{A} = \mathbf{Q}\begin{pmatrix} \mathbf{R} \\ \mathbf{0} \end{pmatrix}.$$

We can also write $\mathbf{A} = \mathbf{Q}_k\mathbf{R}$, where $\mathbf{Q}_k$ consists of the first k columns of $\mathbf{Q}$.

A.5 Partitioned Matrices

- A standard result on inverses of partitioned matrices is

$$\begin{pmatrix} \mathbf{B}_{11} & \mathbf{B}_{12} \\ \mathbf{B}_{21} & \mathbf{B}_{22} \end{pmatrix}^{-1} = \begin{pmatrix} \mathbf{C}_{11}^{-1} & -\mathbf{B}_{11}^{-1}\mathbf{B}_{12}\mathbf{C}_{22}^{-1} \\ -\mathbf{C}_{22}^{-1}\mathbf{B}_{21}\mathbf{B}_{11}^{-1} & \mathbf{C}_{22}^{-1} \end{pmatrix}, \tag{A.1}$$

where $\mathbf{C}_{11} = \mathbf{B}_{11} - \mathbf{B}_{12}\mathbf{B}_{22}^{-1}\mathbf{B}_{21}$ and $\mathbf{C}_{22} = \mathbf{B}_{22} - \mathbf{B}_{21}\mathbf{B}_{11}^{-1}\mathbf{B}_{12}$.
- A related result on determinants of partitioned matrices is

$$\det\begin{pmatrix} \mathbf{B}_{11} & \mathbf{B}_{12} \\ \mathbf{B}_{21} & \mathbf{B}_{22} \end{pmatrix} = \det \mathbf{B}_{11} \det \mathbf{C}_{22} = \det \mathbf{B}_{11} \det\left(\mathbf{B}_{22} - \mathbf{B}_{21}\mathbf{B}_{11}^{-1}\mathbf{B}_{12}\right). \tag{A.2}$$

- Another standard result on inverses of partitioned matrices is

$$\left(\mathbf{B}_{11} - \mathbf{B}_{12}\mathbf{B}_{22}^{-1}\mathbf{B}_{21}\right)^{-1} = \mathbf{B}_{11}^{-1} + \mathbf{B}_{11}^{-1}\mathbf{B}_{12}\left(\mathbf{B}_{22} - \mathbf{B}_{21}\mathbf{B}_{11}^{-1}\mathbf{B}_{12}\right)^{-1}\mathbf{B}_{21}\mathbf{B}_{11}^{-1}. \tag{A.3}$$

- To illustrate, with $\mathbf{R} = \mathbf{B}_{11}$, $\mathbf{Z} = \mathbf{B}_{12}$, $-\mathbf{Z}' = \mathbf{B}_{21}$, and $\mathbf{D}^{-1} = \mathbf{B}_{22}$, we have

$$\mathbf{V}^{-1} = (\mathbf{R} + \mathbf{ZDZ}')^{-1} = \mathbf{R}^{-1} - \mathbf{R}^{-1}\mathbf{Z}(\mathbf{D}^{-1} + \mathbf{Z}'\mathbf{R}^{-1}\mathbf{Z})^{-1}\mathbf{Z}'\mathbf{R}^{-1}. \tag{A.4}$$

- The related determinant result is

$$\ln\det \mathbf{V} = \ln\det \mathbf{R} - \ln\det \mathbf{D}^{-1} + \ln\det(\mathbf{D}^{-1} + \mathbf{Z}'\mathbf{R}^{-1}\mathbf{Z})^{-1}. \tag{A.5}$$

- Suppose that $\mathbf{A}$ is an invertible $p \times p$ matrix and $\mathbf{c}$ and $\mathbf{d}$ are $p \times 1$ vectors. Then from, for example, Graybill (1969G, Theorem 8.9.3) we have

$$(\mathbf{A} + \mathbf{cd}')^{-1} = \mathbf{A}^{-1} - \frac{\mathbf{A}^{-1}\mathbf{cd}'\mathbf{A}^{-1}}{1 + \mathbf{d}'\mathbf{A}^{-1}\mathbf{c}}. \tag{A.6}$$

To check this result, simply multiply $\mathbf{A} + \mathbf{cd}'$ by the right-hand side to get $\mathbf{I}$, the identity matrix.

- Let $\mathbf{P}$ and $\mathbf{Q}$ be idempotent and orthogonal matrices. Let a and b be positive contants. Then,

$$(a\mathbf{P} + b\mathbf{Q})^c = a^c\mathbf{P} + b^c\mathbf{Q}, \tag{A.7}$$

for scalar c. (Baltagi, 2001E).

A.6 Kronecker (Direct) Product

Let $\mathbf{A}$ be an $m \times n$ matrix and $\mathbf{B}$ be an $m_1 \times n_2$ matrix. The direct product is defined to be

$$\mathbf{A} \otimes \mathbf{B} = \begin{pmatrix} a_{11}\mathbf{B} & a_{12}\mathbf{B} & \cdots & a_{1n}\mathbf{B} \\ a_{21}\mathbf{B} & a_{22}\mathbf{B} & \cdots & a_{2n}\mathbf{B} \\ \vdots & \vdots & \ddots & \vdots \\ a_{m1}\mathbf{B} & a_{m2}\mathbf{B} & \cdots & a_{mn}\mathbf{B} \end{pmatrix},$$

an $mm_1 \times nn_2$ matrix. Direct products have the following properties:

$$(\mathbf{A} \otimes \mathbf{B})(\mathbf{F} \otimes \mathbf{G}) = (\mathbf{AF}) \otimes (\mathbf{BG})$$

and

$$(\mathbf{A} \otimes \mathbf{B})^{-1} = \mathbf{A}^{-1} \otimes \mathbf{B}^{-1}.$$

See Graybill (1969G, Chapter 8).

Appendix B

Normal Distribution

B.1 Univariate Normal Distribution

Recall that the probability density function of $y \sim N(\mu, \sigma^2)$ is given by

$$f(y) = \frac{1}{\sigma\sqrt{2\pi}} \exp\left(-\frac{(y-\mu)^2}{2\sigma^2}\right).$$

If $\mu = 0$ and $\sigma = 1$, then $y \sim N(0, 1)$ is said to be *standard normal*. The standard normal probability density function is

$$\phi(y) = \frac{1}{\sqrt{2\pi}} \exp\left(-\frac{y^2}{2}\right).$$

The corresponding standard normal distribution function is denoted by $\Phi(y) = \int_{-\infty}^{y} \phi(z)dz$.

B.2 Multivariate Normal Distribution

A vector of random variables $\mathbf{y} = (y_1\ y_2 \ldots y_n)'$ is said to be multivariate normal if all linear combinations of the form $\mathbf{a}'\mathbf{y} = \Sigma_i a_i y_i$ are normally distributed, where the a_is are constants. In this case, we write $\mathbf{y} \sim N(\boldsymbol{\mu}, \mathbf{V})$, where $\boldsymbol{\mu} = \mathrm{E}\mathbf{y}$ is the expected value of $\mathbf{y}$ and $\mathbf{V} = \mathrm{Var}\,\mathbf{y}$ is the variance–covariance matrix of $\mathbf{y}$. From the definition, we have that $\mathbf{y} \sim N(\boldsymbol{\mu}, \mathbf{V})$ implies that $\mathbf{a}'\mathbf{y} \sim N(\mathbf{a}'\boldsymbol{\mu}, \mathbf{a}'\mathbf{V}\mathbf{a})$.

The multivariate probability density function of $\mathbf{y} \sim N(\boldsymbol{\mu}, \mathbf{V})$ is given by

$$\begin{aligned} f(\mathbf{y}) &= f(y_1, \ldots, y_n) \\ &= (2\pi)^{-n/2} (\det \mathbf{V})^{-1/2} \exp\left(-\frac{1}{2}(\mathbf{y}-\boldsymbol{\mu})'\mathbf{V}^{-1}(\mathbf{y}-\boldsymbol{\mu})\right). \end{aligned}$$

For mixed linear models, the mean is a function of linear combinations of parameters such that $\boldsymbol{\mu} = \mathbf{X}\boldsymbol{\beta}$. Thus, the probability density function of $\mathbf{y} \sim N(\mathbf{X}\boldsymbol{\beta}, \mathbf{V})$ is given by

$$\begin{aligned} f(\mathbf{y}) &= f(y_1, \ldots, y_n) \\ &= (2\pi)^{-n/2} (\det \mathbf{V})^{-1/2} \exp\left(-\frac{1}{2} (\mathbf{y} - \mathbf{X}\boldsymbol{\beta})' \mathbf{V}^{-1} (\mathbf{y} - \mathbf{X}\boldsymbol{\beta}) \right). \end{aligned}$$

B.3 Normal Likelihood

A logarithmic probability density function evaluated using the observations is known as a *log likelihood*. Suppose that this density depends on the mean parameters $\boldsymbol{\beta}$ and variance components $\boldsymbol{\tau}$. Then, the log likelihood for the multivariate normal can be expressed as

$$L(\boldsymbol{\beta}, \boldsymbol{\tau}) = -\frac{1}{2}(n \ln (2\pi) + \ln (\det \mathbf{V}) + (\mathbf{y} - \mathbf{X}\boldsymbol{\beta})' \mathbf{V}^{-1} (\mathbf{y} - \mathbf{X}\boldsymbol{\beta})). \qquad \text{(B.1)}$$

B.4 Conditional Distributions

Suppose that $(\mathbf{y}_1', \mathbf{y}_2')'$ is a multivariate normally distributed vector such that

$$\begin{pmatrix} \mathbf{y}_1 \\ \mathbf{y}_2 \end{pmatrix} \sim N\left(\begin{pmatrix} \boldsymbol{\mu}_1 \\ \boldsymbol{\mu}_2 \end{pmatrix}, \begin{pmatrix} \boldsymbol{\Sigma}_{11} & \boldsymbol{\Sigma}_{12} \\ \boldsymbol{\Sigma}_{12}' & \boldsymbol{\Sigma}_{22} \end{pmatrix} \right).$$

Then, the conditional distribution of $\mathbf{y}_1 \mid \mathbf{y}_2$ is also normal. Specifically, we have

$$\mathbf{y}_1 \mid \mathbf{y}_2 \sim N\left(\boldsymbol{\mu}_1 + \boldsymbol{\Sigma}_{12}\boldsymbol{\Sigma}_{22}^{-1}(\mathbf{y}_2 - \boldsymbol{\mu}_2),\ \boldsymbol{\Sigma}_{11} - \boldsymbol{\Sigma}_{12}\boldsymbol{\Sigma}_{22}^{-1}\boldsymbol{\Sigma}_{12}' \right). \qquad \text{(B.2)}$$

Thus, $\mathrm{E}(\mathbf{y}_1 \mid \mathbf{y}_2) = \boldsymbol{\mu}_1 + \boldsymbol{\Sigma}_{12}\boldsymbol{\Sigma}_{22}^{-1}(\mathbf{y}_2 - \boldsymbol{\mu}_2)$ and $\mathrm{Var}\,(\mathbf{y}_1 \mid \mathbf{y}_2) = \boldsymbol{\Sigma}_{11} - \boldsymbol{\Sigma}_{12}\boldsymbol{\Sigma}_{22}^{-1}\boldsymbol{\Sigma}_{12}'$.

Appendix C

Likelihood-Based Inference

Begin with a random vector $\mathbf{y}$ whose joint distribution is known up to a vector of parameters $\boldsymbol{\theta}$. This joint probability density (mass) function is denoted as $p(\mathbf{y}; \boldsymbol{\theta})$. The log-likelihood function is $\ln p(\mathbf{y}; \boldsymbol{\theta}) = L(\mathbf{y}; \boldsymbol{\theta}) = L(\boldsymbol{\theta})$, when evaluated at a realization of $\mathbf{y}$. That is, the log-likelihood is a function of the parameters with the data fixed rather than a function of the data with the parameters fixed.

C.1 Characteristics of Likelihood Functions

Two basic characteristics of likelihood functions are as follows:

$$\mathrm{E}\left(\frac{\partial}{\partial \boldsymbol{\theta}} L(\boldsymbol{\theta})\right) = \mathbf{0} \tag{C.1}$$

and

$$\mathrm{E}\left(\frac{\partial^2}{\partial \boldsymbol{\theta} \partial \boldsymbol{\theta}'} L(\boldsymbol{\theta})\right) + \mathrm{E}\left(\frac{\partial L(\boldsymbol{\theta})}{\partial \boldsymbol{\theta}} \frac{\partial L(\boldsymbol{\theta})}{\partial \boldsymbol{\theta}'}\right) = \mathbf{0}. \tag{C.2}$$

The derivative of the log-likelihood function, $\partial L(\boldsymbol{\theta})/\partial \boldsymbol{\theta}$, is called the *score function*. From Equation (C.1), we see that it has mean zero. To see Equation (C.1), under suitable regularity conditions, we have

$$\begin{aligned}
\mathrm{E}\left(\frac{\partial}{\partial \boldsymbol{\theta}} L(\boldsymbol{\theta})\right) &= \mathrm{E}\left(\frac{\frac{\partial}{\partial \boldsymbol{\theta}} p(\mathbf{y}; \boldsymbol{\theta})}{p(\mathbf{y}; \boldsymbol{\theta})}\right) = \int \frac{\partial}{\partial \boldsymbol{\theta}} p(\mathbf{y}; \boldsymbol{\theta})\, d\mathbf{y} = \frac{\partial}{\partial \boldsymbol{\theta}} \int p(\mathbf{y}; \boldsymbol{\theta})\, d\mathbf{y} \\
&= \frac{\partial}{\partial \boldsymbol{\theta}} 1 = \mathbf{0}.
\end{aligned}$$

For convenience, this demonstration assumes a density for $\mathbf{y}$; extensions to mass and mixtures are straightforward. The proof of Equation (C.2) is similar and is omitted. Some "suitable regularity conditions" are required to allow the interchange of the derivative and integral sign.

Using Equation (C.2), we may define the *information matrix*

$$I(\boldsymbol{\theta}) = \mathrm{E}\left(\frac{\partial L(\boldsymbol{\theta})}{\partial \boldsymbol{\theta}}\frac{\partial L(\boldsymbol{\theta})}{\partial \boldsymbol{\theta}'}\right) = -\mathrm{E}\left(\frac{\partial^2}{\partial \boldsymbol{\theta}\partial \boldsymbol{\theta}'}L(\boldsymbol{\theta})\right).$$

This quantity is used in the scoring algorithm for parameter estimation.

Under broad conditions, we have that $\partial L(\boldsymbol{\theta})/\partial \boldsymbol{\theta}$ is asymptotically normal with mean $\mathbf{0}$ and variance $I(\boldsymbol{\theta})$.

C.2 Maximum Likelihood Estimators

Maximum likelihood estimators are values of the parameters $\boldsymbol{\theta}$ that are "most likely" to have been produced by the data. Consider random variables $(y_1, \ldots, y_n)' = \mathbf{y}$ with probability function $p(\mathbf{y}; \boldsymbol{\theta})$. The value of $\boldsymbol{\theta}$, say $\boldsymbol{\theta}_{\mathrm{MLE}}$, that maximizes $p(\mathbf{y}; \boldsymbol{\theta})$ is called the *maximum likelihood estimator*. We may also determine $\boldsymbol{\theta}_{\mathrm{MLE}}$ by maximizing $L(\boldsymbol{\theta})$, the log-likelihood function. In many applications, this can be done by finding roots of the score function, $\partial L(\boldsymbol{\theta})/\partial \boldsymbol{\theta}$.

Under broad conditions, we have that $\boldsymbol{\theta}_{\mathrm{MLE}}$ is asymptotically normal with mean $\boldsymbol{\theta}$ and variance $(I(\boldsymbol{\theta}))^{-1}$. Moreover, maximum likelihood estimators are the most efficient in the following sense. Suppose that $\hat{\boldsymbol{\theta}}$ is an alternative unbiased estimator. The Cramer–Rao theorem states that, under mild regularity conditions, for all vectors $\mathbf{c}$, Var $\mathbf{c}'\boldsymbol{\theta}_{\mathrm{MLE}} \leq$ Var $\mathbf{c}'\hat{\theta}$, for sufficiently large n.

We also note that $2(L(\boldsymbol{\theta}_{\mathrm{MLE}}) - \mathrm{L}(\boldsymbol{\theta}))$ has a chi-square distribution with degrees of freedom equal to the dimension of $\boldsymbol{\theta}$.

Example: One-Parameter Exponential Family Let $y_1, \ldots, y_n$ be independent draws from a one-parameter exponential family distribution as in Equation (9.1),

$$p(y, \theta, \phi) = \exp\left(\frac{y\theta - b(\theta)}{\phi} + S(y, \phi)\right). \tag{C.3}$$

The score function is

$$\frac{\partial L(\theta)}{\partial \theta} = \frac{\partial}{\partial \theta}\sum_{i=1}^{n}\ln p(y_i, \theta, \phi) = \frac{\partial}{\partial \theta}\sum_{i=1}^{n}\left(\frac{y_i\theta - b(\theta)}{\phi} + S(y_i, \phi)\right)$$
$$= \sum_{i=1}^{n}\left(\frac{y_i - b'(\theta)}{\phi}\right) = \frac{n(\bar{y} - b'(\theta))}{\phi}.$$

Thus, setting this equal to zero yields $\bar{y} = b'(\theta_{\mathrm{MLE}})$, or $\theta_{\mathrm{MLE}} = b'^{-1}(\bar{y})$. The information matrix is

$$I(\theta) = -\mathrm{E}\left(\frac{\partial^2}{\partial \theta^2}L(\theta)\right) = \frac{nb''(\theta)}{\phi}.$$

In general, maximum likelihood estimators are determined iteratively. For general likelihoods, two basic procedures are used:

- *Newton–Raphson* uses the iterative algorithm

$$\boldsymbol{\theta}_{\text{NEW}} = \boldsymbol{\theta}_{\text{OLD}} - \left.\left\{ \left(\frac{\partial^2 L}{\partial \boldsymbol{\theta}\, \partial \boldsymbol{\theta}'} \right)^{-1} \frac{\partial L}{\partial \boldsymbol{\theta}} \right\}\right|_{\theta=\boldsymbol{\theta}_{\text{OLD}}}.$$

- *Fisher scoring* uses the iterative algorithm

$$\boldsymbol{\theta}_{\text{NEW}} = \boldsymbol{\theta}_{\text{OLD}} + I(\boldsymbol{\theta}_{\text{OLD}})^{-1} \left.\left\{ \frac{\partial L}{\partial \boldsymbol{\theta}} \right\}\right|_{\theta=\boldsymbol{\theta}_{\text{OLD}}},$$

  where $I(\boldsymbol{\theta})$ is the information matrix.

Example: Generalized Linear Model Let $y_1, \ldots, y_n$ be independent draws from a one-parameter exponential family with distribution in Equation (C.3). Suppose that the random variable y_i has mean μ_i with systematic component $\eta_i = \mathrm{g}(\mu_i) = \mathbf{x}_i'\boldsymbol{\beta}$ and canonical link so that $\eta_i = \theta_i$. Assume that the scale parameter varies by observation so that $\phi_i = \phi / w_i$, where w_i is a known weight function. Then, the score function is

$$\begin{aligned}\frac{\partial L(\boldsymbol{\beta})}{\partial \boldsymbol{\beta}} &= \frac{\partial}{\partial \boldsymbol{\beta}} \sum_{i=1}^{n} \left(\frac{y_i \mathbf{x}_i'\boldsymbol{\beta} - b(\mathbf{x}_i'\boldsymbol{\beta})}{\phi_i} + S(y_i, \phi) \right) \\ &= \sum_{i=1}^{n} w_i \left(\frac{y_i - b'(\mathbf{x}_i'\boldsymbol{\beta})}{\phi} \right) \mathbf{x}_i.\end{aligned}$$

The matrix of second derivatives is

$$\frac{\partial^2 L(\boldsymbol{\beta})}{\partial \boldsymbol{\beta} \partial \boldsymbol{\beta}'} = - \sum_{i=1}^{n} w_i b''(\mathbf{x}_i'\boldsymbol{\beta}) \frac{\mathbf{x}_i \mathbf{x}_i'}{\phi}. \tag{C.4}$$

Thus, the Newton–Raphson algorithm is

$$\boldsymbol{\beta}_{\text{NEW}} = \boldsymbol{\beta}_{\text{OLD}} - \left(\sum_{i=1}^{n} w_i \mathrm{b}''(\mathbf{x}_i'\boldsymbol{\beta}_{\text{OLD}}) \mathbf{x}_i \mathbf{x}_i' \right)^{-1} \left(\sum_{i=1}^{n} w_i (y_i - b'(\mathbf{x}_i'\boldsymbol{\beta}_{\text{OLD}})) \mathbf{x}_i \right).$$

Because the matrix of second derivatives is nonstochastic, we have that

$$\frac{\partial^2 L(\boldsymbol{\beta})}{\partial \boldsymbol{\beta} \partial \boldsymbol{\beta}'} = I(\boldsymbol{\beta})$$

and thus the Newton–Raphson algorithm is equivalent to the Fisher scoring algorithm.

C.3 Iterated Reweighted Least Squares

Continue with the prior example concerning the generalized linear model and define an "adjusted dependent variable"

$$y_i^*(\boldsymbol{\beta}) = \mathbf{x}_i'\boldsymbol{\beta} + \frac{y_i - b'(\mathbf{x}_i'\boldsymbol{\beta})}{b''(\mathbf{x}_i'\boldsymbol{\beta})}.$$

This has variance

$$\mathrm{Var}[y_i^*(\beta)] = \frac{\mathrm{Var[y_i]}}{(b''(\mathbf{x}_i'\beta))^2} = \frac{\phi_i b''(\mathbf{x}_i'\beta)}{(b''(\mathbf{x}_i'\beta))^2} = \frac{\phi / w_i}{b''(\mathbf{x}_i'\beta)}.$$

Use the new weight as the reciprocal of the variance, $w_i(\beta) = w_i b''(\mathbf{x}_i'\beta)/\phi$. Then, with the expression

$$w_i(y_i - b'(\mathbf{x}_i'\beta)) = w_i b''(\mathbf{x}_i'\beta)(y_i^*(\beta) - \mathbf{x}_i'\beta) = \phi w_i(\beta)\,(y_i^*(\beta) - \mathbf{x}_i'\beta),$$

from the Newton–Raphson iteration, we have

$$\begin{aligned}
\beta_{\mathrm{NEW}} &= \beta_{\mathrm{OLD}} - \left(\sum_{i=1}^{n} w_i b''(\mathbf{x}_i'\beta_{\mathrm{OLD}})\mathbf{x}_i\mathbf{x}_i'\right)^{-1}\left(\sum_{i=1}^{n} w_i(y_i - b'(\mathbf{x}_i'\beta_{\mathrm{OLD}}))\mathbf{x}_i\right)\\
&= \beta_{\mathrm{OLD}} - \left(\sum_{i=1}^{n} \phi w_i(\beta_{\mathrm{OLD}})\mathbf{x}_i\mathbf{x}_i'\right)^{-1}\\
&\quad\times\left(\sum_{i=1}^{n} \phi w_i(\beta_{\mathrm{OLD}})(y_i^*(\beta_{\mathrm{OLD}}) - \mathbf{x}_i'\beta_{\mathrm{OLD}})\mathbf{x}_i\right)\\
&= \beta_{\mathrm{OLD}} - \left(\sum_{i=1}^{n} w_i(\beta_{\mathrm{OLD}})\mathbf{x}_i\mathbf{x}_i'\right)^{-1}\\
&\quad\times\left(\sum_{i=1}^{n} w_i(\beta_{\mathrm{OLD}})\mathbf{x}_i y_i^*(\beta_{\mathrm{OLD}}) - \sum_{i=1}^{n} w_i(\beta_{\mathrm{OLD}})\mathbf{x}_i\mathbf{x}_i'\beta_{\mathrm{OLD}}\right)\\
&= \left(\sum_{i=1}^{n} w_i(\beta_{\mathrm{OLD}})\mathbf{x}_i\mathbf{x}_i'\right)^{-1}\left(\sum_{i=1}^{n} w_i(\beta_{\mathrm{OLD}})\mathbf{x}_i y_i^*(\beta_{\mathrm{OLD}})\right).
\end{aligned}$$

Thus, this provides a method for iteration using weighted least squares.

C.4 Profile Likelihood

Split the vector of parameters into two components, say, $\boldsymbol{\theta} = (\boldsymbol{\theta}_1', \boldsymbol{\theta}_2')'$. Here, interpret $\boldsymbol{\theta}_1$ to be the parameters of interest, whereas $\boldsymbol{\theta}_2$ are auxiliary, or "nuisance," parameters of secondary interest. Let $\boldsymbol{\theta}_{2,\mathrm{MLE}}(\boldsymbol{\theta}_1)$ be the maximum likelihood estimator of $\boldsymbol{\theta}_2$ for a fixed value of $\boldsymbol{\theta}_1$. Then, the profile likelihood for $\boldsymbol{\theta}_1$ is $p(\boldsymbol{\theta}_1, \boldsymbol{\theta}_{2,\mathrm{MLE}}; \mathbf{y}) = \sup_{\boldsymbol{\theta}_2} p(\boldsymbol{\theta}_1, \boldsymbol{\theta}_2; \mathbf{y})$. This is not a likelihood in the usual sense. As an illustration, it is straightforward to check that Equation (C.1) does not hold for $p(\boldsymbol{\theta}_1, \boldsymbol{\theta}_{2,\mathrm{MLE}}; \mathbf{y})$.

C.5 Quasi-Likelihood

Suppose that y is distributed according to the one-parameter exponential family in Equation (C.3). Then, $\mathrm{E}y = \mu = b'(\theta)$ and $\mathrm{Var}\ y = \phi b''(\theta)$. Thus,

$$\frac{\partial \mu}{\partial \theta} = \frac{\partial b'(\theta)}{\partial \theta} = b''(\theta) = \mathrm{Var}\ y/\phi.$$

Using the chain rule, we have

$$\frac{\partial}{\partial \mu} \ln p(y,\theta,\phi) = \frac{\partial \theta}{\partial \mu}\frac{\partial}{\partial \theta} \ln p(y,\theta,\phi) = \frac{\phi}{\text{Var } y}\frac{(y-b'(\theta))}{\phi} = \frac{y-\mu}{\text{V}(\mu)}.$$

Here, we have explicitly denoted the variance of y as a function of the mean μ by using the notation Var $y = \text{V}(\mu)$. We write $t(y,\mu) = (y-\mu)/\text{V}(\mu)$, a function of y and μ. Similar to the score function, this function has the following properties:

$$\text{E}\, t(y,\mu) = 0$$

and

$$-\text{E}\frac{\partial}{\partial \mu} t(y,\mu) = \frac{1}{\text{V}(\mu)} = \text{Var } t(y,\mu).$$

Since these properties are the ones that are critical for the asymptotics of likelihood analysis, we may think of $Q(y,\mu) = \int_y^\mu t(y,s)ds$ as a quasi–log likelihood. The parameter μ is known up to a finite dimension vector of parameters, $\boldsymbol{\beta}$, and thus we write $\mu(\boldsymbol{\beta})$ for μ. Thus, the quasi-score function is

$$\frac{\partial Q(y,\mu)}{\partial \boldsymbol{\beta}} = t(y,\mu(\boldsymbol{\beta}))\frac{\partial \mu}{\partial \boldsymbol{\beta}}.$$

Estimation proceeds as in the likelihood case. That is, in many applications we assume that $\{y_1,\ldots,y_n\}$ are n independent observations with mean μ_i and variance $\text{V}(\mu_i)$. Then, the quasi-score function is

$$\sum_{i=1}^{n} \frac{\partial Q(y_i,\mu_i)}{\partial \boldsymbol{\beta}} = \sum_{i=1}^{n} t(y_i,\mu_i(\boldsymbol{\beta}))\frac{\partial \mu_i}{\partial \boldsymbol{\beta}}.$$

C.6 Estimating Equations

An alternative method for parameter estimation uses the notion of an "estimating equation," which extends the idea of moment estimation. In the following, we summarize treatments due to McCullagh and Nelder (1989G, Chapter 9) and Diggle et al. (2002S). From an econometrics perspective where this procedure is known as the "generalized method of moments," see Hayashi (2000E).

An estimating function is a function $\mathbf{g}(.)$ of $\mathbf{y}$, an $n \times 1$ vector of random variables, and $\boldsymbol{\theta}$, a $p \times 1$ vector of parameters such that

$$\text{E}\mathbf{g}(\mathbf{y};\boldsymbol{\theta}) = \mathbf{0}_n \quad \text{for all } \boldsymbol{\theta} \text{ in a parameter space of interest,} \tag{C.5}$$

where $\mathbf{0}_n$ is an $n \times 1$ vector of zeros. For example, in many applications, we take $\mathbf{g}(\mathbf{y};\boldsymbol{\theta}) = \mathbf{y} - \boldsymbol{\mu}(\boldsymbol{\theta})$, where $\text{E}\mathbf{y} = \boldsymbol{\mu}(\boldsymbol{\theta})$. The choice of the $\mathbf{g}$ function is critical in applications. In econometrics, Equation (C.5) is known as the "moment condition."

Let $\mathbf{H}$ be an $n \times p$ matrix and define the estimator as the solution of the equation

$$\mathbf{H}'\mathbf{g}(\mathbf{y};\boldsymbol{\theta}) = \mathbf{0}_p,$$

denoted as $\boldsymbol{\theta}_{\mathrm{EE}}$. What is the best choice of $\mathbf{H}$? Using a Taylor-series expansion, we see that

$$\mathbf{0}_p = \mathbf{H}'\mathbf{g}(\mathbf{y}; \boldsymbol{\theta}_{\mathrm{EE}}) \approx \mathbf{H}'\left\{\mathbf{g}(\mathbf{y}; \boldsymbol{\theta}) + \frac{\partial \mathbf{g}(\mathbf{y}; \boldsymbol{\theta})}{\partial \boldsymbol{\theta}}(\boldsymbol{\theta}_{\mathrm{EE}} - \boldsymbol{\theta})\right\}$$

so that

$$\boldsymbol{\theta}_{\mathrm{EE}} - \boldsymbol{\theta} \approx \left(-\mathbf{H}'\frac{\partial \mathbf{g}(\mathbf{y}; \boldsymbol{\theta})}{\partial \boldsymbol{\theta}}\right)^{-1} \mathbf{H}'\mathbf{g}(\mathbf{y}; \boldsymbol{\theta}).$$

Thus, the asymptotic variance is

$$\mathrm{Var}\, \boldsymbol{\theta}_{\mathrm{EE}} = \left(\mathbf{H}'\frac{\partial \mathbf{g}(\mathbf{y}; \boldsymbol{\theta})}{\partial \boldsymbol{\theta}}\right)^{-1} \mathbf{H}' \left(\mathrm{Var}\, \mathbf{g}(\mathbf{y}; \boldsymbol{\theta})\right) \mathbf{H} \left(\frac{\partial \mathbf{g}(\mathbf{y}; \boldsymbol{\theta})'}{\partial \boldsymbol{\theta}} \mathbf{H}\right)^{-1}.$$

The choice of $\mathbf{H}$ that yields the most efficient estimator is $-\left(\mathrm{Var}\, \mathbf{g}(\mathbf{y}; \boldsymbol{\theta})\right)^{-1} \partial \mathbf{g}(\mathbf{y}; \boldsymbol{\theta})/\partial \boldsymbol{\theta}$. This yields

$$\mathrm{Var}\, \boldsymbol{\theta}_{\mathrm{EE}} = \left(\frac{\partial \mathbf{g}(\mathbf{y}; \boldsymbol{\theta})'}{\partial \boldsymbol{\theta}} \left(\mathrm{Var}\, \mathbf{g}(\mathbf{y}; \boldsymbol{\theta})\right)^{-1} \frac{\partial \mathbf{g}(\mathbf{y}; \boldsymbol{\theta})}{\partial \boldsymbol{\theta}}\right)^{-1}.$$

For the case $\mathbf{g}(\mathbf{y}; \boldsymbol{\theta}) = \mathbf{y} - \boldsymbol{\mu}(\boldsymbol{\theta})$, we have $\mathbf{H} = \mathbf{V}^{-1}\partial \boldsymbol{\mu}(\boldsymbol{\theta})/\partial \boldsymbol{\theta}$, where $\mathbf{V} = \mathrm{Var}\, \mathbf{y}$ and

$$\mathrm{Var}\, \boldsymbol{\theta}_{\mathrm{EE}} = \left(\frac{\partial \boldsymbol{\mu}(\boldsymbol{\theta})'}{\partial \boldsymbol{\theta}} \mathbf{V}^{-1} \frac{\partial \boldsymbol{\mu}(\boldsymbol{\theta})}{\partial \boldsymbol{\theta}}\right)^{-1}.$$

In this case, the estimating equations estimator $\boldsymbol{\theta}_{\mathrm{EE}}$ is the solution of the equation

$$\mathbf{0}_p = \frac{\partial \boldsymbol{\mu}(\boldsymbol{\theta})'}{\partial \boldsymbol{\theta}} \mathbf{V}^{-1}(\mathbf{y} - \boldsymbol{\mu}(\boldsymbol{\theta})).$$

For independent observations, this representation is identical to the quasi-likelihood estimators presented in the previous section.

As another example, suppose that $(\mathbf{w}_i, \mathbf{x}_i, y_i)$ are i.i.d. and let $\mathbf{g}(\mathbf{y}; \boldsymbol{\theta}) = (\mathbf{g}_1(\mathbf{y}; \boldsymbol{\theta}), \ldots, \mathbf{g}_n(\mathbf{y}; \boldsymbol{\theta}))'$, where

$$\mathbf{g}_i(\mathbf{y}; \boldsymbol{\theta}) = \mathbf{w}_i(y_i - \mathbf{x}_i'\boldsymbol{\theta}),$$

and $\mathrm{E}(y_i \mid \mathbf{w}_i, \mathbf{x}_i) = \mathbf{x}_i'\boldsymbol{\theta}$. Assume that $\mathrm{Var}\ \mathbf{g}_i(\mathbf{y}; \boldsymbol{\theta}) = \mathrm{Var}\ (\mathbf{w}_i(y_i - \mathbf{x}_i'\boldsymbol{\theta})) = \sigma^2\, \mathrm{E}(\mathbf{w}_i\mathbf{w}_i') = \sigma^2\boldsymbol{\Sigma}_{\mathbf{w}}$. Thus,

$$\mathbf{H} = \left(\mathrm{Var}\, \mathbf{g}(\mathbf{y}; \boldsymbol{\theta})\right)^{-1} \frac{\partial \mathbf{g}(\mathbf{y}; \boldsymbol{\theta})}{\partial \boldsymbol{\theta}} = \left(\mathbf{I}_n \otimes \left(\sigma^2 \boldsymbol{\Sigma}_{\mathbf{w}}\right)\right)^{-1} \begin{pmatrix} -\mathbf{w}_1\mathbf{x}_1' \\ \vdots \\ -\mathbf{w}_n\mathbf{x}_n' \end{pmatrix}$$

$$= \sigma^{-2} \begin{pmatrix} -\boldsymbol{\Sigma}_{\mathbf{w}}^{-1}\mathbf{w}_1\mathbf{x}_1' \\ \vdots \\ -\boldsymbol{\Sigma}_{\mathbf{w}}^{-1}\mathbf{w}_n\mathbf{x}_n' \end{pmatrix}.$$

Therefore, the estimator is a solution of

$$\mathbf{0} = \sigma^{-2}\left(-\mathbf{x}_1 \mathbf{w}_1' \boldsymbol{\Sigma}_{\mathbf{w}}^{-1} \quad \cdots \quad -\mathbf{x}_n \mathbf{w}_n' \boldsymbol{\Sigma}_{\mathbf{w}}^{-1}\right) \begin{pmatrix} \mathbf{w}_1(y_1 - \mathbf{x}_1'\boldsymbol{\theta}) \\ \vdots \\ \mathbf{w}_n(y_n - \mathbf{x}_n'\boldsymbol{\theta}) \end{pmatrix}$$

$$= -\sigma^{-2} \sum_{i=1}^{n} \mathbf{x}_i \mathbf{w}_i' \boldsymbol{\Sigma}_{\mathbf{w}}^{-1} \mathbf{w}_i (y_i - \mathbf{x}_i'\boldsymbol{\theta}).$$

This yields

$$\boldsymbol{\theta}_{\mathrm{EE}} = \left(\sum_{i=1}^{n} \mathbf{x}_i \mathbf{w}_i' \boldsymbol{\Sigma}_{\mathbf{w}}^{-1} \mathbf{w}_i \mathbf{x}_i'\right)^{-1} \sum_{i=1}^{n} \mathbf{x}_i \mathbf{w}_i' \boldsymbol{\Sigma}_{\mathbf{w}}^{-1} \mathbf{w}_i y_i .$$

Using $n^{-1}\sum_{i=1}^{n} \mathbf{w}_i \mathbf{w}_i'$ in place of $\boldsymbol{\Sigma}_{\mathbf{w}}$ yields the *instrumental variable* estimator.

For the case of longitudinal data mixed models, we will assume that the data vector can be decomposed as $\mathbf{y} = (\mathbf{y}_1, \ldots, \mathbf{y}_n)'$, where $\mathrm{E}\, \mathbf{y}_i = \boldsymbol{\mu}_i(\boldsymbol{\theta})$ and $\mathrm{Var}\, \mathbf{y}_i = \mathbf{V}_i = \mathbf{V}_i(\boldsymbol{\theta}, \boldsymbol{\tau})$. Here, the $r \times 1$ vector $\boldsymbol{\tau}$ is our vector of variance components. Assuming independence among subjects, we consider

$$G_{\theta}(\boldsymbol{\theta}, \boldsymbol{\tau}) = \sum_{i=1}^{n} \frac{\partial \boldsymbol{\mu}_i(\boldsymbol{\theta})'}{\partial \boldsymbol{\theta}} \mathbf{V}_i^{-1} (\mathbf{y}_i - \boldsymbol{\mu}_i(\boldsymbol{\theta})). \tag{C.6}$$

The "estimating equations" estimator of $\boldsymbol{\theta}$, denoted by $\boldsymbol{\theta}_{\mathrm{EE}}$, is the solution of the equation $\mathbf{0}_p = \mathrm{G}_{\theta}(\boldsymbol{\theta}, \boldsymbol{\tau})$, where $\mathbf{0}_p$ is a $p \times 1$ vector of zeros.

To compute $\boldsymbol{\theta}_{\mathrm{EE}}$, we require estimators of the variance components $\boldsymbol{\tau}$. For second moments, we will use $\mathbf{y}_i \mathbf{y}_i'$ as our primary data source. For notation, let vech($\mathbf{M}$) denote the column vector created by stacking the columns of the matrix $\mathbf{M}$. Thus, for example,

$$\mathrm{vech}(\mathbf{y}_i \mathbf{y}_i') = \left(y_{i1}^2, y_{i2} y_{i1}, \ldots, y_{i,T_i+1} y_{i1}, y_{i1} y_{i2}, \ldots, y_{i,T_i+1} y_{i,T_i+1}\right)'.$$

Thus, we use $\mathbf{h}_i = (\mathrm{vech}(\mathbf{y}_i \mathbf{y}_i'))$ as our data vector and let $\boldsymbol{\eta}_i = \mathrm{E}\, \mathbf{h}_i$. Then, the estimating equation for $\boldsymbol{\tau}$ is

$$G_{\tau}(\boldsymbol{\theta}, \boldsymbol{\tau}) = \sum_{i=1}^{n} \frac{\partial \boldsymbol{\eta}_i'}{\partial \boldsymbol{\tau}} (\mathbf{h}_i - \boldsymbol{\eta}_i). \tag{C.7}$$

The "estimating equations" estimator of $\boldsymbol{\tau}$, denoted by $\boldsymbol{\tau}_{\mathrm{EE}}$, is the solution of the equation $\mathbf{0}_r = G_{\tau}(\boldsymbol{\theta}, \boldsymbol{\tau})$. To summarize, we first compute initial estimators of $(\boldsymbol{\theta}, \boldsymbol{\tau})$, say $(\boldsymbol{\theta}_{0,\mathrm{EE}}, \boldsymbol{\tau}_{0,\mathrm{EE}})$, typically using basic moment conditions. Then, at the nth stage, recursively, perform the following:

1. Use $\boldsymbol{\tau}_{n,\mathrm{EE}}$ and Equation (C.6) to update the estimator of $\boldsymbol{\theta}$; that is, $\boldsymbol{\theta}_{n+1,\mathrm{EE}}$ is the solution of the equation $G_{\theta}(\boldsymbol{\theta}, \boldsymbol{\tau}_{n,\mathrm{EE}}) = \mathbf{0}_p$.
2. Use $\boldsymbol{\theta}_{n+1,\mathrm{EE}}$ and Equation (C.7) to update the estimator of $\boldsymbol{\tau}$; that is, $\boldsymbol{\tau}_{n+1,\mathrm{EE}}$ is the solution of the equation $G_{\tau}(\boldsymbol{\theta}_{n+1,\mathrm{EE}}, \boldsymbol{\tau}) = \mathbf{0}_r$.
3. Repeat steps 1 and 2 until convergence.

Under mild regularity conditions, $(\boldsymbol{\theta}_{\mathrm{EE}}, \boldsymbol{\tau}_{\mathrm{EE}})$ is consistent and asymptotically normal; see, for example, Diggle et al. (2002S). Under mild regularity conditions, Gourieroux,

Monfort, and Trognon (1984E) show that the estimator $\boldsymbol{\theta}_{\mathrm{EE}}$, calculated using the estimated $\boldsymbol{\tau}$, is just as efficient asymptotically as if $\boldsymbol{\tau}$ were known. Liang and Zeger (1986B) also provide the following estimator of the asymptotic variance–covariance matrix of $(\boldsymbol{\theta}_{\mathrm{EE}}, \boldsymbol{\tau}_{\mathrm{EE}})$:

$$\left(\sum_{i=1}^{n} \mathbf{G}_{1i}\mathbf{G}_{2i}\right)^{-1} \left(\sum_{i=1}^{n} \mathbf{G}_{1i} \begin{pmatrix} \mathbf{y}_i - \boldsymbol{\mu}_i \\ \mathbf{h}_i - \boldsymbol{\eta}_i \end{pmatrix} \begin{pmatrix} \mathbf{y}_i - \boldsymbol{\mu}_i \\ \mathbf{h}_i - \boldsymbol{\eta}_i \end{pmatrix}' \mathbf{G}'_{1i}\right) \left(\sum_{i=1}^{n} \mathbf{G}'_{2i}\mathbf{G}'_{1i}\right)^{-1}, \qquad \text{(C.8)}$$

where

$$\mathbf{G}_{1i} = \begin{pmatrix} \dfrac{\partial \boldsymbol{\mu}'_i}{\partial \boldsymbol{\theta}} \mathbf{V}_i^{-1} & \mathbf{0}_i \\ \mathbf{0}_i & \dfrac{\partial \boldsymbol{\eta}'_i}{\partial \boldsymbol{\tau}} \end{pmatrix}$$

and

$$\mathbf{G}_{2i} = \begin{pmatrix} \dfrac{\partial \boldsymbol{\mu}_i}{\partial \boldsymbol{\theta}} & \dfrac{\partial \boldsymbol{\mu}_i}{\partial \boldsymbol{\tau}} \\ \dfrac{\partial \boldsymbol{\eta}_i}{\partial \boldsymbol{\theta}} & \dfrac{\partial \boldsymbol{\eta}_i}{\partial \boldsymbol{\tau}} \end{pmatrix}.$$

C.7 Hypothesis Tests

We consider testing the null hypothesis $\mathrm{H}_0 : h(\boldsymbol{\theta}) = \mathbf{d}$, where $\mathbf{d}$ is a known vector of dimension $r \times 1$ and $h(.)$ is known and differentiable.

There are three widely used approaches for testing the null hypothesis: the *likelihood ratio, Wald*, and *Rao* tests. The Wald approach evaluates a function of the likelihood at $\boldsymbol{\theta}_{\mathrm{MLE}}$. The likelihood ratio approach uses $\boldsymbol{\theta}_{\mathrm{MLE}}$ and $\boldsymbol{\theta}_{\mathrm{Reduced}}$, where $\boldsymbol{\theta}_{\mathrm{Reduced}}$ is the value of $\boldsymbol{\theta}$ that maximizes $L(\boldsymbol{\theta})$ under the restriction that $h(\boldsymbol{\theta}) = \mathbf{d}$. The Rao approach also uses $\boldsymbol{\theta}_{\mathrm{Reduced}}$ but determines it by maximizing $L(\boldsymbol{\theta}) - \boldsymbol{\lambda}'(h(\boldsymbol{\theta}) - \mathbf{d})$, where $\boldsymbol{\lambda}$ is a vector of Lagrange multipliers. Hence, Rao's test is also called the *Lagrange multiplier test.* The three statistics are:

- Likelihood ratio: $\mathrm{LRT} = 2(L(\boldsymbol{\theta}_{\mathrm{MLE}}) - L(\boldsymbol{\theta}_{\mathrm{Reduced}}))$,
- Wald : $TS_W(\boldsymbol{\theta}_{\mathrm{MLE}})$, where
  $TS_W(\boldsymbol{\theta}) = (h(\boldsymbol{\theta}) - \mathbf{d})'\{\nabla h(\boldsymbol{\theta})'(-I(\boldsymbol{\theta}))^{-1}\nabla h(\boldsymbol{\theta})\}^{-1}(h(\boldsymbol{\theta}) - \mathbf{d})$, and
- Rao : $TS_R(\boldsymbol{\theta}_{\mathrm{Reduced}})$, where $TS_R(\boldsymbol{\theta}) = \nabla \mathrm{L}(\boldsymbol{\theta})'(-I(\boldsymbol{\theta}))^{-1}\nabla \mathrm{L}(\boldsymbol{\theta})$.

$\nabla h(\boldsymbol{\theta}) = \partial h(\boldsymbol{\theta})/\partial \boldsymbol{\theta}$ is the gradient of $h(\boldsymbol{\theta})$ and $\nabla L(\boldsymbol{\theta}) = \partial L(\boldsymbol{\theta})/\partial \boldsymbol{\theta}$ is the gradient of $L(\boldsymbol{\theta})$, the score function.

The main advantage of the Wald statistic is that it only requires computation of $\boldsymbol{\theta}_{\mathrm{MLE}}$ and not $\boldsymbol{\theta}_{\mathrm{Reduced}}$. Similarly, the main advantage of the Rao statistic is that it only requires computation of $\boldsymbol{\theta}_{\mathrm{Reduced}}$ and not $\boldsymbol{\theta}_{\mathrm{MLE}}$. In many applications, computation of $\boldsymbol{\theta}_{\mathrm{MLE}}$ is onerous.

Under broad conditions, all three test statistics are asymptotically chi-square with r degrees of freedom under H_0. All asymptotic methods work well when the number of parameters is bounded and the null hypothesis specifies that $\boldsymbol{\theta}$ is on the interior of the parameter space.

In the usual fixed-effects model, the number of individual-specific parameters is of the same order as the number of subjects. Here, the number of parameters tends to infinity as the number of subjects tends to infinity and the usual asymptotic approximations are not valid. Instead, special *conditional* maximum likelihood estimators enjoy the asymptotic properties similar to maximum likelihood estimators.

When a hypothesis specifies that $\boldsymbol{\theta}$ is on the boundary, then the asymptotic distribution is no longer valid without corrections. An example is $H_0 : \theta = \sigma_\alpha^2 = 0$. Here, the parameter space is $[0, \infty)$. By specifying the null hypothesis at 0, we are on the boundary. Self and Liang (1987S) provide some corrections that improve the asymptotic approximation.

C.8 Goodness-of-Fit Statistics

In linear regression models, the most widely cited goodness-of-fit statistic is the R^2 measure, which is based on the decomposition

$$\sum_i (y_i - \bar{y})^2 = \sum_i (y_i - \hat{y}_i)^2 + \sum_i (\hat{y}_i - \bar{y})^2 + 2\sum_i (y_i - \hat{y}_i)(\hat{y}_i - \bar{y}).$$

In the language of Section 2.3, this decomposition is

$$\textit{Total SS} = \textit{Error SS} + \textit{Regression SS} + 2 \times \textit{Sum of Cross Products}.$$

The difficulty with nonlinear models is that the *Sum of Cross Products* term rarely equals zero. Thus, one gets different statistics when defining R^2 as [(*Regression SS*)/(*Total SS*)] compared to [1 − (*Error SS*)/(*Total SS*)].

An alternative and widely cited goodness-of-fit measure is the *Pearson chi-square* statistic. To define this statistic, suppose that $\mathrm{E}y_i = \mu_i$, $\mathrm{Var}\, y_i = \mathrm{V}(\mu_i)$ for some function V(.), and $\hat{\mu}_i$ is an estimator of μ_i. Then, the Pearson chi-square statistic is defined as $\sum_i (y_i - \hat{\mu}_i)^2/\mathrm{V}(\hat{\mu}_i)$. For Poisson models of count data, this formulation reduces to the form $\sum_i (y_i - \hat{\mu}_i)^2/\hat{\mu}_i$.

In the context of generalized linear models, a goodness-of-fit measure is the *deviance statistic*. To define this statistic, suppose $\mathrm{E}\mathbf{y} = \boldsymbol{\mu} = \boldsymbol{\mu}(\boldsymbol{\theta})$ and write $L(\hat{\boldsymbol{\mu}})$ for the log-likelihood evaluated at $\hat{\boldsymbol{\mu}} = \boldsymbol{\mu}(\hat{\boldsymbol{\theta}})$. The *scaled* deviance statistic is defined as $\mathrm{D}^*(\mathbf{y}, \hat{\boldsymbol{\mu}}) = 2(L(\mathbf{y}) - L(\hat{\boldsymbol{\mu}}))$. In linear exponential families, we multiply by the scaling factor ϕ to define the deviance statistic, $\mathrm{D}(\mathbf{y}, \hat{\boldsymbol{\mu}}) = \phi \mathrm{D}^*(\mathbf{y}, \hat{\boldsymbol{\mu}})$. This multiplication actually removes the variance scaling factor from the definition of the statistic.

Using Appendix 9A, it is straightforward to check that the deviance statistic reduces to the following forms for three important distributions:

- Normal: $\mathrm{D}(\mathbf{y}, \hat{\boldsymbol{\mu}}) = \sum_i (y_i - \hat{\mu}_i)^2$,

- Bernoulli: $\mathrm{D}(\mathbf{y}, \hat{\boldsymbol{\pi}}) = \sum_i \{y_i \ln \frac{y_i}{\hat{\pi}_i} + (1 - y_i) \ln \frac{1-y_i}{1-\hat{\pi}_i}\}$, and
- Poisson: $\mathrm{D}(\mathbf{y}, \hat{\boldsymbol{\mu}}) = \sum_i \{y_i \ln \frac{y_i}{\hat{\mu}_i} + (y_i - \hat{\mu}_i)\}$.

Here, we use the convention that $y \ln y = 0$ when $y = 0$.

C.9 Information Criteria

Likelihood ratio tests are useful for choosing between two models that are *nested*, that is, where one model is a subset of the other. How do we compare models when they are not nested? One way is to use the following information criteria.

The distance between two probability distributions given by probability density functions g and f_θ can be summarized by

$$\mathrm{KL}(g, f_\theta) = \mathrm{E}_g \ln \frac{g(y)}{f_\theta(y)}.$$

This is the *Kullback–Leibler distance*, which turns out to be nonnegative. Here, we have indexed f by a vector of parameters $\boldsymbol{\theta}$. Picking the function g to be f_{θ_0} and then minimizing $\mathrm{KL}(f_{\theta_0}, f_\theta)$ is equivalent to the maximum likelihood principle.

In general, we have to estimate $g(.)$. A reasonable alternative is to minimize

$$AIC = -2L(\boldsymbol{\theta}_{\mathrm{MLE}}) + 2(\text{number of parameters}),$$

known as *Akaike's Information Criterion*. This statistic is used when comparing several alternative (nonnested) models. One picks the model that minimizes AIC. If the models under consideration have the same number of parameters, this is equivalent to choosing the model that maximizes the log likelihood. We remark that, in time-series analysis, the AIC is rescaled by the number of parameters.

The statistic AIC is also useful in that it reduces to the C_p, a statistic that is widely used in regression analysis. This statistic minimizes a bias and variance trade-off when selecting among linear regression models.

An alternative criterion is the *Bayesian Information Criterion*, defined as

$$BIC = -2L(\boldsymbol{\theta}_{\mathrm{MLE}}) + \ln(\text{number of parameters}).$$

This measure gives greater weight to the number of parameters. That is, other things being equal, BIC will suggest a more parsimonious model than AIC.

Appendix D

State Space Model and the Kalman Filter

D.1 Basic State Space Model

Consider the observation equation

$$\mathbf{y}_t = \mathbf{W}_t \boldsymbol{\delta}_t + \boldsymbol{\epsilon}_t, \qquad t = 1, \ldots, T, \tag{D.1}$$

where $\mathbf{y}_t$ is an $n_t \times 1$ vector and $\boldsymbol{\delta}_t$ is an $m \times 1$ vector. The transition equation is

$$\boldsymbol{\delta}_t = \mathbf{T}_t \boldsymbol{\delta}_{t-1} + \boldsymbol{\eta}_t. \tag{D.2}$$

Together, Equations (D.1) and (D.2) define the *state space model*. To complete the specification, define $\mathrm{Var}_{t-1} \boldsymbol{\epsilon}_t = \mathbf{H}_t$ and $\mathrm{Var}_{t-1} \boldsymbol{\eta}_t = \mathbf{Q}_t$, where Var_t is a variance conditional on information up to and including time t, that is, $\{\mathbf{y}_1, \ldots, \mathbf{y}_t\}$. Similarly, let E_t denote the conditional expectation and assume that $\mathrm{E}_{t-1} \boldsymbol{\epsilon}_t = \mathbf{0}$ and $\mathrm{E}_{t-1} \boldsymbol{\eta}_t = \mathbf{0}$. Further define $\mathbf{d}_0 = \mathrm{E}\boldsymbol{\delta}_0$, $\mathbf{P}_0 = \mathrm{Var}\, \boldsymbol{\delta}_0$, and $\mathbf{P}_t = \mathrm{Var}_t\, \boldsymbol{\delta}_t$. Assume that $\{\boldsymbol{\epsilon}_t\}$ and $\{\boldsymbol{\eta}_t\}$ are mutually independent.

In subsequent sections, it will be useful to summarize Equation (D.1). Thus, we define

$$\mathbf{y} = \begin{pmatrix} \mathbf{y}_1 \\ \mathbf{y}_2 \\ \vdots \\ \mathbf{y}_T \end{pmatrix} = \begin{pmatrix} \mathbf{W}_1 \boldsymbol{\delta}_1 \\ \mathbf{W}_2 \boldsymbol{\delta}_2 \\ \vdots \\ \mathbf{W}_T \boldsymbol{\delta}_T \end{pmatrix} + \begin{pmatrix} \boldsymbol{\epsilon}_1 \\ \boldsymbol{\epsilon}_2 \\ \vdots \\ \boldsymbol{\epsilon}_T \end{pmatrix}$$
$$= \begin{pmatrix} \mathbf{W}_1 & \mathbf{0} & \cdots & \mathbf{0} \\ \mathbf{0} & \mathbf{W}_2 & \cdots & \mathbf{0} \\ \vdots & \vdots & \ddots & \vdots \\ \mathbf{0} & \mathbf{0} & \cdots & \mathbf{W}_T \end{pmatrix} \begin{pmatrix} \boldsymbol{\delta}_1 \\ \boldsymbol{\delta}_2 \\ \vdots \\ \boldsymbol{\delta}_T \end{pmatrix} + \begin{pmatrix} \boldsymbol{\epsilon}_1 \\ \boldsymbol{\epsilon}_2 \\ \vdots \\ \boldsymbol{\epsilon}_T \end{pmatrix} = \mathbf{W}\boldsymbol{\delta} + \boldsymbol{\epsilon}. \tag{D.3}$$

With the notation $N = \sum_{t=1}^{T} n_t$, we have that $\mathbf{y}$ is an $N \times 1$ vector of random variables, $\boldsymbol{\delta}$ is a $Tm \times 1$ vector of state variables, $\mathbf{W}$ is an $N \times Tm$ matrix of known variables, and $\boldsymbol{\epsilon}$ is an $N \times 1$ vector of disturbance terms.

D.2 Kalman Filter Algorithm

Taking a conditional expectation and variance of Equation (D.2) yields the "prediction equations"

$$\mathbf{d}_{t/t-1} = \mathrm{E}_{t-1}\boldsymbol{\delta}_t = \mathbf{T}_t\mathbf{d}_{t-1} \quad \text{(D.4a)}$$

and

$$\mathbf{P}_{t/t-1} = \mathrm{Var}_{t-1}\boldsymbol{\delta}_t = \mathbf{T}_t\mathbf{P}_{t-1}\mathbf{T}_t' + \mathbf{Q}_t. \quad \text{(D.4b)}$$

Taking a conditional expectation and variance of Equation (D.1) yields

$$\mathrm{E}_{t-1}\mathbf{y}_t = \mathbf{W}_t\mathbf{d}_{t/t-1} \quad \text{(D.5a)}$$

and

$$\mathbf{F}_t = \mathrm{Var}_{t-1}\mathbf{y}_t = \mathbf{W}_t\mathbf{P}_{t/t-1}\mathbf{W}_t' + \mathbf{H}_t. \quad \text{(D.5b)}$$

The "updating equations" are

$$\mathbf{d}_t = \mathbf{d}_{t/t-1} + \mathbf{P}_{t/t-1}\mathbf{W}_t'\mathbf{F}_t^{-1}(\mathbf{y}_t - \mathbf{W}_t\mathbf{d}_{t/t-1}) \quad \text{(D.6a)}$$

and

$$\mathbf{P}_t = \mathbf{P}_{t/t-1} - \mathbf{P}_{t/t-1}\mathbf{W}_t'\mathbf{F}_t^{-1}\mathbf{W}_t\mathbf{P}_{t/t-1}. \quad \text{(D.6b)}$$

The updating equations can be motivated by assuming that $\boldsymbol{\delta}_t$ and $\mathbf{y}_t$ are jointly normally distributed. With this assumption, and Equation (B.2) of Appendix B, we have

$$\begin{aligned}\mathrm{E}_t\boldsymbol{\delta}_t &= \mathrm{E}_{t-1}(\boldsymbol{\delta}_t \mid \mathbf{y}_t) \\ &= \mathrm{E}_{t-1}\boldsymbol{\delta}_t + \mathrm{Cov}_{t-1}(\boldsymbol{\delta}_t, \mathbf{y}_t)(\mathrm{Var}_{t-1}\mathbf{y}_t)^{-1}(\mathbf{y}_t - \mathrm{E}_{t-1}\mathbf{y}_t)\end{aligned}$$

and

$$\mathrm{Var}_t\boldsymbol{\delta}_t = \mathrm{Var}_{t-1}\boldsymbol{\delta}_t - \mathrm{Cov}_{t-1}(\boldsymbol{\delta}_t, \mathbf{y}_t)(\mathrm{Var}_{t-1}\mathbf{y}_t)^{-1}\mathrm{Cov}_{t-1}(\boldsymbol{\delta}_t, \mathbf{y}_t)'.$$

These expressions yield the updating equations immediately.

For computational convenience, the Kalman filter algorithm in Equations (D.4)–(D.6) can be expressed more compactly as

$$\mathbf{d}_{t+1/t} = \mathbf{T}_{t+1}\mathbf{d}_{t/t-1} + \mathbf{K}_t(\mathbf{y}_t - \mathbf{W}_t\mathbf{d}_{t/t-1}) \quad \text{(D.7)}$$

and

$$\mathbf{P}_{t+1/t} = \mathbf{T}_{t+1}\left(\mathbf{P}_{t/t-1} - \mathbf{P}_{t/t-1}\mathbf{W}_t'\mathbf{F}_t^{-1}\mathbf{W}_t\mathbf{P}_{t/t-1}\right)\mathbf{T}_{t+1}' + \mathbf{Q}_{t+1}, \quad \text{(D.8)}$$

where $\mathbf{K}_t = \mathbf{T}_{t+1}\mathbf{P}_{t/t-1}\mathbf{W}_t'\mathbf{F}_t^{-1}$ is known as the *gain matrix*. To start these recursions, from (D.4) we have $\mathbf{d}_{1/0} = \mathbf{T}_1\mathbf{d}_0$ and $\mathbf{P}_{1/0} = \mathbf{T}_1\mathbf{P}_0\mathbf{T}_1' + \mathbf{Q}_1$.

D.3 Likelihood Equations

Assuming the conditional variances, $\mathbf{Q}_t$ and $\mathbf{H}_t$, and initial conditions, $\mathbf{d}_0$ and $\mathbf{P}_0$, are known, Equations (D.7) and (D.8) allow one to recursively compute $\mathrm{E}_{t-1}\mathbf{y}_t$ and $\mathrm{Var}_{t-1}\mathbf{y}_t = \mathbf{F}_t$ These quantities are important because they allow us to directly evaluate the likelihood of $\{\mathbf{y}_1, \ldots, \mathbf{y}_T\}$. That is, assume that $\{\mathbf{y}_1, \ldots, \mathbf{y}_T\}$ are jointly multivariate normal and let f be used for the joint and conditional density. Then, with Equations (B.1) and (B.2) of Appendix B, the logarithmic likelihood is

$$\begin{aligned} L &= \ln f(\mathbf{y}_1, \ldots, \mathbf{y}_T) = \ln f(\mathbf{y}_1) + \sum_{t=2}^{T} f(\mathbf{y}_t \mid \mathbf{y}_1, \ldots, \mathbf{y}_{t-1}) \\ &= -\frac{1}{2}\left[N \ln 2\pi + \sum_{t=1}^{T} \ln \det(\mathbf{F}_t) + \sum_{t=1}^{T} (\mathbf{y}_t - \mathrm{E}_{t-1}\mathbf{y}_t)' \mathbf{F}_t^{-1} (\mathbf{y}_t - \mathrm{E}_{t-1}\mathbf{y}_t) \right]. \end{aligned} \tag{D.9}$$

From the Kalman filter algorithm in Equations (D.4)–(D.6) , we see that $\mathrm{E}_{t-1}\mathbf{y}_t$ is a *linear* combination of $\{\mathbf{y}_1, \ldots, \mathbf{y}_{t-1}\}$. Thus, we may write

$$\mathbf{Ly} = \mathbf{L}\begin{pmatrix} \mathbf{y}_1 \\ \mathbf{y}_2 \\ \vdots \\ \mathbf{y}_T \end{pmatrix} = \begin{pmatrix} \mathbf{y}_1 - \mathrm{E}_0\mathbf{y}_1 \\ \mathbf{y}_2 - \mathrm{E}_1\mathbf{y}_2 \\ \vdots \\ \mathbf{y}_T - \mathrm{E}_{T-1}\mathbf{y}_T \end{pmatrix}, \tag{D.10}$$

where $\mathbf{L}$ is an $N \times N$ lower triangular matrix with ones on the diagonal. Elements of the matrix $\mathbf{L}$ do not depend on the random variables. The advantages of this transformation are that the components of the right-hand side of Equation (D.10) are mean zero and are mutually uncorrelated.

D.4 Extended State Space Model and Mixed Linear Models

To handle the longitudinal data model described in Section 8.5, we now extend the state space model to mixed linear models, so that it handles fixed and random effects. By incorporating these effects, we will also be able to introduce initial conditions that are estimable.

Specifically, consider the observation equation

$$\mathbf{y}_t = \mathbf{X}_t\boldsymbol{\beta} + \mathbf{Z}_t\boldsymbol{\alpha} + \mathbf{W}_t\boldsymbol{\delta}_t + \boldsymbol{\epsilon}_t \tag{D.11}$$

where $\boldsymbol{\beta}$ is a $K \times 1$ vector of unknown parameters, called "fixed effects," and $\boldsymbol{\alpha}$ is a $q^* \times 1$ random vector, known as "random effects." We assume that $\boldsymbol{\alpha}$ has mean $\mathbf{0}$, has a variance–covariance matrix $\sigma^2\mathbf{B} = \mathrm{Var}\,\boldsymbol{\alpha}$, and is independent of $\{\boldsymbol{\delta}_t\}$ and $\{\boldsymbol{\epsilon}_t\}$. The transition equation is as in Equation (D.2); that is, we assume

$$\boldsymbol{\delta}_t = \mathbf{T}_t\boldsymbol{\delta}_{t-1} + \boldsymbol{\eta}_t. \tag{D.12}$$

Similar to Equation (D.3), we summarize Equation (D.11) as

$$\mathbf{y} = \mathbf{X}\boldsymbol{\beta} + \mathbf{Z}\boldsymbol{\alpha} + \mathbf{W}\boldsymbol{\delta} + \boldsymbol{\epsilon}. \tag{D.13}$$

Here, $\mathbf{y}, \mathbf{W}, \boldsymbol{\delta}$, and $\boldsymbol{\epsilon}$ are defined in Equation (D.3), and $\mathbf{X} = (\mathbf{X}_1', \mathbf{X}_2', \ldots, \mathbf{X}_T')'$ and $\mathbf{Z} = (\mathbf{Z}_1', \mathbf{Z}_2', \ldots, \mathbf{Z}_T')'$.

D.5 Likelihood Equations for Mixed Linear Models

To handle fixed and random effects, begin with the transformation matrix $\mathbf{L}$ defined in Equation (D.10). In developing the likelihood, the important point to observe is that the transformed sequence of random variables, $\mathbf{y} - (\mathbf{X}\boldsymbol{\beta} + \mathbf{Z}\boldsymbol{\alpha})$, has the same properties as the basic set of random variables $\mathbf{y}$ in Section D.1. Specifically, with the transformation matrix $\mathbf{L}$, the T components of the vector

$$\mathbf{L}(\mathbf{y} - (\mathbf{X}\boldsymbol{\beta} + \mathbf{Z}\boldsymbol{\alpha})) = \mathbf{v} = \begin{pmatrix} \mathbf{v}_1 \\ \mathbf{v}_2 \\ \vdots \\ \mathbf{v}_T \end{pmatrix}$$

are mean zero and mutually uncorrelated. Further, conditional on $\boldsymbol{\alpha}$ and $\{\mathbf{y}_1, \ldots, \mathbf{y}_{t-1}\}$, the tth component of this matrix, $\mathbf{v}_t$, has variance $\mathbf{F}_t$.

With the Kalman filter algorithm in (D.7) and (D.8), we define the transformed variables $\mathbf{y}^* = \mathbf{L}\mathbf{y}$, $\mathbf{X}^* = \mathbf{L}\mathbf{X}$, and $\mathbf{Z}^* = \mathbf{L}\mathbf{Z}$. To illustrate, for the jth column of $\mathbf{X}$, say $\mathbf{X}^{(j)}$, one would recursively calculate

$$\mathbf{d}_{t+1/t}(\mathbf{X}^{(j)}) = \mathbf{T}_{t+1}\mathbf{d}_{t/t-1}(\mathbf{X}^{(j)}) + \mathbf{K}_t\left(\mathbf{X}_t^{(j)} - \mathbf{W}_t\mathbf{d}_{t/t-1}(\mathbf{X}^{(j)})\right) \tag{D.7*}$$

in place of Equation (D.7). We begin the recursion in Equation (D.7*) with $\mathbf{d}_{1/0}(\mathbf{X}^{(j)}) = \mathbf{0}$. Equation (D.8) remains unchanged. Then, analogous to expression (D.5a), the tth component of $\mathbf{X}^*$ is

$$\mathbf{X}_t^{*(j)} = \mathbf{X}_t^{(j)} - \mathbf{W}_t\mathbf{d}_{t/t-1}(\mathbf{X}^{(j)}).$$

With these transformed variables, we may express the transformed random variables as

$$\mathbf{y}^* = \mathbf{L}\mathbf{y} = \mathbf{v} + \mathbf{X}^*\boldsymbol{\beta} + \mathbf{Z}^*\alpha.$$

Recall that $\sigma^2\mathbf{B} = \mathrm{Var}\,\boldsymbol{\alpha}$ and note that

$$\mathrm{Var}\,\mathbf{v} = \mathrm{Var}\begin{pmatrix} \mathbf{v}_1 \\ \mathbf{v}_2 \\ \vdots \\ \mathbf{v}_T \end{pmatrix} = \begin{pmatrix} \mathbf{F}_1 & \mathbf{0} & \cdots & \mathbf{0} \\ \mathbf{0} & \mathbf{F}_2 & \cdots & \mathbf{0} \\ \vdots & \vdots & \ddots & \vdots \\ \mathbf{0} & \mathbf{0} & \cdots & \mathbf{F}_T \end{pmatrix} = \sigma^2\boldsymbol{\Lambda}. \tag{D.14}$$

This yields $\mathrm{E}\mathbf{y}^* = \mathbf{X}^*\boldsymbol{\beta}$ and $\mathrm{Var}\,\mathbf{y}^* = \sigma^2(\boldsymbol{\Lambda} + \mathbf{Z}^*\mathbf{B}\mathbf{Z}^{*\prime}) = \sigma^2\mathbf{V}$. We use $\boldsymbol{\tau}$ to denote the vector of (unknown) quantities that parameterize $\mathbf{V}$.

From Equation (B.1) of Appendix B, the logarithmic likelihood is

$$L(\boldsymbol{\beta}, \sigma^2, \boldsymbol{\tau}) = -\frac{1}{2}\{N \ln 2\pi + N \ln \sigma^2 + \sigma^{-2}(\mathbf{y}^* - \mathbf{X}^*\boldsymbol{\beta})'\mathbf{V}^{-1}(\mathbf{y}^* - \mathbf{X}^*\boldsymbol{\beta}) + \ln \det \mathbf{V}\}. \tag{D.15}$$

The corresponding restricted log-likelihood is

$$L_{\mathrm{R}}(\boldsymbol{\beta}, \sigma^2, \boldsymbol{\tau}) = -\frac{1}{2}\{\ln \det(\mathbf{X}^{*\prime}\mathbf{V}^{-1}\mathbf{X}^*) - K \ln \sigma^2\} + L(\boldsymbol{\beta}, \sigma^2, \boldsymbol{\tau}) + \text{constant}. \tag{D.16}$$

Either (D.15) or (D.16) can be maximized to determine an estimator of $\boldsymbol{\beta}$. The result is equivalent to the generalized least-squares estimator

$$\mathbf{b}_{\mathrm{GLS}} = (\mathbf{X}^{*\prime}\mathbf{V}^{-1}\mathbf{X}^*)^{-1}\mathbf{X}^{*\prime}\mathbf{V}^{-1}\mathbf{y}^*. \tag{D.17}$$

Using $\mathbf{b}_{\mathrm{GLS}}$ for $\boldsymbol{\beta}$ in Equations (D.16) and (D.17) yields concentrated likelihoods. To determine the REML estimator of σ^2, we maximize $L_{\mathrm{R}}(\mathbf{b}_{\mathrm{GLS}}, \sigma^2, \boldsymbol{\tau})$ (holding $\boldsymbol{\tau}$ fixed) to get

$$s^2_{\mathrm{REML}} = (N - K)^{-1}(\mathbf{y}^* - \mathbf{X}^*\mathbf{b}_{\mathrm{GLS}})'\mathbf{V}^{-1}(\mathbf{y}^* - \mathbf{X}^*\mathbf{b}_{\mathrm{GLS}}). \tag{D.18}$$

Thus, the logarithmic likelihood evaluated at these parameters is

$$L\left(\mathbf{b}_{\mathrm{GLS}}, s^2_{\mathrm{REML}}, \boldsymbol{\tau}\right) = -\frac{1}{2}\left\{N \ln 2\pi + N \ln s^2_{\mathrm{REML}} + N - K + \ln \det \mathbf{V}\right\}. \tag{D.19}$$

The corresponding restricted logarithmic likelihood is

$$L_{\mathrm{REML}} = -\frac{1}{2}\left\{\ln \det(\mathbf{X}^{*\prime}\mathbf{V}^{-1}\mathbf{X}^*) - K \ln s^2_{\mathrm{REML}}\right\} + L\left(\mathbf{b}_{\mathrm{GLS}}, s^2_{\mathrm{REML}}, \boldsymbol{\tau}\right) + \text{constant}. \tag{D.20}$$

The likelihood expressions in Equations (D.19) and (D.20) are intuitively straightforward. However, because of the number of dimensions, they can be difficult to compute. We now provide alternative expressions that, although more complex, are simpler to compute with the Kalman filter algorithm. From Equations (A.3) and (A.5) of Appendix A, we have

$$\mathbf{V}^{-1} = \boldsymbol{\Lambda}^{-1} - \boldsymbol{\Lambda}^{-1}\mathbf{Z}^*(\mathbf{B}^{-1} + \mathbf{Z}^{*\prime}\boldsymbol{\Lambda}^{-1}\mathbf{Z}^*)^{-1}\mathbf{Z}^{*\prime}\boldsymbol{\Lambda}^{-1} \tag{D.21}$$

and

$$\ln \det \mathbf{V} = \ln \det \boldsymbol{\Lambda} - \ln \det \mathbf{B}^{-1} + \ln \det(\mathbf{B}^{-1} + \mathbf{Z}^{*\prime}\boldsymbol{\Lambda}^{-1}\mathbf{Z}^*). \tag{D.22}$$

With Equation (D.21), we immediately have the expression for $\mathbf{b}_{\mathrm{GLS}}$ in Equation (6.32). From Equation (D.18), the restricted maximum likelihood estimator of σ^2 can be expressed as

$$s^2_{\mathrm{REML}} = (N - K)^{-1}\left\{\mathbf{y}^*\mathbf{V}^{-1}\mathbf{y}^* - \mathbf{y}^{*\prime}\mathbf{V}^{-1}\mathbf{X}^*\mathbf{b}_{\mathrm{GLS}}\right\},$$

which is sufficient for Equation (8.36). This, Equations (D.20) and (D.22) are sufficient for Equation (8.37).

Appendix E
Symbols and Notation

Symbol	Description	Chapter defined
$\mathbf{1}_i$	$T_i \times 1$ vector of ones	2
$a_{i,\text{BLUP}}$	BLUP of α_i in the one-way error-components model	4
$a_{i,\text{MLE}}$	maximum likelihood estimator of α_i	9
$a_i, a_{i,j}, \mathbf{a}_i$	estimators of $\alpha_i, \alpha_{i,j}, \boldsymbol{\alpha}_i$	2
$\mathbf{B}, \boldsymbol{\Gamma}$	regression coefficients for endogenous and exogenous regressors, respectively, in the simultaneous equations model	6
$\mathbf{b}_{\text{IV}}$	instrumental variable estimator of $\boldsymbol{\beta}$	6
$\mathbf{b}_{\text{EC}}$	generalized least squares (GLS) estimator of $\boldsymbol{\beta}$ in the error-components model	3
$\mathbf{b}_{\text{EE}}, \boldsymbol{\tau}_{\text{EE}}$	estimating equations estimators of $\boldsymbol{\beta}$ and $\boldsymbol{\tau}$	9
$\mathbf{b}_{\text{GLS}}$	GLS estimator of $\boldsymbol{\beta}$ in the mixed linear model	3
$\mathbf{b}_{\text{MLE}}$	maximum likelihood estimator (MLE) of $\boldsymbol{\beta}$ in the mixed linear model	3
$\mathbf{b}_{i,\text{OLS}}$	ordinary least-squares estimator of $\boldsymbol{\alpha}_i + \boldsymbol{\beta}$ in the random coefficients model	3
$\mathbf{b}_{\text{W}}$	weighted least-squares estimator of $\boldsymbol{\beta}$ in the mixed linear model	3
$b_j, \mathbf{b}$	estimators of $\beta_j, \boldsymbol{\beta}$	2
$\mathbf{D}$	variance–covariance matrix of subject-specific effects in the longitudinal data mixed model, Var $\boldsymbol{\alpha}_i$	3
$\mathbf{D}_{\text{SWAMY}}$	Swamy's estimator of $\mathbf{D}$	3
$\mathbf{e}_i$	$T_i \times 1$ vector of residuals for the ith subject, $\mathbf{e}_i = (e_{i1}, e_{i2}, \ldots, e_{iT_i})'$.	2
e_{it}	residual for the ith subject, tth time period	2
$e_{it,\text{BLUP}}$	BLUP residual, predictor of ε_{it}	4

(continued)

Symbol	Description	Chapter defined
$f(y_1,.., y_T, \mathbf{x}_1, \ldots, \mathbf{x}_T)$	a generic joint probability density (or mass) function $\{y_1, \ldots, y_T, \mathbf{x}_1, \ldots, \mathbf{x}_T\}$	6
$\mathbf{G}_i$	a $T_i \times g$ matrix of "augmented," independent variables	7
$\mathbf{G}_\mu(\boldsymbol{\beta}, \tau)$	matrix of derivatives	9
$\mathbf{G}_{\mathrm{OLS}}$	ordinary least-squares estimator of $\mathbf{\Gamma}$ in the multivariate regression model	6
$\mathbf{H}$	spatial variance matrix, $=$ Var $\boldsymbol{\epsilon}_t$	8
$\mathbf{I}$	Identity matrix, generally of dimension $T \times T$	2
$\mathbf{I}_i$	$T_i \times T_i$ identity matrix	2
i, t	indices for the ith subject, tth time period	1
$\mathbf{i}(k)$	index set; the set of all indices $(i_1, i_2, \ldots, i_k)$ such that $y_{i_1,i_2,\ldots,i_k}$ is observed	5
$i(k)$	a typical element of $\mathbf{i}(k)$ of the form $(i_1, i_2, \ldots, i_k)$	5
$\mathbf{i}(k-s)$	index set; the set of all indices $(i_1, \ldots, i_{k-s})$ such that $y_{i_1,\ldots,i_{k-s},j_{k-s+1},\ldots,j_k}$ is observed for some $\{j_{k-s+1}, \ldots, j_k\}$	5
$i(k-s)$	a typical element of $\mathbf{i}(k-s)$	5
$\mathbf{J}$	matrix of ones, generally of dimension $T \times T$	2
$\mathbf{J}_i$	$T_i \times T_i$ matrix of ones	2
$\mathbf{K}$	a $(T-1) \times T$ upper triangular matrix such that $\mathbf{K}\,\mathbf{1} = \mathbf{0}$	6
K	number of explanatory variables associated with global parameters	2
$L(.)$	logarithmic likelihood of the entire data set	3
L_{R}	logarithmic restricted maximum likelihood of the entire data set	3
LRT	likelihood ratio test statistic	9
$\mathrm{LP}(\varepsilon_i \mid \mathbf{x}_i)$	a linear projection of ε_i on $\mathbf{x}_i$	6
$l_i(.)$	logarithmic likelihood of the ith subject	3
logit(p)	logit function, defined as $\mathrm{logit}(p) = \ln(p/(1-p))$	9
$\mathbf{M}_i$	a $T_i \times T$ matrix used to specify the availability of observations	7
$\mathbf{M}_i$	$T_i \times T_i$ design matrix that describes missing observations	8
$m_{\alpha,\mathrm{GLS}}$	GLS estimator of μ in the one-way random-effects ANOVA model	4
N	total number of observations, $N = T_1 + T_2 + \cdots + T_n$	2
n	number of subjects	1
$\mathbf{o}_{it}, \mathbf{o}_i$	$\mathbf{o}_{it} = (\mathbf{z}_{it}', \mathbf{x}_{it}')'$ is a $(q+K) \times 1$ vector of observed effects, $\mathbf{o}_i = (\mathbf{o}_{i1}', \ldots, \mathbf{o}_{iT_i}')'$.	6
$\mathbf{P}_{\mathbf{W}}$	a projection matrix, $\mathbf{P}_{\mathbf{W}} = \mathbf{W}(\mathbf{W}'\mathbf{W})^{-1}\mathbf{W}'$	6
p_{it}	probability of y_{it} equaling one	9
$p(\,\mid \alpha_i)$	conditional distribution, given the random effect α_i	9
$\mathbf{Q}_i$	matrix that projects a vector of responses to OLS residuals, $\mathbf{Q}_i = \mathbf{I}_i - \mathbf{Z}_i\left(\mathbf{Z}_i'\mathbf{Z}_i\right)^{-1}\mathbf{Z}_i'$	2
$\mathbf{Q}_{Z,i}$	matrix that projects a vector of responses to GLS residuals, $\mathbf{Q}_{Z,i} = \mathbf{I}_i - \mathbf{R}_i^{-1/2}\mathbf{Z}_i(\mathbf{Z}_i'\mathbf{R}_i^{-1}\mathbf{Z}_i)^{-1}\mathbf{Z}_i'\mathbf{R}_i^{-1/2}$	2

Symbol	Description	Chapter defined
q	number of explanatory variables associated with subject-specific parameters	2
$\mathbf{R}$	$T \times T$ variance–covariance matrix, $\mathbf{R} = \mathrm{Var}\,\boldsymbol{\epsilon}$	2
$\mathbf{R}_i$	$T_i \times T_i$ matrix variance–covariance for the ith subject, $\mathbf{R}_i = \mathrm{Var}\,\boldsymbol{\epsilon}_i$	2
$\mathbf{R}_{rs}$	the element in the rth row and sth column of $\mathbf{R}$, $\mathbf{R}_{rs} = \mathrm{Cov}(\varepsilon_r, \varepsilon_s)$	2
$\mathbf{R}_{\mathrm{AR}}(\rho)$	correlation matrix corresponding to an $AR(1)$ process	8
$\mathbf{R}_{\mathrm{RW}}(\rho)$	correlation matrix corresponding to a (generalized) random walk process	8
R^2_{ms}	max-scaled coefficient of determination (R^2)	9
R_{AVE}	the average of Spearman's rank correlations, $R_{\mathrm{AVE}} = \frac{1}{n(n-1)/2} \sum_{\{i<j\}} sr_{ij}$	2
R^2_{AVE}	the average of squared Spearman's rank correlations, $R^2_{\mathrm{AVE}} = \frac{1}{n(n-1)/2} \sum_{\{i<j\}} (sr_{ij})^2$	2
r	number of time-varying coefficients per time period	8
r_{it}	the rank of the tth residual e_{it} from the vector of residuals $\{e_{i,1}, \ldots, e_{i,T}\}$	2
r_{ij}, $\mathbf{r}$	r_{ij} is an indicator of the ijth observation (a "1" indicates the observation is available), $\mathbf{r} = (r_{11}, \ldots, r_{1T}, \ldots, r_{n1}, \ldots, r_{nT})'$	7
s^2	unbiased estimator of σ^2	2
s^2_α	unbiased estimator of σ^2_α	3
$se(\mathbf{b}_{\mathrm{W}})$	standard error of $\mathbf{b}_{\mathrm{W}}$	3
sr_{ij}	Spearman's rank correlation coefficient between the ith and jth subjects, $sr_{ij} = \frac{\sum_{t=1}^{T} (r_{i,t} - (T+1)/2)(r_{j,t} - (T+1)/2)}{\sum_{t=1}^{T} (r_{i,t} - (T+1)/2)^2}$	2
T_i	number of observations for the ith subject	1
$\mathbf{T}_t$	matrix of time-varying parameters for a generalized transition equation	8
TS	a test statistic for assessing homogeneity in the error-components model	3
$\mathbf{U}_i$	a $T_i \times g$ matrix of unobserved, independent variables	7
U_{itj}, V_{itj}	unobserved utility and value of the jth choice for the ith subject at time t	9
u_{it}	utility function for the ith subject at time t	9
$\mathbf{V}_i$	variance–covariance matrix of the ith subject in the longitudinal data mixed model, $\mathbf{V}_i = \mathbf{Z}_i \mathbf{D} \mathbf{Z}_i' + \mathbf{R}_i$	3
$\mathbf{V}_{\mathbf{H}}$	$\sigma^2_\lambda \mathbf{J}_n + \mathbf{H}$	8
$\mathbf{W}$	a matrix of instrumental variables	6
$\mathbf{W}_i$	a $K \times K$ weight matrix for the ith subject, $\mathbf{W}_i = \sum_{t=1}^{T_i} (\mathbf{x}_{it} - \bar{\mathbf{x}}_i)(\mathbf{x}_{it} - \bar{\mathbf{x}}_i)'$	2

(continued)

Symbol	Description	Chapter defined
w	generic random variable to be predicted	4
$\mathbf{w}_i$	a set of predetermined variables	6
w_{BLUP}	BLUP of w	4
$\mathbf{X}$	$N \times K$ matrix of explanatory variables associated with fixed effects in the mixed linear model	3
$\mathbf{X}^*$	a collection of all observed explanatory variables, $\mathbf{X}^* = \{\mathbf{X}_1, \mathbf{Z}_1, \ldots, \mathbf{X}_n, \mathbf{Z}_n\}$	6
$\mathbf{X}_i$	$T_i \times K$ matrix of explanatory variables associated with global parameters for the ith subject, $\mathbf{X}_i = (\mathbf{x}_{i1}, \mathbf{x}_{i2}, \ldots, \mathbf{x}_{iT_i})'$	2
$\bar{\mathbf{x}}_i$	$K \times 1$ vector of averages, over time, of the explanatory variables associated with the global parameters from the ith subject, $\mathbf{x}_i = T_i^{-1} \sum_{t=1}^{T_i} \mathbf{x}_{it}$	2
$\mathbf{x}_{it}$	$K \times 1$ vector of explanatory variables associated with global parameters for the ith subject, tth time period, $\mathbf{x}_{it} = (x_{it,1}, x_{it,2}, \ldots, x_{it,K})'$	2
$x_{it,j}$	jth explanatory variable associated with global parameters, for the ith subject, tth time period	2
$\mathbf{Y}$	$\mathbf{Y}$ is the vector of all potentially observed responses, $\mathbf{Y} = (y_{11}, \ldots, y_{1T}, \ldots, y_{n1}, \ldots, y_{nT})'$	7
$\mathbf{Y}, \mathbf{X}, \mathbf{\Gamma}$	responses, explanatory variables, and parameters in the multivariate regression model	6
$\mathbf{y}$	$N \times 1$ vector of responses in the mixed linear model	3
$\mathbf{y}_i$	$T_i \times 1$ vector of responses for the ith subject, $\mathbf{y}_i = (y_{i1}, y_{i2}, \ldots, y_{iT_i})'$	2
$\bar{y}_i$	average, over time, of the responses from the ith subject, $\bar{y}_i = T_i^{-1} \sum_{t=1}^{T_i} y_{it}$	2
$\bar{y}_{i,s}$	shrinkage estimator of $\mu + \alpha_i$ in the one-way random-effects ANOVA model	4
$\bar{y}_{i,\mathrm{BLUP}}$	best linear unbiased predictor (BLUP) of $\mu + \alpha_i$ in the one-way random-effects ANOVA model	4
$y_{i_1,i_2,\ldots,i_k}$	a typical dependent variable in a k-level model	5
y_{it}	response for the ith subject, tth time period	1
$\mathbf{Z}$	$N \times q$ matrix of explanatory variables associated with random effects in the mixed linear model	3
$\mathbf{Z}_i$	$T_i \times q$ matrix of explanatory variables associated with subject-specific parameters for the ith subject, $\mathbf{Z}_i = (\mathbf{z}_{i1}, \mathbf{z}_{i2}, \ldots, \mathbf{z}_{iT_i})'$	2
$\mathbf{Z}^{(g)}_{i(k+1-g)}$ $\mathbf{X}^{(g)}_{i(k+1-g)}$	level-g covariates matrices in the high-order multilevel model, analogous to Chapter 2 and 3 $\mathbf{Z}_i$ and $\mathbf{X}_i$ matrices	5
$\mathbf{z}_{it}$	$q \times 1$ vector of explanatory variables associated with subject-specific parameters for the ith subject, tth time period, $\mathbf{z}_{it} = (z_{it,1}, z_{it,2}, \ldots, z_{it,q})'$	2

Symbol	Description	Chapter defined
$z_{it,j}$	jth explanatory variable associated with subject-specific parameters, for the ith subject, tth time period	2
$z_{\alpha,it,j}$, $\mathbf{z}_{\alpha,it}$, $\mathbf{Z}_{\alpha,i}\mathbf{Z}_{\alpha}$	explanatory variables associated with $\boldsymbol{\alpha}_i$; same as Chapter 2 $z_{it,j}$, $\mathbf{z}_{it}$, $\mathbf{Z}_i$, and $\mathbf{Z}$	8
$z_{\lambda,it,j}$, $\mathbf{z}_{\lambda,it}$, $\mathbf{Z}_{\lambda,i}$, $\mathbf{Z}_{\lambda}$	explanatory variables associated with $\boldsymbol{\lambda}_t$; similar to $z_{\alpha,it,j}$, $\mathbf{z}_{\alpha,it}$, $\mathbf{Z}_{\alpha,i}$, and $\mathbf{Z}_{\alpha}$	8
$\boldsymbol{\alpha}$	$q \times 1$ vector of random effects in the mixed linear model	3
α_i	subject-specific intercept parameter for the ith subject	2
$\boldsymbol{\alpha}_i$	$q \times 1$ vector of subject-specific parameter for the ith subject, $\boldsymbol{\alpha}_i = (\alpha_{i1}, \ldots, \alpha_{iq})'$	2
$\alpha_{i,j}$	jth subject-specific parameter for the ith subject, associated with $z_{it,j}$	2
$\boldsymbol{\beta}$	$K \times 1$ vector of global parameters, $\boldsymbol{\beta} = (\beta_1 \beta_2 \cdots \beta_K)'$	2
$\boldsymbol{\beta}$	$K \times 1$ vector of fixed effects in the mixed linear model	3
$\beta_{i,j}$	level-2 dependent variable in a three-level model	5
β_j	jth global parameter, associated with $x_{it,j}$	2
$\boldsymbol{\beta}_{i(k-g)}^{(g)}$ $\boldsymbol{\beta}_g$	level-g parameter matrices in the high-order multilevel model, analogous to Chapter 2 and 3 $\boldsymbol{\alpha}_i$ and $\boldsymbol{\beta}$	5
$\boldsymbol{\beta}_{i(k+1-g)}^{(g-1)}$ $\boldsymbol{\epsilon}_{i(k+1-g)}^{(g)}$	level-g response and disturbance terms in the high-order multilevel model, analogous to Chapter 2 and 3 $\mathbf{y}_i$ and $\boldsymbol{\epsilon}_i$	5
$\boldsymbol{\gamma}$	a $g \times 1$ vector of parameters corresponding to $\mathbf{U}_i$	7
γ	parameter associated with lagged dependent-variable model	8
γ_i	level-3 dependent variable in a three-level model	5
Δ	first difference operator, for example, $\Delta y_{it} = y_{it} - y_{i,t-1}$	6
δ_g	a group effect variable	3
$\boldsymbol{\delta}_t$	vector of unobservables in Kalman filter algorithm, $\boldsymbol{\delta}_t = (\boldsymbol{\lambda}_t', \boldsymbol{\xi}_t')'$	8
ε_{it}	error term for the ith subject, tth time period	2
$\boldsymbol{\epsilon}$	$N \times 1$ vector of disturbances in the mixed linear model	3
$\boldsymbol{\epsilon}_i$	$T_i \times 1$ vector of error terms for the ith subject, $\boldsymbol{\epsilon}_i = (\varepsilon_{i1}, \varepsilon_{i2}, \ldots, \varepsilon_{iT_i})'$	2
ζ_i	weighting (credibility) factor used to compute the shrinkage estimator	4
$\boldsymbol{\eta}_i$	a $T_i \times 1$ i.i.d. noise vector	7
$\boldsymbol{\eta}_i$	reduced-form disturbance term in the simultaneous equations model, $\boldsymbol{\eta}_i = (\mathbf{I} - \mathbf{B})^{-1}\boldsymbol{\epsilon}_i$	6
$\boldsymbol{\eta}_i$, $\boldsymbol{\epsilon}_i$	latent explanatory variable and disturbance term, respectively, in the $\mathbf{y}$-measurement equation of simultaneous equations with latent variables	6
$\boldsymbol{\Theta}_{\varepsilon}$	variance of $\boldsymbol{\epsilon}_i$, Var $\boldsymbol{\epsilon}_i = \boldsymbol{\Theta}_{\varepsilon}$	6
$\boldsymbol{\theta}$, $\boldsymbol{\psi}$	vectors of parameters for $f(.)$	6

(*continued*)

Symbol	Description	Chapter defined
λ_t	time-specific parameter	2
$\boldsymbol{\lambda}_t, \boldsymbol{\lambda}$	vectors of time-varying coefficients, $\boldsymbol{\lambda}_t = (\lambda_{t,1}, \ldots, \lambda_{t,r})'$, $\boldsymbol{\lambda} = (\boldsymbol{\lambda}_1, \ldots, \boldsymbol{\lambda}_T)'$	8
μ_{it}	mean of the ith subject at time t, $\mu_{it} = \mathrm{E}y_{it}$	9
$\boldsymbol{\mu}_\xi$	expected value of $\boldsymbol{\xi}_i$, $\mathrm{E}\,\boldsymbol{\xi}_i = \boldsymbol{\mu}_\xi$	6
$\boldsymbol{\mu}_i$	$T_i \times 1$ vector of means, $\boldsymbol{\mu}_i = (\mu_{i1}, \ldots, \mu_{iT_i})'$	9
$\boldsymbol{\xi}_t$	$np \times 1$ vector of unobservables in Kalman filter algorithm	8
$\boldsymbol{\xi}_i, \boldsymbol{\delta}_i$	latent explanatory variable and disturbance term, respectively, in the **x**-measurement equation of simultaneous equations with latent variables	6
$\boldsymbol{\Pi}$	reduced-form regression coefficients in the simultaneous equations model, $\boldsymbol{\Pi} = (\mathbf{I} - \mathbf{B})^{-1}\boldsymbol{\Gamma}$	6
$\pi(z)$	distribution function for binary dependent variables	9
ρ	correlation between two error terms, $\rho = \mathrm{corr}(\varepsilon_{ir}, \varepsilon_{is})$	2
$\rho(u)$	correlation function for spatial data	8
$\boldsymbol{\Sigma}$	variance of the response in the multivariate regression model	6
$\boldsymbol{\Sigma}_{\mathrm{IV}}$	variance–covariance matrix used to compute the variance of $\mathbf{b}_{\mathrm{IV}}$, $\boldsymbol{\Sigma}_{\mathrm{IV}} = \mathrm{E}\left(\mathbf{W}_i'\mathbf{K}\,\boldsymbol{\epsilon}_i\boldsymbol{\epsilon}_i'\mathbf{K}'\mathbf{W}_i\right)$	6
σ^2	variance of the error term under the homoscedastic model, $\sigma^2 = \mathrm{Var}\,\varepsilon_{it}$	2
σ_α^2	variance of the subject-specific intercept in the one-way error-components model, $\mathrm{Var}\,\alpha_i$	3
$\boldsymbol{\tau}$	vector of variance components	2
$\boldsymbol{\tau}_x, \boldsymbol{\Lambda}_x$	regression coefficients in the **x**-measurement equation of simultaneous equations with latent variables	6
$\boldsymbol{\tau}_y, \boldsymbol{\Lambda}_y$	regression coefficients in the **y**-measurement equation of simultaneous equations with latent variables	6
$\boldsymbol{\tau}_\eta, \mathbf{B}, \boldsymbol{\Gamma}$	regression coefficients in the structural equation of simultaneous equations with latent variables	6
$\boldsymbol{\Phi}_{1t}$	matrix of time-varying parameters for a transition equation	8
$\boldsymbol{\Phi}_2$	matrix representation of $\phi_1, \ldots, \phi_p$	8
$\boldsymbol{\Phi}, \boldsymbol{\Theta}_\delta$	variances of $\boldsymbol{\xi}_i$ and $\boldsymbol{\delta}_i$, respectively: $\mathrm{Var}\,\boldsymbol{\xi}_i = \boldsymbol{\Phi}$ and $\mathrm{Var}\,\boldsymbol{\delta}_i = \boldsymbol{\Theta}_\delta$	6
$\Phi(.)$	standard normal probability distribution function	7
$\phi(.)$	standard normal probability density function	7
$\phi_1, \ldots, \phi_p$	parameters of the $AR(p)$ process, autoregressive of order p	8
$\chi^2_{(q)}$	chi-square random variable with q degrees of freedom	5
$\boldsymbol{\Omega}$	variance of the reduced-form disturbance term in the simultaneous equations model	6

Appendix F

Selected Longitudinal and Panel Data Sets

In many disciplines, longitudinal and panel data sets can be readily constructed using a traditional pre- and postevent study; that is, subjects are observed prior to some condition of interest as well as after this condition or event. It is also common to use longitudinal and panel data methods to examine data sets that follow subjects observed at an aggregate level, such as a government entity (state, province, or nation, for example) or firm.

Longitudinal and panel data sets are also available that follow individuals over time. However, these data sets are generally expensive to construct and are conducted through the sponsorship of a government agency. Thus, although the data are available, data providers are generally bound by national laws requiring some form of user agreement to protect confidential information regarding the subjects. Because of the wide interest in these data sets, most data providers make information about the data sets available on the Internet.

Despite the expense and confidentiality requirements, many countries have conducted, or are in the process of conducting, household panel studies. Socio-demographic and economic information is collected about a household as well as individuals within the household. Information may relate to income, wealth, education, health, geographic mobility, taxes, and so forth. To illustrate, one of the oldest ongoing national panels, the U.S. Panel Study of Income Dynamics (PSID), collects 5,000 variables. Table F.1 cites some major international household panel data sets.

Education is another discipline that has a long history of interest in longitudinal methods. Table F.2 cites some major educational longitudinal data sets. Similarly, because of the dynamic nature of aging and retirement, analysts rely on longitudinal data for answers to important social science questions concerned with these issues. Table F.3 cites some major aging and retirement longitudinal data sets.

Longitudinal and panel data methods are used in many scientific disciplines. Table F.4 cites some other widely used data sets. The focus of Table F.4 is on political science through election surveys and longitudinal surveys of the firm.

Table F.1. *International household panel studies*

Panel study (years available) and Web site	Sample description
Australian Household, Income and Labor Dynamics (2001–) www.melbourneinstitute.com/hilda	First wave was scheduled to be collected in late 2001.
Belgian Socioeconomic Panel (1985–) www.ufsia.ac.be/csb/sep_nl.htm	A representative sample of 6,471 Belgian households in 1985, 3,800 in 1988, 3,800 in 1992 (which includes 900 new households) and 4632 in 1997 (which includes 2375 new households).
British Household Panel Survey (1991–) www.irc.essex.ac.uk/bhps	Annual survey of private households in Britain. A national representative sample of 5,000 households.
Canadian Survey of Labor Income Dynamics (1993–) www.statcan.ca	Approximately 35,000 households
Dutch Socio-Economic Panel (1984–1997) center.uvt.nl/research/facilities/datares.html	A national representative sample of 5,000 households.
French Household Panel (1985–1990) www.ceps.lu/paco/pacofrpa.htm	Baseline of 715 households increased to 2,092 in the second wave.
German Social Economic Panel (1984–) www.diw.de/soep	First wave collected in 1984, included 5,921 West German households consisting of 12,245 individuals.
Hungarian Household Panel (1992–1996) www.tarki.hu	Sample contains 2,059 households.
Indonesia Family Life Survey (1993–) www.rand.org/FLS/IFLS/	In 1993, 7,224 households were interviewed.
Japanese Panel Survey on Consumers (1994–) www.kakeiken.or.jp	National representative sample of 1,500 women age 24–34 in 1993; in 1997, 500 women were added.
Korea Labor and Income Panel Study (1998–) www.kli.re.kr/klips	Sample contains 5,000 households.
Luxembourg Panel Socio-Economique (1985–) www.ceps.lu/psell/pselpres.htm	Representative sample of 2,012 households and 6,110 individuals, 1985–1994. In 1994, it expanded to 2,978 households and 8,232 individuals.
Mexican Family Life Survey (2001–)	Will contain about 8,000 households. Plans are to collect data at two points in time, 2001 and 2004.

Panel study (years available) and Web site	Sample description
Polish Household Panel (1987–1990) www.ceps.lu/paco/pacopopa.htm	Four waves available of a sample of persons living in private households, excluding police officers, military personnel, and members of the "nomenklatura."
Russian Longitudinal Monitoring Survey (1992–) www.cpc.unc.edu/projects/rlms/home.html	Sample contains 7,200 households.
South African KwaZulu–Natal Income Dynamics Study (1993–1998) www.ifpri.cgiar.org/themes/mp17/safkzn2.htm	1,400 households from 70 rural and urban communities were surveyed in 1993 and reinterviewed in 1998.
Swedish Panel Study Market and Nonmarket Activities (1984–) www.nek.uu.se/faculty/klevmark/hus.htm	Sample contains 2,000 individuals surveyed in 1984.
Swiss Household Panel (1999–) www.swisspanel.ch/	First wave in 1999 consists of 5,074 households comprising 7,799 individuals.
Taiwanese Panel Study of Family Dynamics (1999–) srda.sinica.edu.tw	First wave in 1999 consists of individuals aged 36–45. Second wave in 2000 includes individuals aged 46–65.
U.S. Panel Study of Income Dynamics (1968–) psidonline.isr.umich.edu/	Began with 4,802 families, with an oversampling of poor families. Annual interviews were conducted with over 5,000 variables collected on roughly 31,000 individuals.

Sources: Institute for Social Research, University of Michigan: psidonline.isr.umich.edu/guide/relatedsites.html; Haisken-DeNew (2001E).

Table F.2. *Youth and education*

Study (years available) and Web site	Sample description
Early Childhood Longitudinal Study (1998–) www.nces.ed.gov/ecls/	Includes a kindergarten cohort and a birth cohort. The kindergarten cohort consists of a nationally representative sample of approximately 23,000 kindergartners from about 1,000 kindergarten programs. The birth cohort includes a nationally representative sample of approximately 15,000 children born in the calendar year 2000.
High School and Beyond (1980–1992) www.nces.ed.gov/surveys/hsb/	Includes two cohorts: the 1980 senior class and the 1980 sophomore class. Both cohorts were surveyed every two years through 1986, and the 1980 sophomore class was also surveyed again in 1992.
National Educational Longitudinal Survey: 1988 (1988–) www.nces.ed.gov/surveys/nels88/	Survey of 24,599 8th graders in 1988. In the first follow-up in 1990, only 19,363 were subsampled owing to budgetary reasons. Subsequent follow-ups were conducted in 1992, 1994, and 2000.
National Longitudinal Study of the High School Class of 1972 (1972–86) www.nces.ed.gov/surveys/nls72/	Followed the 1972 cohort of high school seniors through 1986. The original sample was drawn in 1972; follow-up surveys were conducted in 1973, 1974, 1976, 1979, and 1986.
The National Longitudinal Survey of Youth 1997 (NLS) www.bls.gov/nls/nlsy97.htm	A nationally representative sample of approximately 9,000 youths who were 12- to 16-years old in 1996.

Table F.3. *Elderly and retirement*

Study (years available) and Web site	Sample description
Framingham Heart Study (1948–) www.nhlbi.nih.gov/about/framingham/index.html	In 1948, 5,209 men and women between the ages of 30 and 62 were recruited to participate in this heart study. They are monitored every other year. In 1971, 5,124 of the original participants' adult children and their spouses were recruited to participate in similar examinations.
Health and Retirement Study (HRS) hrsonline.isr.umich.edu/	The original HRS cohort born 1931–1941 and first interviewed in 1992 (ages 51–61). The AHEAD cohort born before 1923 and first interviewed in 1993 (ages 70 and above). Spouses were included, regardless of age. These cohorts were merged in the 1998 wave and include over 21,000 participants. Variables collected include income, employment, wealth, health conditions, health status, and health insurance coverage.
Longitudinal Retirement History Study (1969–1979) Social Security Administration www.icpsr.umich. edu/index.html	Survey of 11,153 men and nonmarried women age 58–63 in 1969. Follow-up surveys were administered at two-year intervals in 1971, 1973, 1975, 1977, and 1979.
The National Longitudinal Surveys of Labor Market Experience (NLS) www.bls.gov/nls/	Five distinct labor markets are followed: • 5,020 older men (between 45 and 49 in 1966), • 5,225 young men (between 14 and 24 in 1966), • 5,083 mature women (between 30 and 44 in 1967), • 5,159 young women (between 14 and 21 in 1968), and • 12,686 youths (between 14 and 24 in 1979, with additional cohorts added in 1986 and 1997). The list of variables is in the thousands, with an emphasis on the supply side of the labor market.

Table F.4. *Other longitudinal and panel studies*

Study (years available) plus Web site	Sample description
National Election Studies 1956, 1958, 1960 American Panel Study www.umich.edu/~nes/studyres/nes56_60/nes56_60.htm	A sample of 1,514 voters who were interviewed at most five times.
National Election Studies 1972, 1974, 1976 Series File www.umich.edu/~nes/studyres/nes72_76/nes72_76.htm	A sample of 4,455 voters who were interviewed at most five times.
National Election Studies 1980 Panel Study www.umich.edu/~nes/studyres/nes80pan/nes80pan.htm	Over 1,000 voters were interviewed four times over the course of the 1980 presidential election.
National Election Studies 1990–1992 Full Panel File www.umich.edu/~nes/studyres/nes90_92/nes90_92.htm	Voter opinions are traced to follow the fortunes of the Bush presidency.
Census Bureau Longitudinal Research Database (1980–) www.census.gov/pub/econ/www/ma0800.html	Links establishment-level data from several censuses and surveys of manufacturers, and can respond to diverse economic research priorities.
Medical Expenditure Panel Survey (1996–) www.meps.ahrq.gov	Surveys of households, medical care providers, as well as business establishments and governments on health, care use and costs.
National Association of Insurance Commissioners (NAIC) www.naic.org/	Maintains annual and quarterly data for more than 6,000 Life/Health, Property/Casualty, Fraternal, Health, and Title companies.

References

Biological Sciences Longitudinal Data References (B)

Gibbons, R. D., and D. Hedeker (1997). Random effects probit and logistic regression models for three-level data. *Biometrics* 53, 1527–37.

Grizzle, J. E., and M. D. Allen (1969). Analysis of growth and dose response curves. *Biometrics* 25, 357–81.

Henderson, C. R. (1953). Estimation of variance components. *Biometrics* 9, 226–52.

Henderson, C. R. (1973). Sire evaluation and genetic trends. In *Proceedings of the Animal Breeding and Genetics Symposium in Honor of Dr. Jay L. Lush*, 10–41. Amer. Soc. Animal Sci. – Amer. Dairy Sci. Assoc. Poultry Sci. Assoc., Champaign, IL.

Henderson, C. R. (1984). *Applications of Linear Models in Animal Breeding*, University of Guelph, Guelph, ON, Canada.

Hougaard, P. (2000). *Analysis of Multivariate Survival Data.* Springer-Verlag, New York.

Jennrich, R. I., and M. D. Schlucter (1986). Unbalanced repeated-measures models with structured covariance matrices. *Biometrics* 42, 805–20.

Kenward, M. G., and J. H. Roger (1997). Small sample inference for fixed effects from restricted maximum likelihood. *Biometrics* 53, 983–97.

Klein, J. P., and M. L. Moeschberger (1997). *Survival Analysis: Techniques for Censored and Truncated Data.* Springer-Verlag, New York.

Laird, N. M. (1988). Missing data in longitudinal studies. *Statistics in Medicine* 7, 305–15.

Laird, N. M., and J. H. Ware (1982). Random-effects models for longitudinal data. *Biometrics* 38, 963–74.

Liang, K.-Y., and S. L. Zeger (1986). Longitudinal data analysis using generalized linear models. *Biometrika* 73, 12–22.

Palta, M., and T.-J. Yao (1991). Analysis of longitudinal data with unmeasured confounders. *Biometrics* 47, 1355–69.

Palta, M., T.-J. Yao, and R. Velu (1994). Testing for omitted variables and nonlinearity in regression models for longitudinal data. *Statistics in Medicine* 13, 2219–31.

Patterson, H. D., and R. Thompson (1971). Recovery of inter-block information when block sizes are unequal. *Biometrika* 58, 545–54.

Potthoff, R. F., and S. N. Roy (1964). A generalized multivariate analysis of variance model useful especially for growth curve problems. *Biometrika* 51, 313–26.

Prentice, R. L. (1988). Correlated binary regression with covariates specific to each binary observation. *Biometrics* 44, 1033–48.

Rao, C. R. (1965). The theory of least squares when the parameters are stochastic and its application to the analysis of growth curves. *Biometrika* 52, 447–58.

Rao, C. R. (1987). Prediction of future observations in growth curve models. *Statistical Science* 2, 434–71.

Stiratelli, R., N. Laird, and J. H. Ware (1984). Random effects models for serial observations with binary responses. *Biometrics* 40, 961–71.

Sullivan, L. M., K. A. Dukes, and E. Losina (1999). An introduction to hierarchical linear modelling. *Statistics in Medicine* 18, 855–88.

Wishart, J. (1938). Growth-rate determinations in nutrition studies with the bacon pig, and their analysis. *Biometrka* 30, 16–28.

Wright, S. (1918). On the nature of size factors. *Genetics* 3, 367–74.

Zeger, S. L., and K.-Y. Liang (1986). Longitudinal data analysis for discrete and continuous outcomes. *Biometrics* 42, 121–30.

Zeger, S. L., K.-Y. Liang, and P. S. Albert (1988). Models for longitudinal data: A generalized estimating equation approach. *Biometrics* 44, 1049–60.

Econometrics Panel Data References (E)

Amemiya, T. (1985). *Advanced Econometrics*. Harvard University Press, Cambridge, MA.

Anderson, T. W., and C. Hsiao (1982). Formulation and estimation of dynamic models using panel data. *Journal of Econometrics* 18, 47–82.

Andrews, D. W. K. (2001). Testing when a parameter is on the boundary of the maintained hypothesis. *Econometrica* 69, 683–734.

Arellano, M. (1993). On the testing of correlated effects with panel data. *Journal of Econometrics* 59, 87–97.

Arellano, M. (2003). *Panel Data Econometrics*. Oxford University Press, Oxford, UK.

Arellano, M., and S. Bond (1991). Some tests of specification for panel data: Monte Carlo evidence and an application to employment equations. *Review of Economic Studies* 58, 277–97.

Arellano, M., and O. Bover (1995). Another look at the instrumental-variable estimation of error components models. *Journal of Econometrics* 68, 29–51.

Arellano, M., and B. Honoré (2001). Panel data models: Some recent developments. In *Handbook of Econometrics*, Vol. 5, ed. J. J. Heckman and E. Leamer, pp. 3231–96. Elsevier, New York.

Ashenfelter, O. (1978). Estimating the effect of training programs on earnings with longitudinal data. *Review of Economics and Statistics* 60, 47–57.

Avery, R. B. (1977). Error components and seemingly unrelated regressions. *Econometrica* 45, 199–209.

Balestra, P., and M. Nerlove (1966). Pooling cross-section and time-series data in the estimation of a dynamic model: The demand for natural gas. *Econometrica* 34, 585–612.

Baltagi, B. H. (1980). On seemingly unrelated regressions and error components. *Econometrica* 48, 1547–51.

Baltagi, B. H. (2001). *Econometric Analysis of Panel Data*, Second Edition. Wiley, New York.

Baltagi, B. H., and Y. J. Chang (1994). Incomplete panels: A comparative study of alternative estimators for the unbalanced one-way error component regression model. *Journal of Econometrics* 62, 67.

Baltagi, B. H., and Q. Li (1990). A Lagrange multiplier test for the error components model with incomplete panels. *Econometric Reviews* 9, 103–107.

Baltagi, B. H., and Q. Li (1992). Prediction in the one-way error components model with serial correlation. *Journal of Forecasting* 11, 561–67.

Becker, R., and V. Henderson (2000). Effects of air quality regulations on polluting industries. *Journal of Political Economy* 108, 379–421.

Bhargava, A., L. Franzini, and W. Narendranathan (1982). Serial correlation and the fixed effects model. *Review of Economic Studies* 49, 533–49.

Blundell, R., and S. Bond (1998). Initial conditions and moment restrictions in dynamic panel data models. *Journal of Econometrics* 87, 115–43.

Bound, J., D. A. Jaeger, and R. M. Baker (1995). Problems with instrumental variables estimation when the correlation between the instruments and endogenous explanatory variables is weak. *Journal of the American Statistical Assocication* 90, 443–50.

Breusch, T. S., and A. R. Pagan (1980). The Lagrange multiplier test and its applications to model specification in econometrics. *Review of Economic Studies* 47, 239–53.

Cameron, A. C., and P. K. Trivedi (1998). *Regresson Analysis of Count Data*. Cambridge University Press, Cambridge, UK.

Card, D. (1995). Using geographic variation in college proximity to estimate the return to schooling. In *Aspects of Labour Market Behavior: Essays in Honour of John Vanderkamp*, ed. L. N. Christophides, E. K. Grant, and R. Swidinsky, pp. 201–222. University of Toronto Press, Toronto, Canada.

Chamberlain, G. (1980). Analysis of covariance with qualitative data. *Review of Economic Studies* 47, 225–38.

Chamberlain, G. (1982). Multivariate regression models for panel data. *Journal of Econometrics* 18, 5–46.

Chamberlain, G. (1984). Panel data. In *Handbook of Econometrics*, ed. Z. Griliches and M. Intrilligator, pp. 1247–1318. North-Holland, Amsterdam.

Chamberlain, G. (1992). Comment: Sequential moment restrictions in panel data. *Journal of Business and Economic Statistics* 10, 20–26.

Chib, S., E. Greenberg, and R. Winkelman (1998). Posterior simulation and Bayes factor in panel count data models. *Journal of Econometrics* 86, 33–54.

Davidson, R., and J. G. MacKinnon (1990). Specification tests based on artificial regressions. *Journal of the American Statistical Association* 85, 220–27.

Engle, R. F., D. F. Hendry, and J. F. Richard (1983). Exogeneity. *Econometrica* 51, 277–304.

Feinberg, S. E., M. P. Keane, and M. F. Bognano (1998). Trade liberalization and "delocalization": New evidence from firm-level panel data. *Canadian Journal of Economics* 31, 749–77.

Frees, E. W. (1995). Assessing cross-sectional correlations in longitudinal data. *Journal of Econometrics* 69, 393–414.

Glaeser, E. L., and D. C. Maré (2001). Cities and skills. *Journal of Labor Economics* 19, 316–42.

Goldberger, A. S. (1962). Best linear unbiased prediction in the generalized linear regression model. *Journal of the American Statistical Association* 57, 369–75.

Goldberger, A. S. (1972). Structural equation methods in the social sciences. *Econometrica* 40, 979–1001.

Goldberger, A. S. (1991). *A Course in Econometrics*. Harvard University Press, Cambridge, MA.

Gourieroux, C., A. Monfort, and A. Trognon (1984). Pseudo-maximum likelihood methods: Theory. *Econometrica* 52, 681–700.

Greene, W. H. (2002), *Econometric Analysis*, Fifth Edition. Prentice-Hall, Englewood Cliffs, NJ.

Haavelmo, T. (1944). The probability approach to econometrics. *Supplement to Econometrica* 12, iii–vi, 1–115.

Haisken-DeNew, J. P. (2001). A hitchhiker's guide to the world's household panel data sets. *The Australian Economic Review* 34(3), 356–66.

Hausman, J. A. (1978). Specification tests in econometrics. *Econometrica* 46, 1251–71.

Hausman, J. A., B. H. Hall, and Z. Griliches (1984). Econometric models for count data with an application to the patents–R&D relationship. *Econometrica* 52, 909–38.

Hausman, J. A., and W. E. Taylor (1981). Panel data and unobservable individual effects. *Econometrica* 49, 1377–98.

Hausman, J. A., and D. Wise (1979). Attrition bias in experimental and panel data: The Gary income maintenance experiment. *Econometrica* 47, 455–73.

Hayashi, F. (2000). *Econometrics*. Princeton University Press, Princeton, NJ.

Heckman, J. J. (1976). The common structure of statistical models of truncation, sample selection and limited dependent variables, and a simple estimator for such models. *Ann. Econ. Soc. Meas* 5, 475–92.

Heckman, J. J. (1981). Statistical models for discrete panel data. In *Structural Analysis Of Discrete Data with Econometric Applications*, ed. C. F. Manski and D. McFadden, pp. 114–78. MIT Press, Cambridge, MA.

Hoch, I. (1962). Estimation of production function parameters combining time-series and cross-section data. *Econometrica* 30, 34–53.

Hsiao, C. (2002). *Analysis of Panel Data*, Second Edition. Cambridge University Press, Cambridge, UK.

Johnson, P. R. (1960). Land substitutes and changes in corn yields. *Journal of Farm Economics* 42, 294–306.

Judge, G. G., W. E. Griffiths, R. C. Hill, H. Lütkepohl, and T. C. Lee (1985). *The Theory and Practice of Econometrics*. Wiley, New York.

Keane, M. P., and D. E. Runkle (1992). On the estimation of panel data models with serial correlation when instruments are not strictly exogenous. *Journal of Business and Economic Statistics* 10, 1–9.

Kiefer, N. M. (1980). Estimation of fixed effects models for time series of cross sections with arbitrary intertemporal covariance. *Journal of Econometrics* 14, 195–202.

Kuh, E. (1959). The validity of cross-sectionally estimated behavior equation in time series application. *Econometrica* 27, 197–214.

Lancaster, T. (1990). *The Econometric Analysis of Transition Data*. Cambridge University Press, New York.

Maddala, G. S. (1971). The use of variance components models in pooling cross section and time series data. *Econometrica* 39, 341–58.

Manski, C. (1992). Comment: The impact of sociological methodology on statistical methodology by C. C. Clogg. *Statistical Science* 7(2), 201–203.

Mátyás, L., and, P. Sevestre eds. (1996). *The Econometrics of Panel Data: Handbook of Theory and Applications*. Kluwer Academic Publishers, Dordrecht.

McFadden, D. (1974). Conditional logit analysis of qualitative choice behavior. In *Frontiers of Econometrics*, ed. P. Zarembka, pp. 105–142. Academic Press, New York.

McFadden, D. (1978). Modeling the choice of residential location. In *Spatial Interaction Theory and Planning Models*, ed. A Karlqvist et al., pp. 75–96. North-Holland, Amsterdam.

McFadden, D. (1981). Econometric models of probabilistic choice. In *Structural Analysis of Discrete Data with Econometric Applications*, ed. C. Manski and D. McFadden, pp. 198–272. MIT Press, Cambridge, MA.

Mundlak, Y. (1961). Empirical production function free of management bias. *Journal of Farm Economics* 43, 44–56.

Mundlak, Y. (1978a). On the pooling of time series and cross-section data. *Econometrica* 46, 69–85.

Mundlak, Y. (1978b). Models with variable coefficients: Integration and extensions. *Annales de L'Insee* 30–31, 483–509.

Neyman, J., and E. L. Scott (1948). Consistent estimates based on partially consistent observations. *Econometrica* 16, 1–32.

Polachek, S. W., and M. Kim (1994). Panel estimates of the gender earnings gap. *Journal of Econometrics* 61, 23–42.

Swamy, P. A. V. B. (1970). Efficient inference in a random coefficient regression model. *Econometrica* 38, 311–23.

Theil, H., and A. Goldberger (1961). On pure and mixed estimation in economics. *International Economic Review* 2, 65–78.

Valletta, R. G. (1999). Declining job security. *Journal of Labor Economics* 17, S170-S197.

Wallace, T., and A. Hussain (1969). The use of error components in combining cross section with time series data. *Econometrica* 37, 55–72.

White, H. (1980). A heteroskedasticity-consistent covariance matrix estimator and a direct test for heteroskedasticity. *Econometrica* 48, 817–38.

White, H. (1984). *Asymptotic Theory for Econometricians*. Academic Press, Orlando, FL.

Wooldridge, J. M. (2002). *Econometric Analysis of Cross Section and Panel Data*. MIT Press, Cambridge, MA.

Zellner, A. (1962). An efficient method of estimating seemingly unrelated regression and tests for aggregation bias. *Journal of the American Statistical Association* 57, 348–68.

Educational Science and Psychology References (EP)

Baltes, P. B., and J. R. Nesselroade (1979). History and rational of longitudinal research. In *Longitudinal Research in the Study of Behavior and Development*, ed. J. R. Nesselroade and P. B. Baltes, pp. 1–39. Academic Press, New York.

Bollen, K. A. (1989). *Structural Equations with Latent Variables*. Wiley, New York.

de Leeuw, J., and I. Kreft (2001). Software for multilevel analysis. In *Multilevel Modeling of Health Statistics* ed. A. H. Leyland and H. Goldstein, pp. 187–204. Wiley, New York.

Duncan, O. D. (1969). Some linear models for two wave, two variable panel analysis. *Psychological Bulletin* 72, 177–82.

Duncan, T. E., S. C. Duncan, L. A. Stryakar, F. Li, and A. Alpert (1999). *An Introduction to Latent Variable Growth Curve Modeling*. Lawrence Erlbaum, Mahwah, NJ.

Frees, E. W. and J.-S. Kim (2004). Multilevel model prediction. UW working paper, available at http://research.bus.wisc.edu/jFrees/.

Goldstein, H. (2002). *Multilevel Statistical Models*, Third Edition. Edward Arnold, London.

Guo, S., and D. Hussey (1999). Analyzing longitudinal rating data: A three-level hierarchical linear model. *Social Work Research* 23(4), 258–68.

Kreft, I., and J. de Leeuw (1998). *Introducing Multilevel Modeling*. Sage, New York.

Lee, V. (2000). Using hierarchical linear modeling to study social contexts: The case of school effects. *Educational Psychologist* 35(2), 125–41.

Lee, V., and J. B. Smith (1997). High school size: Which works best and for whom? *Educational Evaluation and Policy Analysis* 19(3), 205–27.

Longford, N. T. (1993). *Random Coefficient Models*. Clarendon Press, Oxford, UK.

Rasch, G. (1961). On general laws and the meaning of measurement in psychology. *Proc. Fourth Berkeley Symposium* 4, 434–58.

Raudenbush, S. W., and A. S. Bryk (2002). *Hierarchical Linear Models: Applications and Data Analysis Methods*, Second Edition. Sage, London.

Rubin, D. R. (1974). Estimating causal effects of treatments in randomized and nonrandomized studies. *Journal of Educational Psychology* 66, 688–701.

Singer, J. D. (1998). Using SAS PROC MIXED to fit multilevel models, hierarchical models, and individual growth models. *Journal of Educational and Behavioral Statistics* 27, 323–55.

Singer, J. D., and J. B. Willett (2003). *Applied Longitudinal Data Analysis: Modeling Change and Event Occurrence*. Oxford University Press, Oxford, UK.

Toon, T. J. (2000). *A Primer in Longitudinal Data Analysis*. Sage, London.

Webb, N. L., W. H. Clune, D. Bolt, A. Gamoran, R. H. Meyer, E. Osthoff, and C. Thorn. (2002). Models for analysis of NSF's systemic initiative programs – The impact of the urban system initiatives on student achievement in Texas, 1994–2000. Wisconsin Center for Education Research Technical Report, July. Available at http://facstaff.wcer.wisc.edu/normw/technical_reports.htm.

Willett, J. B., and A. G. Sayer (1994). Using covariance structure analysis to detect correlates and predictors of individual change over time. *Psychological Bulletin* 116, 363–81.

Other Social Science References (O)

Ashley, T., Y. Liu, and S. Chang (1999). Estimating net lottery revenues for states. *Atlantic Economics Journal* 27, 170–78.

Bailey, A. (1950). Credibility procedures: LaPlace's generalization of Bayes' rule and the combination of collateral knowledge with observed data. *Proceedings of the Casualty Actuarial Society* 37, 7–23.

Beck, N., and J. N. Katz (1995). What to do (and not to do) with time-series cross-section data. *American Political Science Review* 89, 634–47.

Brader, T., and J. A. Tucker (2001). The emergence of mass partisanship in Russia, 1993–1996. *American Journal of Political Science* 45, 69–83.

Bühlmann, H. (1967). Experience rating and credibility. *ASTIN Bulletin* 4, 199–207.

Bühlmann, H., and E. Straub (1970). Glaubwürdigkeit für Schadensätze. *Mitteilungen der Vereinigung Schweizerischer Versicherungsmathematiker* 70, 111–33.

Dannenburg, D. R., R. Kaas, and M. J. Goovaerts (1996). *Practical Actuarial Credibility Models*. Institute of Actuarial Science and Econometrics, University of Amsterdam, Amsterdam.

Dielman, T. E. (1989). *Pooled Cross-Sectional and Time Series Data Analysis*. Marcel Dekker, New York.

Frees, E. W., and T. W. Miller (2004). Sales forecasting using longitudinal data models. *International Journal of Forecasting* 20, 97–111.

Frees, E. W., V. Young, and Y. Luo (1999). A longitudinal data analysis interpretation of credibility models. *Insurance: Mathematics and Economics* 24, 229–47.

Frees, E. W., V. Young, and Y. Luo (2001). Case studies using panel data models. *North American Actuarial Journal* 4(4), 24–42.

Frischmann, P. J., and E. W. Frees (1999). Demand for services: Determinants of tax preparation fees. *Journal of the American Taxation Association* 21, Supplement 1–23.

Green, R. K., and S. Malpezzi (2003). *A Primer on U.S. Housing Markets and Policy*. Urban Institute Press, Washington, DC.

Haberman, S., and E. Pitacco (1999). *Actuarial Models for Disability Insurance*. Chapman and Hall/CRC, Boca Raton.

Hachemeister, C. A. (1975). Credibility for regression models with applications to trend. In *Credibility: Theory and Applications*, ed. P. M. Kahn, pp. 129–163 Academic Press, New York.

Hickman, J. C., and L. Heacox (1999). Credibility theory: The cornerstone of actuarial science. *North American Actuarial Journal* 3(2), 1–8.

Jain, D. C., N. J. Vilcassim, and P. K. Chintagunta (1994). A random-coefficients logit brand choice model applied to panel data. *Journal of Business and Economic Statistics* 12, 317–28.

Jewell, W. S. (1975). The use of collateral data in credibility theory: A hierarchical model. *Giornale dell'Istituto Italiano degli Attuari* 38, 1–16. (Also, Research Memorandum 75–24 of the International Institute for Applied Systems Analysis, Laxenburg, Austria.)

Kim, Y.-D., D. R. Anderson, T. L. Amburgey, and J. C. Hickman (1995). The use of event history analysis to examine insurance insolvencies. *Journal of Risk and Insurance* 62, 94–110.

Klugman, S. A. (1992). *Bayesian Statistics in Actuarial Science: With Emphasis on Credibility*. Kluwer Academic, Boston.

Klugman, S., H. Panjer, and G. Willmot (1998). *Loss Models: From Data to Decisions*. Wiley, New York.

Kung, Y. (1996). *Panel Data with Serial Correlation*. Unpublished Ph.D. thesis, University of Wisconsin, Madison.

Lazarsfeld, P. F., and M. Fiske (1938). The panel as a new tool for measuring opinion. *Public Opinion Quarterly* 2, 596–612.

Ledolter, J., S. Klugman, and C.-S. Lee (1991). Credibility models with time-varying trend components. *ASTIN Bulletin* 21, 73–91.

Lee, H. D. (1994). *An Empirical Study of the Effects of Tort Reforms on the Rate of Tort Filings*. Unpublished Ph.D. thesis, University of Wisconsin, Madison.

Lee, H. D., M. J. Browne, and J. T. Schmit (1994). How does joint and several tort reform affect the rate of tort filings? Evidence from the state courts. *Journal of Risk and Insurance* 61(2), 595–616.

Lintner, J. (1965). The valuation of risky assets and the selection of risky investments in stock portfolios and capital budgets. *Review of Economics and Statistics* 47, 13–37.

Luo, Y., V. R. Young, and E. W. Frees (2004). Credibility ratemaking using collateral information. To appear in *Scandinavian Actuarial Journal*.

Malpezzi, S. (1996). Housing prices, externalities, and regulation in U.S. metropolitan areas. *Journal of Housing Research* 7(2), 209–41.

Markowitz, H. (1952). Portfolio selection. *Journal of Finance* 7, 77–91.

Mowbray, A. H. (1914). How extensive a payroll exposure is necessary to give a dependable pure premium. *Proceedings of the Casualty Actuarial Society* 1, 24–30.

Norberg, R. (1980). Empirical Bayes credibility. *Scandinavian Actuarial Journal* 1980, 177–94.

Norberg, R. (1986). Hierarchical credibility: Analysis of a random effect linear model with nested classification. *Scandinavian Actuarial Journal* 1986, 204–22.

Paker, B. S. (2000). *Corporate Fund Raising and Capital Structure in Japan during the 1980s and 1990s*. Unpublished Ph.D. thesis, University of Wisconsin, Madison.

Pinquet, J. (1997). Allowance for cost of claims in bonus-malus systems. *ASTIN Bulletin* 27, 33–57.

Sharpe, W. (1964). Capital asset prices: A theory of market equilibrium under risk. *Journal of Finance* 19, 425–42.

Shumway, T. (2001). Forecasting bankruptcy more accurately: A simple hazard model. *Journal of Business* 74, 101–24.

Stimson, J. (1985). Regression in space and time: A statistical essay. *American Journal of Political Science* 29, 914–47.

Stohs, M. H., and D. C. Mauer (1996). The determinants of corporate debt maturity structure. *Journal of Business* 69(3), 279–312.

Taylor, G. C. (1977). Abstract credibility. *Scandinavian Actuarial Journal* 1977, 149–68.

Venter, G. (1996). Credibility. In *Foundations of Casualty Actuarial Science*, Third Edition, ed. I. K. Bass et al. Casualty Actuarial Society, Arlington, VA.

Zhou, X. (2000). Economic transformation and income inequality in urban China: Evidence from panel data. *American Journal of Sociology* 105, 1135–74.

Statistical Longitudinal Data References (S)

Banerjee, M., and E. W. Frees (1997). Influence diagnostics for linear longitudinal models. *Journal of the American Statistical Association* 92, 999–1005.

Chen, Z., and L. Kuo (2001). A note on the estimation of multinomial logit model with random effects. *American Statistician* 55, 89–95.

Conaway. M. R. (1989). Analysis of repeated categorical measurements with conditional likelihood methods. *Journal of the American Statistical Association* 84, 53–62.

Corbeil, R. R., and S. R. Searle (1976a). Restricted maximum likelihood (REML) estimation of variance components in the mixed model. *Technometrics* 18, 31–38.

Corbeil, R. R., and S. R. Searle (1976b). A comparison of variance components estimators. *Biometrics* 32, 779–91.

Davidian, M., and D. M. Giltinan (1995). *Nonlinear Models for Repeated Measurement Data*. Chapman–Hall, London.

Diggle, P. J., P. Heagarty, K.-Y. Liang, and S. L. Zeger (2002). *Analysis of Longitudinal Data*, Second Edition. Oxford University Press, Oxford, UK.

Fahrmeir, L., and G. Tutz (2001). *Multivariate Statistical Modelling Based on Generalized Linear Models*, Second Edition. Springer-Verlag, New York.

Frees, E. W. (2001). Omitted variables in panel data models. *Canadian Journal of Statistics* 29(4), 1–23.

Frees, E. W., and C. Jin (2004). Empirical standard errors for longitudinal data mixed linear models. To appear in *Computational Statistics*.

Ghosh, M., and J. N. K. Rao (1994). Small area estimation: An appraisal. *Statistical Science* 9, 55–93.

Harvey, A. C. (1989). *Forecasting, Structural Time Series Models and the Kalman Filter*. Cambridge University Press, Cambridge, UK.

Harville, D. A. (1974). Bayesian inference for variance components using only error contrasts. *Biometrika* 61, 383–85.

Harville, D. (1976). Extension of the Gauss–Markov theorem to include the estimation of random effects. *Annals of Statistics* 2, 384–95.

Harville, D. (1977). Maximum likelihood estimation of variance components and related problems. *Journal of the American Statistical Association* 72, 320–40.

Herbach, L. H. (1959). Properties of model II type analysis of variance tests, A: Optimum nature of the *F*-test for model II in the balanced case. *Annals of Mathematical Statistics* 30, 939–59.

Hildreth, C., and C. Houck (1968). Some estimators for a linear model with random coefficients. *Journal of the American Statistical Association* 63, 584–95.

Jones, R. H. (1993). *Longitudinal Data with Serial Correlation: A State-Space Approach*. Chapman and Hall, London.

Jöreskog, K. G., and A. S. Goldberger (1975). Estimation of a model with multiple indicators and multiple causes of a single latent variable. *Journal of the American Statistical Association* 70, 631–39.

Kackar, R. N., and D. Harville (1984). Approximations for standard errors of estimators of fixed and random effects in mixed linear models. *Journal of the American Statistical Association* 79, 853–62.

Lin, X., and N. E. Breslow (1996). Bias correction on generalized linear mixed models with multiple components of dispersion. *Journal of the American Statistical Association* 91, 1007–16.

Littell, R. C, G. A. Milliken, W. W. Stroup, and R. D. Wolfinger (1996). *SAS System for Mixed Models*. SAS Institute, Cary, NC.

Parks, R. (1967). Efficient estimation of a system of regression equations when disturbances are both serially and contemporaneously correlated. *Journal of the American Statistical Association* 62, 500–509.

Pinheiro, J. C., and D. M. Bates (2000). *Mixed-Effects Models in S and S-Plus*. Springer-Verlag, New York.

Rao, C. R. (1965). The theory of least squares when the parameters are stochastic and its application to the analysis of growth curves. *Biometrika* 52, 447–58.

Rao, C. R. (1970). Estimation of variance and covariance components in linear models. *Journal of the American Statistical Association* 67, 112–15.

Reinsel, G. C. (1982). Multivariate repeated-measurement or growth curve models with multivariate random-effects covariance structure. *Journal of the American Statistical Association* 77, 190–95.

Reinsel, G. C. (1984). Estimation and prediction in a multivariate random effects generalized linear model. *Journal of the American Statistical Association* 79, 406–14.

Robinson, G. K. (1991). The estimation of random effects. *Statistical Science* 6, 15–51.

Searle, S. R., G. Casella, and C. E. McCulloch (1992). *Variance Components*. Wiley, New York.

Self, S. G., and K. Y. Liang (1987). Asymptotic properties of maximum likelihood estimators and likelihood ratio tests under nonstandard conditions. *Journal of the American Statistical Association* 82, 605–10.

Stram, D. O., and J. W. Lee (1994). Variance components testing in the longitudinal mixed effects model. *Biometrics* 50, 1171–77.

Swallow, W. H., and S. R. Searle (1978). Minimum variance quadratic unbiased estimation (MIVQUE) of variance components. *Technometrics* 20, 265–72.

Tsimikas, J. V., and J. Ledolter (1994). REML and best linear unbiased prediction in state space models. *Communications in Statistics: Theory and Methods* 23, 2253–68.

Tsimikas, J. V., and J. Ledolter (1997). Mixed model representations of state space models: New smoothing results and their application to REML estimation. *Statistica Sinica* 7, 973–91.

Tsimikas, J. V., and J. Ledolter (1998). Analysis of multi-unit variance components models with state space profiles. *Annals of the Institute of Statist Math* 50(1), 147–64.

Verbeke, G., and G. Molenberghs (2000). *Linear Mixed Models for Longitudinal Data*. Springer-Verlag, New York.

Vonesh, E. F., and V. M. Chinchilli (1997). *Linear and Nonlinear Models for the Analysis of Repeated Measurements*. Marcel Dekker, New York.

Ware, J. H. (1985). Linear models for the analysis of longitudinal studies. *The American Statistician* 39, 95–101.

Wolfinger, R., R. Tobias, and J. Sall (1994). Computing Gaussian likelihoods and their derivatives for general linear mixed models. *SIAM Journal of Scientific Computing* 15(6), 1294–310.

Zeger, S. L., and M. R. Karim (1991). Generalized linear models with random effects: A Gibbs sampling approach. *Journal of the American Statistical Association* 86, 79–86.

General Statistics References (G)

Agresti, A. (2002). *Categorical Data Analysis*. Wiley, New York.

Andersen, E. B. (1970). Asymptotic properties of conditional maximum-likelihood estimators. *Journal of the Royal Statistical Society B* 32, 283–301.

Anderson, T. W. (1958). *An Introduction to Multivariate Statistical Analysis*. Wiley, New York.

Angrist, J. D, G W. Imbens, and D. B. Rubin (1996). Identification of causal effects using instrumental variables. *Journal of the American Statistical Association* 91, 444–72.

Becker, R. A., W. S. Cleveland, and M.-J. Shyu (1996). The visual design and control of trellis graphics displays. *Journal of Computational and Graphical Statistics* 5, 123–56.

Bickel, P. J., and Doksum, K. A. (1977). *Mathematical Statistics*. Holden-Day, San Francisco.

Box, G. E. P. (1979). Robustness in the strategy of scientific model building. In *Robustness in Statistics*, ed. R. Launer and G. Wilderson, pp. 201–236. Academic Press, New York.

Box, G. E. P., G. M. Jenkins, and G. Reinsel. (1994). *Time Series Analysis*. Prentice Hall, Englewood Cliffs, NJ.

Carroll, R. J., and D. Ruppert (1988). *Transformation and Weighting in Regression*. Chapman and Hall, London.

Cleveland, W. S. (1993). *Visualizing Data*. Hobart Press, Summit, NJ.

Congdon, P. (2003). *Applied Bayesian Modelling*. Wiley, New York.

Cook, D., and S. Weisberg (1982). *Residuals and Influence in Regression*. Chapman and Hall, London.

Draper, N., and H. Smith (1981). *Applied Regression Analysis*, Second Edition. Wiley, New York.

Friedman, M. (1937). The use of ranks to avoid the assumption of normality implicit in the analysis of variance. *Journal of the American Statistical Association* 89, 517–25.

Fuller, W. A., and G. E. Battese (1973). Transformations for estimation of linear models with nested error structure. *Journal of the American Statistical Association* 68, 626–32.

Gelman, A., J. B. Carlin, H. S. Stern, and D. B. Rubin (2004). *Bayesian Data Analysis*, Second Edition. Chapman and Hall, New York.

Gill, J. (2002). *Bayesian Methods for the Social and Behavioral Sciences*. Chapman and Hall, New York.

Graybill, F. A. (1969). *Matrices with Applications in Statistics*, Second Edition. Wadsworth, Belmont CA.

Hocking, R. (1985). *The Analysis of Linear Models*. Wadsworth and Brooks/Cole, Pacific Grove, CA.

Hosmer, D. W., and S. Lemeshow (2000). *Applied Logistic Regression*. Wiley, New York.

Hougaard, P. (1987). Modelling multivariate survival. *Scandinavian Journal of Statistics* 14, 291–304.

Huber, P. J. (1967). The behaviour of maximum likelihood estimators under non-standard conditions. *Proceedings of the Fifth Berkeley Symposium on Mathematical Statistics and Probability* 1, ed. L. M. LeCam and J. Neyman, pp. 221–33, University of California Press, Berkeley.

Hutchinson, T. P., and C. D. Lai (1990). *Continuous Bivariate Distributions, Emphasising Applications*. Rumsby Scientific Publishing, Adelaide, Australia.

Johnson, R. A., and D. Wichern (1999). *Applied Multivariate Statistical Analysis*. Prentice Hall, Englewood Cliffs, NJ.

Kalman, R. E. (1960). A new approach to linear filtering and prediction problems. *Journal of Basic Engineering* 82, 34–45.

Layard, M. W. (1973). Robust large sample tests for homogeneity of variance. *Journal of the American Statistical Association* 68, 195–198.

Lehmann, E. (1991). *Theory of Point Estimation*. Wadsworth and Brooks/Cole, Pacific Grove, CA.

Little, R. J. (1995). Modelling the drop-out mechanism in repeated-measures studies. *Journal of the American Statistical Association* 90, 1112–21.

Little, R. J., and D. B. Rubin (1987). *Statistical Analysis with Missing Data*. Wiley, New York.

McCullagh, P., and J. A. Nelder (1989). *Generalized Linear Models*, Second Edition. Chapman and Hall, London.

McCulloch, C. E., and S. R. Searle (2001). *Generalized, Linear, and Mixed Models*. Wiley, New York.

Nelder, J. A., and R. W. Wedderburn (1972). Generalized linear models. *Journal of the Royal Statistical Society Ser. A* 135, 370–84.

Neter, J., and W. Wasserman (1974). *Applied Linear Statistical Models*. Irwin, Homewood, IL.

Rubin, D. R. (1976). Inference and missing data. *Biometrika* 63, 581–92.

Rubin, D. R. (1978). Bayesian inference for causal effects. *Annals of Statistics* 6, 34–58.

Rubin, D. R. (1990). Comment: Neyman (1923) and causal inference in experiments and observational studies. *Statistical Science* 5, 472–80.

Scheffé, H. (1959). *The Analysis of Variance*. Wiley, New York.

Searle, S. R. (1971). *Linear Models*. Wiley, New York.

Searle, S. R. (1987). *Linear Models for Unbalanced Data*. Wiley, New York.

Seber, G. A. (1977). *Linear Regression Analysis*. Wiley, New York.

Serfling, R. J. (1980). *Approximation Theorems of Mathematical Statistics*. Wiley, New York.

Stigler, S. M. (1986). *The History of Statistics: The Measurement of Uncertainty before 1900*. Harvard University Press, Cambridge, MA.

Tufte, E. R. (1997). *Visual Explanations*. Graphics Press, Cheshire, CT.

Tukey, J. W. (1977). *Exploratory Data Analysis*. Addison-Wesley, Reading, MA.

Venables, W. N., and B. D. Ripley (1999). *Modern Applied Statistics with S-PLUS*, Third Edition. Springer-Verlag, New York.

Wachter, K. W., and J. Trusell (1982). Estimating historical heights. *Journal of the American Statistical Association* 77, 279–301.

Wedderburn, R. W. (1974). Quasi-likelihood functions, generalized linear models and the Gaussian method. *Biometrika* 61, 439–47.

Wong, W. H. (1986). Theory of partial likelihood. *Annals of Statistics* 14, 88–123.

Index